D. Gilbon
-Madison
istry

Bacteria (bought since 1998)

Cell-Cell Signaling in Bacteria

Cell-Cell Signaling in Bacteria

Edited by

Gary M. Dunny
Department of Microbiology, University of Minnesota
Minneapolis, Minnesota 55455

Stephen C. Winans
Section of Microbiology, Cornell University
Ithaca, New York 14853

ASM PRESS WASHINGTON, D.C.

Cover: Aggregation of myxobacteria during early stages of fruiting body formation. Drawings by Roland Thaxter, 1892 ("On the *Myxobacteriaceae*, a new order of *Schizomycetes*," *Botanical Gazette* **17**:389–406).

Library of Congress Cataloging-in-Publication Data

Cell-cell signaling in bacteria / edited by Gary M. Dunny, Stephen C. Winans.

     p.   cm.
    Includes bibliographical references and index.
    ISBN 1-55581-149-3
    1. Cell interaction.   2. Bacteria—Physiology.   3. Symbiosis.
4. Genetic transformation.   5. Cellular signal transduction.
I. Dunny, Gary M.   II. Winans, Stephen Carlyle.
    [DNLM: 1. Bacteria—cytology.   2. Cell Communication.
3. Bacteria—growth & development.   4. Gene Expression Regulation,
Bacterial.   5. Symbiosis.   QW 52 C3928 1999]
    QR96.5.C45   1994
    579.3—dc21
    DNLM/DLC
    for Library of Congress   98-44808
                                      CIP

CONTENTS

CONTRIBUTORS

Bonnie L. Bassler
Department of Molecular Biology, Princeton University, Princeton, New Jersey
08544-1014

Susanne Beck von Bodman
Department of Plant Science and Department of Molecular and Cell Biology, University
of Connecticut, 1376 Storrs Road, U-67, Storrs, Connecticut 06269-4067

Don B. Clewell
Department of Biologic and Materials Science, School of Dentistry, and Department of
Microbiology and Immunology, School of Medicine, The University of Michigan,
Ann Arbor, Michigan 48109

Willem M. de Vos
Microbial Ingredients Section, NIZO Food Research, Kernhemseweg 2, 6710 BA Ede,
The Netherlands

Gary M. Dunny
Department of Microbiology, University of Minnesota Medical School, 1460 Mayo
Building/Box 196, 420 Delaware Street, S.E., Minneapolis, Minnesota 55455-0312

Anatol Eberhard
Department of Chemistry, Ithaca College, Ithaca, New York 14850

Vincent G. H. Eijsink
Laboratory of Microbial Gene Technology, Department of Biotechnological Sciences,
Agricultural University of Norway, P.O. Box 5051, N-1432 Ås, Norway

Clay Fuqua
Department of Biology, Trinity University, San Antonio, Texas 78212

E. P. Greenberg
Department of Microbiology, University of Iowa, Iowa City, Iowa 52242

Alan D. Grossman
Department of Biology, Building 68-530, Massachusetts Institute of Technology,
Cambridge, Massachusetts 02139

Leiv Sigve Håvarstein
Department of Biotechnological Sciences, Seksjon Mikrobiologi, Fellesbygget, Agricultural University of Norway, P.O. Box 5040, 1342 Ås, Norway

Sueharu Horinouchi
Department of Biotechnology, Graduate School of Agriculture and Life Sciences, The University of Tokyo, Bunkyo-ku, Tokyo 113-8657, Japan

Barbara H. Iglewski
Department of Microbiology and Immunology, University of Rochester School of Medicine and Dentistry, University of Rochester, Rochester, New York 14642

Dale Kaiser
Biochemistry Department, Stanford University, Stanford, California 94305

Heidi B. Kaplan
Department of Microbiology and Molecular Genetics, University of Texas Medical School at Houston, 6431 Fannin, Houston, Texas 77030

Michiel Kleerebezem
Microbial Ingredients Section, NIZO Food Research, Kernhemseweg 2, 6710 BA Ede, The Netherlands

Oscar P. Kuipers
Microbial Ingredients Section, NIZO Food Research, Kernhemseweg 2, 6710 BA Ede, The Netherlands

Beth A. Lazazzera
Department of Biology, Building 68-530, Massachusetts Institute of Technology, Cambridge, Massachusetts 02139

Bettina A. B. Leonard
Department of Microbiology and Immunology, Temple University School of Medicine, 3400 N. Broad Street, Philadelphia, Pennsylvania 19140

Margret I. Moré
Section of Microbiology, Wing Hall, Cornell University, Ithaca, New York 14853

Donald A. Morrison
Laboratory for Molecular Biology, Department of Biological Sciences, University of Illinois at Chicago, 900 South Ashland Avenue, Chicago, Illinois 60607

Kenneth H. Nealson
Jet Propulsion Laboratory, California Institute of Technology, 4800 Oak Grove Drive, Pasadena, California 91109

Ingolf F. Nes
Laboratory of Microbial Gene Technology, Department of Biotechnological Sciences, Agricultural University of Norway, P.O. Box 5051, N-1432 Ås, Norway

Richard P. Novick
Skirball Institute of Biomolecular Medicine, New York University Medical Center, 540 First Avenue, New York, New York 10016

Tanya Palmer
Department of Biology, Massachusetts Institute of Technology, Cambridge, Massachusetts 02139

Marta Perego
Department of Molecular and Experimental Medicine, Division of Cellular Biology, The Scripps Research Institute, 10550 N. Torrey Pines Road, NX-1, La Jolla, California 92037

Everett C. Pesci
Department of Microbiology and Immunology, School of Medicine, East Carolina University, Greenville, North Carolina 27858

Leland S. Pierson III
Department of Plant Pathology, 204 Forbes Hall, University of Arizona, Tucson, Arizona 85721-0036

Lynda Plamann
Division of Cell Biology and Biophysics, School of Biological Sciences, 2411 Holmes, University of Missouri-Kansas City, Kansas City, Missouri 64108

Andreas Podbielski
Department of Medical Microbiology and Hygiene, University of Ulm Clinic, D-89081 Ulm, Germany

John Quisel
Department of Biology, Massachusetts Institute of Technology, Cambridge, Massachusetts 02139

Edward G. Ruby
Pacific Biomedical Research Center, University of Hawaii, 41 Ahui Street, Honolulu, Hawaii 96813

Lawrence J. Shimkets
Department of Microbiology, 527 Biological Sciences Building, University of Georgia, Athens, Georgia 30602-2605

Ann M. Stevens
Department of Biology, Virginia Polytechnic Institute and State University, Blacksburg, Virginia 24061

Gordon S. A. B. Stewart
School of Pharmaceutical Sciences, University of Nottingham, Nottingham NG7 2RD, United Kingdom

Simon Swift
Institute of Infections and Immunity, Queen's Medical Centre, University of Nottingham, Nottingham NG7 2UH, and School of Pharmaceutical Sciences, University of Nottingham, Nottingham NG7 2RD, United Kingdom

Karen L. Visick
Department of Microbiology and Immunology, Loyola University of Chicago, 2160 South First Avenue, Building 105, Maywood, Illinois 60153

Paul Williams
Institute of Infections and Immunity, Queen's Medical Centre, University of Nottingham, Nottingham NG7 2UH, and School of Pharmaceutical Sciences, University of Nottingham, Nottingham NG7 2RD, United Kingdom

Stephen C. Winans
Section of Microbiology, Wing Hall, Cornell University, Ithaca, New York 14853

Derek W. Wood
Department of Plant Pathology, 204 Forbes Hall, University of Arizona, Tucson, Arizona 85721-0036

Jun Zhu
Section of Microbiology, Wing Hall, Cornell University, Ithaca, New York 14853

ACKNOWLEDGMENTS

The editors would like to thank Pete Greenberg and Barbara Iglewski for providing a great deal of the initial encouragement and support for the idea of putting together a book on bacterial cell-cell signaling. Greg Payne, of ASM Press, provided the initial "organizational encouragement" and facilitated the process of nuturing this project from the planning stages right through to the finished text. At ASM Press, Ellie Tupper handled the book production process with competence and efficiency.

Virtually all chapters in the book were reviewed for scientific content by one of the editors and one additional reader. Individuals who helped with this task included Don Clewell, Donald Morrison, Alan Grossman, Heidi Kaplan, Patrick Schlievert, David Sherman, Martin Dworkin, and Kurt Fredrick. We are grateful to Shirley Cramer and Melodie Bahan for secretarial help. We also want to acknowledge the tremendous efforts and enthusiasm of the authors of all the chapters who put forth their best efforts to produce high-quality contributions, and who also completed their chapters in a timely fashion.

Finally, we acknowledge the efforts of Ed Atkeson, of Berg Design, in adapting the beautiful drawings of myxobacterial cell-cell interactions by Roland Thaxter to the cover of the book.

BACTERIAL LIFE: NEITHER LONELY NOR BORING

Gary M. Dunny and Stephen C. Winans

The paradigm of the asocial existence of the bacterial cell has been a major intellectual force driving research in modern microbiology. In this view of prokaryotic life, organisms live a unicellular existence, with responses to external stimuli limited to the detection of chemical and physical signals of environmental origin. Perhaps this point of view was most clearly elucidated by François Jacob (1973) when he stated, "What, then, could be the aim of the bacterium? What does it want to produce that justifies its existence, determines its organization, and underlies its work? There is apparently only one answer to this question. A bacterium continuously strives to produce two bacteria. This seems to be its one project, its sole ambition." No one could argue that the growth of a microbial cell into two progeny is not an important bacterial activity. However, it is now clear that certain developmental processes and other forms of multicellular behavior not directly related to vegetative growth are critical elements in the biology of microorganisms.

Over 60 years ago the literature contained snippets of data suggesting the possibility of bacterial group behavior with the implication of requisite intercellular signaling mechanisms (see chapters 2 and 18). The significance of such reports was not widely appreciated. Indeed, the prevailing view was that each member of the population of a microbial culture in balanced growth conditions was essentially identical to all others, and most importantly, also identical to a single cell grown in the same medium. Again, in the words of Jacob (1973), "It is perfectly possible to imagine a rather boring universe without sex, without hormones and without nervous systems; a universe peopled only by individual cells reproducing *ad infinitum*. This universe, in fact, exists. It is the one formed by a culture of bacteria."

In the 1960s and 1970s, sufficient genetic and biochemical tools became available to allow for detailed investigations of phenomena that challenged the unicellular paradigm. During this time period, studies of three groups of microorganisms (the gram-positive cocci, the myxobacteria, and the luminescent marine vibrios) provided convincing evidence for multicellular forms of behavior employing sophisticated chemical communication systems to coordinate the activities of individuals within a population.

Gary M. Dunny, Department of Microbiology, University of Minnesota Medical School, 1460 Mayo Bldg./Box 196, 420 Delaware St., SE, Minneapolis, MN 55455-0312. *Stephen C. Winans*, Section of Microbiology, Wing Hall, Cornell University, Ithaca, NY 14853.

Cell-Cell Signaling in Bacteria, Edited by Gary M. Dunny and Stephen C. Winans
©1999 American Society for Microbiology, Washington, D.C.

In the case of the myxobacteria, early observations suggesting the possibility of bacterial signaling during fruiting body formation were given concrete support by genetic studies (McVittie et al., 1962) demonstrating extracellular complementation of defects in fruiting body formation between two mutant strains of *Myxococcus xanthus*. Dworkin (1991b) summarized additional information about the early evidence for myxobacterial signaling, while chapters 5 and 6 of this volume present the current status of this field.

Tomasz (1965) presented the first direct experimental evidence for release of a signal molecule into growth medium. He demonstrated that previous observations of variable expression of pneumococcal competence, ascribed to a shift in the physiological state of the organisms (Hotchkiss, 1954; Tomasz and Hotchkiss, 1964), were due to density-dependent accumulation of an extracellular, competence-activating substance. He drew an analogy between this factor and hormones in higher organisms and made a number of predictions about the factor and its mode of action. Although the initial hypothesis that the factor was a fairly large protein turned out to be incorrect, most of the other predictions were ultimately verified 30 years later when its purification and molecular characterization were completed (chapter 2).

Within a few years, models invoking cell-cell communication by chemical signals were also proposed in two additional systems that have developed into important experimental model systems for detailed study of bacterial communication. These include the control of expression of bioluminescence in marine vibrios (chapters 14, 15, 17, 18), and conjugative plasmid transfer in enterococci (chapters 4 and 20). Although these systems are presently recognized as important paradigms for ubiquitous microbial communication systems, the idea of group behavior controlled by cell-cell signaling in bacteria was not always accepted with immediate enthusiasm (chapter 18). The pioneering work on induction of bioluminescence by the homoserine lactone (HSL)

autoinducers of the marine *Vibrio* spp. (reviewed in chapter 18) included the first identification of a bacterial extracellular signal. This class of molecules is now known to function in a wide range of microbial processes carried out by gram-negative bacteria (chapters 7, 8, 10, 20). The enterococcal sex pheromones (chapters 4 and 20) were the first peptide signals identified, and peptides are now known to comprise the predominant mode of signaling in gram-positive bacteria.

The level of understanding both of specific signaling mechanisms and of the biological processes regulated by such mechanisms has increased to a remarkable degree. Perhaps the most notable development in the area of microbial signaling in recent years is the explosion of data from the studies of diverse bacteria. These results indicate that a well-defined set of signaling mechanisms has been very widely disseminated among prokaryotes. The extent to which cell-cell signaling has been found to regulate important microbial processes suggests that signaling and group behavior are actually the typical modus operandi for many if not all bacteria. Studies of cell-cell interactions in bacterial colonies on agar medium (Shapiro, 1997) suggest that this group of microbes may well include the laboratory workhorse strains of *Escherichia coli*, although the molecular basis for these interactions is not yet understood in this organism.

The general importance of multicellular behavior in many bacteria was first explicitly acknowledged by a pair of conferences (ASM Conferences on Multicellular Behavior of Bacteria, Woods Hole, Mass., 1990 and 1993) and books on this subject which were organized and edited by M. Dworkin and J. Shapiro (Dworkin, 1991a; Shapiro and Dworkin, 1997). Subsequently, molecular and genetic analyses of signaling mechanisms controlling multicellular behaviors, and the identification of such mechanisms in additional organisms, have reached the stage where they are widely appreciated as one of the most important areas of current research. In a review of the microbial communication systems using homoserine lactone signal molecules, the term "quorum sensing" was used to describe density-dependent phenomena controlled by extracellular signals (Fuqua et al., 1994). This term (see box) has become prominent in the microbiological literature, since so many microbial activities have been found to be controlled by this kind of mechanism. The explosion of knowledge in this area, and the realization that work being done in each of the signaling systems being studied may have important implications for other organisms not closely related by phylogeny or ecological niche, stimulated the generation of this book.

In terms of extracellular signaling mechanisms, a limited number of systems have apparently evolved, with each one having been adopted by many organisms to regulate a variety of biological processes (Table 1). In gram-negative bacteria the most commonly identified signaling mechanism uses homoserine lactones for extracellular communication, with the signal entering the cell via diffusion and interacting with intracellular effectors (see chapter 15). Although a butyrolactone-signal is used to regulate antibiotic production in postexponential cultures of *Streptomyces* (see chapter 13), homoserine lactone–based signaling has not been found to date in the gram-positive bacteria. In these organisms, peptide-based signaling seems to be the preferred mode of communication. Interestingly, the signaling mechanism can either involve import of the signal and subsequent interaction with an intracellular effector (chapters 3, 4, 15, 17 and 20), or transduction across the membrane via a two-component sensing system (chapters 2, 3, 9, 11, 12). Peptide-based systems have also been identified in gram-negative organisms, but in the myxobacterial A-signal system, the active peptides seem to be processed to amino acids, which appear to constitute the actual signal (chapter 5). These organisms have developed a plethora of extracellular communication circuits, including some dependent on direct contact between the communicating cells (chapter 6), where the actual signal molecule remains unknown.

The primary interest of many current researchers who study microorganisms is to elucidate the molecular basis for a particular biological process by using the experimental advantages offered by microbial systems. It is therefore not surprising that a considerable portion of the current research in microbial communication, as well as a significant portion of this book, is devoted to the molecular biology of signaling processes. However, bacteria are also tremendously important from an organismal point of view. Much of the initial interest in the study of microorganisms was due to their importance in interactions with higher organisms (especially in disease situations). Of course, bacteria also play a vital role in global ecology and are widely used in the production of fermented foods, medicines, and other useful products. Each of these areas continues to be of major importance and is turning out to have microbial cell-cell signaling as a major component of its biology. To emphasize this fact, we have chosen to organize the chapters primarily according to the nature of the biological phenomenon affected by signaling. Such phenomena include symbioses, genetic transfer and microbial development, and production of antimicrobial compounds. The molecular basis of signaling is covered to some extent in most of the chapters, but is emphasized in detail in section IV.

On the Genesis of "Quorum Sensing"

The term "quorum sensing" was coined over Thanksgiving dinner, when one of the authors (S.C.W.) asked members of his extended family to help find a lively new phrase to describe the autoinducer-type pheromones. It was agreed that the terms "communulin" and "gridlockin" did not resonate. At some point between turkey and pie, a nonscientist in the group, Rob Johnston, mused that a high cell density could be viewed as a bacterial quorum. He then exclaimed, "Got it! Quormone!" In hindsight, it seems that the venerable terms "pheromone" and "autoinducer" suffice after all, but that "quorum" appears to have answered some etymological need within the microbiology community.

TABLE 1 Cell-cell communication systems in bacteria

Genus (genera)	Process(es) regulated	Signal molecule(s)	Mechanism(s)[a]	Chapter(s)
Vibrio	Bioluminescence	HSLs[b]	Import via diffusion	14, 15, 17, 18, 21
Streptococcus	Competence, virulence	Peptides	Transduction or import	2, 20
Myxococcus	Development	Peptides, amino acids, others??	Transduction, others?	5, 6
Bacillus	Competence, development	Peptides	Import or transduction	3, 16
Erwinia, Ralstonia, Rhizobium, Xanthomonas, Pseudomonas	Plant symbiosis, plant pathogenesis	(HSLs)	Import via diffusion	7, 19
Agrobacterium	Conjugation	HSLs	Import via diffusion	8, 14
Pseudomonas	Mammalian pathogenesis	HSLs	Import via diffusion	10, 19
Enterococcus	Conjugation, plasmid maintenance, pathogenesis	Peptides	Import	4, 20
Staphylococcus	Pathogenesis	Peptides	Transduction	9
Lactococcus and other lactic acid bacteria	Bacteriocin production	Peptides	Transduction	11, 12
Streptomyces	Antibiotic production	γ-Butyrolactone	Import via diffusion?	13

[a] HSLs enter the cell via diffusion, while some peptides are imported by a dedicated transport system. Both types of signal then interact with intracellular effectors. Other peptides appear to transduce a signal via a two-component sensing system. Direct evidence for these mechanisms has only been obtained for a few systems. The others are inferred via sequence homology.
[b] HSLs, homoserine lactones.

The final section of the book contains four chapters that provide some perspective on where the field has been and where it may be going. Obviously, a number of chapters could have been placed in more than one section. We hope that the arrangement chosen will help the readers gain a higher level of appreciation for the scope of microbial biology that is controlled by intercellular communication as they page through the book on their way to the chapters on their favorite microbes.

It must be emphasized that an enormous amount remains to be learned about microbial signaling in terms of its basic mechanisms, its application for useful purposes, and the extent to which these mechanisms really pervade the microbial world. For example, there is no coverage of the Archaea in this book, even though there is every reason to believe that cell-cell communication processes occur in these organisms. Even as this book goes to press, new examples have emerged of extracellular control of important microbial processes that have been of interest to researchers for years. These include the determination of the location of the heterocyst cell in growing chains of *Anabaena* (Golden, 1998) and the development of biofilms (Davies et al., 1998).

Many short open reading frames in bacterial DNA that could encode peptide signals are generally dismissed as too short to be of biological significance. The potential importance of these kinds of Orfs is exemplified by the ComX pheromone of *Bacillus subtilis*, which is encoded by a small gene that had been sequenced for several years before any biological function was suspected (chapter 3).

It has also become clear that at least some well-studied signal molecules may serve more than one function (see chapters 3, 4, 17, and 20). Which function actually came first? What were the key evolutionary steps involved in appropriation of signal molecules and sensing machinery for new purposes? As pointed out in chapter 17, there are also signaling systems whose molecular biology is quite well worked out, but where the role of the system in the ecology of the organism is still elusive. In terms of applications, it is obvious that manipulation of systems that control expression of useful compounds such as nisin holds great potential for increasing production of these compounds as well as for developing heterologous expression systems (chapters 11 through 13). The signaling systems controlling expression of virulence determinants are promising targets for development of new vaccines and antimicrobial agents.

Finally, even in the signaling systems that have been subjected to the most extensive experimental scrutiny, new fundamental insights into the cellular and molecular biology of bacteria are yet to come. For example, in a pure culture of competent *B. subtilis*, only a fraction of the cells acquire the ability to transform, even though all the cells in the culture are, to the best of our knowledge, genetically identical. All these cells are growing in exactly the same chemical environment in the presence of the same concentration of competence pheromone (chapter 3). This means that, even in a controlled laboratory situation, a process regulated by a well-studied pheromone signaling mechanism still contains unsolved mysteries. This suggests that the assertion of Jacob about the "boring universe" of the bacterial culture is clearly a considerable oversimplification.

ACKNOWLEDGMENT

We thank Marty Dworkin for many helpful insights on the importance of signaling and multicellularity.

REFERENCES

Davies, D. G., M. R. Parsek, J. P. Pearson, B. H. Iglewski, J. W. Costerton, and E. P. Greenberg. 1998. The involvement of cell-cell signals in the development of a bacterial biofilm. *Science* **280**:295–298.

Dworkin, M. (ed.). 1991a. *Microbial Cell-Cell Interactions*. American Society for Microbiology, Washington, D.C.

Dworkin, M. 1991b. Cell-cell interactions in myxobacteria, p. 179–216. *In* M. Dworkin (ed.), *Microbial Cell-Cell Interactions*. American Society for Microbiology, Washington, D.C.

Fuqua, W. C., S. C. Winans, and E. P. Greenberg. 1994. Quorum sensing in bacteria: the LuxR-LuxI family of cell density-responsive transcriptional activators. *J. Bacteriol.* **176**:269–275.

Golden, J. 1998. Personal communication.

Hotchkiss, R. D. 1954. Cyclical behavior in pneumococcal growth and transformability occasioned by environmental changes. *Proc. Natl. Acad. Sci. USA* **40**:49–55.

Jacob, F. 1973. *The Logic of Living Systems: A History of Heredity*. English translation by Betty E. Spillman. Alan Lane (Div. of Penguin Books, Ltd.), London.

McVittie, A., F. Messik, and S. A. Zahler. 1962. Developmental biology of *Myxococcus*. *J. Bacteriol.* **84**:546–551.

Shapiro, J. A. 1997. Multicellularity: the rule, not the exception, p. 14–49. *In* J. A. Shapiro and M. Dworkin (ed.), *Bacteria as Multicellular Organisms*. Oxford University Press, New York.

Shapiro, J. A., and M. Dworkin. 1997. *Bacteria as Multicellular Organisms*. Oxford University Press, New York.

Tomasz, A. 1965. Control of the competent state in *Pneumococcus* by a hormone-like cell product: an example of a new type of regulatory mechanism in bacteria. *Nature* **208**:155–159.

Tomasz, A., and R. D. Hotchkiss. 1964. Regulation of the transformability of pneumococcal cultures by macromolecular cell products. *Proc. Natl. Acad. Sci. USA* **51**:480–487.

GENE TRANSFER AND MICROBIAL DEVELOPMENT

I

QUORUM SENSING AND PEPTIDE PHEROMONES IN STREPTOCOCCAL COMPETENCE FOR GENETIC TRANSFORMATION

Leiv Sigve Håvarstein and Donald A. Morrison

2

It has been pointed out that the technique employed in the foregoing experiments varied in certain particulars from that which had been adopted in previous unsuccessful in vitro experiments. One of these variations consisted in the use of very small seedings of the R culture . . . These experiments demonstrated the importance of employing small amounts of the R culture and afforded an explanation for the failure of many previous attempts to secure transformation of type by in vitro methods. . . . Large inocula usually resulted in the growth of R forms only. The growth of large numbers of R pneumococci apparently created conditions unfavorable for transformation. No adequate explanation of this finding is offered (Dawson and Sia, 1931).

Cultures of certain pneumococcal strains can develop a cellular condition during growth in which most of the individual bacteria are capable of reacting with DNA molecules present in the environment: the DNA molecules are absorbed and may be incorporated into the genome of the recipient cells. Recent investigations on this "competent" or "transformable" condition revealed that pneumococci in the competent state can "communicate" this property to incompetent cells. A key component of this system, a macromolecular activator substance, was separated from competent bacteria and was found to be responsible for the cell-to-cell transfer of competence (Tomasz and Mosser, 1966).

Streptococcal transformation has long been linked to cell-to-cell signaling. The roots of the recognition that genetic transformation is controlled in *Streptococcus pneumoniae* (pneumococcus) by a quorum-sensing mechanism can be traced to the work of Dawson and Sia (1931), which showed that the "transformation of types," which had previously been observed only as a transfer of alleles occurring in a living mouse host (Griffith, 1928), could also occur in vitro, but that competence was not a constant property of pneumococcus. Cultures grown from small inocula to high density in the presence of donor cell extracts sometimes contained recombinants at the conclusion of culture growth. Previous attempts to demonstrate transformation in vitro had apparently been frustrated by the use of inocula at a cell density above that at which a pneumococcal quorum-sensing mechanism operates. This method for in vitro genetic transformation led in time to the classic experiments identifying the active donor agent as DNA, studies in which it was also shown that both active growth and other favorable conditions were required for the appearance

Leiv Sigve Håvarstein, Department of Biotechnological Sciences, Seksjon Mikrobiologi, Fellesbygget, Agricultural University of Norway, P. O. Box 5040, 1342 Ås, Norway.   *Donald A. Morrison*, Laboratory for Molecular Biology, Department of Biological Sciences, University of Illinois at Chicago, Chicago, IL 60607.

Cell-Cell Signaling in Bacteria, Edited by Gary M. Dunny and Stephen C. Winans
©1999 American Society for Microbiology, Washington, D.C.

of high levels of competence for genetic trans-formation (Avery et al., 1944). It was two dec-ades later, however, that analysis of this pneumococcal phenomenon (Tomasz and Hotchkiss, 1964; Tomasz, 1965; Tomasz and Mosser, 1966), and of a parallel one in *Strep-tococcus sanguis* (Pakula and Walczak, 1963; Dobrzanski and Osowiecki, 1967), showed that a coordinated induction of competence occurs at a particular cell density. These anal-yses implicated a proteinaceous extracellular factor, termed competence activator (CF), that could substitute for other cells in causing the induction. The evocative term quorum sensor was later coined to describe such behavior (Fuqua et al., 1994). More recently, purifica-tion of a 17-residue basic peptide with the properties of CF has led to the characterization of a regulatory circuit including peptide-processing and peptide receptor proteins (Hå-varstein et al., 1995a, 1996). Natural genetic transformation has been reported in other streptococcal species, but only recently, with identification of conserved features of the quorum-sensing system, has it become possi-ble to screen easily for similar systems in many new species or strains. The results already show that competence for genetic transfor-mation, regulated by a peptide-mediated quorum-sensing system, is common within a large group of closely related streptococci. The competence-regulating quorum-sensing genes share ancestors with certain peptide bac-teriocin regulatory genes, including a family of peptide-secretion ABC transporter proteins and a family of histidine protein kinases.

Regulation of transformability seems at first to be an unusual role for quorum sensing. Most quorum-sensing systems described in this book were identified from roles in coor-dinating the release of products whose effect-iveness for the producer cell would be compromised by dilution. In the case of trans-formation, there is no known released product required for DNA recognition or uptake at all, and dilute competent cells take up DNA just as efficiently as do concentrated cells (Håvar-stein et al., 1995a). However, the one obvious extracellular component of transformation is DNA; if transformation evolved as a mecha-nism of intraspecies transfer of alleles, then a sensor for potential gene donors of the same species makes sense. In contrast, other pro-posed major roles of transformation, such as nutrition, response to stress, or interspecies gene transfer (Dubnau, 1991), are more diffi-cult to reconcile with regulation of compe-tence through a quorum-sensing mechanism.

In this chapter, we describe a peptide com-ponent of the competence quorum-sensing system, survey the variety of streptococcal competence-regulator peptides, consider the components of the regulatory system identi-fied to date, and outline a model for the in-teraction of those components.

PEPTIDE COMPETENCE PHEROMONES: INDUCTION OF COMPETENCE IN NATURALLY TRANSFORMABLE STREPTOCOCCI BY PEPTIDE SIGNALS

Purification of the Competence-Stimulating Peptide (CSP) from *S. pneumoniae* Rx

As described above, researchers discovered more than three decades ago that an activator present in the culture supernatants of com-petent pneumococci could substitute for other cells in causing competence induction. At-tempts to characterize the physicochemical nature of this substance demonstrated that it was highly cationic and sensitive to proteases. Estimates of its molecular weight ranged from that of a peptide to the size of a small protein. Unfortunately, the activating substance turned out to be highly refractory to further charac-terization. Biological activity was lost during purification, and it was widely believed that it might be too unstable to purify by conven-tional techniques. Recently, however, two in-dependent discoveries gave a clue to the nature of the CF. First, an ABC transporter (ComA) was identified and implicated in se-cretion of the activator by genetic evidence (Hui and Morrison, 1991; Zhou et al., 1995). A few years later, a new family of ABC transporters, dedicated to the export of anti-microbial peptides, was shown to contain N-terminal proteolytic domains (Håvarstein et al., 1995b). Concomitant with export, these proteolytic domains remove so-called double-glycine leader sequences from their peptide substrates. By homology searches, an N-terminal domain of this type was found in ComA, suggesting that its substrate, the CF, might be a peptide synthesized with a Gly-Gly leader sequence. The new insight was used to design a purification protocol for activator, based upon principles developed for peptide purification. This strategy proved to be successful, making it possible to isolate enough activator from *S. pneumoniae* Rx to determine its primary structure by Edman degradation (Håvarstein et al., 1995a). The purified competence-inducing factor, termed CSP, turned out to be a cationic peptide consisting of 17 amino acid residues (Fig. 1). Subsequent sequence determination of the corresponding structural gene (*comC*) revealed that CSP is synthesized as a precursor containing a Gly-Gly leader sequence, as predicted (Håvarstein et al., 1995a). Chemically synthesized CSP was shown to possess biological activity indistinguishable from that of the activator substance, strongly indicating that the natural peptide is unmodified. This is very fortunate, because with synthetic CSP at hand, competence can be induced without the elaborate protocols that have been used to support competence induction by endogenous activator.

Identification of a Competence Regulation Operon

By sequencing upstream and downstream of *comC*, it was soon established that the CSP structural gene is located on the same operon as two larger open reading frames (Fig. 2). Homology searches identified their products as a histidine kinase (ComD) and a response regulator (ComE), which together constitute a two-component regulatory system (Pestova et al., 1996). Signal transduction pathways of this kind are common in bacteria, where they play a central role in adapting bacterial growth to changing environmental conditions (Stock et al., 1995). Histidine kinases, usually situated in the plasma membrane, serve as environmental sensors. They modulate (by a phosphotransfer mechanism) the activity of their cognate cytoplasmic response regulators, which in most cases activate transcription of appropriate target genes. On the basis of the organization of the *comCDE* operon, it seemed likely that the two component system (ComDE) acts downstream of CSP in the signaling chain leading to competence. Genetic evidence showed those genes to be required in some way for the response to CSP (Pestova et al., 1996). Indeed, by tracing a CSP strain specificity of the competence response to strain-specific alleles of *comD*, evidence has been obtained which demonstrates that the histidine kinase ComD is the CSP receptor (Håvarstein et al., 1996). The DNA sequences of 10 different *comD* alleles, representing different pheromone specificities (pherotypes), have now been published (Håvarstein et al., 1996, 1997). Prediction of the possible topology of these receptors shows in all cases that they contain five, six, or most probably seven transmembrane segments. Comparisons of the primary structures of pairs of closely related *comD* alleles revealed that amino acid substitutions are clustered within their N-terminal region (Håvarstein et al., 1996; Håvarstein, unpublished results). This region, which encompasses the first two to three transmembrane segments (Fig. 3), must be involved in recognizing and binding the competence pheromone. Studies further elucidating the role of the two-component regulatory system are discussed below.

CSPs from Other Streptococci

After identification of the *comCDE* genes of *S. pneumoniae* a few years ago, it became possible to look for the presence of these genes in other species as well. In pneumococcus, where the widest surveys have been done, two different alleles of *comC* are common. They encode CSP-1, which corresponds to the first

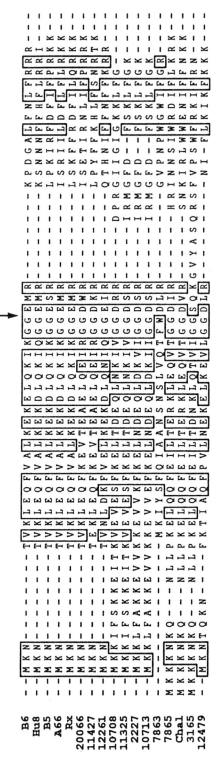

FIGURE 1 Comparison of the primary structure of deduced CSP precursors from different streptococcal strains. The N-terminal Gly-Gly leaders are cleaved from the mature pheromones, as indicated by the vertical arrow. Boxes indicate identical or similar amino acid residues in at least half of the compared peptides. 10708, *S. milleri* NCTC 10708; 10713, *S. anginosus* NCTC 10713; 11325, *S. constellatus* NCTC 11325; 11427, *S. oralis* NCTC 11427; 12261, *S. mitis* NCTC 12261; 12479, *S. crista* NCTC 12479; 20066, *S. mitis* DSM 20066; 2227, *S. intermedius* NCDO 2227; 3165, *S. gordonii* NCTC 3165; 7863, *S. sanguis* NCTC 7863; 7865, *S. gordonii* NCTC 7865; A66, *S. pneumoniae* A66; B5, *S. mitis* B5; B6, *S. mitis* B6; Chal, *S. gordonii* Challis; Hu8, *S. mitis* Hu8; Rx, *S. pneumoniae* Rx. (From Håvarstein et al., 1997).

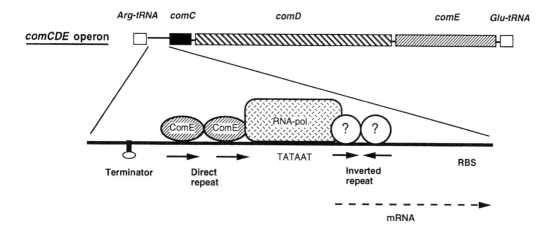

Promoter region

FIGURE 2 Location of possible binding sites for transcriptional regulators in the promoter region of the competence pheromone operon of *S. pneumoniae* Rx. Preliminary results show that ComE binds to the direct repeat as indicated. A sequence believed to be the Pribnow box (TATAAT) is situated about 30 bp downstream of the direct repeat. Seven basepairs further downstream there is an inverted repeat which, judging from its proximity to the Pribnow box, may bind a repressor.

peptide isolated from strain Rx, and CSP-2, a closely related peptide identified in strain A66 by Pozzi et al. (1996). Among 42 CSP-1- or CSP-2-carrying strains identified in this study, 20 respond to one of the pheromones, demonstrating that responsiveness to cognate peptide is quite common. An extensive survey conducted by Ramirez et al. (1997) described the distribution of competence genes within a large array of pneumococcal strains. Among 214 strains examined from various countries and contexts of isolation, 100% carried the

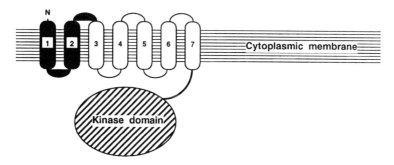

FIGURE 3 Model depicting the domain organization of CSP receptors (ComD). Comparisons of closely related ComD sequences representing different pheromone types (phenotypes) show that amino acid substitutions altering phenotype specificity are located to the N-terminal part of the membrane domain (indicated by solid fills). Thus, this part of the receptor must be essential for CSP recognition.

genes *comA* and *comC*. In 50 strains in which the sequence of *comC* was determined, all but 1 carried one of the two *comC* alleles mentioned above; a single instance of a third allele is of unknown significance. In a recent study (Håvarstein et al., 1997), PCR was used to look for the *comCDE* operon in a number of streptococcal species, to map the prevalence of this operon in the genus as a whole. The PCR amplification was carried out with primers complementary to conserved tRNA genes previously shown to flank the competence regulation operon in *S. pneumoniae* and *Streptococcus gordonii* strains (Pestova et al., 1996; Håvarstein et al., 1996). Using this approach, it was found that species belonging to the mitis and anginosus groups in general possess the *comCDE* operon, whereas it was not detected in strains assigned to the salivarius, bovis, mutans, or pyogenic groups. These results do not necessarily mean that strains found to be lacking the *comCDE* operon are noncompetent, as competence in these strains could be regulated by an entirely different mechanism, or their *comCDE* operons might not be amplified by the set of primers used.

So far, the primary structures of 15 different CSPs have been determined. The alignment in Fig. 1 shows that all of them are cationic peptides consisting of 14 to 23 amino acid residues. Although the mechanism of action of CSP at the receptor is not known, comparison of related peptides may provide some clues. An alanine replacement scan of the peptide CSP-1 revealed that some sites were more important for its potency than others (Table 1), with residual activities of the CSP analogs covering a wide range (Coomaraswamy, 1996). Certain terminal or subterminal residues were especially important. *N*-Acetylation of the N-terminal glutamate, or substitution by alanine, reduced activity severely, while replacement of the conserved arginine at position 3 abolished activity. At the C terminus, a positive charge was essential, although loss of only one or two of the three basic residues had modest effects. Certain hydrophobic residues were also important, especially the phenylalanines at positions 7, 8, and 11. Two additional important hydrophobic residues (I12 and L4) share with the phenylalanines a single surface of the peptide when it is arranged in an alpha-helix with a 3.6-residue pitch, while less important, and mainly hydrophilic, residues occupy the opposite face. Finally, comparison with Fig. 1 also shows that just those features most sensitive to replacement in CSP-1 are conserved strongly in the known streptococcal competence pheromones, despite a wide sequence variation among them. All have a positive charge at position 3 and most have a positive charge at the C terminus and a negative charge at the N terminus. All seem to contain as well a central core capable of forming a 3- or 4-turn amphipathic helix. The congruence of conserved structures with the most important residues identified in CSP-1 may be more than coincidence and suggests that peptide binds to receptor in a helical form, with the most critical residues limited to one hydrophobic face of the helix and to the charged termini.

PHEROTYPE SPECIFICITY

At the peak of competence the culture supernatant of streptococci from the mitis and anginosus groups contains a competence-inducing factor. If supernatant is harvested from a particular competent isolate, and a small volume is added to a noncompetent culture of the same bacterium, competence can be induced. However, corresponding samples will rarely induce competence when added to noncompetent cultures of other isolates (Perry and Slade, 1966; Gaustad, 1979). The mechanism behind this remarkable specificity can now be explained. Streptococci secrete CSPs with different primary structures, which interact specifically with their cognate histidine kinase receptors (ComD). Therefore, a CSP produced by one species will in general not be recognized by the receptor domain of the receptor from another species. Even different strains/isolates of the same species often have distinct pherotypes, as demonstrated by the comparison of CSPs in Fig. 1. Interestingly, three species from the anginosus group, *Streptococcus anginosus* NCTC

TABLE 1 Structural analysis of peptide pheromone: activity of replacement and deletion derivatives of pneumococcal CSP-1

Position in peptide	CSP-1 residue	Substitution in analog	Relative activity of analog (%)[a]
1	E	A	1.6
2	M	A	4.2
3	R	A	<0.1
4	L	A	2
5	S	A	17
6	K	A	4
7	F	A	0.4
8	F	A	0.4
9	R	A	5.3
10	D	A	16
11	F	A	0.4
12	I	A	1.9
13	L	A	3.6
14	Q	A	8
15	R	A	30
16	K	A	26
17	K	A	88
15,16,17	RKK	RK	28
15,16,17	RKK	R	20
15,16,17	RKK		0.88
CSP			100

[a] Activity is proportional to the inverse of the concentration required to elicit 30% of a maximal response in CP1214 (comA) (data from Coomaraswamy [1996]). Actual value at 100% relative activity was 125 ml/μg.

10713, *Streptococcus constellatus* NCTC11325, and *Streptococcus intermedius* NCDO 2227, differ from the others. These species produce and respond to identical CSPs (Fig. 1) and therefore belong to the same pherotype. A comparison of the nucleic acid sequences of the *comCDE* operons of *S. anginosus*, *S. constellatus*, and *S. intermedius* shows that the sequences are highly similar, indicating that the *comCDE* genes have been transferred horizontally between the species relatively recently. Further evidence of horizontal gene transfer comes from the discovery of mosaic structures in the *comCDE* genes of some other strains (Fig. 4). Nucleic acid sequence comparisons of different alleles of these genes show that blocks of low homology (53 to 62% identity) are inserted between flanking sequences that are 92 to 100% identical (Håvarstein et al., 1997). These mosaic structures most likely result from interspecies recombinational exchanges. In all cases the inserted sequences encompass a region encoding the CSP and the part of the receptor postulated to be involved in CSP binding. Consequently, the recombination events appear to have led to switches in pherotypes for the strains involved in a single step. The possible biological significance of such changes is not known, but judging from the frequency of mosaic structures found so far, switching of pherotypes seems to be selected under certain growth conditions.

Considering the number of different CSPs already identified, many new pherotypes will certainly be found as the number of isolates examined grows. What is the biological significance of such a plethora of pherotypes?

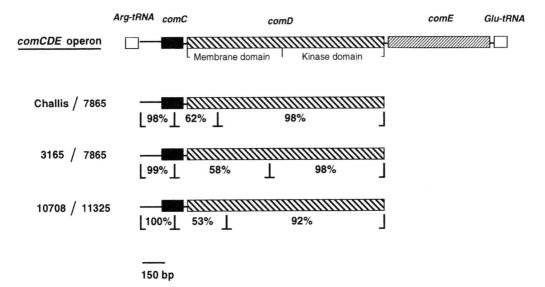

FIGURE 4 Mosaic structure of *comC* and *comD* genes. The nucleic acid sequences (approx. 1,600 bp) between the Arg-tRNA and *comE* genes of several related streptococcal strains were compared in pairs as indicated: Challis/7865, *S. gordonii* Challis against *S. gordonii* NCTC 7865; 3165/7865, *S. gordonii* NCTC 3165 against *S. gordonii* NCTC 7865; 10708/11325, *S. milleri* NCTC 10708 against *S. constellatus* NCTC 11325. Regions of high and low homology are indicated by brackets and different fill patterns. Percentage identity is calculated for each region. The low-homology regions are believed to have been introduced by recombination. (From Håvarstein et al., 1997.)

Pherotypes divide competent streptococci into separate populations, which in most cases will consist of only closely related bacteria. Members of each population are "deaf" to signals from bacteria other than those from their own group and develop competence only after the pherotype population to which they belong reaches a critical density. It is possible, though, that highly similar pheromones exhibit low cross-inducing activity. Perhaps natural transformation is a mechanism for exchange of DNA between bacteria belonging to the same pherotype, and the diversity of pherotypes has evolved to ensure that a bacterium does not develop competence when surrounded by more distantly related bacteria. This hypothesis requires that some or all bacteria in a competent population act as donors of DNA. Indeed, experiments indicating that small amounts of DNA are released in pneumococcal cultures during competence development have been reported (Ottolenghi and Hotchkiss, 1960, 1962). Thus, natural transformation in streptococci may serve the same function that sex is believed to serve in higher organisms, namely, to create greater genetic diversity for natural selection to act upon, by promoting genetic recombination between closely related individuals. Gene exchanges between less related individuals might not in general be beneficial, because homologous recombination would more frequently give rise to mosaic genes encoding inferior or nonfunctional gene products. This hypothesis, however, does not account well for the observed pherotype switching or for the fact that three different species from the anginosus group produce and respond to identical pheromones.

THE BIOLOGY OF TRANSFORMATION AND COMPETENCE: GLOBAL REGULATION RESPONDS TO POPULATION DENSITY BUT ALSO TO OTHER ASPECTS OF METABOLISM OR ENVIRONMENT

Despite the caveat that pure in vitro cultures must reflect behavior in vivo imperfectly, competence induction appears to offer a particularly clear case of quorum-sensing regulation. In the typical example of competence induction by endogenous signals, there is a very sudden onset of transformability at the first appearance of competence during growth of a culture (Hotchkiss, 1954; Tomasz, 1965, 1966; Chen and Morrison, 1987). This first "wave" of competence subsides nearly as rapidly, limiting the period of maximal competence to 10 to 20 min. One or two subsequent waves of competence are often observed, peaking at intervals of ~1.5 doubling times as the culture continues exponential growth, but involving smaller fractions of the population. Approach to the stationary-growth phase terminates this competence pattern, whenever it was first triggered. The requirement for actively growing cells and the suppression of competence in later growth phases seem to be sufficient to explain the failure of the early efforts to detect transformation in vitro mentioned above and the success achieved when inocula were reduced (Dawson and Sia, 1931). The "lore" of streptococcal competence indicates that numerous additional factors affect whether this train of events is initiated, but they are not generally well defined (McCarty et al., 1946). Thus, incorporation of $CaCl_2$ and bovine serum albumin into the growth medium favors competence, while substitution of maltose for glucose or replacement of the PO_4 supplement with another buffer prevents competence induction. Similarly, reducing the initial growth medium pH from 7.8 to 6.8 blocks competence induction (Tomasz and Hotchkiss, 1964), while at intermediate pH values, the critical cell density for endogenous competence induction varies continuously over this pH range (Chen and Morrison,

1987), as if either the production of a signal or the sensitivity to the signal varied sensitively with pH. Endogenous induction also varies with cell genotype: some strains become competent more readily than others (Ramirez et al., 1997), and endogenous induction of competence is generally suppressed by the presence of a polysaccharide capsule (Yother et al., 1986; Pozzi et al., 1996; and references therein). Some strains are temperature sensitive for some part of this process and will complete competence development well only if cooled below the usual culture temperature of 37°C (Fox and Hotchkiss, 1957; Lacks and Greenberg, 1973). Peptide supply may affect competence induction (Pakula and Ihler, 1969; Alloing et al., 1996), and zinc starvation can apparently interfere with later steps of transformation (Dintilhac et al., 1997). Additional links, to cell-wall metabolism, are suggested by competence phenotypes of mutations affecting regulatory proteins originally identified as penicillin-resistance mutations (Guenzi et al., 1994). Studies using a bioassay for CF have shown that its level increases strongly with the onset of competence, whether provoked by endogenous signals or by a small dose of CSP, suggesting that the suddenness of induction reflects an autoinducible signal (Tomasz, 1966; Morrison, 1981). The response requires protein synthesis (Tomasz, 1970) and in fact entails a synchronized global switch of protein synthesis (Morrison and Baker, 1979; Morrison, 1981) and acquisition of a highly efficient pathway for transfer of genetic material from external DNA to the cell genome.

This constellation of properties suggests that pneumococcal competence may offer a valuable model for dissection of bacterial quorum sensing. The response to the synthetic peptide signal is dramatic, rapid, global, and yet self-limiting. The entire process of induction, response, and attenuation of the response can occur in a short time, entirely during exponential-growth phase, disassociated from the complexity of the transition to stationary phase. In the induction of competence by endogenously produced signals, the presence of other pneumococcal cells is sensed at rather low densities, where unrelated effects of high cell density, such as nutrient limitation, may be expected to be minimal; yet the response to attainment of a critical cell density is rapid and pervasive, affecting all cells in a culture.

EVIDENCE LINKING CSP AND ITS GENE comC TO QUORUM SENSING

Evidence that the peptide CSP participates in pneumococcal competence regulation combines several types of data, including (i) properties of mutants, (ii) response of cells to synthetic CSP peptide, (iii) response of genes to synthetic peptide, and (iv) biochemical observations.

The first indications that CSP plays a role in regulation of competence were the results of experiments in which CSP induced competence development in cells that were not otherwise destined to become competent, either because the cell number was too low, because the pH was inappropriate for competence, or because the strain was an activator-deficient mutant (Håvarstein et al., 1995a). A strong additional indication that CSP is, or effectively substitutes for, the normal competence-stimulating signal is the observation that synthetic CSP elicits a competence response that qualitatively and quantitatively reflects many aspects of the competence cycle generated by endogenous signals. Thus, a rapid and synchronous first wave of competence, affecting 100% of the cells present, follows addition of CSP to a culture of (dilute or concentrated) exponentially growing noncompetent cells (Håvarstein et al., 1995a; Morrison, 1997); it is followed by a shutdown of competence and then by further waves of competence as growth proceeds, the waves soon merging into a low-level continuum of competence that involves only a minority of the cells at any specific moment (Morrison, 1997). The well-known pH modulation of competence is also reflected directly in the CSP response, with sensitivity to the peptide decreasing with decreasing pH: for example, amounts of CSP eliciting half-maximal response in a non-CSP-producing strain were 2 ng/ml at pH 8, but 20 ng/ml at pH 6.1 (Håvarstein et al., 1995a).

A distinct indication of intimate involvement of CSP in competence regulation is obtained from analysis of the kinetics of competence induction. Early experiments using a bioassay of activator in culture supernatants suggested that the signal acted autocatalytically to stimulate an explosive accumulation of activator once a critical cell density was reached, effectively synchronizing competence throughout a culture (Tomasz, 1966; Morrison, 1981). With identification of CSP and its gene, comC, fusion of a lacZ reporter to comC made it possible to follow transcription of the CSP gene directly in a variety of culture conditions. The reporter showed that expression of this gene is detectable at a low level in growing, noncompetent cells and that the expression is greatly increased after addition of CSP to such a culture (Pestova, 1997). The strong response of the peptide precursor gene to synthetic peptide shows directly that comC is induced at competence. Indeed, this behavior may offer an explanation of the otherwise puzzling observation (Coomaraswamy, 1996) that a strain (comA) that cannot release CSP is less sensitive to CSP than an isogenic Com+ strain: this would be expected if CSP production were activated by lower levels of CSP than are required for the full competence response.

During the initial characterization of CSP, a few micrograms of the peptide were purified directly from an S. pneumoniae Rx culture, and the structure was determined by sequential Edman degradation (Håvarstein et al., 1995a). However, by itself, that experiment did not provide evidence for an association of CSP production with competence; it only showed that this strain can produce CSP. Furthermore, there is one report (Cheng et al., 1997) that the comC gene in a particular reporter-fusion derivative of strain R6x was not induced by CSP, although attempts to reproduce this result in strain R6 have not been successful (Alloing et al., 1998). To describe the behavior of *comC* more thoroughly, and to ask whether induction of transcription of the *comC* gene does actually correlate with accumulation of high levels of CSP, a more detailed examination of the early steps of competence induction included an enzyme-linked immunosorbent assay (ELISA) for CSP (Coomaraswamy, 1996). The kinetics of the response of competence-proficient Rx cells to a low dose of CSP (Fig. 5) show that *comC* expression precedes the appearance of DNA transport capacity, but that there is also a delay of approximately 5 min in expression of *comC*, a period during which some early steps of signal transmission may occur. The ELISA showed that a burst of CSP release coincided with *comC* induction and also preceded development of competence for DNA uptake by several minutes. Although the sensitivity of this assay was not high enough to demonstrate the basal level of CSP production, it clearly reveals the induction of CSP release in response to a small CSP dose, and the accumulation of a high level of CSP coincident with competence. The simplest way to explain that CSP acts autocatalytically would be to assume that ComE is a transcriptional activator that binds to its own promoter. Inspection of the promoter region of the

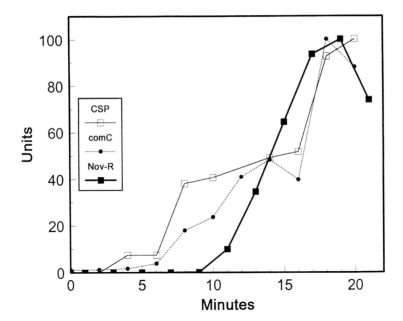

FIGURE 5 Kinetics of competence induction and *comC* expression in *S. pneumoniae*. The response was initiated by addition of 10 ng of CSP per ml to 6×10^7 CP1251 cells per ml growing in a complex medium (Chen and Morrison, 1987) at 37°C. Competence was assayed at the indicated times by 2-min exposures to 10 μg of Nv-R DNA per ml. *comC* induction was assayed as β-galactosidase activity from a *comC-lacZ* fusion made by insertion of pXF520 to duplicate part of the *comCDE* locus in strain CP1250, as described by Pestova (1996). The level of CSP in culture supernatants was measured by a sandwich ELISA using rabbit anti-CSP and a horseradish peroxidase-conjugated second antibody (Coomaraswamy, 1996). All three parameters remained at background levels when synthetic CSP was withheld. Ordinate values represent units of 10,700 Nv-R/ml (■), 4.2 Miller units (●), and 2.7 ng/ml CSP (□). Some of these data are from Coomaraswamy (1996).

comCDE operon suggests some apparent regulatory sites: an inverted repeat close to the *comC* initiation codon and a direct repeat further upstream (Fig. 2). Studies on protein-DNA interactions by electrophoretic mobility shift assays show that ComE is a DNA-binding protein and that it most likely recognizes the direct repeat (Ween et al., unpublished results). In addition, ComE seems to bind a similar motif in the promoter of the *comAB* operon, suggesting that this response regulator activates transcription of the genes encoding the CSP secretion apparatus as well. The position of the inverted repeat, relative to the binding site of ComE and the −10 region, suggests that the protein postulated to bind to this site will inhibit transcription of the *comCDE* operon.

A SIMPLE MODEL FOR A PEPTIDE QUORUM-SENSING CIRCUIT

In pneumococcus, five genes have been implicated in the peptide-mediated regulatory circuit. The products of the five genes fit naturally, on the basis of similarities to proteins of established function in other species, into the following model for a peptide-dependent quorum-sensing system, as outlined in Fig. 6. This simplified model of competence induction represents a combination of the relevant evidence for their involvement available at the time of this writing and some of the speculations discussed in this chapter.

- Peptide precursor, made by translation of *comC*, is processed and exported by ComAB, releasing the mature peptide extracellularly.
- ComC expression is normally maintained at a basal level, allowing production of peptide in proportion to cell numbers.
- ComD acts as a membrane-bound receptor/kinase and acts through a response regulator, ComE, to transmit a signal reflecting the (extracellular) abundance of CSP to responder genes.
- A low-threshold response to CSP involves binding of the pheromone to ComD, sub-

sequent phosphorylation of ComE, and binding of this response regulator to the *comAB* and *comCDE* promoters, generating the autoinduction behavior typical of this system.
- Subsequently, at higher levels of CSP, a coordinated induction of the late genes required to assemble the DNA-processing transformation machinery constitutes a high-threshold response to CSP.

ARE ALL STREPTOCOCCI NATURALLY COMPETENT?

Both direct and indirect evidence indicates that natural competence is more widespread in the genus *Streptococcus* than the results of the PCR study mentioned above might suggest. By using an in vitro transformation assay, Perry and Kuramitsu (1981) were able to demonstrate competence development in three strains of *Streptococcus mutans* (HS6, GS5, and MT557), indicating that at least some strains assigned to the mutans group are naturally competent, even though the *comCDE* operon was not detected in the *S. mutans* strains (NCTC 10449, NCTC 10832, and NCTC 10919) examined by PCR (Håvarstein et al., 1997). Analyses of the data released from the nearly completed *Streptococcus pyogenes* genome sequencing project have revealed interesting new information regarding natural competence in this species (Håvarstein, unpublished results). Searches in the *S. pyogenes* genome for homologs of the ComABCDE proteins, involved in regulation of competence development in the mitis and anginosus groups, indicate that *S. pyogenes* does not produce and secrete a signal molecule related to the CSP. As mentioned above, this pheromone is secreted by ComAB, a dedicated export machinery belonging to a subfamily of ABC transporters specific for peptides with double-glycine type leader sequences. No open reading frames encoding proteins corresponding to ComAB were identified in the *S. pyogenes* genome. In contrast, an operon encoding proteins closely related to ComDE, the two-component regulatory system acting

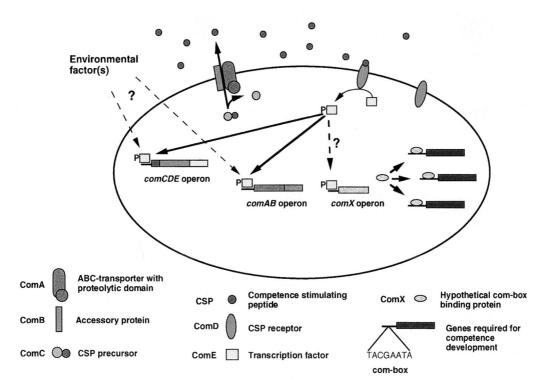

FIGURE 6 Schematic representation of signal transduction and information flow in the pathway leading to induction of competence in *S. pneumoniae* and related streptococci. Solid arrows indicate processing steps or transcriptional activation that have been shown to take place or where some evidence supports the model. Dashed arrows indicate hypothetical parts of the signal transduction chain where evidence supporting the model is lacking.

downstream of CSP in the signal transduction pathway leading to competence induction, is present; in fact, two alleles of the *comD* gene precede the gene encoding the response regulator (*comE*). The predicted topology of the *S. pyogenes* ComD-like histidine kinases shows that the proteins possess about seven trans-membrane alpha-helices, a topology typical for members of the family of membrane-anchored histidine kinases to which the CSP receptors belong. By sensing the concentration of secreted peptide pheromones, members of this family regulate competence, virulence, and bacteriocin production in different gram-positive bacteria in a cell density-dependent manner (Dunny and Leonard, 1997). In the mitis and anginosus groups, the CSP structural gene (*comC*) is always located immediately upstream of the histidine kinase

gene (*comD*) (Håvarstein et al., 1997). However, there is no open reading frame corresponding to *comC* adjacent to the genes encoding the ComDE homologs in *S. pyogenes*, in accordance with the finding that a secretion apparatus specific for ComC is missing. Thus, if *S. pyogenes* is found to be naturally competent, the regulation of competence induction in this species could be different from the peptide-pheromone-triggered quorum-sensing system controlling development of competence in the mitis and anginosus groups. Such differences in regulation could help explain why all attempts in the laboratory to make *S. pyogenes* competent for natural transformation have failed. Searches for the presence of late competence genes revealed that the genome of *S. pyogenes* encodes proteins with high homology to competence-specific gene products from the *comE* and *comG* operons of *Bacillus subtilis*. These products, which have been demonstrated to be essential for transformation in *B. subtilis* (Albano et al., 1989; Hahn et al., 1993; Inamine and Dubnau, 1995), *S. pneumoniae* (Pestova, 1997), and *S. gordonii* (Lunsford and Roble, 1997), most likely constitute parts of the DNA binding and uptake apparatus of naturally competent bacteria. Furthermore, the promoter regions of the *S. pyogenes* genes encoding this uptake apparatus contain a conserved sequence element (TACGAATA) also found to be present in the promoters of late-competence genes in *S. pneumoniae* (Fig. 7) (Pestova, 1997; Campbell, et al., 1998). It is assumed that binding of an unknown transcriptional activator to this "com-box" coordinates transcriptional activation of the late-competence genes. Considering the po-

sition of the com-box relative to the start codon and the ribosome binding site, the postulated com-box-binding protein could be a competence-specific sigma factor. Altogether, these data indicate that strains from two additional major clusters within the genus *Streptococcus*, the pyogenic and the mutans groups, are naturally competent. In addition, the streptococcal phylogenetic tree shows that the apparently competent pyogenic and mutans streptococci share common ancestors, with streptococci assigned to the bovis and salivarius groups, suggesting that members of these groups could be competent as well. If competence for natural transformation really is a more general property of streptococci than has been recognized, it will have significant implications for our understanding of the epidemiology and pathogenesis of these important human pathogens.

comE	GGATATACTT**TACGAATA**AAAAGATGAGGTAAAAAA***ATG***
recA	GAGATCAGAT**TACGAATA**GATAAGTAGGTGGGGGAA***ATG***
comG	TTTTTTTGTT**TACGAATA**AATGAGTAAGGAGGAATT***ATG***
SSB	CCACAATTTT**TACGAATA**ATAAGATAAGGAGGTCAGT***ATG***
orf1	TTTTCTTTTT**TACGAATA**AATGAAGTAAGGAGGTAACTTA***GTG***
orf2	CTGCTAAGAT**TACGAATA**GATGGGGACAATAA***ATG***
orf3	TCCTGTAAAA**TACGAATA**AATAAGTAGGAGGGTAAAA***ATG***

COM-BOX

FIGURE 7 Competence-related genes in *S. pyogenes*. A conserved-sequence motif termed the com-box (underlined) is present in the promoter regions, immediately upstream of the start codons (bold), of several late-competence gene homologs identified in the genome of *S. pyogenes*. As corresponding sequences have been discovered in promoters of late genes from *S. pneumoniae* (Campbell, et al., 1998; Pestova, 1997), the com-box is believed to bind a protein that activates transcription of the late genes. comE and comG, genes encoding proteins highly homologous to ComEA and ComGA, which are involved in DNA binding and uptake in competent *B. subtilis*. recA and SSB, the RecA protein plays an essential role in general recombination, whereas SSBs (single-strand binding proteins) may protect single-stranded DNA taken up during competence from the action of nucleases. It has been shown that the transcription rate of *recA* and production of an SSB increase upon induction of competence in *S. pneumoniae*. orf1-3, open reading frames adjacent to com-boxes, and therefore possibly transcriptionally activated upon competence induction.

OPEN QUESTIONS

The existence of numerous streptococcal CSPs and their utility in doing genetics in the species that possess them are very well established, and related roles for some accessory genes in competence regulation have been partially identified. However, much remains unknown about these quorum-sensing systems, particularly about genes in parts of the circuit prior to CSP synthesis and after its recognition. It is not known what, if anything, modulates the expression of *comC*, beyond its response to the mature pheromone, an apparently autostimulatory behavior. Competence is perhaps better known, among laboratory workers, for its absence than for its presence; it is not known why, but some combinations of strains and media and growth phase lead to competence, but many others do not. The basis for this variability has not been explored or defined. There are hints of influences from other genes in the cell, including *ciaHR* (Guenzi et al., 1994), from oligopeptide permeases, and from encapsulation, as well as influences from environmental signals like temperature, calcium, and zinc ions. Possible influence of the upstream tRNA gene is especially interesting, as linkage of basal expression to transcriptional readthrough could provide the mechanism forbidding competence during stationary phase. The possibility that various aspects of cell metabolism are monitored in regulating the basal synthesis of CSP is suggested by the fact that high phosphate is required for competence, but this requirement is bypassed in some mutants of the *phoP* homolog *ciaH*, as well as by the fact that CSP can bypass "media" limitations in some cases. Finally, it has never been clear what aspect of encapsulation interferes with competence. Early hypotheses suggested that DNA access was blocked by the polysaccharide of capsules, or that CF was similarly blocked from access to the cell surface. As a majority of the encapsulated strains tested to date (>250) respond to CSP and transform, it is clear that, for many encapsulated strains, such barriers are not insurmountable; thus, the capsule appears to exert its effect earlier in the signaling pathway.

Since the early investigations of competence regulation were done with incompletely purified activator preparations or crude culture supernatants or cell extracts, it could not be determined how many signal molecules were involved, but working models were based on the simplifying hypothesis of a single pheromone entity. With CSP in hand, it has become possible to determine the activities of one signal in isolation. Thus, for example, the idea of an "autocatalytic" signal rested on the hypothesis of a single signal species. If there were, in fact, two signal species, one could act to stimulate production of the other, with neither acting autocatalytically. Does CSP provide a complete quorum-sensing mechanism? Or is it possible that CSP is just one (albeit an important one) in a chain of extracellular signals? If it is the last extracellular signal, is its autoinduction important? That is, does CSP up-regulate *comC* during endogenous induction, or is another signal sent and sensed first? A hint of a possible second signal may be found in a mixed-culture experiment reported by Cheng et al. (1997). If CSP does amplify a second, prior, signal, there should exist mutations that abolish basal CSP synthesis but allow normal competence control. Perhaps the one classical property of activator not yet reflected by experience with CSP is its association with cell walls in certain strains of pneumococcus (Tomasz and Mosser, 1966; Morrison, 1981; Yother et al., 1986; Cheng et al., 1997); it will be interesting to know how CSP is processed or presented in those strains. Ultimately, the existence of a preceding intercellular signal will only be established by identifying that signal and demonstrating its activity. Meanwhile, as the *comABCDE* system itself displays sufficient properties to explain all major features of the known biology of competence regulation, it is reasonable to propose that there is a unitary activator molecule and that CSP plays that role.

The nature of the interaction of CSP with the receptor is unknown, except that pherotype-specific residues cluster at one end of the receptor protein. How that interaction creates a signal transmitted to, presumably, ComE is equally unknown. While it has been proposed that proteins of the transformation/DNA-processing pathway are synthesized specifically on induction of competence, and it has been shown that several genes implicated in one way or another in DNA transport are induced at competence, the catalog of competence-induced genes is not nearly complete. Nor is it clear what proportion of genes essential for transformation are induced specifically at competence or what fraction of induced genes are involved in or important for transformation. Despite decades of study of the transformation process, we are still ignorant of the biologically significant source of donor DNA. Yet, the ubiquity of competence in certain species strongly suggests the importance of the mechanism for the species.

Is competence the only function regulated in *S. pneumoniae* by a peptide-mediated quorum-sensing mechanism? The circuit described here seems to offer a rather flexible device for linking cell activity to aspects of the cell's environment. Adaptation of the structure of the response regulator, the receptor, and the peptide, or modulation of their synthesis rates, could in principle create a collection of circuits to activate specific sets of genes at different cell densities, and in different environments. It would be valuable to explore such possibilities for any other Gly-Gly precursors visible in the pneumococcal genome sequence.

ACKNOWLEDGMENTS

Parts of the recent work discussed in this chapter were supported by the National Science Foundation (MCB-9506785). Participation of Mary Ann Yucius in the kinetic assays shown in Fig. 5 is gratefully acknowledged. We acknowledge also the access to unpublished data generously provided by the Streptococcal Genome Sequencing Project (B. A. Roe, S. Clifton, M. McShan, and J. Ferretti; supported by USPHS/NIH grant #AI38406).

REFERENCES

Albano, M., R. Breitling, and D. Dubnau. 1989. Nucleotide sequence and genetic organization of the *Bacillus subtilis comG* operon. *J. Bacteriol.* **171:** 5386–5404.

Alloing, G., C. Granadel, D. A. Morrison, and J.-P. Claverys. 1996. Competence pheromone, oligopeptide permease, and induction of competence in *Streptococcus pneumoniae*. *Mol. Microbiol.* **21:** 471–478.

Alloing, G., B. Martin, C. Granadel, and J. P. Claverys. 1998. Development of competence in *Streptococcus pneumoniae*: pheromone autoinduction and control of quorum sensing by the oligopeptide permease. *Mol. Microbiol.* **29:**75–83.

Avery, O. T., C. M. MacLeod, and M. McCarty. 1944. Studies on the chemical nature of the substance inducing transformation of pneumococcal types. Induction of transformation by a desoxyribonucleic acid fraction isolated from pneumococcus type III. *J. Exp. Med.* **79:**137–157.

Campbell, E. A., S. Y. Choi, and H. R. Masure. 1998. A competence regulon in *Streptococcus pneumoniae* revealed by genomic analysis. *Mol. Microbiol.* **27:**929–939.

Chen, J. D., and D. A. Morrison. 1987. Modulation of competence for genetic transformation in *Streptococcus pneumoniae*. *J. Gen. Microbiol.* **133:** 1959–1967.

Cheng, Q., E. A. Campbell, A. M. Naughton, S. Johnson, and H. R. Masure. 1997. The Com locus controls genetic transformation in *Streptococcus pneumoniae*. *Mol. Microbiol.* **23:**683–692.

Coomaraswamy, G. 1996. Induction of genetic transformation in *Streptococcus pneumoniae* by a pheromone peptide and its synthetic analogues. Ph.D. thesis. University of Illinois, Chicago.

Dawson, M. H., and R. H. Sia. 1931. In vitro transformation of pneumococcal types. I. A technique for inducing transformation of pneumococcal types in vitro. *J. Exp. Med.* **54:**681–699.

Dintilhac, A., G. Alloing, C. Granadel, and J.-P. Claverys. 1997. Competence and virulence of *Streptococcus pneumoniae*—adc and psaA mutants exhibit a requirement for Zn and Mn resulting from inactivation of putative ABC metal permeases. *Mol. Microbiol.* **25:**727–739.

Dobrzanski, W. T., and H. Osowiecki. 1967. Isolation and some properties of the competence factor from Group H Streptococcus strain Challis. *J. Gen. Microbiol.* **48:**299–304.

Dubnau, D. 1991. Genetic competence in *Bacillus subtilis*. *Microbiol. Rev.* **55:**395–424.

Dunny, G. M., and B. A. B. Leonard. 1997. Cell-cell communication in Gram-positive bacteria. *Annu. Rev. Microbiol.* **51:**527–564.

Fox, M. S., and R. D. Hotchkiss. 1957. Initiation of bacterial transformation. *Nature* **179:**1322–1325.

Fuqua, W. C., S. C. Winans, and E. P. Greenberg. 1994. Quorum sensing in bacteria: the

LuxR-LuxI family of cell density-responsive transcriptional regulators (minireview). *J. Bacteriol.* **176:**269–75.

Gaustad, P. 1979. Genetic transformation in *Streptococcus sanguis*: distribution of competence and competence factors in a collection of strains. *Acta Pathol. Microbiol. Scand. Sect. B* **87:**123–128.

Griffith, F. 1928. The significance of pneumococcal types. *J. Hyg.* **27:**108–159.

Guenzi, E., A.-M. Gasc, M. A. Sicard, and R. Hakenbeck. 1994. A two-component signal-transducing system is involved in competence and penicillin susceptibility in laboratory mutants of *Streptococcus pneumoniae*. *Mol. Microbiol.* **12:**505–515.

Hahn, J., G. Inamine, Y. Kozlov, and D. Dubnau. 1993. Characterization of comE, a late competence operon of *Bacillus subtilis* required for the binding and uptake of transforming DNA. *Mol. Microbiol.* **10:**99–111.

Håvarstein, L. S. Unpublished results.

Håvarstein, L. S., G. Coomaraswamy, and D. A. Morrison. 1995a. An unmodified heptadecapeptide pheromone induces competence for genetic transformation in *Streptococcus pneumoniae*. *Proc. Natl. Acad. Sci. USA* **92:**11140–11144.

Håvarstein, L. S., D. B. Diep, and I. F. Nes. 1995b. A family of bacteriocin ABC transporters carry out proteolytic processing of their substrates concomitant with export. *Mol. Microbiol.* **16:**229–240.

Håvarstein, L. S., P. Gaustad, I. F. Nes, and D. A. Morrison. 1996. Identification of the streptococcal competence pheromone receptor. *Mol. Microbiol.* **21:**863–869.

Håvarstein, L. S., R. Hakenbeck, and P. Gaustad. 1997. Natural competence in the genus *Streptococcus*: evidence that streptococci can change pheroperotype by interspecies recombinational exchanges. *J. Bacteriol.* **179:**6589–6594.

Hotchkiss, R. D. 1954. Cyclical behavior in pneumococcal growth and transformability occasioned by environmental changes. *Proc. Natl. Acad. Sci. USA* **40:**49–55.

Hui, F., and D. A. Morrison. 1991. Genetic transformation in *Streptococcus pneumoniae*: nucleotide sequence analysis shows *comA*, a gene required for competence induction, to be a member of the bacterial ATP- dependent transport protein family. *J. Bacteriol.* **173:**372–381.

Inamine, G. S., and D. Dubnau. 1995. ComEA, a *Bacillus subtilis* integral membrane protein required for genetic transformation, is needed for both DNA binding and transport. *J. Bacteriol.* **177:**3045–3051.

Lacks, S. A., and B. Greenberg. 1973. Competence for deoxyribonucleic acid uptake and deoxyribonuclease action external to cells in the genetic transformation of *Diplococcus pneumoniae*. *J. Bacteriol.* **114:**152–163.

Lunsford, R. D., and A. G. Roble. 1997. *comYA*, a gene similar to *comGA* of *Bacillus subtilis*, is essential for competence-factor-dependent DNA transformation in *Streptococcus gordonii*. *J. Bacteriol.* **179:**3122–3126.

McCarty, M., H. E. Taylor, and O. T. Avery. 1946. Biochemical studies of environmental factors essential in transformation of pneumococcal types. *Cold Spring Harbor Symp. Quant. Biol.* **11:**177–183.

Morrison, D. A. 1981. Competence-specific protein synthesis in *Streptococcus pneumoniae*, p. 39–53. *In* M. Polsinelli and G. Mazza (ed.), *Transformation 1980: Proceedings of Fifth European Meeting on Bacterial Transformation and Transfection*. Cotswold Press, Oxford.

Morrison, D. A. 1997. Streptococcal competence for genetic transformation: regulation by peptide pheromones. *Microbial Drug Res.* **3:**27–38.

Morrison, D. A., and M. F. Baker. 1979. Competence for genetic transformation in pneumococcus depends on synthesis of a small set of proteins. *Nature* **282:**215–217.

Ottolenghi, E., and R. D. Hotchkiss. 1960. Appearance of genetic transforming activity in pneumococcal cultures. *Science* **132:**1257–1258.

Ottolenghi, E., and R. D. Hotchkiss. 1962. Release of genetic transforming agent from pneumococcal cultures during growth and disintegration. *J. Exp. Med.* **116:**491–519.

Pakula, R., and D. Ihler. 1969. Peptides for transformability of a streptococcus. *Can. J. Microbiol.* **15:**649–650.

Pakula, R., and W. Walczak. 1963. On the nature of competence of transformable streptococci. *J. Gen. Microbiol.* **31:**125–133.

Perry, D., and H. K. Kuramitsu. 1981. Genetic transformation of *Streptococcus mutans*. *Infect. Immun.* **32:**1295–1297.

Perry, D., and H. D. Slade. 1966. Transformation of streptococcus to streptomycin resistance. *J. Bacteriol.* **83:**443–449.

Pestova, E. V. 1997. *Streptococcus pneumoniae Transformation-Specific Genes and Elements of the Quorum-Sensing Pathway*. Ph.D. thesis. University of Illinois, Chicago.

Pestova, E. V., L. S. Håvarstein, and D. A. Morrison. 1996. Regulation of competence for genetic transformation in *Streptococcus pneumoniae* by an auto-induced peptide pheromone and a two-component regulatory system. *Mol. Microbiol.* **21:**853–862.

Pozzi, G., L. Masala, F. Iannelli, R. Manganelli, L. S. Håvarstein, L. Piccoli, D. Simon, and D. Morrison. 1996. Competence for genetic transformation in encapsulated strains of *Streptococcus pneumoniae*: two allelic variants of the peptide pheromone. *J. Bacteriol.* **178:**6087–6090.

Ramirez, M., D. A. Morrison, and A. Tomasz. 1997. Ubiquitous distribution of the competence-related genes *comA* and *comC* among isolates of *Streptococcus pneumoniae*. *Microb. Drug Res.* **3**:39–52.

Stock, J. B., M. G. Surette, M. Levit, and P. Park. 1995. Two-component signal transduction systems: structure-function relationships and mechanism of catalysis, p. 25–51. *In* J. A. Hoch and T. J. Silhavy (ed.), *Two-Component Signal Transduction*. American Society for Microbiology, Washington, D.C.

Tomasz, A. 1965. Control of the competent state in Pneumococcus by a hormone-like cell product: an example for a new type of regulatory mechanism in bacteria. *Nature* **208**:155–159.

Tomasz, A. 1966. Model for the mechanism controlling the expression of competent state in *Pneumococcus* cultures. *J. Bacteriol.* **91**:1050–1061.

Tomasz, A. 1970. Cellular metabolism in genetic transformation of pneumococci: requirement for protein synthesis during induction of competence. *J. Bacteriol.* **101**:860–871.

Tomasz, A., and R. D. Hotchkiss. 1964. Regulation of the transformability of pneumococcal cultures by macromolecular cell products. *Proc. Natl. Acad. Sci. USA* **51**:480–487.

Tomasz, A., and J. L. Mosser. 1966. On the nature of the pneumococcal activator substance. *Proc. Natl. Acad. Sci. USA* **55**:58–66.

Ween, O., P. Gaustad, and L. S. Håvarstein. Unpublished results.

Yother, J., L. S. McDaniel, and D. E. Briles. 1986. Transformation of encapsulated *Streptococcus pneumoniae*. *J. Bacteriol.* **168**:1463–1465.

Zhou, L., F. M. Hui, and D. A. Morrison. 1995. Competence for genetic transformation in *Streptococcus pneumoniae*: organization of a regulatory locus with homology to two lactococcin A secretion genes. *Gene* **153**:25–31.

CELL DENSITY CONTROL OF GENE EXPRESSION AND DEVELOPMENT IN *BACILLUS SUBTILIS*

Beth A. Lazazzera, Tanya Palmer, John Quisel, and Alan D. Grossman

3

Many organisms have mechanisms for sensing and responding to population density. In *Bacillus subtilis*, high cell density (quorum sensing) contributes to the regulation of at least two different developmental processes, the development of genetic competence (the natural ability to bind and take up DNA) and sporulation (the formation of environmentally resistant endospores). In both cases, high cell density is indicated by the accumulation of peptide signaling molecules in the growth medium. Some of these peptide signals appear to control a general physiological response to high cell density, and parts of that response involve stimulation of competence and sporulation.

CELL DENSITY PHENOMENA IN *B. SUBTILIS*

Both competence development and sporulation have long been known to be more efficient in cells at high rather than low cell densities (Akrigg and Ayad, 1970; Akrigg et al., 1967; Grossman and Losick, 1988; Ireton et al., 1993; Joenje et al., 1972; Vasantha and Freese, 1979; Waldburger et al., 1993). In general, such cell density effects suggest that there might be cell-cell signaling. However, nutritional conditions change as cells grow to high density, especially in complex medium. It can be difficult to distinguish nutrient effects from more obvious forms of cell-cell signaling. For both competence development and sporulation in *B. subtilis*, it is clear that specific cell-cell signaling peptides are involved in regulation by cell density.

Cell density control of competence may serve to indicate the likely presence of exogenous DNA, which could arise from cell lysis. In addition, competence development mediated by cell-cell signaling will probably favor species-specific transfer of genetic material (Solomon and Grossman, 1996). If the signaling factors are species specific, then competence development will occur only when cells are crowded by similar cells. It is interesting to note that of the four well-studied organisms that are naturally transformable, the two that take up DNA independently of sequence, *B. subtilis* and *Streptococcus pneumoniae*, use cell-cell signaling to regulate competence development. In contrast, the two organisms that have sequence-specific DNA uptake, *Neisseria gonorrhoeae* and *Haemophilus influenzae*, do not use cell-cell signaling to regulate competence development.

Beth A. Lazazzera, Tanya Palmer, John Quisel, and Alan D. Grossman, Department of Biology, Building 68-530, Massachusetts Institute of Technology, Cambridge, MA 02139.

Cell-Cell Signaling in Bacteria, Edited by Gary M. Dunny and Stephen C. Winans
©1999 American Society for Microbiology, Washington, D.C.

We suspect that the cell density effect on sporulation is used to indicate crowding and possible competition for scarce nutrients. Sporulation is most efficient upon nutrient depletion when cells are crowded. *B. subtilis* is found in the environment growing in soil, usually as a colony. If cells are starving in a crowded colony, it may be advantageous to sporulate rather than compete for additional nutrients. If cells are dispersed (i.e., at low cell density), then the chances of finding additional nutrients might be higher and sporulation might be less desirable.

In this chapter, we summarize what is known about cell density regulation in *B. subtilis*, with the main focus on cell density control of competence and sporulation.

OVERVIEW OF COMPETENCE DEVELOPMENT IN *B. SUBTILIS*

Natural genetic competence in *B. subtilis* (and many other bacteria) is the ability of cells to bind and take up exogenous DNA (reviewed by Dubnau, 1991, 1993; Grossman, 1995; Solomon and Grossman, 1996). In *B. subtilis*, competence develops in a subpopulation of cells, typically 1 to 10%, and requires over 40 genes. Competent cells are truly differentiated since they have a different buoyant density than noncompetent cells and have reduced metabolic activity.

The key regulatory event in competence development is the activation of the transcription factor encoded by *comK* (Fig. 1) (Hahn et al., 1994, 1996; Kong and Dubnau, 1994; Turgay et al., 1997; van Sinderen et al., 1994, 1995; van Sinderen and Venema, 1994). ComK activates expression of all of the known competence genes that encode the DNA processing and uptake machinery. ComK also increases expression of *recA*, presumably to stimulate recombination between the incoming DNA and the chromosome (Haijema et al., 1996). ComK directly activates its own transcription, an autoregulatory loop that ensures a rapid regulatory response and probably contributes to committing cells to the competence pathway (Hahn et al., 1994, 1996; van Sinderen et al., 1995; van Sinderen and Venema, 1994). Transcription of *comK* is also controlled by additional regulatory proteins, the products of *abrB*, *sinR*, *degU*, and *codY* (Hahn et al., 1996; Serror and Sonenshein, 1996). This complex regulation probably serves to integrate multiple physiological signals that contribute to the decision to become competent.

The activity of ComK protein is inhibited by the *mecA* and *clpC* (*mecB*) gene products (Fig. 1), and mutations in either *mecA* or *clpC* cause high, constitutive levels of ComK and of competence (Dubnau and Roggiani, 1990; Kong et al., 1993; Msadek et al., 1994; Roggiani et al., 1990). MecA binds directly to ComK, inhibiting its activity, and this inhibition requires ClpC, which interacts with MecA to stimulate binding to ComK (Kong and Dubnau, 1994; Msadek et al., 1994; Tur-

FIGURE 1 Regulation of ComK activity and synthesis. The transcription factor ComK activates its own expression in a positive autoregulatory loop. AbrB at low concentrations, DegU, and SinR also activate transcription of *comK*. AbrB at high concentrations and CodY inhibit transcription of *comK*. The activity of the ComK protein is inhibited by MecA and ClpC (MecB), and inhibition is antagonized by interaction of ComS with MecA. Positive or stimulatory interactions between proteins or genes are indicated by arrows (→); negative or inhibitory interactions are indicated by lines with bars (——|).

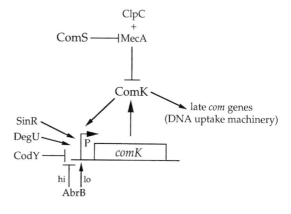

gay et al., 1997). The inhibitory effect of MecA/ClpC on ComK activity is overcome by the *comS* gene product (Turgay et al., 1997), and expression of *comS* is controlled by cell density (Magnuson et al., 1994).

CELL DENSITY CONTROL OF COMPETENCE DEVELOPMENT

In *B. subtilis*, competence develops at high cell density, during late exponential growth or after the onset of stationary phase, depending on the culture medium. In defined minimal medium, in the absence of complex mixtures of amino acids, competence develops in a subpopulation of cells during exponential growth of the culture. Under these conditions, the main signals that initiate competence development are two peptide pheromones, ComX pheromone and competence and sporulation factor (CSF), which accumulate in culture supernatants and serve as signals of cell density.

Both ComX pheromone and CSF stimulate transcription of the *srfA* operon, and this stimulatory activity was used in assays to purify the two extracellular factors (Magnuson et al., 1994; Solomon et al., 1995, 1996). *srfA* is required for production of the peptide antibiotic surfactin (Nakano et al., 1988, 1991a). The only part of the large *srfA* operon that is required for competence development is *comS*, a small open reading frame (ORF) (internal to a larger ORF) encoding a 46-amino-acid peptide (D'Souza et al., 1994; Hamoen et al., 1995). ComS is required for activation of the transcription factor ComK (Turgay et al., 1997). Transcription from the *srfA* (*comS*) promoter increases as cells grow to high density owing to the accumulation of ComX pheromone and CSF in the culture medium (Magnuson et al., 1994; Solomon et al., 1995, 1996).

The transcription of *srfA* is activated by the phosphorylated form of the response regulator, ComA (Nakano et al., 1991b; Nakano and Zuber, 1989, 1991; Roggiani and Dubnau, 1993; Weinrauch et al., 1989). ComX pheromone and CSF appear to stimulate *srfA* expression by increasing the level of ComA~P (Fig. 2). Briefly, ComX pheromone appears to activate a kinase ComP, which donates phosphate to ComA (Magnuson et al., 1994; Solomon et al., 1995; Weinrauch et al., 1990). CSF, in contrast, appears to inhibit the activity of a phosphatase RapC that is proposed to dephosphorylate ComA~P (Lazazzera et al., 1997; Solomon et al., 1996).

THE ComA REGULON AND QUORUM SENSING

It seems certain that the primary role of the ComA response regulator is to mediate responses to high cell density. Activation of *srfA* (*comS*) and competence is only one part of that response, and *srfA* is the only target of ComA that is required for competence development (Hahn and Dubnau, 1991; Nakano and Zuber, 1991). In addition to *srfA*, three other loci have been shown to be controlled by ComA: *degQ*, *rapA* (*gsiA*), and *rapC*, and each is involved in a cell density-regulated process. *degQ* encodes a small protein that contributes to production of extracellular degradative enzymes at high cell density (Msadek et al., 1991; Yang et al., 1986). *rapA* encodes a response regulator, aspartyl-phosphate phosphatase (Perego et al., 1994), that inhibits sporulation (Mueller et al., 1992; Mueller and Sonenshein, 1992; Perego et al., 1994; Perego and Hoch, 1996). *rapC* encodes a negative regulator of competence (Solomon et al., 1996), most likely a phosphatase for ComA~P (see below).

To identify additional targets of ComA regulation and to broaden our understanding of cell density-regulated processes, we searched the *B. subtilis* genome sequence (Kunst et al., 1997) for potential ComA-binding sites. A consensus ComA binding site, 5'-TTGCGGNNNNCCGCAA-3' (N represents any base) was derived from a comparison of the four known ComA~P targets (Mueller et al., 1992). Allowing two mismatches, this consensus sequence is found approximately 420 times in the entire *B. subtilis* genome. For simplicity, we will consider only the 17 sites in noncoding regions within a few

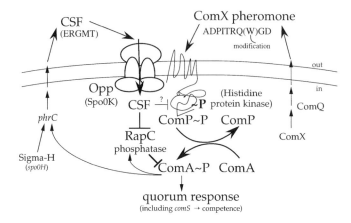

FIGURE 2 Production of and response to CSF and ComX pheromone. The cell membrane is shown with the two competence factors, CSF and ComX pheromone, outside the cell. ComX is the 55-amino-acid precursor for the peptide portion of the ComX pheromone, and ComQ is thought to be required for processing or modification. Response to ComX pheromone requires the membrane-bound histidine protein kinase ComP, and ComX pheromone probably interacts directly with ComP to activate the kinase. ComP donates phosphate to ComA, and ComA~P mediates the quorum response, part of which is to activate transcription of *comS* (*srfA*), which leads to competence development. *phrC* encodes the precursor of CSF, and its transcription is activated by sigma-H (*spo0H*) from one promoter and ComA~P from another. The oligopeptide permease Opp (Spo0K) transports CSF into the cell. CSF probably directly inhibits the activity of the phosphatase RapC, which is a negative regulator of ComA~P. ComA~P activates expression of *rapC*, creating a homeostatic regulatory loop. High concentrations of CSF inhibit expression of ComA-controlled genes, perhaps by inhibiting activity of ComP.

hundred base pairs upstream from an ORF or operon (Table 1). Based on this analysis, we estimate that ComA directly regulates transcription of at least 21 genes.

Of the 21 ORFs with putative ComA binding sites, 4 were the previously identified ComA-regulated genes, 7 were ORFs with no known function, and 10 were genes with known or inferred function (Table 1). These last include *pelA*, encoding pectate lyase; *psd*, encoding phosphatidylserine decarboxylase; *yodJ*, encoding D-alanine D-alanyl carboxypeptidase; *ylaM*, encoding glutaminase; *deoD*, encoding purine nucleoside dephosphorylase; *yurI*, encoding RNase; *infA*, encoding translation initiation factor 1; *tsf*, encoding elon-

gation factor Ts; and *yurDCB*, encoding a putative 4-hydroxy-benzoyl-coenzyme A reductase.

It is remarkable that pectate lyase production may be regulated by cell density in *B. subtilis*. Production of pectate lyase by the gram-negative plant pathogen *Erwinia carotovora* is also induced by a putative cell density-sensing, *N*-acyl homoserine lactone signaling system (Jones et al., 1993; Pirhonen et al., 1993), similar to the LuxI/LuxR system of *Vibrio harveyi*. Utilization of complex polymers (e.g., pectin) by bacteria is usually more efficient at high population densities owing to high concentrations of extracellular degradative enzymes. This may explain why such ev-

TABLE 1 Potential ComA binding sites identified in the *B. subtilis* genome

Gene	Function	ComA binding site[a] (position)[b]
Known ComA-regulated genes		
degQ[c]	Regulator of degradative enzyme production	TTGCGG TGTC ACGCAG (-55)
rapA	Protein phosphatase	TTGCGG TTAG CCGAAA (-60)
rapC	Protein phosphatase	TTGCGT GCTC CCGAAA
srfA, comS	Surfactin biosynthesis; competence	TTGCGG CATC CCGCAA (-103)
		TTTCGG CATC CCGCAT (-59)
Putative ComA-regulated genes		
deoD-yodJ	Purine nucleoside phosphorylase (*deoD*); D-alanine-D-alanyl carboxypeptidase (*yodJ*)	CTGCGG AAAC CCGCAG
infA	Translation initiation factor 1	TTGCGT ATCT CCGGAA
pelA	Pectate lyase	TTGCAG AATG CGGCAA
psd	Phosphatidylserine decarboxylase	TTTCGG GGAT CGGCAA
tsf	Elongation factor Ts	TTGCGG AAGC TCGCAA
ydfS	Unknown	TTCCGG CCAA TCGCAA
ydiQ	Unknown	TTTCGG GTCT CGGCAA
ylaM	Glutaminase	TTGCGG TCAA ACGAAA
yugM	Unknown	TTGCGT TACT CGGCAA
yurFEDCB	Unknown (*yurFE*); 4-hydroxy-benzoyl-coenzyme A reductase subunits (*yurDCB*)	TTGCGG AACT CCGCAA
yurI[d]	RNase	TGCCGG CAGC CCGCAA
yxjI	Unknown	TCGCCG CAAA CCGCAA

[a] The consensus ComA binding site, 5'-TTGCGG NNNN CCGCAA-3', was derived from the known sites. Putative sites were identified in the *B. subtilis* genome sequence using the consensus and allowing up to two mismatches.

[b] When the transcription start site is known, then the position of the 3' end of the ComA binding site is indicated relative to the transcription start site. In genes where the transcription start site has not been mapped, no position is indicated. We estimate that these potential binding sites are typically between 60 and 160 bp upstream from probable -10 regions (for sigma-A promoters).

[c] The ComA binding site upstream of *degQ* may also regulate transcription from the *yuzC* (unknown function) promoter directed in the opposite direction.

[d] The putative ComA binding site upstream of *yurI* may also regulate transcription from the *yurHG* promoter directed in the opposite direction. *yurHG* encodes products similar to N-carbamyl-L-amino acid aminohydrolase and aspartate aminotransferase, respectively.

olutionarily divergent bacteria as *E. carotovora* and *B. subtilis* both appear to regulate synthesis of pectate lyase in response to high cell density. Perhaps several of the genes that are regulated by cell density signals will be similar across a wide range of bacteria.

The significance of the probable cell density regulation of *psd*, *yodJ*, *ylaM*, *deoD*, *yurI*, *infA*, and *tsf* remains obscure. It is intriguing to speculate that cells adjust to growth at high cell density by adjusting membrane lipid composition, fine-tuning translation rates, and inducing the synthesis of degradative enzymes that could be used to provide the cell with additional energy, but further research will be needed to address these points.

Through the regulation of ComA~P, ComX pheromone and CSF may mediate a general physiological response to cell density. The following sections will focus on the production and mechanism of response to these peptide factors.

ComX PHEROMONE

Identification of ComX Pheromone

ComX pheromone was purified from culture supernatant based on its ability to stimulate transcription of *srfA* when added to wild-type cells at low cell density (Magnuson et al., 1994). Although its name reflects the fact that it was initially discovered as a factor regulating competence development, ComX pheromone appears to be a general indicator of high cell density and is probably involved in the regulation of many genes.

ComX pheromone is a 10-amino-acid peptide (Fig. 2 and 3) containing a modified tryptophan (Magnuson et al., 1994). The modification is hydrophobic and required for pheromone activity. If the tryptophan side chain is intact, then the modification is 206 Da, based on mass analysis. The peptide moiety of the pheromone corresponds to the C-terminal 10 amino acids of the 55-amino-acid peptide encoded by *comX* (Fig. 3). ComX pheromone accumulates to ~50 nM in culture medium as cells approach stationary phase, and the half-maximal response to pheromone occurs at ~5 to 10 nM (Magnuson et al., 1994). Mutants unable to produce ComX pheromone are reduced approximately 100-fold in transcription of *srfA* (Magnuson et al., 1994; Solomon et al., 1995).

Production of ComX Pheromone

Production of mature, active ComX pheromone requires expression (transcription and translation) of *comX*, processing of the 55-amino-acid precursor to 10 amino acids, modification of the tryptophan residue 3 amino acids from the C terminus, and export (Fig. 3). We suspect that there are multiple proteins involved in the production of ComX pheromone; however, an extensive hunt for mutants with decreased production of ComX pheromone led to the isolation of 18 independent mutations, all located in *comQ* or *comX* (Palmer and Grossman, manuscript in preparation). *comQ* (Msadek et al., 1991) is required for pheromone production and is immediately upstream from and partly overlapping with *comX* (Magnuson et al., 1994). These findings indicate that, if there are genes other than *comQ* and *comX* required for production of ComX pheromone, then they are likely to be redundant and/or essential.

ComQ is a good candidate for the protein that modifies ComX pheromone. ComQ has several stretches of similarity to regions conserved in isoprenyl transferases, and the mass and hydrophobic properties of the modification are consistent with farnesylation. Farnesylation is a common modification in

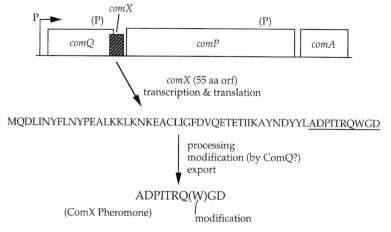

FIGURE 3 Production of ComX pheromone. The *comQXPA* genes are indicated. Putative promoters are indicated by (P). The known promoter upstream of *comQ* is also indicated. *comX* encodes a 55-amino-acid (aa) peptide, the precursor of ComX pheromone, and the *comQ* gene product is probably involved in the modification of the pheromone. The C-terminal 10 amino acids of ComX are in the active pheromone. orf, open reading frame.

eukaryotes that most notably occurs on Ras proteins and on the yeast pheromone **a**-factor (for reviews see Caldwell et al., 1995; Zhang and Casey, 1996). However, in these cases, a cysteine is farnesylated, not a tryptophan. Modification of tryptophan in peptides is relatively rare, and identification of the modification and characterization of the role of ComQ should reveal interesting biochemical reactions.

It is not yet known how ComX pheromone is exported and processed. ComX contains neither an obvious Sec-dependent signal sequence nor a recognizable proteolytic cleavage site, suggesting that export and processing probably require specialized proteins. ComX also lacks the signature double-glycine-type leader peptide found in several peptides that are exported via ATP-binding cassette (ABC) exporters with proteolytic processing activity (Håvarstein et al., 1995b; van Belkum et al., 1997). There appear to be approximately 77 ABC transporters encoded in the *B. subtilis* genome (Kunst et al., 1997). Of these, six are noticeably similar to ABC exporters required for secretion of peptides or small proteins in other organisms (Palmer and Grossman, unpublished results). Perhaps several of these are involved in export of ComX pheromone (or a precursor).

Response to ComX Pheromone
Response to ComX pheromone requires the two-component system encoded by *comP* and *comA*. ComP is a membrane-bound histidine protein kinase (Weinrauch et al., 1990) and is almost certainly the receptor for ComX pheromone. *comP* null mutants are unable to respond to exogenous pheromone, and double mutant analysis indicates that ComP and ComX pheromone are on the same genetic pathway (Magnuson et al., 1994; Solomon et al., 1995). Once activated, ComP autophosphorylates on a histidine residue and transfers phosphate to an aspartate residue in the cognate response regulator ComA. ComA~P then directly activates transcription from several promoters, including that of the *srfA* operon (Nakano et al., 1991b; Nakano and Zuber, 1989, 1991; Roggiani and Dubnau, 1993; Weinrauch et al., 1989).

Peptide Signaling Cassettes
The *comQXPA* genes form a peptide signaling cassette, a paradigm for quorum sensing in gram-positive organisms (Fig. 4) (Kleerebezem et al., 1997). Similar cassettes have been identified in a variety of organisms, including *Streptococcus* sp. (Håvarstein et al., 1996; Pestova et al., 1996), *Lactobacillus* sp. (Diep et al., 1995), *Carnobacterium piscicola* (Quadri et al., 1997), and *Staphylococcus aureus* (Ji et al., 1995). In all cases, there is a gene encoding the precursor of an extracellular peptide signaling factor, followed by genes for a membrane-bound histidine protein kinase and a response regulator that mediate response to the extracellular peptide. If the extracellular peptide is modified (*B. subtilis* and *S. aureus*), then upstream of the gene encoding the peptide precursor there is a gene required for factor production, probably modification (Fig. 4). In addition to having roles in the processes for which they were originally discovered, we suspect that many of the peptide signaling cassettes are involved in a general regulatory response to cell density. Several of these peptide signaling cassettes are described in more detail in other chapters in this volume.

CSF
The competence and sporulation factor CSF is one of a likely family of peptides used for signaling in *B. subtilis*. CSF accumulates in culture supernatant as cells grow to high density and is involved in controlling expression of genes that are regulated by cell density. Below, we summarize what is known about CSF, how it is produced, the cellular response, and its role in cell-cell signaling.

Identification of CSF
Whereas ComX pheromone is the major extracellular competence factor of *B. subtilis*, there is a second factor CSF that modulates competence development as cells grow to

FIGURE 4 Peptide signaling cassettes. In each case, the gene encoding the pheromone precursor is directly upstream of the gene for the histidine protein kinase. If the pheromone is modified, the upstream gene (*comQ*, *agrB*) is required for production (modification?). The gene downstream from the kinase gene encodes the cognate response regulator. *Lactobacillus plantarum* has two response regulator genes (*plnC*, *plnD*). The direction of transcription is from left to right.

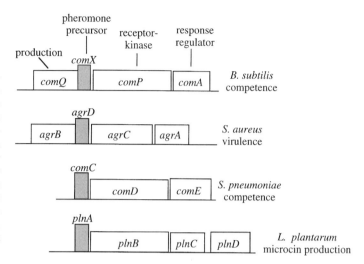

high density. CSF was purified from culture supernatant based on its ability to stimulate expression of a *srfA-lacZ* fusion in cells at low density (Solomon et al., 1996). Analysis of the purified material revealed that CSF is an unmodified 5-amino-acid peptide of the sequence Glu-Arg-Gly-Met-Thr (ERGMT) (Solomon et al., 1996).

Production of CSF

The sequence of the CSF peptide corresponds to the C-terminal 5 amino acids of the *phrC* gene product (Fig. 5) (Solomon et al., 1996). A *phrC* null mutant does not make CSF and has reduced and delayed expression of *srfA* (Solomon et al., 1996). *phrC* encodes a 40-amino-acid precursor peptide with a Sec-dependent signal sequence and putative peptidase cleavage sites (Perego et al., 1996). A peptide between 11 and 25 amino acids in length is predicted to be secreted from the cell through the Sec-dependent pathway. How this precursor peptide is processed to the mature 5-amino-acid form that is found in culture supernatants is unknown. There may be a specific extracellular protease responsible for processing CSF.

Response to CSF

The response of cells to CSF is more complicated than that to ComX pheromone (Fig. 6).

At relatively low extracellular concentrations (1 to 5 nM), CSF stimulates the expression of *srfA*, and at higher extracellular concentrations (>20 nM), CSF inhibits the expression of *srfA* and stimulates sporulation (Solomon et al., 1996). This section will focus on the role of CSF in regulating *srfA* and other ComA-controlled genes. The following section will discuss the role of CSF in cell density control of sporulation.

Experiments characterizing the mechanisms by which CSF regulates ComA-controlled genes (Lazazzera et al., 1997; Solomon et al., 1995, 1996) have led to the following model (Fig. 2). Extracellular CSF binds to the oligopeptide permease Opp (Spo0K) and is transported into the cell. Once inside, CSF binds to intracellular targets to regulate the level of ComA~P. ComA~P levels are increased because CSF binds to and inhibits the activity of the RapC phosphatase, which dephosphorylates ComA~P. The intracellular target responsible for the inhibition of *srfA* expression in response to high concentrations of CSF has not yet been identified, but one model is that CSF inhibits the histidine protein kinase ComP, thereby reducing accumulation of ComA~P. Thus, the role of CSF in cell density regulation is to modulate the level of ComA~P, increasing the level during

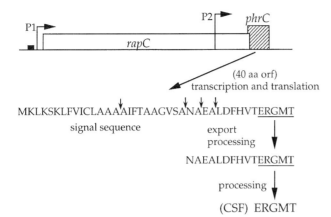

MKLKSKLFVICLAAAAIFTAAGVSANAEALDFHVT<u>ERGMT</u>
signal sequence

export
processing

NAEALDFHVT<u>ERGMT</u>

processing

(CSF) <u>ERGMT</u>

FIGURE 5 Production of CSF. P1 is the promoter upstream of *rapC* and is activated by ComA~P. The ComA~P binding site is indicated by a little black box. P2 is upstream of *phrC*, internal to *rapC*, and is controlled by RNA polymerase containing sigma-H. *phrC* encodes a 40-amino-acid (aa) peptide, the precursor of CSF. The first steps in production probably involve the normal secretion pathway, and four potential peptidase cleavage sites are indicated by arrows above the 40-amino-acid sequence. The C-terminal 5 amino acids of PhrC constitute active CSF. orf, open reading frame.

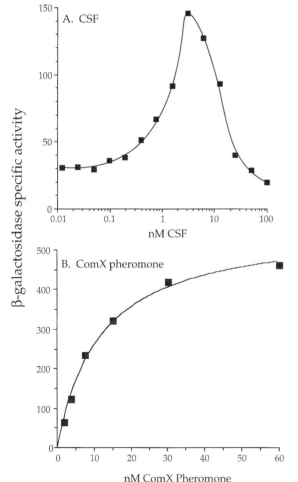

FIGURE 6 Response of cells to different amounts of CSF and ComX pheromone. Cells containing a *srfA-lacZ* fusion (Magnuson et al., 1994) were grown in defined minimal medium. Cells at low cell density (OD_{600} ~0.1) were mixed with the indicated amount of peptide, incubated for 70 min at 37°C, and assayed for β-galactosidase specific activity. (A) The response of wild-type cells (*srfA-lacZ*) to different amounts of CSF. In the absence of CSF, β-galactosidase specific activity was 28 units. (B) The response of cells unable to make ComX pheromone (*comQ*, *srfA-lacZ*) to different amounts of ComX pheromone. In the absence of ComX pheromone, β-galactosidase specific activity was 8 units.

growth and possibly decreasing the level near the onset of stationary phase when CSF levels are high.

CSF Functions Intracellularly

CSF is a member of an emerging class of cell-cell signaling peptides in prokaryotes that function intracellularly, requiring active transport by an oligopeptide permease and subsequent interaction with an intracellular target (Lazazzera et al., 1997; Perego, 1997). Cell-cell signaling peptides were generally thought to interact with membrane receptors, such as histidine protein kinases, that transduce a signal across the membrane in response to the extracellular peptide (Dunny and Leonard, 1997; Kleerebezem et al., 1997). Other cell-cell signaling molecules that function intracellularly, such as the homoserine lactones, freely cross the cell membrane (Fuqua et al., 1994, 1996; Swift et al., 1996), are not peptide based, and are not subject to degradation by peptidases.

Other bacterial species also utilize signaling peptides that probably function intracellularly to mediate cell-cell signaling. Like the requirement for the oligopeptide permease for competence and sporulation in *B. subtilis*, these organisms also utilize an oligopeptide permease in signaling. Mating in *Enterococcus faecalis* requires an oligopeptide permease, apparently to transport mating pheromones (Leonard et al., 1996). Competence regulation in *S. pneumoniae* is modulated by several oligopeptide binding proteins and an oligopeptide permease. While these are not required for response to the major competence pheromone, they do play a role in modulating competence development and suggest a possible role for additional peptide signals during competence development in *S. pneumoniae* (Alloing et al., 1996; Pearce et al., 1994). Development of aerial mycelium in *Streptomyces coelicolor* also requires an oligopeptide permease (Nodwell et al., 1996). It seems likely that these oligopeptide permeases import peptide signaling molecules that interact with intra-cellular receptors to mediate a physiological response.

For a signaling peptide to function intra-cellularly, it needs to be transported into the cell at the low extracellular concentrations (pM to nM) at which these peptides usually are active (Diep et al., 1995; Dunny, 1990; Håvarstein et al., 1995a; Kuipers et al., 1995; Magnuson et al., 1994; Solomon et al., 1996). In *B. subtilis*, CSF is indeed transported into the cell by the oligopeptide permease at concentrations required to elicit a physiological response (Lazazzera et al., 1997). However, knowing that a signaling peptide is transported into the cell does not indicate whether the peptide actually functions intracellularly. For CSF, it was shown that expressing a mature 5-amino-acid form of CSF intracellularly functioned to regulate expression of *srfA* (Lazazzera et al., 1997). In this case, a mutant form of CSF (ARGMT instead of ERGMT), which is able to inhibit but not stimulate *srfA* expression, was used (Lazazzera et al., 1997). CSF has also been shown in vitro to interact with RapB, a cytoplasmic protein that is involved in stimulating sporulation in response to CSF (see below) (Perego, 1997).

The Oligopeptide Permease and the Regulation of ComA~P Levels

Oligopeptide permeases are members of the widespread ABC family of transporters (Higgins, 1992). Oligopeptide permeases generally transport peptides of three to five amino acids relatively nonspecifically, making them ideal for transporting small, unmodified peptide signaling molecules like CSF. Some ABC transporters are known to have roles in signal transduction in addition to their roles in transport (e.g., Cangelosi et al., 1990; Cox et al., 1988; Manson et al., 1986; Wanner, 1993). Because CSF can function intracellularly (Lazazzera et al., 1997) and interacts with at least one intracellular protein (Perego, 1997), it seems unlikely that Opp serves as a receptor for CSF in addition to its role in transport. However, the role of Opp in regulating

ComA-controlled genes is more complicated than simply transporting CSF. ComA-controlled gene expression is much more severely reduced in a Δ*opp* mutant than in a Δ*phrC* mutant (Lazazzera et al., 1997; Solomon et al., 1996). This additional role for Opp is poorly understood and could be related to transport of other extracellular peptide signaling molecules similar to CSF (see below).

The RapC Phosphatase and Response to CSF

rapC, the gene immediately upstream of and overlapping with *phrC*, encodes a negative regulator of ComA-controlled gene expression (Solomon et al., 1996). RapC shares approximately 43% amino acid identity to two biochemically characterized protein aspartyl-phosphate phosphatases, RapA and RapB, of *B. subtilis* (Perego et al., 1996). RapA and RapB dephosphorylate Spo0F~P, a response regulator required for the initiation of sporulation (Perego et al., 1994). This has led to the suggestion that RapC negatively regulates ComA-controlled genes by dephosphorylating ComA~P, a response regulator. Since RapC is required for CSF to stimulate *srfA* expression (Solomon et al., 1996), CSF likely increases the levels of ComA~P by binding to and inhibiting the activity of RapC.

Autoregulation of *rapC* and *phrC*

The promoter immediately upstream of *rapC* is preceded by a DNA sequence that matches the consensus binding site for ComA. Indeed, the transcription of the *rapC phrC* operon increases with increasing cell density owing to activation of ComA (Lazazzera et al., manuscript in preparation). This regulation by ComA~P allows RapC and CSF to regulate the transcription of their own genes. RapC negatively regulates its own synthesis, and CSF positively regulates its own synthesis by inhibiting RapC activity (Fig. 2).

In addition to this regulation by ComA, immediately upstream of *phrC* is a promoter controlled by the sigma factor sigma-H (Car-

ter et al., 1991). Sigma-H is required for sporulation and becomes more active during the transition from exponential to stationary phase (Healy et al., 1991; Weir et al., 1991). Sigma-H-dependent regulation of *phrC* causes an increase in extracellular CSF concentrations to approximately 100 nM shortly after entry into stationary phase (Lazazzera et al., manuscript in preparation). This high concentration of CSF is probably important for the ability of CSF to inhibit *srfA* expression and to stimulate sporulation (see below).

rapC and *phrC*, Members of a Family of Phosphatase and Regulator Pairs

RapC shares amino acid similarity with eight putative phosphatases, in addition to the two characterized phosphatases, RapA and RapB (Fig. 7) (Kunst et al., 1997; Perego et al., 1996). Whereas RapC is a negative regulator of competence development (Solomon et al., 1996), RapA and RapB negatively regulate the initiation of sporulation (Perego et al., 1994), and RapE is reported to inhibit sporulation when overexpressed (Perego et al., 1996).

Seven of these *rap* genes are in operons with a downstream, usually overlapping *phr* gene that appears to encode a small extracellular peptide (Kunst et al., 1997). By analogy to PhrA/RapA and PhrC/RapC, these Phr pheromones are likely to negatively regulate their cognate Rap. It has been shown that a 5-amino-acid peptide derived from the C terminus of PhrA inhibits the activity of the RapA phosphatase in vitro (Perego, 1997). CSF appears to inhibit RapC and has been shown in vitro to inhibit the activity of RapB, indicating that a Phr peptide may regulate more than one phosphatase, and that a Rap phosphatase without a cognate Phr may still be subject to regulation (Perego, 1997). Whether the uncharacterized Rap/Phr pairs will play roles in regulating cellular processes in response to cell density, as CSF does for competence development and sporulation, awaits experimental analysis.

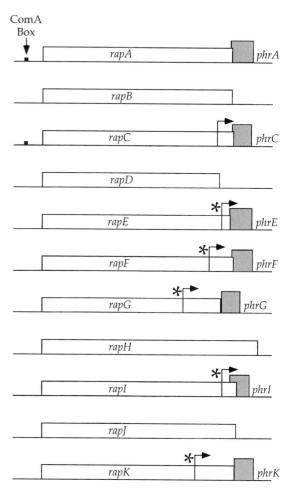

FIGURE 7 *rap* and *phr* family of genes. Eleven genes encoding products homologous to the RapA and RapB phosphatases have been identified in the *B. subtilis* genome. Six of these are followed by small ORFs, each encoding the precursor for a putative extracellular peptide. Binding sites for ComA~P (ComA boxes) are found upstream of *rapA* and *rapC*, and both of these operons are activated by ComA~P in vivo. Upstream of *phrC* is a promoter controlled by RNA polymerase containing sigma-H (sigma-H is the *spo0H* gene product). We have found potential sigma-H-dependent promoters upstream of each of the other *phr* genes, with the exception of *phrA*. These putative promoters are indicated by an arrow with an asterisk (*).

QUORUM SENSING AND THE CONTROL OF SPORULATION

Under conditions of nutrient depletion, *B. subtilis* can undergo a process of development that leads to the formation of dormant, environmentally resistant spores. Sporulation involves an asymmetric cell division, generating two distinct cell types with different patterns of gene expression and different developmental fates (reviewed by Errington, 1993; Grossman, 1995; Stragier and Losick, 1996).

Sporulation is inefficient in cells at low cell density, even upon starvation, indicating that nutrient depletion is not sufficient for efficient sporulation (Grossman and Losick, 1988; Ireton et al., 1993; Vasantha and Freese, 1979; Waldburger et al., 1993). This cell density effect on sporulation is mediated by signaling molecules (spore factors) that accumulate extracellularly as cells grow to high density and become crowded.

The cell density control of sporulation is complicated, probably requiring several extracellular peptide factors. Some sporulation factors may be important only under some conditions, and there may be redundancy in the function of other extracellular spore factors. Under some conditions, ComX pheromone (Magnuson et al., 1994) and CSF (Lazzazera et al., 1997; Solomon et al., 1996) can function as spore factors. Furthermore, during the purification of CSF, an additional spore factor activity was identified (Solomon et al., 1996), and a spore factor activity in culture supernatant that is insensitive to proteases has been reported (Waldburger et al., 1993). The biochemical identity of these last two spore factors has not been determined. Analysis of the complete *B. subtilis* genome sequence might facilitate the identification of genes encoding or involved in production of additional extracellular spore factors.

Extracellular Spore Factors and the Level of Spo0A~P

The *spo0A* gene product is the key transcription factor required for the initiation of sporulation (reviewed by Grossman, 1995; Hoch, 1993), and its activity is regulated, in part, by cell density signals (Ireton et al., 1993). Spo0A is a member of the response regulator family of proteins, but unlike most other members of this family, Spo0A does not receive phosphate directly from a histidine-protein kinase. There are two phosphorelay proteins, Spo0F and Spo0B, that transfer phosphate from one of three histidine-protein kinases, KinA, KinB, and KinC, to Spo0A (Fig. 8) (Antoniewski et al., 1990; Burbulys et al., 1991; Kobayashi et al., 1995; LeDeaux and Grossman, 1995; LeDeaux et al., 1995; Perego et al., 1989; Trach and Hoch, 1993). The accumulation of Spo0A~P is negatively regulated by the activities of at least three phosphatases. Spo0F~P is the target of two phosphatases, RapA and RapB (Perego et al., 1994), and Spo0A~P is dephosphorylated by the phosphatase Spo0E (Ohlsen et al., 1994).

The physiological function of the phosphorelay is to integrate the many signals that control the initiation of sporulation, including nutrient depletion, DNA damage, DNA replication, tricarboxylic acid cycle signals, and cell density (Grossman, 1995; Ireton and Grossman, 1992, 1994; Ireton et al., 1993, 1995). Constitutively active forms of Spo0A, which bypass the requirement for phosphorylation for activity, bypass the need for high cell density for efficient sporulation (Ireton et al., 1993). Many gene products affect the phosphorylation of Spo0A, and these proteins are potential targets for the cell density signals.

CSF and Regulation of the RapB Phosphatase

The RapB phosphatase appears to regulate the level of Spo0A~P in response to cell density. The ability of RapB to dephosphorylate Spo0F is inhibited in vitro by CSF (Perego, 1997). In a particular mutant background (see below), CSF is required for efficient sporulation, indicating that CSF functions as an extracellular spore factor (Lazazzera et al., 1997; Solomon et al., 1996). RapB provides the link between control of Spo0A~P levels and CSF.

The concentrations of extracellular CSF required to stimulate sporulation are similar to the concentrations of CSF required to inhibit *srfA* expression (>20 nM) (Lazazzera et al., 1997; Solomon et al., 1996). Despite the overlap in the concentrations of CSF, the two processes are not related. Different amino ac-

FIGURE 8 The phosphorelay and phosphatases. Activation (phosphorylation) of the transcription factor Spo0A is controlled by at least three histidine protein kinases (KinA, B, C). The kinases autophosphorylate on a histidine residue and donate phosphate to the single-domain response regulator Spo0F. Phosphate is transferred from Spo0F~P to Spo0B, and finally to Spo0A. RapA and RapB are phosphatases for Spo0F~P, and Spo0E is a phosphatase for Spo0A~P. RapA and RapB are inhibited by the PhrA and CSF (PhrC) pentapeptides, respectively. It is not known what, if anything, controls activity of the phosphatase Spo0E.

ids of CSF are required for the two processes (Lazazzera et al., 1997), and RapB is not required for CSF to inhibit *srfA* expression (Lazazzera, unpublished data).

The PhrA Peptide and Regulation of the RapA Phosphatase

Another member of the Rap family that regulates sporulation is RapA. *rapA* is followed by a small gene *phrA* that encodes the precursor of a 5-amino-acid peptide that inhibits the activity of RapA (Perego, 1997). It has been proposed that the PhrA peptide serves as a timing mechanism for the control of the initiation of sporulation (Perego, 1997). The synthesis of RapA and PhrA is induced by ComA~P in response to high cell density. After synthesis, the PhrA precursor is secreted by the normal secretory apparatus. Like PhrC, it is not known how PhrA is cleaved to the 5-amino-acid form. The PhrA pentapeptide is then imported via the oligopeptide permease and acts inside the cell (Lazazzera et al., 1997) to inhibit the activity of the phosphatase RapA (Perego, 1997).

It is not known if the PhrA peptide accumulates in culture supernatants as cells grow to high density, a prerequisite for its ability to serve as a cell density signal. However, the sporulation defect of *phrA* mutant cells is rescued by growth in the presence of wild-type cells (Perego and Hoch, 1996), indicating that the PhrA peptide can serve as a cell-cell signaling molecule. In addition, because PhrA and RapA synthesis is controlled by ComA~P, and therefore CSF and ComX pheromone, PhrA and RapA may indirectly mediate cell density control of sporulation.

The Sporulation Sigma Factor Sigma-H and Regulation of Extracellular Spore Factors

A search for mutants defective in the production of extracellular spore factors identified a mutant designated *ecs191* with a missense mutation in *spo0H*, the gene encoding the sporulation-specific sigma factor sigma-H (Lazazzera et al., 1997). This mutant sporulates poorly, but the sporulation frequency can be largely rescued by growing the *ecs191* mutant in the presence of wild-type cells or by adding conditioned medium from wild-type cells (Lazazzera et al., 1997). This indicates that sigma-H is responsible for transcribing a gene(s) required for production of at least one extracellular spore factor.

phrC, the gene encoding CSF, is transcribed in part from a sigma-H-dependent promoter (Fig. 2 and 7). However, the sporulation frequency of an *ecs191* mutant cannot be fully rescued by the addition of exogenous CSF, indicating that the production of at least one other spore factor is affected in the *ecs191* mutant. Candidate genes that have putative sigma-H promoters that could encode spore factors include the Phr family of peptide regulators. Five other members of this family have been identified in the *B. subtilis* genome, in addition to PhrA and PhrC (Kunst et al., 1997). All five of these *phr* genes have putative sigma-H promoters (Fig. 7). The targets of the cognate Rap phosphatases of these Phrs are unknown. It is possible that these Phrs may contribute to the cell density control of sporulation in a manner similar to that of CSF.

The involvement of these Phr peptides in the cell density control of sporulation may be difficult to dissect if these peptide factors are redundant. Analysis of the role of CSF in the regulation of sporulation has indicated that its function may be redundant with one or more extracellular spore factors. Introduction of a ΔphrC mutation into the *ecs191* mutant background resulted in a 10-fold decrease in sporulation, indicating that CSF is required for efficient sporulation (Lazazzera et al., 1997). However, a ΔphrC mutation had no significant effect on the level of sporulation of an otherwise wild-type strain. This suggests that there may be redundancy in extracellular spore factors. The removal of one spore factor may not significantly reduce the sporulation frequency, but if the levels of a number of spore factors are reduced, as may be the case in the *ecs191* mutant, then removing one spore factor may have a large effect on sporulation.

The Oligopeptide Permease and Response to Extracellular Spore Factors

Opp is required for efficient sporulation (Perego et al., 1991; Rudner et al., 1991). This is due to decreased levels of Spo0A~P in a Δ*opp* mutant. Increasing the level of Spo0A~P, either by overexpressing the histidine protein kinase KinC or by eliminating the RapA and RapB phosphatases, bypasses the sporulation defect of a Δ*opp* mutant (LeDeaux and Grossman, 1995; Perego and Hoch, 1996).

At least part of the requirement of Opp for sporulation can be explained by its role in import of the extracellular Phr peptides, CSF and PhrA (Lazazzera et al., 1997; Perego and Hoch, 1996). Consistent with Opp's role in transporting the cell density signal CSF, Δ*opp* mutants are partly defective for the cell density control of sporulation. Sporulation of Δ*opp* mutants is still decreased at low cell density, indicating that there is at least one extracellular spore factor that is not transported by Opp.

Phr Peptides: Signals of Cell Density and Possibly More

Sequence gazing indicates that there are at least seven *phr* genes that encode potential extracellular peptides (Fig. 7). Whereas extracellular CSF, the pentapeptide produced from the *phrC* gene product, is clearly an indicator of cell density, the physiological function of the other Phr peptides is less clear. In addition to its role in cell density signaling, CSF (and perhaps other Phr peptides) is an indicator of other physiological conditions. Since activity of sigma-H increases when cells sense nutrient depletion, control of transcription of several *phr* genes by sigma-H serves to indicate, in part, nutritional status or energy stress (Palmer and Grossman, manuscript in preparation). In this way, extracellular CSF is an indicator of more than just cell density, and could contribute to the integration of multiple physiological signals.

It is also possible that some of the Phr signaling peptides function in a cell-autonomous manner. For this to happen, a signaling peptide would have to preferentially act on the producing cell and probably could not be freely diffusible in culture medium. Such a role has been suggested for the PhrA pentapeptide (Perego, 1997), although it clearly can act in cell-cell signaling (Lazazzera et al., 1997; Perego and Hoch, 1996). The notion that a secreted peptide functions in a cell-autonomous manner raises additional questions concerning the physiological function of the peptide. As suggested previously (Perego, 1997), perhaps production of and response to the peptide serves as a timing mechanism to coordinate complex regulatory responses.

We also speculate that successful production of and response to some of the Phr peptides, whether or not they function in a cell-autonomous manner, might be used to indicate a functioning secretion pathway. The secretory pathway is required for the initiation of sporulation (Asai et al., 1997, and references therein), and several of the Phr peptides (CSF, PhrA, and perhaps others) regulate the initiation of sporulation. Perhaps the ability to properly secrete the Phr peptides, and then reinternalize them to inhibit Rap phosphatases, is a signal that the secretory pathway is functioning and sporulation can be initiated.

SUMMARY

Cell-cell signaling mediates cell density control of sporulation, the development of genetic competence, and possibly other physiological processes in *B. subtilis*. Small extracellular peptide factors are used to mediate this cell density control. There is obvious conservation among gram-positive bacteria for one mechanism of cell-cell signaling. Analogs of the *B. subtilis* *comQXPA* cassette, required for the production of and response to an extracellular peptide, have been found in at least five different bacterial genera thus far. There is also evidence in other organisms, for signaling by small peptides that are actively transported into the cell, in a manner similar to CSF in *B. subtilis*. Monitoring the crowding state of a population before committing to a complicated developmental process must be advantageous for *B. subtilis*. For competence, this probably indicates that there is a greater likelihood of DNA from its own species being present. For sporulation, monitoring crowding may indicate the degree of competition for a new source of food.

ACKNOWLEDGMENTS

Work in the Grossman lab is supported by Public Health Service grants GM50895 and GM41934. B.A.L. was supported, in part, by Fellowship DRG1384 of the Cancer Research Fund of the Damon Runyon-Walter Winchell Foundation. T.P. and J.Q. were supported, in part, by predoctoral fellowships from the National Science Foundation.

REFERENCES

Akrigg, A., and S. R. Ayad. 1970. Studies on the competence-inducing factor of *Bacillus subtilis*. *Biochem. J.* **117:**397–403.

Akrigg, A., S. R. Ayad, and G. R. Barker. 1967. The nature of a competence-inducing factor in *Bacillus subtilis*. *Biochem. Biophys. Res. Commun.* **28:** 1062–1067.

Alloing, G., C. Granadel, D. A. Morrison, and J.-P. Claverys. 1996. Competence pheromone, oligopeptide permease, and induction of competence in *Streptococcus pneumoniae*. *Mol. Microbiol.* **21:** 471–478.

Antoniewski, C., B. Savelli, and P. Stragier. 1990. The *spoIIJ* gene, which regulates early developmental steps in *Bacillus subtilis*, belongs to a class of environmentally responsive genes. *J. Bacteriol.* **172:**86–93.

Asai, K., F. Kawamura, Y. Sadaie, and H. Takahashi. 1997. Isolation and characterization of a sporulation initiation mutation in the *Bacillus subtilis secA* gene. *J. Bacteriol.* **179:**544–547.

Burbulys, D., K. A. Trach, and J. A. Hoch. 1991. Initiation of sporulation in *B. subtilis* is controlled by a multicomponent phosphorelay. *Cell* **64:**545–552.

Caldwell, G., F. Naider, and J. Becker. 1995. Fungal lipopeptide mating pheromones: a model system for the study of protein prenylation. *Microbiol. Rev.* **59:**406–422.

Cangelosi, G. A., R. G. Ankenbauer, and E. W. Nester. 1990. Sugars induce the *Agrobacterium* virulence genes through a periplasmic binding protein and a transmembrane signal protein. *Proc. Natl. Acad. Sci. USA* **87:**6708–6712.

Carter, H. L., III, K. M. Tatti, and C. P. Moran, Jr. 1991. Cloning of a promoter used by sigma-H RNA polymerase in *Bacillus subtilis*. *Gene* **96:**101–105.

Cox, G. B., D. Webb, J. Godovac-Zimmermann, and H. Rosenberg. 1988. Arg-220 of the PstA protein is required for phosphate transport through the phosphate-specific transport system in *Escherichia coli* but not for alkaline phosphatase repression. *J. Bacteriol.* **170:**2283–2286.

Diep, D. B., L. S. Håvarstein, and I. F. Nes. 1995. A bacteriocin-like peptide induces bacteriocin synthesis in *Lactobacillus plantarum* C11. *Mol. Microbiol.* **18:**631–639.

D'Souza, C., M. M. Nakano, and P. Zuber. 1994. Identification of *comS*, a gene of the *srfA* operon that regulates the establishment of genetic competence in *Bacillus subtilis*. *Proc. Natl. Acad. Sci. USA* **91:**9397–9401.

Dubnau, D. 1991. Genetic competence in *Bacillus subtilis*. *Microbiol. Rev.* **55:**395–424.

Dubnau, D. 1993. Genetic exchange and homologous recombination, p. 555–584. *In* A. L. Sonenshein, J. A. Hoch, and R. Losick (ed.), *Bacillus subtilis and other Gram-positive bacteria: Biochemistry, Physiology, and Molecular Genetics*. American Society for Microbiology, Washington, D.C.

Dubnau, D., and M. Roggiani. 1990. Growth medium-independent genetic competence mutants of *Bacillus subtilis*. *J. Bacteriol.* **172:**4048–4055.

Dunny, G. M. 1990. Genetic functions and cell-cell interactions in the pheromone-inducible plasmid transfer system of *Enterococcus faecalis*. *Mol. Microbiol.* **4:**689–696.

Dunny, G. M., and B. A. B. Leonard. 1997. Cell-cell communication in Gram-positive bacteria. *Annu. Rev. Microbiol.* **51:**527–564.

Errington, J. 1993. *Bacillus subtilis* sporulation: regulation of gene expression and control of morphogenesis. *Microbiol. Rev.* **57:**1–33.

Fuqua, W. C., S. C. Winans, and E. P. Greenberg. 1994. Quorum sensing in bacteria: the LuxR-LuxI family of cell density-responsive transcriptional regulators. *J. Bacteriol.* **176:**269–275.

Fuqua, C., S. C. Winans, and E. P. Greenberg. 1996. Census and consensus in bacterial ecosystems: the LuxR-LuxI family of quorum-sensing transcriptional regulators. *Annu. Rev. Microbiol.* **50:** 727–751.

Grossman, A. D. 1995. Genetic networks controlling the initiation of sporulation and the development of genetic competence in *Bacillus subtilis*. *Annu. Rev. Genet.* **29:**477–508.

Grossman, A. D., and R. Losick. 1988. Extracellular control of spore formation in *Bacillus subtilis*. *Proc. Natl. Acad. Sci. USA* **85:**4369–4373.

Hahn, J., and D. Dubnau. 1991. Growth stage signal transduction and the requirements for *srfA* induction in development of competence. *J. Bacteriol.* **173:**7275–7282.

Hahn, J., L. Kong, and D. Dubnau. 1994. The regulation of competence transcription factor synthesis constitutes a critical control point in the regulation of competence in *Bacillus subtilis*. *J. Bacteriol*. **176:**5753–5761.

Hahn, J., A. Luttinger, and D. Dubnau. 1996. Regulatory inputs for the synthesis of ComK, the competence transcription factor of *Bacillus subtilis*. *Mol. Microbiol*. **21:**763–775.

Haijema, B. J., D. van Sinderen, K. Winterling, J. Kooistra, G. Venema, and L. Hamoen. 1996. Regulated expression of the *dinR* and *recA* genes during competence development and SOS induction in *Bacillus subtilis*. *Mol. Microbiol*. **22:**75–85.

Hamoen, L. W., H. Eshuis, J. Jongbloed, G. Venema, and D. van Sinderen. 1995. A small gene, designated *comS*, located within the coding region of the fourth amino acid-activation domain of *srfA*, is required for competence development in *Bacillus subtilis*. *Mol. Microbiol*. **15:**55–63.

Håvarstein, L. S., G. Coomaraswamy, and D. Morrison. 1995a. An unmodified heptadecapeptide pheromone induces competence for genetic transformation in *Streptococcus pneumoniae*. *Proc. Natl. Acad. Sci. USA* **92:**11140–11144.

Håvarstein, L. S., D. B. Diep, and I. F. Nes. 1995b. A family of bacteriocin ABC transporters carry out proteolytic processing of their substrates concomitant with export. *Mol. Microbiol*. **16:**229–240.

Håvarstein, L. S., P. Gaustad, I. F. Nes, and D. A. Morrison. 1996. Identification of the streptococcal competence-pheromone receptor. *Mol. Microbiol*. **21:**863–869.

Healy, J., J. Weir, I. Smith, and R. Losick. 1991. Post-transcriptional control of a sporulation regulatory gene encoding transcription factor σ^H in *Bacillus subtilis*. *Mol. Microbiol*. **5:**477–487.

Higgins, C. F. 1992. ABC transporters: from microorganisms to man. *Annu. Rev. Cell Biol*. **8:**67–113.

Hoch, J. A. 1993. Regulation of the phosphorelay and the initiation of sporulation in *Bacillus subtilis*. *Annu. Rev. Microbiol*. **47:**441–465.

Ireton, K., and A. D. Grossman. 1992. Coupling between gene expression and DNA synthesis early during development in *Bacillus subtilis*. *Proc. Natl. Acad. Sci. USA* **89:**8808–8812.

Ireton, K., and A. D. Grossman. 1994. A developmental checkpoint couples the initiation of sporulation to DNA replication in *Bacillus subtilis*. *EMBO J*. **13:**1566–1573.

Ireton, K., D. Z. Rudner, K. J. Siranosian, and A. D. Grossman. 1993. Integration of multiple developmental signals in *Bacillus subtilis* through the Spo0A transcription factor. *Genes Dev*. **7:**283–294.

Ireton, K., S. F. Jin, A. D. Grossman, and A. L. Sonenshein. 1995. Krebs cycle function is required for activation of the Spo0A transcription factor in *Bacillus subtilis*. *Proc. Natl. Acad. Sci. USA* **92:**2845–2849.

Ji, G., R. C. Beavis, and R. P. Novick. 1995. Cell density control of staphylococcal virulence mediated by an octapeptide pheromone. *Proc. Natl. Acad. Sci. USA* **92:**12055–12059.

Joenje, H., M. Gruber, and G. Venema. 1972. Stimulation of the development of competence by culture fluids in *Bacillus subtilis* transformation. *Biochim. Biophys. Acta* **262:**189–199.

Jones, S., B. Yu, J. J. Bainton, M. Birdsall, B. W. Bycroft, S. R. Chabra, A. J. R. Cox, P. Golby, P. J. Reeves, S. Stephens, M. K. Winson, G. P. Salmond, G. S. Stewart, and P. Williams. 1993. The *lux* autoinducer regulates the production of exoenzyme virulence determinants in *Erwinia carotovora* and *Pseudomonas aeruginosa*. *EMBO J*. **12:**2477–2482.

Kleerebezem, M., L. E. N. Quadri, O. P. Kuipers, and W. M. de Vos. 1997. Quorum sensing by peptide pheromones and two-component signal-transduction systems in Gram-positive bacteria. *Mol. Microbiol*. **24:**895–904.

Kobayashi, K., K. Shoji, T. Shimizu, K. Nakano, T. Sato, and Y. Kobayashi. 1995. Analysis of a suppressor mutation *ssb* (*kinC*) of *sur0B20* (*spo0A*) mutation in *Bacillus subtilis* reveals that *kinC* encodes a histidine protein kinase. *J. Bacteriol*. **177:**176–182.

Kong, L., and D. Dubnau. 1994. Regulation of competence-specific gene expression by Mec-mediated protein-protein interaction in *Bacillus subtilis*. *Proc. Natl. Acad. Sci. USA* **91:**5793–5797.

Kong, L., K. J. Siranosian, A. D. Grossman, and D. Dubnau. 1993. Sequence and properties of *mecA*, a negative regulator of genetic competence in *Bacillus subtilis*. *Mol. Microbiol*. **9:**365–373.

Kuipers, O. P., M. M. Beerthuyzen, P. G. G. A. de Ruyter, E. J. Luesink, and W. M. de Vos. 1995. Autoregulation of nisin biosynthesis in *Lactococcus lactis* by signal transduction. *J. Biol. Chem*. **270:**27299–27304.

Kunst, F., N. Ogasawara, I. Moszer, A. M. Albertini, G. Alloni, V. Azevedo, et al. 1997. The complete genome sequence of the Gram-positive bacterium *Bacillus subtilis*. *Nature* **390:**249–256.

Lazazzera, B. A. Unpublished data.

Lazazzera, B. A., J. M. Solomon, and A. D. Grossman. 1997. An exported peptide functions intracellularly to contribute to cell density signaling in *B. subtilis*. *Cell* **89:**917–925.

Lazazzera, B. A., I. Kurtser, and A. D. Grossman. Manuscript in preparation.

LeDeaux, J. R., and A. D. Grossman. 1995. Isolation and characterization of *kinC*, a gene that encodes a sensor kinase homologous to the sporulation sensor kinases KinA and KinB in *Bacillus subtilis*. *J. Bacteriol.* **177:**166–175.

LeDeaux, J. R., N. Yu, and A. D. Grossman. 1995. Different roles for KinA, KinB, and KinC in the initiation of sporulation in *Bacillus subtilis*. *J. Bacteriol.* **177:**861–863.

Leonard, B. A. B., A. Podbielski, P. J. Hedberg, and G. M. Dunny. 1996. *Enterococcus faecalis* pheromone binding protein, PrgZ, recruits a chromosomal oligopeptide permease system to import sex pheromone cCF10 for induction of conjugation. *Proc. Natl. Acad. Sci. USA* **93:**260–264.

Magnuson, R., J. Solomon, and A. D. Grossman. 1994. Biochemical and genetic characterization of a competence pheromone from *B. subtilis*. *Cell* **77:**207–216.

Manson, M. D., V. Blank, G. Brade, and C. F. Higgins. 1986. Peptide chemotaxis in *E. coli* involves the Tap signal transducer and the dipeptide permease. *Nature* **321:**253–256.

Msadek, T., F. Kunst, A. Klier, and G. Rapoport. 1991. DegS-DegU and ComP-ComA modulator-effector pairs control expression of the *Bacillus subtilis* pleiotropic regulatory gene *degQ*. *J. Bacteriol.* **173:**2366–2377.

Msadek, T., F. Kunst, and G. Rapoport. 1994. MecB of *Bacillus subtilis*, a member of the ClpC ATPase family, is a pleiotropic regulator controlling competence gene expression and growth at high temperature. *Proc. Natl. Acad. Sci. USA* **91:**5788–5792.

Mueller, J. P., and A. L. Sonenshein. 1992. Role of the *Bacillus subtilis gsiA* gene in regulation of early sporulation gene expression. *J. Bacteriol.* **174:**4374–4383.

Mueller, J. P., G. Bukusoglu, and A. L. Sonenshein. 1992. Transcriptional regulation of *Bacillus subtilis* glucose starvation-inducible genes: control of *gsiA* by the ComP-ComA signal transduction system. *J. Bacteriol.* **174:**4361–4373.

Nakano, M. M., M. A. Marahiel, and P. Zuber. 1988. Identification of a genetic locus required for biosynthesis of the lipopeptide antibiotic surfactin in *Bacillus subtilis*. *J. Bacteriol.* **170:**5662–5668.

Nakano, M. M., and P. Zuber. 1989. Cloning and characterization of *srfB*, a regulatory gene involved in surfactin production and competence in *Bacillus subtilis*. *J. Bacteriol.* **171:**5347–5353.

Nakano, M. M., and P. Zuber. 1991. The primary role of ComA in establishment of the competent state in *Bacillus subtilis* is to activate expression of *srfA*. *J. Bacteriol.* **173:**7269–7274.

Nakano, M. M., M. A. Marahiel, and P. Zuber. 1988. Identification of a genetic locus required for biosynthesis of the lipopeptide antibiotic surfactin in *Bacillus subtilis*. *J. Bacteriol.* **170:**5662–5668.

Nakano, M. M., R. Magnuson, A. Meyers, J. Curry, A. D. Grossman, and P. Zuber. 1991a. *srfA* is an operon required for surfactin production, competence development, and efficient sporulation in *Bacillus subtilis*. *J. Bacteriol.* **173:**1770–1778.

Nakano, M. M., L. Xia, and P. Zuber. 1991b. Transcription initiation region of the *srfA* operon, which is controlled by the ComP-ComA signal transduction system in *Bacillus subtilis*. *J. Bacteriol.* **173:**5487–5493.

Nodwell, J. R., K. McGovern, and R. Losick. 1996. An oligopeptide permease responsible for the import of an extracellular signal governing aerial mycelium formation in *Streptomyces coelicolor*. *Mol. Microbiol.* **22:**881–893.

Ohlsen, K. L., J. K. Grimsley, and J. A. Hoch. 1994. Deactivation of the sporulation transcription factor Spo0A by the Spo0E protein phosphatase. *Proc. Natl. Acad. Sci. USA* **91:**1756–1760.

Palmer, T., and A. D. Grossman. Manuscript in preparation.

Palmer, T., and A. D. Grossman. Unpublished results.

Pearce, B. J., A. M. Naughton, and H. R. Masure. 1994. Peptide permeases modulate transformation in *Streptococcus pneumoniae*. *Mol. Microbiol.* **12:**881–892.

Perego, M. 1997. A peptide export-import control circuit modulating bacterial development regulates protein phosphatases of the phosphorelay. *Proc. Natl. Acad. Sci. USA* **94:**8612–8617.

Perego, M., and J. A. Hoch. 1996. Cell-cell communication regulates the effects of protein aspartate phosphatases on the phosphorelay controlling development in *Bacillus subtilis*. *Proc. Natl. Acad. Sci. USA* **93:**1549–1553.

Perego, M., S. P. Cole, D. Burbulys, K. Trach, and J. A. Hoch. 1989. Characterization of the gene for a protein kinase which phosphorylates the sporulation-regulatory proteins Spo0A and Spo0F of *Bacillus subtilis*. *J. Bacteriol.* **171:**6187–6196.

Perego, M., C. F. Higgins, S. R. Pearce, M. P. Gallagher, and J. A. Hoch. 1991. The oligopeptide transport system of *Bacillus subtilis* plays a role in the initiation of sporulation. *Mol. Microbiol.* **5:**173–185.

Perego, M., C. Hanstein, K. M. Welsh, T. Djavakhishvili, P. Glaser, and J. A. Hoch. 1994. Multiple protein-aspartate phosphatases provide a mechanism for the integration of diverse signals in the control of development in *B. subtilis*. *Cell* **79:**1047–1055.

Perego, M., P. Glaser, and J. A. Hoch. 1996. Aspartyl-phosphate phosphatases deactivate the response regulator components of the sporulation signal transduction system in *Bacillus subtilis*. *Mol. Microbiol.* **19:**1151–1157.

Pestova, E. V., L. S. Håvarstein, and D. A. Morrison. 1996. Regulation of competence for genetic transformation in *Streptococcus pneumoniae* by an auto-induced peptide pheromone and a two-component regulatory system. *Mol. Microbiol.* **21:**853–862.

Pirhonen, M., D. Flego, R. Heikinheimo, and E. T. Palva. 1993. A small diffusible signal molecule is responsible for the global control of virulence and exoenzyme production in the plant pathogen *Erwinia carotovora*. *EMBO J.* **12:**2467–2476.

Quadri, L. E. N., M. Kleerebezem, O. P. Kuipers, W. M. de Vos, K. L. Roy, J. C. Vederas, and M. E. Stiles. 1997. Characterization of a locus from *Carnobacterium piscicola* LV17B involved in bacteriocin production and immunity: evidence for global inducer-mediated transcriptional regulation. *J. Bacteriol.* **179:**6163–6171.

Roggiani, M., and D. Dubnau. 1993. ComA, a phosphorylated response regulator protein of *Bacillus subtilis*, binds to the promoter region of *srfA*. *J. Bacteriol.* **175:**3182–3187.

Roggiani, M., J. Hahn, and D. Dubnau. 1990. Suppression of early competence mutations in *Bacillus subtilis* by *mec* mutations. *J. Bacteriol.* **172:**4056–4063.

Rudner, D. Z., J. R. LeDeaux, K. Ireton, and A. D. Grossman. 1991. The *spo0K* locus of *Bacillus subtilis* is homologous to the oligopeptide permease locus and is required for sporulation and competence. *J. Bacteriol.* **173:**1388–1398.

Serror, P., and A. L. Sonenshein. 1996. CodY is required for nutritional repression of *Bacillus subtilis* genetic competence. *J. Bacteriol.* **178:**5910–5915.

Solomon, J. M., and A. D. Grossman. 1996. Who's competent and when: regulation of natural genetic competence in bacteria. *Trends Genet.* **12:**150–155.

Solomon, J. M., R. Magnuson, A. Srivastava, and A. D. Grossman. 1995. Convergent sensing pathways mediate response to two extracellular competence factors in *Bacillus subtilis*. *Genes Dev.* **9:**547–558.

Solomon, J. M., B. A. Lazazzera, and A. D. Grossman. 1996. Purification and characterization of an extracellular peptide factor that affects two different developmental pathways in *Bacillus subtilis*. *Genes Dev.* **10:**2014–2024.

Stragier, P., and R. Losick. 1996. Molecular genetics of sporulation in *Bacillus subtilis*. *Annu. Rev. Genet.* **30:**297–341.

Swift, S., J. P. Throup, P. Williams, G. P. C. Salmond, and G. S. A. B. Stewart. 1996. Quorum sensing: a population-density component in the determination of bacterial phenotype. *Trends Biochem.* **21:**214–219.

Trach, K. A., and J. A. Hoch. 1993. Multisensory activation of the phosphorelay initiating sporulation in *Bacillus subtilis*: identification and sequence of the protein kinase of the alternate pathway. *Mol. Microbiol.* **8:**69–79.

Turgay, K., L. W. Hamoen, G. Venema, and D. Dubnau. 1997. Biochemical characterization of a molecular switch involving the heat shock protein ClpC, which controls the activity of ComK, the competence transcription factor of *Bacillus subtilis*. *Genes Dev.* **11:**119–128.

van Belkum, M., R. Worobo, and M. Stiles. 1997. Double-glycine-type leader peptides direct secretion of bacteriocins by ABC transporters: colicin V secretion in *Lactococcus lactis*. *Mol. Microbiol.* **23:**1293–1301.

van Sinderen, D., and G. Venema. 1994. *comK* acts as an autoregulatory control switch in the signal transduction route to competence in *Bacillus subtilis*. *J. Bacteriol.* **176:**5762–5770.

van Sinderen, D., A. ten Berge, B. J. Hayema, L. Hamoen, and G. Venema. 1994. Molecular cloning and sequence of *comK*, a gene required for genetic competence in *Bacillus subtilis*. *Mol. Microbiol.* **11:**695–703.

van Sinderen, D., A. Luttinger, L. Kong, D. Dubnau, G. Venema, and L. Hamoen. 1995. *comK* encodes the competence transcription factor, the key regulatory protein for competence development in *Bacillus subtilis*. *Mol. Microbiol.* **15:**455–462.

Vasantha, N., and E. Freese. 1979. The role of manganese in growth and sporulation of *Bacillus subtilis*. *J. Gen. Microbiol.* **112:**329–336.

Waldburger, C., D. Gonzalez, and G. H. Chambliss. 1993. Characterization of a new sporulation factor in *Bacillus subtilis*. *J. Bacteriol.* **175:**6321–6327.

Wanner, B. L. 1993. Gene regulation by phosphate in enteric bacteria. *J. Cell. Biochem.* **51:**47–54.

Weinrauch, Y., N. Guillen, and D. Dubnau. 1989. Sequence and transcription mapping of *Bacillus subtilis* competence genes *comB* and *comA*, one of which is related to a family of bacterial regulatory determinants. *J. Bacteriol.* **171:**5362–5375.

Weinrauch, Y., R. Penchev, E. Dubnau, I. Smith, and D. Dubnau. 1990. A *Bacillus subtilis* regulatory gene product for genetic competence and sporulation resembles sensor protein members of the bacterial two-component signal-transduction systems. *Genes Dev.* **4:**860–872.

Weir, J., M. Predich, E. Dubnau, G. Nair, and I. Smith. 1991. Regulation of *spo0H*, a gene coding for the *Bacillus subtilis* s^H factor. *J. Bacteriol.* **173:**521–529.

Yang, M., E. Ferrari, E. Chen, and D. J. Henner. 1986. Identification of the pleiotropic *saqQ* gene of *Bacillus subtilis*. *J. Bacteriol.* **166:**113–119.

Zhang, F. L., and P. J. Casey. 1996. Protein prenylation: molecular mechanisms and functional consequences. *Annu. Rev. Biochem.* **65:**241–269.

SEX PHEROMONE SYSTEMS IN ENTEROCOCCI

Don B. Clewell

4

The number of bacterial species known to make use of pheromones has been growing rapidly, in part because of the multitude of quorum-sensing systems being reported. These systems involve a "self-induction" which occurs when bacteria respond to concentrations of a secreted compound reached only if the cell population collectively represents a critical density. The behavior is exemplified by the phenomenon of competence in species such as *Bacillus subtilis* and *Streptococcus pneumoniae*; the principle is essentially the same with regard to bioluminescence in *Vibrio fischeri* and a variety of other systems, many of which are discussed elsewhere in this volume. In these types of intercellular communication, the organisms involved are generally members of the same population.

In other cases, pheromone phenomena involve communication between organisms of different genotypes. Examples of this include mating systems in *Enterococcus faecalis*, in which a given peptide pheromone may have no known effect on the bacterial producer, even at high density, but induces a conjugative mating response in a nearby nonidentical bacterium of the same or closely related species. From a broader perspective, signaling phenomena related to conjugation could include the stimulation of transfer of "tumor-inducing" DNA from strains of *Agrobacterium tumefaciens* into plants by phenolic compounds emitted from injured plant tissue. The plant tumors (e.g., crown galls) that result from such infections in turn secrete compounds (e.g., opines or arginine analogs encoded by bacterial DNA integrated in the plant genome) which the bacteria outside can use as a carbon/nitrogen source. Interestingly, these substances also act as "aphrodisiacs" able to stimulate interbacterial transfer of plasmid DNA via the activation of a homoserine lactone quorum-sensing system (see chapter 8 of this volume). Thus far, the enterococci are the only bacterial group known to secrete sex pheromones able to activate conjugation systems in certain plasmid-containing donor cells; such systems will constitute the focus of this review. For earlier reviews see Clewell (1993a, 1993b), Dunny and Leonard (1997), Dunny et al., (1995), and Wirth (1994).

The enterococci are gram-positive bacteria that normally inhabit the human intestine and are opportunistic pathogens that can be asso-

Don B. Clewell, Department of Biologic and Materials Sciences, School of Dentistry, and Department of Microbiology and Immunology, School of Medicine, The University of Michigan, Ann Arbor, MI 48109.

Cell-Cell Signaling in Bacteria, Edited by Gary M. Dunny and Stephen C. Winans
©1999 American Society for Microbiology, Washington, D.C.

ciated with bacteremia, urinary tract infections, and infective endocarditis (Jett et al., 1994; Moellering, 1992; Murray, 1990). They are among the top three types of bacteria involved in nosocomial infections in the United States (Lewis and Zervos, 1990). These organisms frequently harbor multiple transferable plasmids as well as conjugative transposons and probably represent a significant reservoir of genetic information for a variety of bacterial species in the gut. Multiple antibiotic resistance, including resistance to the "last resort" antibiotic vancomycin, is common in enterococci; resistance as well as certain virulence traits can be associated with highly conjugative plasmids (Clewell, 1990).

PHEROMONE SYSTEMS OF *E. FAECALIS*

A given plasmid-free (recipient) strain of *E. faecalis* secretes a variety of peptide sex pheromones (Table 1); each stimulates a mating response by donor cells bearing a member of a specific plasmid family (Dunny et al., 1979). (Not all conjugative plasmids in enterococci involve pheromones.) The plasmid-encoded response enables relatively efficient transfer in broth matings (e.g., 10^{-3} to 10^{-1} per donor). When a copy of the plasmid is acquired, the transconjugant ceases to produce detectable pheromone and becomes a responder, but the cell continues to secrete other host-determined pheromones active on donors carrying different plasmids. Studies on bacteria harboring two different pheromone-responding plasmids have shown that the conjugation system activated is specific for the particular pheromone to which the cells are exposed (Ehrenfeld et al., 1986).

In all cases examined, the plasmid that determines a pheromone response also encodes a peptide that is secreted and acts as a competitive inhibitor of pheromone (Clewell et al., 1990; Ike et al., 1983; Mori et al., 1986b) (Table 1). These peptides correspond to the carboxyl-terminal 7- to 8-amino-acid residues of 21- to 23- amino-acid precursor structures that appear designed to ensure exportation. The inhibitor is believed to prevent the self-induction of donors by endogenous pheromone that may not have been totally shut down. In at least one known case (pCF10), there is little if any reduction of endogenous pheromone, and the inhibitor plays a major role in countering self-induction (Nakayama et al., 1995d). In other cases (e.g., pAD1) very little pheromone is secreted from donor cells, and the inhibitor is in significant excess (Nakayama et al., 1995d); however, if production of the inhibitor (iAD1) is mutationally eliminated, self-induction does occur (Bastos et al., 1997; Tanimoto, personal communication). Generally, if there is an equal number of donors and recipients in a broth culture, the

TABLE 1

Pheromone or inhibitor	Peptide structure	Plasmid system	Reference
cAD1	LFSLVLAG	pAD1	Mori et al., 1984
iAD1	LFVVTLVG		Mori et al., 1986
cPD1	FLVMFLSG	pPD1	Suzuki et al., 1984
iPD1	ALILTLVS		Mori et al., 1987
cCF10	LVTLVFV	pCF10	Mori et al., 1988
iCF10	AITLIFI		Nakayama et al., 1984
cAM373	AIFILAS(A)[a]	pAM373	Mori et al., 1986; Nakayama et al., 1996
iAM373	SIFTLVA		Nakayama et al., 1995c
cOB1	VAVLVLGA	pOB1	Nakayama et al., 1995b

[a] In *S. aureus* the A is present in place of the S.

recipient-produced pheromone will be in excess and induction will occur. The structures of a number of pheromones and their inhibitors have been determined biochemically by the Suzuki group in Japan (Table 1). Synthetic peptides have been generated in most cases and are fully active.

The first examples of pheromone response systems in *E. faecalis* involved the bacteriocin-encoding plasmid pPD1, originally identified in strain 39-5 (Dunny et al., 1978; Yagi et al., 1983), and the cytolysin (hemolysin/bacteriocin)-encoding pAD1 of the multiply resistant strain DS16 (Clewell et al., 1982a; Dunny et al., 1979). The most detailed genetic analyses have related to pAD1 and the tetracycline resistance-encoding plasmid pCF10 from the isolate SF-7 (Dunny et al., 1981); however, studies conducted in Japan are generating increasing information on pPD1 (Fujimoto et al., 1995; Nakayama et al., 1995a). The discussion below will deal primarily with these three plasmids and will focus on regulatory aspects of the conjugation phenomena.

Aggregation Substance

When donor and recipient bacteria are mixed together in broth during log phase, the culture will be seen to "clump" within a few hours as a result of the donors' synthesizing a plasmid-encoded surface protein designated aggregation substance (Galli et al., 1989). The protein is involved in initiating contact between donors and recipients by binding to enterococcal binding substance on the recipient surface, and aggregation requires the presence of phosphate and a divalent cation such as Mg^{2+} (Yagi et al., 1983). Enterococcal binding substance, which may in part consist of lipoteichoic acid (Bensing and Dunny, 1993; Ehrenfeld et al., 1986), is also present on the donor surface; in the absence of recipients, donors can be observed to self-clump if exposed to a culture filtrate of recipients or to synthetic pheromone. Indeed, experiments have shown that significant mating can occur between donors during such clumping, although the frequency is usually at least an order of magnitude lower than that which occurs between donors and plasmid-free recipients owing to a plasmid-encoded surface exclusion function (Clewell and Brown, 1980; Dunny et al., 1985). The phenomenon of donor clumping serves as the basis of a convenient microtiter assay for quantifying pheromone activity (Dunny et al., 1979).

Results involving immunoelectron microscopy of induced cells have been reported for the pPD1 (Yagi et al., 1983), pAD1 (Galli et al., 1989; Hirt et al., 1993; Wanner et al., 1989) and pCF10 (Olmsted et al., 1993) systems, revealing a microfibrillar substance on the donor surface. In the case of pAD1 (Hirt et al., 1993) and pCF10 (Olmsted et al., 1993), the aggregation substances Asa1 and Asc10, respectively, did not coat cells uniformly; interestingly, the aggregation substance appeared mainly on "older" sections of the surface of dividing cells (Olmsted et al., 1993; Wanner et al., 1989). Scanning electron microscopy of cells in which both the pCF10 aggregation substance and the plasmid-encoded surface exclusion protein are immunologically labeled has also been reported (Olmsted et al., 1993).

Upon exposure to pheromone, donor cells such as those carrying pAD1 require 60 to 90 min to generate a maximum mating potential, which can be $\geq 100,000$-fold greater than that of unexposed cells (Ike and Clewell, 1984). The level of induction can be easily measured by determining the mating potential during a brief (e.g., 10-min) mixing of donors and recipients. Such short exposures to recipients are not long enough to result in a significant pheromone response by donors; thus, they are useful in estimating the mating potential of cells previously induced. Pheromones are active at concentrations as low as 5×10^{-11} M, and there is a report relating to pCF10 estimating that donors are able to detect just a few molecules of the cognate peptide cCF10 (Mori et al., 1988).

Genetic Organization

Figure 1 compares current maps of pAD1, pCF10, and pPD1 with regard to a key region

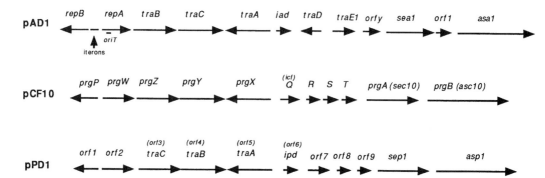

FIGURE 1 Comparison of the organization of selected determinants on pAD1, pCF10, and pPD1. The products, their properties, and related functions are listed in Table 2.

that includes regulatory determinants and the gene for aggregation substance. Table 2 lists related products with respect to function. With regard to the genes for aggregation substance (e.g., *asa1*) and surface exclusion (e.g., *sea1*), there are significant similarities in general organization and sequence content (Hirt et al., 1996). Indeed, based on Southern blot analyses, aggregation substance determinants of numerous pheromone-responding plasmids exhibit strong homology; an exception, however, is the plasmid pAM373 (Galli and Wirth, 1991; Hirt et al., 1996). In the case of pAD1,

asa1 is preceded by an open reading frame (ORF) (*orf1*) of unknown function (Galli et al., 1990). Interestingly, pPD1 has been shown to have an insertion element sequence located within the 3′ end of its surface exclusion determinant *sep1* (Hirt et al., 1996).

Additional structural genes relating to conjugation, including those involved in DNA transfer, are located over a 15-kb span (at least for pAD1) downstream of the determinant for aggregation substance (not shown in Fig. 1) (Ehrenfeld and Clewell, 1987; Pontius and Clewell, 1991). In the case of pAD1, most if

TABLE 2 Products, properties, and functions of pAD1, pCF10, and pPD1

Function	pAD1		pCF10		pPD1	
	Protein	No. of amino acids[a]	Protein	No. of amino acids[a]	Protein	No. of amino acids[a]
Aggregation	Asa1	1,296	Asc10 (PrgB)	1,305	Asp1	1,306
Surface exclusion	Sea1	890	Sec10 (PrgA)	891	Sep1	?
Inhibitor	iAD1	22	iCF10 (PrgQ)	23	iPD1 (Orf6)	21
Positive regulation	TraE1	119	PrgQ	RNA	Orf7?	125
			PrgR	121	Orf8?	48
			PrgS	73	Orf9?	101
			PrgT	62		
Negative regulation	TraA	320	PrgX?	317	TraA	321
	TraB	389	PrgY	383	TraB	384
Pheromone uptake	TraC	543	PrgZ	545	TraC	545
Plasmid maintenance	RepA	336	PrgW	333	Orf2	333
	RepB	287	PrgP	309		
	RepC	123	PrgO	110		

[a] Sizes represent unprocessed proteins.

not all of the downstream determinants are in the same orientation as *asa1* and are probably organized into several operons. Appropriately oriented Tn*917lac* insertions scattered through this region all show LacZ induction upon pheromone exposure. Transposon insertions over an additional 8 kb further downstream, where no apparent phenotype is evident, are also inducible with pheromone. Thus, pheromone exposure controls expression of genes over at least 23 kb of DNA downstream of *asa1*, a span that comes to within 1 to 2 kb of the plasmid's hemolysin/bacteriocin determinant.

Hirt et al. (1996) have sequenced more than 6 kb beyond *asa1* and have resolved 12 ORFs, all oriented similarly. No significant homology was observed with sequences in the DNA database, although regional hybridization experiments involving pCF10 and pPD1 showed homology in some locations. No homology was seen with a determinant *traF* (PD78) located downstream of *asp1* and involved in transfer of pPD1 (Nakayama et al., 1995e). A number of other plasmids that encode pheromone responses were probed and found to have extensive homology in some cases and very little in others (Hirt et al., 1996).

The region on pAD1 involving *traA-iad-traD-traE1* and the corresponding regions on pCF10 and pPD1 (Fig. 1) are involved in regulation of expression of conjugation functions and the response to pheromone exposure. There are significant differences among the three plasmids with regard to this region, and these probably relate to some basic differences in the molecular control mechanisms (discussed below). *iad*, *icf* (*prgQ*), and *ipd* represent determinants for the corresponding inhibitor peptides iAD1, iCF10, and iPD1, respectively.

traC on pAD1 is equivalent to *traC* on pPD1 and *prgZ* on pCF10; these determinants exhibit strong homology (Fujimoto et al., 1995; Nakayama et al., 1995a; Ruhfel et al., 1993; Tanimoto et al., 1993). The positions of *traC* and *traB* on pAD1 are reversed in the case of pPD1 and pCF10. The related prod-

ucts are located on the cell surface and are involved in pheromone binding; there is good evidence that the proteins participate together with a host-encoded peptide transport system to facilitate pheromone uptake (Leonard et al., 1996). Mutations in these determinants result in cells that require exposure to higher concentrations of pheromone in order to generate induction. It has been reported that *traC* of pAD1 is able to complement *traC* mutants of pPD1 (Nakayama and Suzuki, 1997), implying that binding to the pheromone peptides is not necessarily highly specific. The binding proteins were found to resemble lipoproteins corresponding to oligopeptide binding proteins from *Escherichia coli* and *B. subtilis*.

traB on pAD1 (An and Clewell, 1994; Ike and Clewell, 1984) and pPD1 (Fujimoto et al., 1995; Nakayama et al. 1995a), as well as *prgY* on pCF10 (Dunny and Leonard, 1997; Ruhfel et al., 1993) are closely related and involve the ability to prevent self-induction by endogenous pheromone. The sequence indicates a membrane protein, and in the case of *traB* of pAD1, and apparently also for pPD1, its function involves shutdown of secretion of endogenous pheromone. In the case of pCF10, in which cCF10 appears unaffected, PrgY appears able to somehow block what has been suggested as resembling an autocrine circuit (Dunny and Leonard, 1997).

repA and *repB* of pAD1 relate to plasmid maintenance (Weaver and Clewell, 1989; Weaver et al., 1993). *repA* is probably involved in the initiation of vegetative replication; whereas *repB* is involved in plasmid copy number (i.e., the frequency of initiation). Between the diverging *repA* and *repB* are a series of iteron sequences. These correspond to repeats of 13 and 12 contiguous octanucleotide sequences separated by a 78-bp "spacer." Iterons of this nature do not appear to be present to this extent in the corresponding regions of pCF10 (i.e., between *prgW* and *prgP*) and pPD1 (between *orf2* and *orf1*). *repA* exhibits homology with *prgW* of pCF10 (Ruhfel et al. 1993) and *orf2* of pPD1 (Fujimoto et al., 1995). *repB* exhibits some homology with *orf1* of pPD1 (Fujimoto et al., 1995) and *prgP* of pCF10 (Hedberg et al., 1996). Interestingly, the origin of conjugative transfer (*oriT*) is located within the *repA* determinant (An and Clewell, 1997), raising the possibility that there may be a connection between conjugative and replication functions of this element. (*oriT* sites have not yet been identified in the case of pPD1 and pCF10.) The product of the homologous determinant *prgW* of pCF10 also has an intriguing link with conjugation control, namely, the ability to bind pheromone (Leonard et al., 1996).

Regulation of the pAD1 Pheromone Response

Two primary components of the pAD1-encoded response to cAD1 are TraE1, a positive regulator of all or most of the conjugation-related structural genes (Pontius and Clewell, 1992a; Weaver and Clewell, 1988, 1990), and TraA, which negatively regulates TraE1 expression (Pontius and Clewell, 1992b) (Fig. 2). TraA protein has been found to bind to DNA between *iad* and *traA* and to reduce transcription from the *iad* promoter (Tanimoto and Clewell, 1993). Northern blot analyses have indicated that a significant amount of transcription occurs from this location even in the uninduced state, and termination occurs about 500 nucleotides (nt) downstream where two transcription termination/pause sites, t1 and t2, are located 128 nt apart (Pontius and Clewell, 1992a; Tanimoto and Clewell, 1993). The two sites resemble intrinsic terminators, but calculations indicate that t2 is significantly weaker than t1. TraA is believed to facilitate termination in the t1-t2 region, although it is not clear if it enhances termination by simply reducing transcription from the *iad* promoter. Interestingly, a sequence TTATTTTATTT located within the *iad* promoter and containing the site where TraA has been observed to bind (Tanimoto and Clewell, 1993) is also located partially within the 3′ end of t2 (Pontius and Clewell, 1992a), suggesting that TraA may also interact here and influence (prevent?)

transcription beyond t2. However, TraA has not been observed to bind to DNA at this site in vitro under the conditions for which it was observed to bind at the *iad* promoter.

Exposure of cells to pheromone results in a significant increase in transcription from the *iad* promoter, some of which is able to pass through t1–t2 and proceed through *traE1* and even *sea1* (Galli et al., 1992; Tanimoto and Clewell, 1993), as illustrated in Fig. 2. Once TraE1 is made, it is able to positively regulate expression of the various structural genes. Muscholl et al. (1993) reported that TraE1 could, in *trans*, positively regulate expression of aggregation substance (Asa1) that was being transcribed from the promoter of the adjacent upstream *orf1*. There is evidence that TraE1 also up-regulates itself at its own promoter located within t2 (Tanimoto and Clewell, 1993); a Tn*917lac* insertion in *traE1* that exhibits pheromone-inducible LacZ does not initiate RNA synthesis from the *traE1* promoter, in contrast to the wild type (Fujimoto et al., 1997; Bastos et al., in press). Interestingly, there is an 11-nt sequence TTAAAAAATAT, located 21 nt upstream of the translational start site of *asa1*, that is also located 18 nt upstream of the translational start site of *traE1* (Galli et al., 1990; Pontius and Clewell, 1992a). It is conceivable, therefore, that this sequence corresponds to the target of TraE1; however, because *asa1* is thought to be transcribed from the further-upstream *orf1* promoter, the regulation of *asa1* would appear to involve facilitating extension of transcription beyond an already-transcribing *orf1*. (Northern blot studies [Tomita, personal communication)] suggest that *orf1* is indeed transcribed in the absence of exposure to pheromone.) In the case of *traE1*'s own promoter, the target sequence is 12 nt downstream of the apparent transcriptional start site. And the first two nucleotides (TT) at its 5′ end correspond to the last two nucleotides at the 3′ end of the above-noted possible TraA target. (Together they appear as 5′ - TTATTTTATTTAAAAAATAT - 3′.) Thus, TraA binding, if it does occur here, might interfere with TraE1 binding or vice versa.

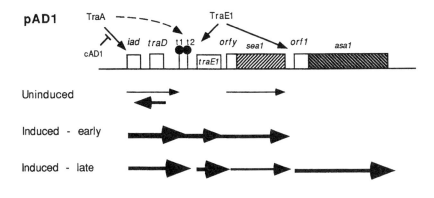

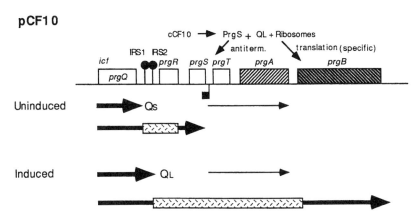

FIGURE 2 Models relating to regulation of the pheromone response of pAD1 and pCF10. The thickness of the arrow lines qualitatively reflects relative amounts of transcription. The hatched areas within arrows indicate degradation and/or processing. In the case of pAD1, the pheromone (cAD1) acts by binding to the negatively regulating TraA, resulting in enhanced transcription from the *iad* promoter along with an antitermination at t1-t2. Once TraE1 is synthesized, it activates itself at its own promoter within t2 and is also able to activate transcription of *asa1*. In the case of pCF10, the cCF10 pheromone is thought to bind to ribosomes in such a way as to enhance translation of PrgS RNA. PrgS protein then facilitates or stabilizes binding of QL RNA to ribosomes so that PrgB (Asc10) RNA can be specifically translated. PrgS or the ribosome complex is also believed to be involved in transcription antitermination at the terminator ("flag") between *prgS* and *prgT*. Although the Dunny group has not previously used the term *icf* for the determinant of iCF10, it is placed here simply for comparison to *iad* of pAD1.

Biochemical studies have shown that cAD1 is able to bind TraA directly near its carboxyl-terminal end and eliminate its ability to bind to DNA (Fujimoto et al., 1997; Fujimoto and Clewell, 1998). This result is consistent with earlier genetic data showing that, while most

transposon insertions in *traA* result in constitutive expression of conjugation functions, insertions near the 3′ end of the determinant result in little if any derepression and exhibit an insensitivity to pheromone (Ike and Clewell, 1984; Pontius and Clewell, 1992b). The analyses of late showed that elimination of 19 or more amino acid residues from the carboxyl terminus of the 319-residue TraA resulted in a protein that, in contrast to wild-type protein, would maintain its binding to plasmid DNA in the presence of cAD1; modifications closer to the carboxyl terminus affected in vitro peptide binding specificity (Fujimoto and Clewell, 1998).

The *traD* determinant, located downstream of *iad* and in the opposite orientation, appears to encode a 23-amino-acid peptide (Bastos et al., 1997). That this determinant plays a regulatory role was initially suggested by the generation of a thermolabile point mutation resulting in a mating response when cells were shifted from 30 to 42°C. Studies have implied that it is *traD* RNA— not the encoded peptide—that plays the key functional role, since the temperature sensitive mutation could be complemented in *trans* by a translationally defective form of the transcript (Bastos et al., in press). *traD* RNA corresponds to a 200-nt product that, based on computer analysis, could exhibit extensive secondary structure; calculations revealed that the thermolabile derivative is unable to assume the most energetically stable structures. It is expressed at a relatively high level when cells are in an uninduced state and is greatly reduced upon induction of cells by pheromone. The RNA would appear to act together with TraA to influence the level of transcription at the *iad* promoter, or at least the ability to transcribe through t1/t2. Interestingly, a Tn917lac insertion 43 nt downstream of the *iad* translational stop codon and within a *traD* untranslated region results in expression of β-galactosidase that is much higher than the expression in the case of an insertion close to, but preceding, t1 and upstream of the transcription start site of *traD* (Pontius and Clewell, 1992a). Because this mutant was not complemented in *trans* by *traD* RNA, it is possible that the apparent RNA effector operating at the *iad* promoter could involve both *traD* RNA and *iad* RNA, since the latter would also be affected by the insertion. It is also noteworthy that a 70-nt segment of the 5′ end of *traD* RNA has strong complementarity with untranslated RNA between *orf1* and *asa1*; thus, it is very possible that

a localized annealing of the two plays a role in control of *asa1* expression.

The general working hypothesis for the initial stages of the pheromone induction circuit is currently viewed as follows. Internalized cAD1 binds directly to TraA, neutralizing its affinity for DNA and resulting in enhancement of transcription from the *iad* promoter. (Since *traA* mutants that transcribe from the *iad* promoter at a maximal level secrete only a twofold increase in iAD1, there would appear to be factors [e.g., RNA secondary structure?] that limit translation.) The significantly increased transcription interferes with the promoter activity of the oppositely oriented *traD*, resulting in a reduction of TraD RNA. If *traD* RNA interacts with TraA or an RNA polymerase complex to down-regulate the *iad* promoter, then reduction of *traD* RNA would lead to even further up-regulation of that promoter. If TraD RNA also acts to down-regulate *asa1* expression (see above), then its reduction during pheromone exposure would reverse this activity.

The pCF10 Response and Comparisons with pAD1

The pCF10 response differs in some respects from the pAD1 response. There does not appear to be a homolog of the TraE1 positive regulator of pAD1; whereas *prgX* of pCF10 exhibits some sequence similarity with *traA* of pAD1, there is no genetic evidence that it behaves in a like manner. (Mutants of *prgX* have not been generated and may be lethal [Dunny, personal communication].) A group of adjacent determinants designated *prgQ*, *prgR*, and *prgS* (Fig. 2) are required for expression of the aggregation substance PrgB (Asc10) (Chung et al., 1995; Chung and Dunny, 1992, 1995; Kao et al., 1991). For all three, expression is from the *prgQ* promoter, which is also necessary in *cis* for *prgB* expression (Chung and Dunny, 1992). The pCF10-encoded peptide inhibitor iCF10 is encoded within *prgQ* and corresponds to the last seven residues of a 23-amino-acid precursor (Nakayama et al. 1994). Transcription beyond *prgQ* extends through *prgS* and, in the absence of pheromone exposure, appears to terminate at or near a site that would give rise to significant local secondary loop-structures ("flag" in Fig. 2) within the transcript (Bensing et al., 1997). It has been suggested that a major part of the transcript is degraded, resulting in a structure corresponding to the first 400 nt of the RNA (Bensing and Dunny, 1997; Bensing et al., 1997); however, it would seem that these transcripts could also result from the apparent factor-independent terminator corresponding to IRS1. This transcript has been designated Qs. The noninduced state also appears to involve initiation of a transcript from a promoter upstream of *prgT* that extends through the surface exclusion determinant *prgA* but not into *prgB*. Its promoter is within the region ("flag" in Fig. 2) that terminates the transcription from upstream (Bensing et al., 1997).

Expression from the *prgQ* promoter is believed to be constitutive, and induction is viewed as involving an antitermination event between prgS and prgT, resulting in transcription through *prgB* (Bensing and Dunny, 1997; Bensing et al., 1996, 1997). Bensing et al. (1996) have reported biochemical evidence that RNA elongated through *prgB* is subsequently processed. A significant amount of a transcript QL (530 nt), which corresponds to a 130-nt extension of Qs, also accumulates in the induced state; these transcripts would seem to reflect transcription termination or processing at IRS2. Induction is believed to be stimulated by the interaction of the internalized cCF10 peptide with ribosomes in such a way as to initially enhance translation of *prgS* (Bensing and Dunny, 1997; Bensing et al., 1997). The product (PrgS) is then thought to facilitate or stabilize the interaction of QL with ribosomes in such a way as to provide some specificity to ribosomes for translating the *prgB* message. It is possible that PrgS or the modified ribosome also plays a role in the transcriptional antitermination event that occurs between *prgS* and *prgT*. The interaction of cCF10 with ribosomes is supported by biochemical data showing that the peptide specifically inhibited the binding of an 18-kDa

ribosomal S5-like protein to a heparin affinity matrix; by a similar approach, QL RNA was shown to interact with a 23-kDa L6-like ribosomal protein (Bensing and Dunny, 1997; Bensing et al., 1997).

It is interesting that the 3′ end of QL contains inverted repeat sequences, IRS1 and IRS2, which are essentially identical to the t1 and t2 sites, respectively, of pAD1. Indeed, there is near identity between pAD1 and pCF10 with respect to a point just upstream of the translational start site of *traD* (of pAD1) through t2. The related RNA structural features may therefore play similar roles in the two systems. It has been pointed out in the case of pAD1 that there is a small ORF (15 amino acids) starting within t1, and there is a potential ribosome binding site just upstream (Pontius and Clewell, 1992a). (In the case of pCF10 the small ORF reflects 18 residues.) The corresponding IRS1-IRS2 region within QL RNA has been noted to have localized sequences that could hybridize with 16S ribosomal RNA and interact with ribosomal protein L6; indeed, genetic analyses have supported such an interaction (Bensing and Dunny, 1997). It is possible that the role of the small ORF is to help favor the association of ribosomes with the IRS1 region of the transcript, perhaps contributing to antitermination (or decreased pausing) of transcription.

It is important to note that, while there are likely to be some similar regulatory features between pCF10 and pAD1, the pCF10 system is believed to transcribe constitutively from the *prgQ* promoter whereas the pAD1 system involves regulation at the *iad* promoter; an equivalent of the TraA protein and TraD RNA has not been demonstrated for pCF10. It is conceivable that a TraA-like factor (e.g., PrgX?) and a *traD*-like determinant could actually exist, but this has not yet been examined. As noted above, the near identity of IRS1-IRS2 and t1-t2 implies a likelihood that similar regulation involving ribosomal interaction occurs at these sites. While the pCF10 system involves transcription from the *prgQ* promoter, in which induction involves extension of transcription all the way through the aggregation substance gene, a similar extended transcription through *asa1* is apparently not essential in the case of pAD1. Recall that there is an ORF just upstream of *asa1* (i.e., *orf1*) whose promoter provides transcription into its downstream neighbor, and *traE1* is able to activate transcription in *trans* (Muscholl et al., 1993).

The pPD1 Response

Because of the general similarity in structure, pPD1 is likely to have some regulatory features similar to those of pAD1 and pCF10. Indeed, a gene *traA* (Fig. 1) has been demonstrated to encode a protein that negatively controls the pheromone response (Nakayama and Suzuki, 1997; Tanimoto et al., 1996), and there is evidence that TraA binds to cPD1 (Nakayama et al., 1998). The inhibitor peptide iPD1 corresponds to the last 8 residues within a 21-amino-acid precursor (Fujimoto et al., 1995; Nakayama et al., 1995a). Transcriptional analyses have not yet been reported, and an equivalent of the pAD1 *traD* determinant has not been identified. It is worth noting, however, that the sequence downstream of *ipd* includes a 300-nt segment exhibiting significant homology with the corresponding regions in pAD1 and pCF10. This includes a likely transcription terminator site corresponding to t1 and IRS1 of pAD1 and pCF10, respectively. Similarity does not continue beyond that point, and there is no evidence for a second inverted repeat structure corresponding to t2 or IRS2. *orf7* and *orf8* of pPD1 do not exhibit sequence similarity with the corresponding regions of pAD1 or pCF10 (Fig. 1), although it is likely that they play a role in positive regulation.

Phase Variation in the pAD1 System

Cells containing pAD1 exhibit a plasmid-linked phase variation enabling an override of the physiological processes that control the pheromone response (Heath et al., 1995; Pontius and Clewell, 1991). The phenomenon is reversible and occurs at a frequency of 10^{-4} to 10^{-3} per cell per generation. It can be easily observed as a change in colony morphology, as cells that have turned on their conjugation functions give rise to a "dry" colony, compared to the more "watery" colony of cells that are not in this state. Phase variants are thus called Dry^+ and Dry^c ("c" for constitutively dry). Dry^c cells clump when suspended in broth and transfer pAD1 at a relatively high frequency in short matings.

Surprisingly, the basis of phase variation did not turn out to involve a structural change within the *traE1-traD-iad-traA* region, as one might have anticipated; rather, variants exhibited changes in the number of iterons located between *repA* and *repB* that are involved in plasmid replication and maintenance (Heath et al., 1995). It has been speculated that the structural changes may affect the expression of RepA and/or RepB and that at least one of these products is able to affect synthesis of TraA (Heath et al., 1995). It was noted that there are some iteron-like sequences within *traA* that might be able to bind a protein such as RepA or RepB, which may have an affinity for iterons.

EPIDEMIOLOGY OF PHEROMONE-RESPONDING PLASMIDS AND POSSIBLE CLINICAL SIGNIFICANCE

Plasmids that encode a pheromone response are widespread among the enterococci (Clewell, 1993b; Dunny et al., 1979; Hirt et al., 1996; Wirth et al., 1992), and there are reports of detectable transfer in vivo (Huycke et al., 1992; Wirth and Marcinek, 1995). Hemolysin-encoding elements closely resembling pAD1 have been identified in the United States, Europe, and Japan; they are commonly harbored by hosts resistant to multiple antibiotics (Borderon et al., 1982; Colmar and Horaud, 1987; Dunny et al., 1979; Huycke et al., 1991; Ike and Clewell, 1992). However, in most cases the resistance determinants are located on co-resident plasmids or are chromosome borne. (The erythromycin-resistance transposon Tn917 [Tomich et al., 1979, 1980] and the tetracycline-resistance-associated conjugative transposon Tn916 [Franke and Clewell, 1981] were both originally identified in the same strain as pAD1.) The pAD1-like, pheromone-responding elements tend to be, but are not always, members of the same incompatibility group and confer a response to the cAD1 peptide (Colmar and Horaud, 1987; Ike and Clewell, 1992). Interestingly, pAD1 and a number of closely related plasmids encode resistance to UV light (Clewell et al., 1986; Ike and Clewell, 1992); characterization of the related determinants has been reported (Ozawa et al., 1997).

The hemolysin of pAD1 is actually a cytolysin that has both hemolytic and bacteriolytic activity (Ike et al., 1990; Tomich et al. 1979), a trait that was recognized a number of years ago for a similar hemolysin plasmid (Brock and Davie, 1963). Gilmore and associates (Booth et al., 1996; Gilmore et al., 1994) have characterized the cytolysin system both genetically and biochemically and have shown that the lytic protein is a two-component lantibiotic. Hemolytic strains are more frequently observed in human parenteral isolates than in fecal specimens (Coque et al., 1995; Huycke and Gilmore, 1995; Ike et al., 1987), and it has been shown that production of hemolysin contributes to virulence in animal models (Chow et al., 1993; Ike et al., 1984; Jett et al., 1992).

It is interesting that the aggregation substance determinants of pAD1, pCF10, and pPD1 all bear two RGD motifs (such motifs are known to bind to integrins on the surface of eukaryotic cells) that might contribute to bacterial binding to human tissue, perhaps facilitating colonization. Indeed, experiments using tissue culture systems support this view (Kreft et al., 1992; Olmsted et al., 1994), although Jett et al. (1998) reported that aggregation substance did not enhance binding to membrane structures within the vitreous in a rabbit endophthalmitis model. It has been reported that a factor in serum is able to induce the synthesis of the pAD1 aggregation substance (Kreft et al., 1992). It could be speculated that the pAD1 phase variation event (see above) plays a role in facilitating adherence to epithelial tissue in the urinary tract in a manner similar to that involving the phase-varied expression of type I pili in E. coli (Eisenstein, 1981).

Using probes corresponding to asa1 and a component of the hemolysin determinant, Huycke and Gilmore (1995) and Coque et al. (1995) screened clinical isolates of E. faecalis associated with bacteremias and endocarditis as well as strains from unrelated stool specimens. When clonality was taken into consideration, there was no consistent support for the view that aggregation substance or hemolysin was specifically associated with endocarditis; there did, however, appear to be a significant correlation between hemolysin production and bacteremia. Schlievert et al. (1998) reported that the aggregation substance of pCF10, as well as E. faecalis host enterococcal binding substance, contributed to virulence in a rabbit endocarditis model.

Potential clinical significance of pheromones or their peptide inhibitors per se has been reported in that some of these molecules are able to induce neutrophil chemotaxis at low (submicromolar) concentrations (Ember and Hugli, 1989; Sannomiya et al., 1990). Although not quite as potent as the well-studied neutrophil activation peptide fMet-Leu-Phe, the fact that these peptides may be released at higher concentrations than fMet-Leu-Phe suggests they could play significant physiological roles in modulating the inflammatory response (see Sannomiya et al., 1990). The iAD1 produced in the case of cells carrying pAD1 exhibits such an activity; however, one would predict that this would favor the host in its defense against infection. In contrast, cAD1 does not exhibit significant activity.

pPD1 was originally identified in a human oral isolate, 39-5, associated with an acute case of periodontitis (Rosan and Williams, 1964). Interestingly, it was one of six co-resident plasmids, among which was an unrelated hemolysin/bacteriocin plasmid pPD5 (Yagi et al., 1983). pPD1 encodes a bacteriocin designated Bac21, which has a wide antibacterial spectrum (Tomita et al., 1997). Analyses have shown that Bac21 has 100% homology with the 70-amino-acid, cyclic, antibiotic protein AS-48 (Martinez-Bueno et al., 1994), which is encoded on the *E. faecalis* conjugative plasmid pMB2 (Martinez-Bueno et al., 1990) and is known to have a broad antibacterial spectrum that includes *Staphylococcus aureus*. Another interesting feature related to pPD1 is that both cPD1 and iPD1 exhibit relatively potent neutrophil chemotaxis activities (Ember and Hugli, 1989; Sannomiya et al., 1990).

An important trait associated with pCF10 is tetracycline resistance, which was shown to correspond to a conjugative transposon, Tn*925* (Christie et al., 1987), which is similar to Tn*916* (see Clewell et al. 1995). While serving as a convenient marker for conjugation experiments involving pCF10, Tn*925* was able to move independently at much lower frequencies than pCF10. It is interesting that a vancomycin-resistant *Enterococcus faecium* strain isolated a number of years later from the same hospital as the *E. faecalis* strain carrying pCF10 has been found to carry the *van* determinant on a plasmid with some significant similarities to pCF10 (Heaton et al., 1996).

CONCLUDING REMARKS

Because of the high efficiency with which pheromone-responding plasmids transfer, these elements are probably involved in the dissemination of a variety of enterococcal traits. In addition to picking up and disseminating resistance transposons (Clewell et al., 1982a; Clewell et al., 1985; Tomich et al., 1979, 1980), they are able to mobilize other plasmids (Clewell et al., 1982a; Dunny and Clewell, 1975) and even chromosomal markers (Franke et al., 1978). Certain mobile elements, including conjugative transposons like Tn*916* that normally transfer poorly in broth (transfer experiments are generally performed on solid surfaces), are able to transfer at much higher frequencies in broth if a pheromone-responding plasmid is also present in the host (Franke and Clewell, 1981). And there is evidence that certain otherwise nonconjugative

plasmids are able to form stable cointegrates (Clewell et al., 1982a; Heaton et al., 1996). The fact that certain plasmids such as pAM373 (Clewell et al., 1985; Muscholl-Silberhorn et al., 1997) and the *E. faecium* plasmid pHKK100 (Handwerger et al., 1990) with cognate pheromone activities also produced by other genera such as *S. aureus* and *Streptococcus gordonii* suggests a potential involvement in the exchange of genetic material with nonenterococcal species. It should be noted that, while most pheromone-responding plasmids are not known for their ability to replicate outside the genus *Enterococcus*, there are numerous enterococcal nonpheromonal elements that exhibit host ranges extending well beyond this genus (Clewell, 1990). Indeed, while pAM373 was not able to establish in *S. aureus*, there is evidence that it may facilitate delivery of a transposon from *E. faecalis* to *S. aureus* (Clewell et al., 1985). It is easily envisioned that there will eventually be transfer of vancomycin resistance from enterococci to *S. aureus* by a mechanism involving a cAM373-like activity.

Insofar as the *E. faecalis* host produces numerous sex pheromones, it is interesting that most bacterial strains are not "full" of acquired pheromone-responding plasmids, especially since plasmids that relate to different pheromones usually are members of different incompatibility groups. While there are at least two examples of bacteria simultaneously carrying three such plasmids (Clewell et al., 1982b; Murray et al., 1988), the majority of strains appear to carry only one or no such element. Conceivably, certain pheromones might play other roles in the cell in addition to their involvement as mating signals. Possibly some could play a role in quorum sensing relating to phenomena not connected to plasmids, although high-cell-density-induced secretion of known pheromones has not been observed. Indeed, plasmids may have simply evolved in such a way as to take advantage of these substances as conjugation signals. It has been possible to generate mutants defective in the production of pheromones (e.g., cAD1 and cPD1 [Clewell et al., 1984; Ike et al., 1983]); however, a small amount (e.g., 2 to 3% of wild type) of pheromone can still be detected in such strains. Certain peptides may be essential to the cell or contribute to survival under specific environmental conditions; in such cases there might be selective pressure to "eliminate" a plasmid that is otherwise shutting down or greatly reducing pheromone production. An interesting feature of the pAD1-related system that probably has some environmental significance is the role of oxygen in cAD1 production. Under anaerobic conditions, the amount of cAD1 secreted is about 20-fold higher than that when cells are vigorously aerated (Weaver and Clewell, 1991); no such effect was observed on the production of two other pheromones, cPD1 and cAM373.

In the case of pCF10, there is evidence that cCF10 is necessary for vegetative plasmid replication, since construction of a plasmid derivative able to provide cCF10 enabled replication in a strain of *Lactococcus* sp. (Dunny and Leonard, 1997). In this regard, it is interesting that a pCF10 protein, PrgW, that is presumed to be involved in plasmid replication has been reported to bind to cCF10 (Leonard et al., 1996).

Finally, whereas little is yet known about the chromosome-borne pheromone determinants, their structures should soon be revealed as a result of the complete *E. faecalis* genome analyses currently under way at the Institute for Genomic Research in Rockville, Md. (Dougherty, personal communication). Interestingly, Firth et al. (1994) have reported that the *traH* gene of conjugative staphylococcal plasmid pSK41 encodes a lipoprotein precursor bearing a signal sequence whose carboxyl-terminal region consists of seven of eight contiguous amino acids identical to those of cAD1. The only difference is a threonine taking the place of a serine. cAD1 activity could actually be detected in culture supernatant of staphylococcal strains harboring pSK41 but not in that of plasmid-free strains; a synthetically prepared octapeptide exhibited activity

(Berg et al., 1997). It would appear that the TraH precursor is processed in such a way that a product consisting of only the active octapeptide is secreted. Similarity to the inhibitor peptides encoded by the *E. faecalis* pheromone plasmids is noteworthy in that the precursor forms appear to resemble lone signal peptides within which the last seven or eight residues correspond to the mature product. It will be interesting to see if the pheromone determinants on the *E. faecalis* genome prove to be parts of signal sequences of various proteins.

NOTE IN PROOF

Since this chapter was submitted, an updated TIGR database has revealed the likely determinants of three pheromones, cPD1, cCF10, and cOB1. Each appears to reflect part of a protein signal sequence.

ACKNOWLEDGMENTS

Work conducted in my laboratory relating to material discussed in this review was supported by National Institutes of Health Grant GM33956. I thank G. Dunny, R. Wirth, A. Suzuki, J. Nakayama, Y. Ike, K. Tanimoto, S. Fujimoto, H. Tomita, K. Weaver, M. Gilmore, E. DeBoever, J. Chow, and F. An for recent reprints/preprints and helpful discussions.

REFERENCES

An, F. Y., and D. B. Clewell. 1994. Characterization of the determinant (*traB*) encoding sex pheromone shutdown by the hemolysin/bacteriocin plasmid pAD1 in *Enterococcus faecalis*. *Plasmid* **31**:215–221.

An, F. Y., and D. B. Clewell. 1997. The origin of transfer (*oriT*) of the Enterococcal, pheromone-responding, cytolysin plasmid pAD1 is located within the *repA* determinant. *Plasmid* **37**:87–94.

Bastos, M. C. F., H. Tomita, K. Tanimoto, and D. B. Clewell. Regulation of the *Enterococcus faecalis* pAD1-related sex pheromone response: analyses of *traD* expression and its role in controlling conjugation functions. *Mol. Microbiol.*, in press.

Bastos, M. C. F., K. Tanimoto, and D. B. Clewell. 1997. Regulation of transfer of the *Enterococcus faecalis* pheromone-responding plasmid pAD1: temperature-sensitive transfer mutants and identification of a new regulatory determinant, *traD. J. Bacteriol.* **179**:3250–3259.

Bensing, B. A., and G. M. Dunny. 1993. Cloning and molecular analysis of genes affecting expression of binding substance, the recipient-encoded receptor(s) mediating mating aggregate formation in *Enterococcus faecalis*. *J. Bacteriol.* **175**:7421–7429.

Bensing, B. A., and G. M. Dunny. 1997. Pheromone-inducible expression of an aggregation protein in *Enterococcus faecalis* requires interaction of a plasmid-encoded RNA with components of the ribosome. *Mol. Microbiol.* **24**:295–308.

Bensing, B. A., B. J. Meyer, and G. M. Dunny. 1996. Sensitive detection of bacterial transcription initiation sites and differentiation from RNA processing sites in the pheromone-induced plasmid transfer system of *Enterococcus faecalis*. *Proc. Natl. Acad. Sci. USA* **93**:7794–7799.

Bensing, B. A., D. A. Manias, and G. M. Dunny. 1997. Pheromone cCF10 and plasmid pCF10-encoded regulatory molecules act post-transcriptionally to activate expression of downstream conjugation functions. *Mol. Microbiol.* **24**:285–294.

Berg, T., N. Firth, and R. A. Skurray. 1997. Enterococcal pheromone-like activity derived from a lipoprotein signal peptide encoded by a *Staphylococcus aureus* plasmid, p. 1041–1044. *In* T. Horaud, M. Sicard, A. Bouve, and H. de Montelos (ed.) *Streptococci and the Host*. Plenum Press, New York.

Booth, M. C., C. P. Bogie, H. Sahl, R. J. Siezen, K. L. Hatter, and M. S. Gilmore. 1996. Structural analysis and proteolytic activation of *Enterococcus faecalis* cytolysin, a novel lantibiotic. *Mol. Microbiol.* **21**:1175–1184.

Borderon, E., G. Bieth, and T. Horodniceanu. 1982. Genetic and physical studies of *Streptococcus faecalis* hemolysin plasmids. *FEMS Microbiol. Lett.* **14**:51–55.

Brock, T., and J. M. Davie. 1963. Probable identity of a group D hemolysin with a bacteriocine. *J. Bacteriol.* **86**:708–712.

Chow, J. W., L. A. Thal, M. B. Perri, J. A. Vazquez, S. M. Donabedian, D. B. Clewell, and M. J. Zervos. 1993. Plasmid-associated hemolysin and aggregation substance production contributes to virulence in experimental enterococcal endocarditis. *Antimicrob. Agents Chemother.* **37**:2474–2477.

Christie, P. J., R. Z. Korman, S. A. Zahler, J. C. Adsit, and G. M. Dunny. 1987. Two conjugation systems associated with *Streptococcus faecalis* plasmid pCF10: identification of a conjugative transposon that transfers between *S. faecalis* and *Bacillus subtilis*. *J. Bacteriol.* **169**:2529–2536.

Chung, J. W., and G. M. Dunny. 1992. Cis-acting, orientation-dependent, positive control system activates pheromone-inducible conjugation functions at distances greater than 10 kilobases upstream from its target in *Enterococcus faecalis*. *Proc. Natl. Acad. Sci. USA* **89**:9020–9024.

Chung, J. W., and G. M. Dunny. 1995. Transcriptional analysis of a region of the *Enterococcus faecalis* plasmid pCF10 involved in positive regulation of conjugative transfer functions. *J. Bacteriol.* **177**:2118–2124.

Chung, J. W., B. A. Bensing, and G. M. Dunny. 1995. Genetic analysis of a region of the *Enterococcus faecalis* plasmid pCF10 involved in positive regulation of conjugative transfer functions. *J. Bacteriol.* **177**:2107–2117.

Clewell, D. B. 1990. Movable genetic elements and antibiotic resistance in enterococci. *Eur. J. Clin. Microbiol. Infect. Dis.* **9**:90–102.

Clewell, D. B. 1993a. Bacterial sex pheromone-induced plasmid transfer. *Cell* **73**:9–12.

Clewell, D. B. 1993b. Sex pheromones and the plasmid-encoded mating response in *Enterococcus faecalis*, p. 349–367. *In* D. B. Clewell (ed.), *Bacterial Conjugation*. Plenum Press, New York.

Clewell, D. B., and B. L. Brown. 1980. Sex pheromone cAD1 in *Streptococcus faecalis*: induction of a function related to plasmid transfer. *J. Bacteriol.* **143**:1063–1065.

Clewell, D. B., P. K. Tomich, M. C. Gawron-Burke, A. E. Franke, Y. Yagi, and F. Y. An. 1982a. Mapping of *Streptococcus faecalis* plasmids pAD1 and pAD2 and studies relating to transposition of Tn*917*. *J. Bacteriol.* **152**:1220–1230.

Clewell, D. B., Y. Yagi, Y. Ike, R. A. Craig, B. L. Brown, and F. An. 1982b. Sex pheromones in *Streptococcus faecalis*: multiple pheromone systems in strain DS5, similarities of pAD1 and pAMγ1, and mutants of pAD1 altered in conjugative properties, p. 97–100. *In* D. Schlessinger (ed.), *Microbiology—1982*, American Society for Microbiology, Washington, D.C.

Clewell, D. B., B. A. White, Y. Ike, and F. Y. An. 1984. Sex pheromones and plasmid transfer in *Streptococcus faecalis*, p. 133–149. *In* R. Losick and L. Shapiro (ed.), *Microbial Development*, Cold Spring Harbor Laboratory, Cold Spring Harbor, N.Y.

Clewell, D. B., F. Y. An, B. A. White, and C. Gawron-Burke. 1985. *Streptococcus faecalis* sex pheromone (cAM373) also produced by *Staphylococcus aureus* and identification of a conjugative transposon (Tn*918*). *J. Bacteriol.* **162**:1212–1220.

Clewell, D. B., E. E. Ehrenfeld, R. E. Kessler, Y. Ike, A. E. Franke, M. Madion, J. H. Shaw, R. Wirth, F. An, M. Mori, C. Kitada, M. Fujino, and A. Suzuki. 1986. Sex-pheromone systems in *Streptococcus faecalis*, p. 131–139. *In* S. B. Levy and R. P. Novick (ed.), *Banbary Report 24: Antibiotic Resistance Genes: Ecology, Transfer and Expression*. Cold Spring Harbor Laboratory, Cold Spring Harbor, N.Y.

Clewell, D. B., L. T. Pontius, F. Y. An, Y. Ike, A. Suzuki, and J. Nakayama. 1990. Nucleotide sequence of the sex pheromone inhibitor (iAD1) determinant of *Enterococcus faecalis* conjugative plasmid pAD1. *Plasmid* **24**:156–161.

Clewell, D. B., S. E. Flannagan, and D. D. Jaworski. 1995. Unconstrained bacterial promiscuity: the Tn*916*-Tn*1545* family of conjugative transposons. *Trends Microbiol.* **3**:229–236.

Colmar, I., and T. Horaud. 1987. *Enterococcus faecalis* hemolysin-bacteriocin plasmids belong to the same incompatibility group. *Appl. Environ. Microbiol.* **53**:567–570.

Coque, T. M., J. E. Patterson, J. M. Steckelberg, and B. E. Murray. 1995. Incidence of hemolysin, gelatinase and aggregation substance among enterococci isolates from patients with endocarditis and other infections, and from feces of hospitalized and community based individuals. *J. Infect. Dis.* **171**:1223–1229.

Dougherty, B. Personal communication.

Dunny, G. Personal communication.

Dunny, G. M., and D. B. Clewell. 1975. Transmissible toxin (hemolysin) plasmid in *Streptococcus faecalis* and its mobilization of a noninfectious drug resistance plasmid. *J. Bacteriol.* **124**:784–790.

Dunny, G. M., and B. A. B. Leonard. 1997. Cell-cell communication in gram-positive bacteria. *Annu. Rev. Microbiol.* **51**:527–564.

Dunny, G. M., B. L. Brown, and D. B. Clewell. 1978. Induced cell aggregation and mating in *Streptococcus faecalis*: evidence for a bacterial sex pheromone. *Proc. Natl. Acad. Sci. USA* **75**:3479–3483.

Dunny, G. M., R. A. Craig, R. L. Carron, and D. B. Clewell. 1979. Plasmid transfer in *Streptococcus faecalis*: production of multiple sex pheromones by recipients. *Plasmid* **2**:454–465.

Dunny, G. M., C. Funk, and J. Adsit. 1981. Direct stimulation of the transfer of antibiotic resistance by sex pheromones in *Streptococcus faecalis*. *Plasmid* **6**:270–278.

Dunny, G. M., D. L. Zimmerman, and M. L. Tortorello. 1985. Induction of surface exclusion by *Streptococcus faecalis* sex pheromones: use of monoclonal antibodies to identify an inducible surface antigen involved in the exclusion process. *Proc. Natl. Acad. Sci. USA* **82**:8582–8586.

Dunny, G. M., Leonard, B. A. B., and P. J. Hedberg. 1995. Pheromone-inducible conjugation in *Enterococcus faecalis*: interbacterial and host-parasite chemical communication. *J. Bacteriol.* **177**:871–876.

Ehrenfeld, E. E., and D. B. Clewell. 1987. Transfer functions of the *Streptococcus faecalis* plasmid pAD1: organization of plasmid DNA encoding response to sex pheromone. *J. Bacteriol.* **169**:3473–3481.

Ehrenfeld, E. E., R. E. Kessler, and D. B. Clewell. 1986. Identification of pheromone-induced surface proteins in *Streptococcus faecalis* and evidence of a role for lipoteichoic acid in formation of mating aggregates. *J. Bacteriol.* **168**:6–12.

Eisenstein, B. I. 1981. Phase variation of type I fimbriae in *Escherichia coli* is under transcriptional control. *Science* **214**:337–339.

Ember, J. A., and T. E. Hugli. 1989. Characterization of the human neutrophil response to sex pheromones from *Streptococcus faecalis*. *Am J. Pathol.* **134**:797–805.

Firth, N. P. D. Fink, L. Johnson, and R. A. Skurray. 1994. A lipoprotein signal peptide encoded by the staphylococcal conjugative plasmid pSK41 exhibits an activity resembling that of *Enterococcus faecalis* pheromone cAD1. *J. Bacteriol.* **176**:5871–5873.

Franke, A. E., and D. B. Clewell. 1981. Evidence for a chromosome-borne resistance transposon (Tn*916*) in *Streptococcus faecalis* that is capable of "conjugal" transfer in the absence of a conjugative plasmid. *J. Bacteriol.* **145**:494–502.

Franke, A. E., G. M. Dunny, B. L. Brown, F. An, D. R. Oliver, S. P. Damle, and D. B. Clewell. 1978. Gene transfer in *Streptococcus faecalis*: evidence for the mobilization of chromosomal determinants by transmissible plasmids, p. 45–47. *In* D. Schlessinger (ed.), *Microbiology—1978*. American Society for Microbiology, Washington, D.C.

Fujimoto, S., and D. B. Clewell. 1998. Regulation of the pAD1 sex pheromone response of *Enterococcus faecalis* by direct interaction between the cAD1 peptide mating signal and the negatively regulating, DNA-binding TraA protein. *Proc. Natl. Acad. Sci. USA* **95**:6430–6435.

Fujimoto, S., H. Tomita, E. Wakamatsu, K. Tanimoto, and Y. Ike. 1995. Physical mapping of the conjugative bacteriocin plasmid pPD1 of *Enterococcus faecalis* and identification of the determinant related to the pheromone response. *J. Bacteriol.* **177**:5574–5581.

Fujimoto, S., M. Bastos, K. Tanimoto, F. An, K. Wu, and D. B. Clewell. 1997. The pAD1 sex pheromone response in *Enterococcus faecalis*, p. 1037–1040. *In* T. Horaud, M. Sicard, A. Bouve, and H. de Montelos (ed.), *Streptococci and the Host*. Plenum, New York .

Galli, D., and R. Wirth. 1991. Comparative analysis of *Enterococcus faecalis* sex pheromone plasmids identifies a single homologous DNA region which codes for aggregation substance. *J. Bacteriol.* **173**:3029–3033.

Galli, D., R. Wirth, and G. Wanner. 1989. Identification of aggregation substances of *Enterococcus faecalis* after induction by sex pheromones. *Arch. Microbiol.* **151**:486–490.

Galli, D., F. Lottspeich, and R. Wirth. 1990. Sequence analysis of *Enterococcus faecalis* aggregation substance encoded by the sex pheromone plasmid pAD1. *Mol. Microbiol.* **4**:895–904.

Galli, D., A. Friesenegger, and R. Wirth. 1992. Transcriptional control of sex-pheromone-inducible genes on plasmid pAD1 of *Enterococcus faecalis* and sequence analysis of a third structural gene for (pPD1-encoded) aggregation substance. *Mol. Microbiol.* **6**:1297–1308.

Gilmore, M. S., R. A. Segarra, M. C. Booth, C. P. Bogie, L. R. Hall, and D. B. Clewell. 1994. Genetic structure of the Enterococcus faecalis plasmid pAD1-encoded cytolytic toxin system and its relationship to lantibiotic determinants. *J. Bacteriol.* **176**:7335–7344.

Handwerger, S., M. J. Pucci, and A. Kolokathis. 1990. Vancomycin resistance is encoded on a pheromone response plasmid in *Enterococcus faecium* 228. *Antimicrob. Agents Chemother.* **34**:358–360.

Heath, D. G., F. Y. An, K. E. Weaver, and D. B. Clewell. 1995. Phase variation of *Enterococcus faecalis* pAD1 conjugation functions relates to changes in iteron sequence region. *J. Bacteriol.* **177**:5453–5459.

Heaton, M. P., L. F. Discotto, M. J. Pucci, and S. Handwerger. 1996. Mobilization of vancomycin resistance by transposon-mediated fusion of a VanA plasmid with an *Enterococcus faecium* sex pheromone-response plasmid. *Gene* **171**:9–17.

Hedberg, P. J., B. A. B. Leonard, R. E. Ruhfel, and G. M. Dunny. 1996. Identification and characterization of the genes of *Enterococcus faecalis* plasmid pCF10 involved in replication and in negative control of pheromone-inducible conjugation. *Plasmid* **35**:46–57.

Hirt, H., G. Wanner, D. Galli, and R. Wirth. 1993. Biochemical, immunological and ultrastructural characterization encoded by *Enterococcus faecalis* sex-pheromone plasmids. *Eur. J. Biochem.* **211**:711–716.

Hirt, H., R. Wirth, and A. Muscholl. 1996. Comparative analysis of 18 sex pheromone plasmids from *Enterococcus faecalis*: detection of a new insertion element on pPD1 and implications for the evolution of this plasmid family. *Mol. Gen. Genet.* **252**:640–647.

Huycke, M. M., and M. S. Gilmore. 1995. Frequency of aggregation substance and cytolysin genes among enterococcal endocarditis isolates. *Plasmid* **34**:152–156.

Huycke, M. M., C. A. Spiegel, and M. S. Gilmore. 1991. Bacteremia caused by hemolytic, high-level gentamicin-resistant *Enterococcus faecalis*. *Antimicrob. Agents Chemother.* **35**:1626–1634.

Huycke, M. M., M. S. Gilmore, B. D. Jett, and J. L. Booth. 1992. Transfer of pheromone-inducible plasmids between *Enterococcus faecalis* in the Syrian hamster gastrointestinal tract. *J. Infect. Dis.* **166**:1188–1191.

Ike, Y., and D. B. Clewell. 1984. Genetic analysis of the pAD1 pheromone response in *Streptococcus faecalis*, using transposon Tn917 as an insertional mutagen. *J. Bacteriol.* **158**:777–783.

Ike, Y., and D. B. Clewell. 1992. Evidence that the hemolysin/bacteriocin phenotype of *Enterococcus faecalis* subsp. *zymogenes* can be determined by plasmids in different incompatibility groups as well as by the chromosome. *J. Bacteriol.* **174**:8172–8177.

Ike, Y., R. C. Craig, B. A. White, Y. Yagi, and D. B. Clewell. 1983. Modification of *Streptococcus faecalis* sex pheromones after acquisition of plasmid DNA. *Proc. Natl. Acad. Sci. USA* **80**:5369–5373.

Ike, Y., H. Hashimoto, and D. B. Clewell. 1984. Hemolysin of *Streptococcus faecalis* subspecies *zymogenes* contributes to virulence in mice. *Infect. Immun.* **45**:528–530.

Ike, Y., H. Hashimoto, and D. B. Clewell. 1987. High incidence of hemolysin production by *Enterococcus* (*Streptococcus*) *faecalis* strains associated with human parenteral infections. *J. Clin. Microbiol.* **25**:1524–1528.

Ike, Y., D. B. Clewell, R. A. Segarra, and M. S. Gilmore. 1990. Genetic analysis of the pAD1 hemolysin/bacteriocin determinant in *Enterococcus faecalis*: Tn917 insertional mutagenesis and cloning. *J. Bacteriol.* **172**:155–163.

Jett, B. D., H. G. Jensen, R. E. Nordquist, and M. S. Gilmore. 1992. Contribution of the pAD1-encoded cytolysin to the severity of experimental *Enterococcus faecalis* endophthalmitis. *Infect. Immun.* **60**:2445–2452.

Jett, B. D., M. M. Huycke, and M. S. Gilmore. 1994. Virulence of enterococci. *Clin. Microbiol. Rev.* **7**:462–478.

Jett, B. D., R. V. Atkuri, and M. S. Gilmore. 1998. *Enterococcus faecalis* localization in experimental endophthalmitis; role of plasmid-encoded aggregation substance. *Infect. Immun.* **66**:843–848.

Kao, S.-M., S. B. Olmsted, A. S. Viksnins, J. C. Gallo, and G. M. Dunny. 1991. Molecular and genetic analysis of a region of plasmid pCF10 containing positive control genes and structural genes encoding surface proteins involved in pheromone-inducible conjugation in *Enterococcus faecalis*. *J. Bacteriol.* **173**:7650–7664.

Kreft, B., R. Marre, U. Schramm, and R. Wirth. 1992. Aggregation substance of *Enterococcus faecalis* mediates adhesion to cultured renal tubular cells. *Infect. Immun.* **60:**25–30.

Leonard, B. A. B., A. Podbielski, P. J. Hedberg, and G. M. Dunny. 1996. *Enterococcus faecalis* pheromone binding protein, PrgZ, recruits a chromosomal oligopeptide permease system to import sex pheromone cCF10 for induction of conjugation. *Proc. Natl. Acad. Sci. USA* **93:**260–264.

Lewis, C. M., and M. J. Zervos. 1990. Clinical manifestations of enterococcal infection. *Eur. J. Clin. Microbiol. Infect. Dis.* **9:**111–117.

Martinez-Bueno, M., A. Galvez, E. Valdivia, and M. Maqueda. 1990. A transferable plasmid associated with AS-48 production in *Enterococcus faecalis. J. Bacteriol.* **172:**2817–2818.

Martinez-Bueno, M., M. Maqueda, A. Galvez, B. Samyn, J. V. Beeumen, J. Coyette, and E. Valdivia. 1994. Determination of the gene sequence and the molecular structure of the enterococcal peptide antibiotic AS-48. *J. Bacteriol.* **176:**6334–6339.

Moellering, R. C. 1992. Emergence of enterococcus as a significant nosocomial pathogen. *Clin. Infect Dis.* **14:**1173–1178.

Mori, M., Y. Sakagami, M. Narita, A. Isogai, M. Fujino, C. Kitada, R. Craig, D. Clewell, and A. Suzuki. 1984. Isolation and structure of the bacterial sex pheromone, cAD1, that induces plasmid transfer in *Streptococcus faecalis. FEBS Lett.* **178:**97–100.

Mori, M., H. Tanaka, Y. Sakagami, A. Isogai, M. Fujino, C. Kitada, B. A. White, F. Y, An, D. B. Clewell, and A. Suzuki. 1986a. Isolation and structure of the *Streptococcus faecalis* sex pheromone, cAM373. *FEBS Lett.* **206:**69–72.

Mori, M., A. Isogai, Y. Sakagami, M. Fujino, C. Kitada, D. B. Clewell, and A. Suzuki. 1986b. Isolation and structure of *Streptococcus faecalis* sex pheromone inhibitor, iAD1, that is excreted by donor strains harboring plasmid pAD1. *Agric. Biol. Chem.* **50:**539–541.

Mori, M., H. Tanaka, Y. Sakagami, A. Isogai, M. Fujino, C. Kitada, D. B. Clewell, and A. Suzuki. 1987. Isolation and structure of the sex pheromone inhibitor, iPD1, excreted by *Streptococcus faecalis* donor strains harboring plasmid pPD1. *J. Bacteriol.* **169:**1747–1749.

Mori, M., Y. Sakagami, Y. Ishii, A. Isogai, C. Kitada, M. Fujino, J. C. Adsit, G. M. Dunny, and A. Suzuki. 1988. Structure of cCF10, a peptide sex pheromone which induces conjugative transfer of the *Streptococcus faecalis* tetracycline resistance plasmid, pCF10. *J. Biol. Chem.* **263:**14574–14578.

Murray, B. E. 1990. The live and times of the enterococcus. *Clin. Microbiol. Rev.* **3:**46–65.

Murray, B. E., F. An, and D. B. Clewell. 1988. Plasmids and pheromone response of the β-lactamase producer *Streptococcus (Enterococcus) faecalis* HH22. *Antimicrob. Agents Chemother.* **32:**547–551.

Muscholl, A., D. Galli, G. Wanner, and R. Wirth. 1993. Sex pheromone plasmid pAD1-encoded aggregation substance of *Enterococcus faecalis* is positively regulated in *trans* by traE1. *Eur. J. Biochem.* **214:**333–338.

Muscholl-Silberhorn, A., E. Samberger, and R. Wirth. 1997. Why does *Staphylococcus aureus* secrete and *Enterococcus faecalis*-specific pheromone? *FEMS Microbiol. Lett.* **157:**261–266.

Nakayama, J., and A. Suzuki. 1997. Genetic analysis of plasmid-specific pheromone signaling encoded by pPD1 in *Enterococcus faecalis. Biosci. Biotechnol. Biochem.* **61:**1796–1799.

Nakayama, J., R. E. Ruhfel, G. M. Dunny, A. Isogai, and A. Suzuki. 1994. The *prgQ* gene of the *Enterococcus faecalis* tetracycline resistance plasmid pCF10 encodes a peptide inhibitor, iCF10. *J. Bacteriol.* **176:**2003–2004.

Nakayama, J., K. Yoshida, H. Kobayashi, A. Isogai, D. B. Clewell, and A. Suzuki. 1995a. Cloning and characterization of a region of *Enterococcus faecalis* plasmid pPD1 encoding pheromone inhibitor (*ipd*), pheromone sensitivity (*traC*), and pheromone shutdown (*traB*) genes. *J. Bacteriol.* **177:**5567–5573.

Nakayama, J. Y. Abe, A. Isogai, and A. Suzuki. 1995b. Isolation and structure of the *Enterococcus faecalis* sex pheromone, cOB1, that induces conjugal transfer of the hemolysin/bacteriocin plasmids, pOB1 and pYI1. *Biosci. Biotechnol. Biochem.* **59:**703–705.

Nakayama, J., Y. Ono, A. Isogai, D. B. Clewell, and A. Suzuki. 1995c. Isolation and structure of the sex pheromone inhibitor, iAM373, of *Enterococcus faecalis. Biosci. Biotechnol. Biochem.* **59:**1358–1359.

Nakayama, J., G. M. Dunny, D. B. Clewell, and A. Suzuki. 1995d. Quantitative analysis for pheromone inhibitor and pheromone shutdown in *Enterococcus faecalis. Dev. Biol. Stand.* **85:**35–38.

Nakayama, J., D. B. Clewell, and A. Suzuki. 1995e. Targeted disruption of the PD78 (*traF*) reduces pheromone-inducible conjugal transfer of the bacteriocin plasmid pPD1 in *Enterococcus faecalis. FEMS Microbiol. Lett.* **128:**283–288.

Nakayama, J., S. Igarashi, H. Nagasawa, D. B. Clewell, F. Y. An, and A. Suzuki. 1996. Isolation and structure of staph-cAM373 produced by *Staphylococcus aureus* that induces conjugal transfer of *Enterococcus faecalis* plasmid pAM373. *Biosci. Biotechnol. Biochem.* **60:**1038–1039.

Nakayama, J., Y. Takanami, T. Horii, S. Sakuda, and A. Suzuki. 1998. Molecular mechanism of peptide-specific pheromone signaling in *Enterococcus faecalis*: functions of pheromone receptor TraA and pheromone-binding protein TraC encoded by plasmid pPD1. *J. Bacteriol.* **180:**449–456.

Olmsted, S. B., S. L. Erlandsen, G. M. Dunny, and C. L. Wells. 1993. High-resolution visualization by field emission scanning electron microscopy of *Enterococcus faecalis* surface proteins encoded by the pheromone-inducible conjugative plasmid pCF10. *J. Bacteriol.* **175:**6229–6237.

Olmsted, S. B., G. M. Dunny, S. L. Erlandsen, and C. L. Wells. 1994. A plasmid-encoded surface protein on *Enterococcus faecalis* augments its internalization by cultured epithelial cells. *J. Infect. Dis.* **170:**1549–1556.

Ozawa, Y., K. Tanimoto, S. Fujimoto, H. Tomita, and Y. Ike. 1997. Cloning and genetic analysis of the UV resistance determinant (*uvr*) encoded on the *Enterococcus faecalis* pheromone-responsive conjugative plasmid pAD1. *J. Bacteriol.* **23:**7468–7475.

Pontius, L. T., and D. B. Clewell. 1991. A phase variation event that activates conjugation functions encoded by the *Enterococcus faecalis* plasmid pAD1. *Plasmid* **26:**172–185.

Pontius, L. T., and D. B. Clewell. 1992a. Conjugative transfer of *Enterococcus faecalis* plasmid pAD1: nucleotide sequence and transcriptional fusion analysis of a region involved in positive regulation. *J. Bacteriol.* **174:**3152–3160.

Pontius, L. T., and D. B. Clewell. 1992b. Regulation of the pAD1-encoded pheromone response in *Enterococcus faecalis*: nucleotide sequence analysis of *traA*. *J. Bacteriol.* **174:**1821–1827.

Rosan, B., and N. Williams. 1964. Hyaluronidase production by oral enterococci. *Arch. Oral Biol.* **9:** 291–298.

Ruhfel, R. E.., D. A. Manias, and G. M. Dunny. 1993. Cloning and characterization of a region of the *Enterococcus faecalis* conjugative plasmid, pCF10, encoding a sex pheromone-binding function. *J. Bacteriol.* **175:**5253–5259.

Sannomiya, P., R. A. Craig, D. B. Clewell, A. Suzuki, M. Fujino, G. O. Till, and W. A. Marasco. 1990. Characterization of a class of nonformylated *Enterococcus faecalis*-derived neutrophil chemotactic peptides: the sex pheromones. *Proc. Natl. Acad. Sci. USA* **87:**66–70.

Schlievert, P. M., P. J. Gahr, A. P. Assimacopoulos, M. M. Dinges, J. A. Stoehr, J. W. Harmala, H. Hirt, and G. M. Dunny. 1998. Aggregation and binding substances enhance pathogenicity in rabbit models of *Enterococcus faecalis* endocarditis. *Infect. Immun.* **66:**218–223.

Suzuki, A., M. Mori, Y. Sakagami, A. Isogai, M. Fujino, C. Kitada, R. A. Craig, and

D. B. Clewell. 1984. Isolation and structure of bacterial sex pheromone, cPD1. *Science* **226:**849–850.

Tanimoto, K. Personal communication.

Tanimoto, K., and D. B. Clewell. 1993. Regulation of the pAD1-encoded sex pheromone response in *Enterococcus faecalis*: expression of the positive regulator TraE1. *J. Bacteriol.* **175:**1008–1018.

Tanimoto, K., F. Y. An, and D. B. Clewell. 1993. Characterization of the *traC* determinant of the *Enterococcus faecalis* hemolysin/bacteriocin plasmid pAD1. Binding of sex pheromone. *J. Bacteriol.* **175:**5260–5264.

Tanimoto, K., H. Tomita, and Y. Ike. 1996. The *traA* gene of the *Enterococcus faecalis* conjugative plasmid pPD1 encodes a negative regulator for the pheromone response. *Plasmid* **36:**55–61.

Tomich, P. K., F. Y. An, S. P. Damle, and D. B. Clewell. 1979. Plasmid related transmissibility and multiple drug resistance in *Streptococcus faecalis* subspecies *zymogenes* strain DS16. *Antimicrob. Agents Chemother.* **15:**828–830.

Tomich, P. K., F. Y. An, and D. B. Clewell. 1980. Properties of erythromycin-inducible transposon Tn*917* in *Streptococcus faecalis*. *J. Bacteriol.* **141:**1366–1374.

Tomita, H. Personal communication.

Tomita, H., S. Fujimoto, K. Tanimoto, and Y. Ike. 1997. Cloning and genetic and sequence analyses of the bacteriocin 21 determinant encoded on the *Enterococcus faecalis* pheromone-responsive conjugative plasmid pPD1. *J. Bacteriol.* **179:**7843–7855.

Wanner, G., H. Formanek, D. Galli, and R. Wirth. 1989. Localization of aggregation substances of *Enterococcus faecalis* after induction by sex pheromones. An ultrastructural comparison using immuno labelling, transmission and high resolution scanning electron microscopic techniques. *Arch. Microbiol.* **151:**491–497.

Weaver, K. E., and D. B. Clewell. 1988. Regulation of the pAD1 sex pheromone response in *Enterococcus faecalis*: construction and characterization of *lacZ* transcriptional fusions in a key control region of the plasmid. *J. Bacteriol.* **170:**4343–4352.

Weaver, K. E., and D. B. Clewell. 1989. Construction of *Enterococcus faecalis* pAD1 miniplasmids: identification of a minimal pheromone response regulatory region and evaluation of a novel pheromone-dependent growth inhibition. *Plasmid* **22:**106–119.

Weaver, K. E., and D. B. Clewell. 1990. Regulation of the pAD1 sex pheromone response in *Enterococcus faecalis*: effects of host strain and *traA*, *traB*, and C region mutants on expression of an E region pheromone-inducible *lacZ* fusion. *J. Bacteriol.* **172:**2633–2641.

Weaver, K. E., and D. B. Clewell. 1991. Control of *Enterococcus faecalis* sex pheromone cAD1 elaboration: effects of culture aeration and pAD1 plasmid-encoded determinants. *Plasmid* **25:**177–189.

Weaver, K. E., D. B. Clewell, and F. An. 1993. Identification, characterization, and nucleotide sequence of a region of *Enterococcus faecalis* pheromone-responsive plasmid pAD1 capable of autonomous replication. *J. Bacteriol.* **175:**1900–1909.

Wirth, R. 1994. The sex pheromone system of *Enterococcus faecalis*. More than just a plasmid-collection mechanism? *Eur. J. Biochem.* **222:**235–246.

Wirth, R., and H. Marcinek. 1995. *In vivo* gene transfer by *Enterococcus faecalis*. *Dev. Biol. Stand.* **85:**51–54.

Wirth, R., A. Friesenegger, and T. Horaud. 1992. Identification of new sex pheromone plasmids in *Enterococcus faecalis*. *Mol. Gen. Genet.* **233:**157–160.

Yagi, Y., R. Kessler, J. Shaw, D. Lopatin, F. An, and D. Clewell. 1983. Plasmid content of *Streptococcus faecalis* strain 39-5 and identification of a pheromone (cPD1)-induced surface antigen. *J. Gen. Microbiol.* **129:**1207–1215.

CELL-DENSITY SENSING DURING EARLY DEVELOPMENT IN *MYXOCOCCUS XANTHUS*

Lynda Plamann and Heidi B. Kaplan

5

Myxobacteria are motile, gram–negative, rod–shaped bacteria that inhabit diverse soil types the world over. These microbes, which belong to the delta subdivision of the proteobacteria, glide over solid surfaces and colonize habitats such as decaying leaves, rotting wood, or the dung of herbivores (Reichenbach, 1993). Although myxobacteria were originally classified as fungi, they are now widely recognized as unique bacteria that exhibit complex social behaviors. Observation of these social behaviors has inspired researchers to investigate the nature of the cell-cell interactions and signaling phenomena that coordinate them.

The selective force behind the social behaviors of *Myxococcus xanthus* is thought to be its unusual mode of cooperative feeding (Dworkin, 1973). Swarms of *M. xanthus* cells, sometimes referred to as "microbial wolf packs," release antibiotics to kill neighboring competing species and break down their cell walls and other macromolecules using a battery of hydrolytic enzymes. The protein contents of these prey microorganisms provide the major source of nitrogen, carbon, and energy for the myxobacteria. Feeding as part of a swarm allows the individual cell to grow more rapidly because the extracellular concentration of hydrolytic enzymes reaches a higher level within a swarm, resulting in a higher concentration of transportable nutrients (Dworkin, 1973; Kaiser, 1984).

Fruiting body formation of *M. xanthus* is the myxobacterial social behavior that has been the most extensively studied. When *M. xanthus* cells are starving and at a high cell density, approximately 100,000 cells glide to aggregation centers where they form multicellular, haystack-shaped fruiting bodies. Cells within nascent fruiting bodies differentiate from rod-shaped cells into spherical, heat- and desiccation-resistant myxospores. Formation of fruiting bodies probably aids in the dispersal of myxospores and ensures that when spores germinate in the presence of a suitable food supply, there will be a sufficiently high number of cells to participate in cooperative feeding (Dworkin and Kaiser, 1985; Kaiser, 1984).

It has long been known that fruiting body formation in *M. xanthus* requires nutrient limitation, a high cell density, and a solid surface.

Lynda Plamann, School of Biological Sciences, Division of Cell Biology and Biophysics, 2411 Holmes, University of Missouri-Kansas City, Kansas City, MO 64108. *Heidi B. Kaplan*, Department of Microbiology and Molecular Genetics, University of Texas Medical School at Houston, 6431 Fannin, Houston, TX 77030.

Cell-Cell Signaling in Bacteria, Edited by Gary M. Dunny and Stephen C. Winans
©1999 American Society for Microbiology, Washington, D.C.

How do cells sense these conditions and then synchronize the initiation of development? How do cells coordinate their movements toward aggregation centers? How do cells achieve specific alignments to construct a fruiting body with the characteristic haystack shape? Although progress has been made toward finding answers to these questions, much remains to be learned.

This chapter covers the advances that have been made in the area of cell-density sensing during early development in *M. xanthus*. It begins with an overview of work on cell-cell signaling that led to the identification of a cell-density signal. This signal, A signal, differs from the acyl homoserine lactones discussed elsewhere in this book. Although the use of *N*-acyl homoserine lactones as cell-density signals in *M. xanthus* cannot be ruled out, these more "traditional" quorum-sensing molecules have not been detected, despite attempts to do so (Shi and Kaplan, personal communication). The A signal is composed of a specific set of extracellular amino acids that are generated in amounts proportional to cell density through the action of extracellular proteases (Kuspa et al., 1992a). The most recent research has focused on understanding the signal transduction mechanisms that regulate A-signal generation and mediate the response to A signal.

ISOLATION OF CELL-CELL SIGNALING MUTANTS

The genetic experiments that opened the door to studies on cell-cell signaling in myxobacteria were carried out by Hagen et al. (1978) in Dale Kaiser's laboratory at Stanford University. These experiments were an extension of earlier work by McVittie et al. (1962), who observed that mixing pairs of particular developmental mutants results in normal sporulation. It was reasoned that if intercellular signals are used by myxobacteria to coordinate fruiting body formation and sporulation, then it should be possible to isolate mutants that fail to release molecular signals and, therefore, do not carry out these processes. Such mutants should be able to fruit and sporulate when mixed with cells that are capable of producing the missing molecules. A simple genetic screen was devised to isolate the conditional, nonsporulating mutants (Hagen et al., 1978). Mutagenized cells were plated at a high cell density on agar plates that promote fruiting. After allowing fruiting bodies to form, any remaining vegetative (heat-sensitive) cells were killed by heating the plates. Myxospores were plated to form single colonies on a low-nutrient medium called clone-fruiting agar. Under these conditions, myxospores germinate and form a thin colony. Once the thin colonies have formed, the cells have exhausted the nutrient supply, and fruiting bodies form within the colony. The conditional mutants—those that sporulate only in the presence of signal-producing cells—fail to form fruiting bodies on these plates.

After isolating approximately 50 nonsporulating, conditional mutants using this protocol, the question arose whether all of these mutants are defective in the same function. To test this, the conditional mutants were placed in pairwise mixtures to assay for sporulation. The prediction is that synergism (sporulation) will occur only in mixtures composed of two mutants that each fail to make different signals. These experiments led to the identification of four groups of mutants (groups A, B, C, and D), with each group predicted to be defective in producing an intercellular signal required for normal fruiting and sporulation. Of the mutants that could be classified in this manner, approximately one-third belong to group A (Hagen et al., 1978). The remainder of this chapter is focused on the group A mutants, which we now know are defective in production of extracellular A signal.

TIMELINE OF DEVELOPMENT

The group A mutants (from here on referred to as the *asg* mutants for A-signal-generating mutants) exhibit defects early in development. On solid clone-fruiting medium, the mutant cells form irregular, loose mounds that do not contain myxospores (LaRossa et al., 1983). This developmental phenotype is similar to that of the group B, C, and D mutants; therefore, visual examination of developing cultures provides little information regarding the relative times of developmental arrest. As a first step in pinpointing more precisely the times of arrest within the developmental pathway, LaRossa et al. (1983) examined the mutants for the presence of certain development-specific molecules: guanosine tetra- and pentaphosphate [(p)ppGpp], protein S, and myxobacterial hemagglutinin (MBHA). There is considerable evidence that (p)ppGpp acts as a starvation signal in *M. xanthus* (Singer and Kaiser, 1995), and it is thought that accumulation of (p)ppGpp may be one of the earliest events in fruiting. All of the conditional mutants tested, including the *asg* mutants, accumulate (p)ppGpp following transfer of the cells from a rich medium to one lacking nutrients, indicating that the mutants are at least able to initiate a response to starvation. However, some aspect of starvation sensing or response appears to be defective in the *asg* mutants. When wild-type cells are placed on solid minimal medium lacking phenylalanine, growth ceases and fruiting bodies form. It appears that limitation for phenylalanine is sufficient to induce development of wild-type *M. xanthus* cells that are at high cell density and on a solid medium. Under these same conditions, the *asg* mutants (with one exception) continue to divide, as if they have lost the ability to respond to nutrient limitation (LaRossa et al., 1983).

Protein S is a spore coat protein that normally is produced 6 h after the initiation of development; MBHA, a lectin, is normally produced approximately 12 to 15 h into development. LaRossa et al. (1983) found that the *asg* mutants delay production of protein S and fail to produce MBHA. These studies placed the point of developmental arrest in the *asg* mutants sometime before 6 h on the 24-h timeline of development. The other groups of conditional mutants appeared to arrest at later points in the developmental pathway.

Although the above results provided a rough estimate of the times of developmental arrest and confirmed the finding that there are at least four classes of conditional mutants, more developmental markers were needed to realize the full potential of these experiments. To this end, Kroos and Kaiser (1984) constructed a transposable promoter probe, Tn*5lac*, which generates transcriptional fusions to *lacZ* when it transposes in the proper orientation into a gene. This advance, along with gene delivery methods utilizing P1 phage (Kuner and Kaiser, 1981; O'Connor and Zusman, 1983; Shimkets et al., 1983), brought the field of myxobacterial development into the age of molecular genetics. Tn*5lac* was used to identify a number of genes that are induced at least threefold at various times during development (Kroos et al., 1986). Expression of these new Tn*5lac* reporter genes was examined in *asg* mutant strains to determine if the *asg* defect altered their expression patterns (Kuspa et al., 1986). Assuming that the developmental pathway is linear and unbranched, the prediction is that those genes normally expressed after the *asg* block would not be expressed in the *asg* mutant background. Twenty-one developmentally expressed reporter genes were separated into two groups following examination of their expression patterns in wild-type and *asg* mutant backgrounds. Three genes termed "A-signal-independent" have the same expression pattern in the wild-type and *asg* mutant backgrounds. These genes increase in expression between 0 and 3 h of development. The expression of the other 18 genes, termed "A-signal-dependent," is reduced or abolished in the *asg* mutant background. These genes increase in expression only after 1.5 h of development. These data indicate that the *asg* block occurs within the first 1 or 2 h of development. Thus, A signal functions very early in *M. xanthus* development.

At this point, the stage was set to test predictions of the original cell-cell signaling hypothesis: if the *asg* mutants are defective in release of an extracellular signal, and the absence of that signal results in failure to express A-signal-dependent Tn*5lac* insertion genes, then expression of these genes in an *asg* mutant should be restored by addition of wild-type cells. Furthermore, if the extracellular signal is indeed released by wild-type cells, then it should be possible to recover the activity from suspensions of developing wild-type cells. This conditioned medium, when added to *asg* mutant cells, should result in restored expression of the A-signal-dependent genes. All of these predictions were found to hold true, and in the process of testing them, the A-signal (also referred to as A factor) bioassay was developed (Kuspa et al., 1986).

BIOASSAY FOR THE A SIGNAL

The "detector" in the bioassay for A-signal activity is an *asg* mutant strain that contains the Tn*5lac* insertion $\Omega4521$. Expression of the *4521* gene initiates at 1.5 h of development in wild-type cells and is abolished in *asg* mutant cells. A-signal activity is detected by measuring the increase in β-galactosidase activity that results when A signal is supplied to the starving detector cells (Kuspa et al., 1986). The detector cells are suspended in buffer and placed in multiwell tissue culture plates. Cells or extracts are added to the wells to test for the presence of A-signal activity. The plates are incubated for 20 h, after which cells are removed and assayed for β-galactosidase activity. Although the cells in this bioassay do not form fruiting bodies, because they are suspended in liquid rather than on a solid surface, early steps in development have been shown to occur under these conditions (Downard and Zusman, 1985; Kuspa et al., 1986). The A-signal bioassay shows a linear response to the addition of A signal, and thus can be used for quantification of A-signal activity (Kuspa et al., 1986).

PURIFICATION OF THE A SIGNAL: PROTEASES AND AMINO ACIDS

With the development of the A-signal bioassay, the tools for identification of the A-signal molecule(s) were in place. Kuspa et al. (1986) showed that A-signal activity is released from wild-type cells beginning an hour or so after cells are placed under starvation conditions, including starvation in shaken suspension. Removing the cells from such a suspension leaves behind a translucent supernatant, which is a convenient starting material for A-signal purification. Early in the A-signal purification, it was noted that some of the activity is rapidly inactivated by heating, while the remaining activity is stable to heating at 100°C for 10 min. The heat-stable activity was found to pass through dialysis tubing with a 3-kDa molecular size cut-off and to be included in gel filtration columns with molecular exclusion limits of 2 kDa. The heat-labile activity was found to be nondialyzable and seemed to copurify with proteolytic activity. Therefore, it was concluded that at least two physically different types of A-signal activity exist (Plamann et al., 1992).

Proteolytic activities were measured throughout the A-signal purification, and it was noted that proteolytic activity consistently copurified with heat-labile A-signal activity. Plamann et al. (1992) found that developing wild-type cells release at least two different proteins that possess both proteolytic and A-signal activities. One of these proteases is a 27-kDa protein with a trypsin-like specificity. The second protease has a molecular mass of approximately 10 kDa and shows a pattern of substrate specificity that differs from that of the 27-kDa protein. Because two proteases with seemingly different substrate specificities were found to possess A-signal activity, it seemed plausible that common laboratory proteases have A-signal activity. Indeed, it was found that pronase, proteinase K, papain, trypsin, and chymotrypsin all have heat-labile A-signal activity (Plamann et al., 1992). Furthermore, it was found that pronase, and to a lesser extent trypsin, rescue fruiting body formation in the *asg* mutants (Kuspa et al., 1992a). Thus, it appeared that heat-labile A signal is a mixture of proteases. Because the cell envelope is remodeled as the rod-shaped cell differentiates to form the ovoid myxospore, it is reasonable to assume that proteins not needed by the myxospore serve as a substrate for the A-signal proteases.

The purification of heat-stable A signal and its identification as a mixture of amino acids and peptides supports the hypothesis that the peptides and amino acids generated as a consequence of extracellular proteolysis act as the true A signal. Fifteen different amino acids were found to have A-signal activity (Kuspa et al., 1992b). Any one of these amino acids, when added to *asg* mutant cells in the A-signal assay, restores Tn*5lac* Ω*4521* β-galactosidase activity to a near-normal level. The amino acids with the highest A-signal activities are tyrosine, proline, phenylalanine, tryptophan, leucine, isoleucine, and alanine; these amino acids can account for approximately half of the heat-stable A-signal activity in the cell-free supernatants. The threshold concentration for these amino acids is approximately 10 μM; below this concentration, *4521* expression is not rescued in an *asg* mutant. Small peptides also have heat-stable A-signal activity. Their activities can be roughly approximated by adding up the activity of the individual amino acids that constitute the peptides. Kuspa et al. (1992b) suggest that the half of heat-stable A signal that is not accounted for by individual amino acids may comprise small peptides with fewer than 100 amino acid residues.

There is considerable evidence to support the hypothesis that A signal is generated by extracellular proteases as they degrade proteins to peptides and finally to amino acids, which are the primary A signal. First, when cell-free conditioned buffer was incubated in vitro, there was a 1.7-fold increase in the levels of free amino acids. Heat-stable A-signal activity also increased by 1.7-fold in these experiments. In addition, developing wild-type cells released amino acids and peptides at the same time that A-signal activity appeared in conditioned buffer, and it has been estimated that the amount of extracellular amino acids generated by wild-type cells during the first 12 h of development is sufficient to account for the observed level of *4521* expression (Kuspa et al., 1992b). Finally, two proteases with differing substrate specificities were isolated from buffer conditioned by developing wild-type cells, and all commercially available proteases that were tested have A-signal activity (Plamann et al., 1992).

THE A SIGNAL AS A CELL-DENSITY SIGNAL

The logic behind the hypothesis that A signal is a cell-density signal for development is as follows. First, *4521* gene expression requires a high cell density (Kuspa et al., 1992a). Second, *4521* is not expressed in an *asg* mutant (Kuspa et al., 1986). Third, *4521* expression in an *asg* mutant is rescued by addition of amino acids (Kuspa et al., 1992a,b). Finally, the concentration of extracellular amino acids is proportional to the concentration of cells undergoing early development (Kuspa et al., 1992). If the above hypothesis is correct, then *4521* expression at low density should be rescued by the addition of amino acids. This is precisely what was observed (Kuspa et al., 1992a). Furthermore, it was found that *asg* mutants, which release 5 to 10% of the normal level of A signal, express *4521* and sporulate when their cell density is raised 10- to 20-fold above the normal density required for development.

Assuming extracellular amino acids induce *4521* expression and development, what prevents these processes from being triggered during growth when the extracellular concentration of amino acids is high? It is likely that the mechanisms for sensing and responding to extracellular A signal are operational only when the cells are starving. Genetic suppression analysis, detailed below, has begun to address this issue.

Why does *M. xanthus* use amino acids as a cell-density signal? Perhaps it is a question of metabolic economics. It makes "economic sense" that extracellular and periplasmic proteins dispensable for fruiting body formation and spore production are recycled during development. Instead of taking on the expense of producing an alternative cell-density signal during early development, *M. xanthus* may have evolved a mechanism to use the extracellular amino acids as both a source of carbon and nitrogen and a signal for development.

THREE *asg* LOCI

In the original screen for cell-cell signaling mutants, Hagen et al. (1978) identified 18 *asg* (group A) mutants, which represents about one-third of the total number of extracellularly complementable mutants isolated. The *asg* mutations map to three unlinked loci: *asgA*, *asgB*, and *asgC* (Kuspa and Kaiser, 1989). The three types of *asg* mutants share a number of phenotypic characteristics in addition to their defects in A-signal generation and fruiting body formation. For example, wild-type cells exhibit tan-to-yellow and yellow-to-tan phase variation, while *asg mutants* are permanently tan. In addition, the *asg* mutants have less tendency to form clumps in liquid culture, and they release less extracellular protein. All of these phenotypes are related in some way to defects in the cell surface or the extracellular matrix, which led to the suggestion that the *asg* mutations in some way alter export (Kuspa and Kaiser, 1989), including export of the proteases that generate A signal.

DNA sequence analyses of the *asgA*, *asgB*, and *asgC* genes resulted in significant advances in our understanding of the cellular roles of these genes (Davis et al., 1995; Plamann et al., 1994, 1995). The *asgA* gene encodes a histidine protein kinase that contains domains that are highly conserved among sensor kinases and response regulators of two-component signal transduction systems. These systems are widely recognized for their roles in sensing and responding to environmental signals in bacteria (Parkinson and Kofoid, 1992) and recently have been identified in eukaryotic systems (Appleby et al., 1996). The paradigm two-component system consists of a membrane-bound sensor that detects an environmental stimulus and a cytoplasmic response regulator protein that alters gene expression in response to signals from the sensor. Commonly, each of the two proteins contains two domains. The sensor contains an input domain that interacts with the environmental signal and a second domain that catalyzes autophosphorylation (histidine protein kinase domain). Typically, the response regulator contains a receiver domain that removes the phosphate from the sensor and attaches it to an aspartate residue and a DNA binding domain that regulates gene expression (Parkinson, 1993; Stock et al., 1989). AsgA is an unusual regulatory protein in that it consists entirely of a receiver domain followed by a histidine protein kinase domain and appears to lack the membrane-spanning, hydrophobic regions that characterize most of the sensors (Plamann et al., 1995). AsgA protein has been purified and shown to have autokinase activity (Li and Plamann, 1996).

As described above, in the paradigm two-component signal transduction system, the histidine protein kinase domain of the sensor and the receiver domain of the response regulator are located on separate polypeptides along with their associated input and output domains, respectively. However, many, if not most "two-component" signal transduction systems exhibit variations on this theme. Some systems are composed of three or more proteins, and the different modules may have been separated from one another or rearranged to give rise to alternative signal transduction circuits (Parkinson and Kofoid, 1992). In *Bacillus subtilis*, a phosphorelay containing proteins homologous to histidine protein kinases and response regulators (as well as other proteins) controls the initiation of sporulation (Burbulys et al., 1991). In this phosphorelay, phosphoryl groups are transferred from ATP to a histidine residue of one of at least two kinases, and from there to an aspartate residue of Spo0F, from there to a histidine residue of Spo0B, and finally, to an aspartate residue of Spo0A. Similar or related His-Asp-His-Asp phosphorelays have been reported, and it has been proposed that such a relay could function to ensure that the signal is of sufficient intensity or duration before embarking on a "costly" response such as sporulation (Appleby et al., 1996). Furthermore, a relay provides additional points at which multiple signals may influence the flow of information and affect the decision to sporulate (Burbulys et al., 1991; Grossman, 1991; Ireton et al.,

1993). Perhaps the role of AsgA in A-signal generation is to function within a phospho-relay that begins with a starvation-sensing autokinase and ends with a transcriptional regulator that affects genes required for A-signal generation (see below and Fig. 1).

The *asgB* gene encodes a 163-amino-acid polypeptide with a potential helix-turn-helix (HTH) DNA binding motif near its C terminus (Plamann et al., 1994). This predicted HTH is highly similar to the HTH found in region 4 of the major sigma factors. Sigma factor region 4 is one of the two most highly conserved regions among the sigma factors and is characterized by its location at the C terminus and a conserved HTH. There is considerable genetic and biochemical evidence to suggest that this HTH directly contacts the −35 region (TTGACA) of promoter sequences (Dombroski et al., 1992; Gardella et al., 1989; Lonetto et al., 1992; Margolis et al., 1991; Siegele et al., 1989). AsgB does not contain sequences that are similar to conserved sigma factor regions 1, 2 or 3; therefore, it is unlikely that AsgB functions as a sigma factor (Plamann et al., 1994). A more likely scenario is that AsgB is a transcription factor, perhaps one that recognizes a DNA sequence closely related to the −35 hexamer. One simple model, which takes into account that *asgB* appears to be essential for growth (Plamann et al., 1994), is that AsgB represses transcription of an early class of developmental genes during growth. In this model, it is assumed that transcription of these early developmental genes is lethal in growing cells and that the mutant containing the *asgB480* point mutation is defective in developmental derepression. An alternative model is that AsgB is part of a multicomponent sigma factor in which the AsgB subunit provides specificity for the −35 region, while other proteins allow interaction with RNA polymerase or other promoter sequences.

The *asgC* gene has been localized to a region that contains genes homologous to *rpsU*, *dnaG*, and *rpoD* of the *Escherichia coli* macromolecular synthesis operon (Davis et al., 1995). These genes encode proteins that are required for the initiation of protein, DNA, and RNA synthesis, respectively (Versalovic et al., 1993). The *asgC767* mutation was identified as two consecutive base substitutions resulting in a glutamate-to-lysine substitution within the *rpoD* homolog (known as *sigA*), which encodes the major sigma factor in *M. xanthus*. This amino acid substitution is at position 598 within sigma factor conserved region 3 (Davis et al., 1995).

One possible explanation for the A-signaling defect observed in the *asgC* (*sigA*) mutant strain is that the mutant sigma subunit fails to interact productively with a transcriptional regulator that affects A-signal generation. Mutations that affect interactions between RNA polymerase and transcription factors have been identified in the genes encoding the alpha and sigma subunits of RNA polymerase (Ishihama, 1993). *E. coli* sigma mutations that affect interactions with the transcription factors PhoB (Makino et al., 1993), CRP (at the P1*gal* promoter) (Kolb et al., 1993), AraC (Hu and Gross, 1985), and the λ cI repressor (Kuldell and Hochschild, 1994; Li et al., 1994) have been localized to region 4. Kumar et al. (1994) suggest, from data provided through a deletional analysis of the C-terminal portions of *rpoD*, that a region extending from at least region 3.2 to upstream of region 4.2 may be involved in association with transcription factors. More recently, Bramucci et al. (1995) identified a mutation in a *B. subtilis* sigma factor gene (*spo0H*) that suppresses the transcriptional defects of a mutant form of the transcription factor Spo0A (Spo0A9V). This *spo0H* mutation is located between the sequences encoding regions 2 and 3 and is proposed to allow Spo0A9V to interact with the mutant sigma-H, restoring transcriptional activation. Similarly, the A-signaling defect observed in the *asgC* (*sigA*) mutant strain may be caused by a failure of the mutant sigma subunit to interact functionally with a transcriptional regulator (AsgB?) necessary for A-signal production. An alternative hypothesis is that the mutant sigma factor has a greater affinity for core polymerase, preventing alternative sigma factors necessary for A-signal generation from gaining access to core.

Hernandez and Cashel (1995) identified *E. coli rpoD* mutations that suppress the defects of a strain unable to synthesize ppGpp. The phenotypes of the *E. coli* mutants suggest that they are hypersensitive to ppGpp, or that their RNA polymerase acts as if ppGpp is already present. Interestingly, the mutations result in substitutions within conserved region 3.1 of *E. coli* RpoD, immediately adjacent to the site of the substitutions in the *M. xanthus asgC* (*sigA*) mutant. It may be that the *M. xanthus asgC* mutant, rather than being hypersensitive to ppGpp, is insensitive to ppGpp, and therefore is unable to alter gene expression in response to starvation. This hypothesis fits well with the phenotype of the *asgC* mutant and does not exclude the transcriptional regulator/sigma factor/interaction hypothesis explained above, because RNA polymerase may respond indirectly to ppGpp, or to ppGpp that is associated with a regulatory protein.

MODEL FOR THE A-SIGNAL-GENERATING PATHWAY

Figure 1 illustrates one of many plausible models for the role of the *asg* gene products in A-signal generation. Given that the *asg* mutants have very similar phenotypes and the *asg* genes encode proteins similar to ones with known regulatory functions, we hypothesize that the *asg* gene products function together in a signal transduction pathway that is required for generation of extracellular A signal. In Fig. 1, AsgA (a histidine protein kinase) is shown interacting with an unknown starvation sensor, which may be another histidine protein kinase, a serine/threonine kinase, or a small-molecule phosphodonor. This interaction results in activation of the AsgA autokinase, and AsgA is converted to AsgA-phosphate. The phosphoryl group is then transferred to an unidentified protein or through a series of proteins, and finally to AsgB (a putative transcription factor). In this model, AsgB

acts as a repressor of early developmental genes when it is unphosphorylated, and/or as an activator of these genes when phosphorylated. Expression of the genes required for A-signal generation requires the wild-type major sigma factor (SigA), perhaps for interaction with AsgB and/or ppGpp. Expression of these genes, which may include genes encoding proteases or secretory machinery, results in release of extracellular proteases that generate A signal. Finally, sensing and transduction of A signal leads to the expression of A-signal-dependent genes such as *4521*. It is easy to imagine several variations of this model. For example, the regulator at the downstream end of the pathway may be a transcriptional activator whose production is regulated by AsgB, or AsgA may function downstream rather than upstream of AsgB.

ANATOMY OF AN A-SIGNAL-DEPENDENT GENE: *4521*

One class of *asg*-dependent developmentally expressed genes appears to be the primary target of A signal (Bowden and Kaplan, 1996). Expression of these genes, identified by the Tn*5lac* insertions $\Omega 4442$, $\Omega 4457$, $\Omega 4494$, and $\Omega 4521$, begins between 1 and 3 h after the initiation of development and is cell density dependent. In wild-type cells, their expression levels are low at low cell densities and rise dramatically as the cell density is increased above 5×10^8 cells per ml. In cells at low density, expression of these genes can be rescued to near-maximum levels per cell if A signal is added.

The best-characterized member of this class is *4521*. Its expression pattern has been analyzed directly by measuring RNA accumulation (Kaplan et al., 1991) and indirectly by measuring the β-galactosidase activity of strains containing the $\Omega 4521$-Tn*5lac* fusion (Kaplan et al., 1991; Kuspa et al., 1986; Plamann et al., 1995). Expression of *4521* requires independent input from both starvation and A signal (Kaplan et al., 1991) and occurs only if the cells are starving and at a high cell density or starving and exposed to extracellular

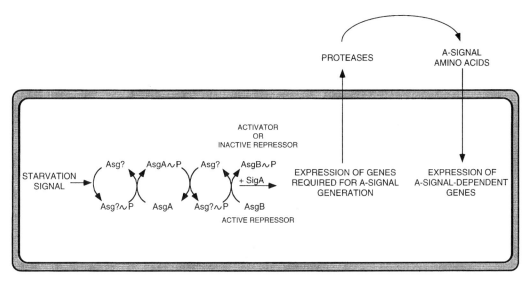

FIGURE 1 A model of the A-signal generating pathway. AsgA is a histidine protein kinase that is predicted to function within a phosphorelay that senses starvation and regulates the expression of genes required for A-signal generation. Unknown proteins (Asg?) are shown interacting upstream and downstream of AsgA in this phosphorelay. AsgB is predicted to be a transcriptional regulator. SigA (AsgC) is the major sigma factor in *M. xanthus*. See the text for a more detailed description.

A signal. In common with the expression of the other genes in this class, *4521* expression can be restored to starving *asg* mutants or starving low-density cells by the addition of exogenous A signal (Kuspa et al., 1986). The expression of *4521* can also be restored by the presence of *asg* suppressor mutations (Kaplan et al., 1991), designated *sas*.

Studies of the *cis*-acting elements controlling *4521* expression have identified the *4521* promoter as a member of the sigma-54 family (Keseler and Kaiser, 1995). However, the sequences upstream of the *4521* transcription start site (TSS) do not perfectly match those present in the sigma-54 consensus promoter (Keseler and Kaiser, 1995). The conserved GC dinucleotides normally present at −12 are located at −14 upstream of the *4521* TSS. The GG dinucleotide normally present at −24 is proposed to correspond with the AG dinucleotide at −26. The *4521* promoter does include the sigma-54 consensus pyrimidine-purine-pyrimidine-purine pattern at −24, a 10-base-pair separation of the −12 and −24

dinucleotides, and the consensus TT at −16. A mutational analysis of the conserved residues indicates that altering bases within the conserved regions greatly reduces *4521* in vivo expression. In addition, changes that alter the spacing between the two conserved regions abolishes promoter activity.

A 5′-deletion analysis of the DNA region upstream of the *4521* TSS determined that the region critical for wild-type *4521* expression during growth and development begins between 125 and 146 bp upstream of the TSS (Gulati et al., 1995). This result is consistent with the finding that all other known sigma-54 promoters require for open complex formation the binding of an activator protein to sites located upstream of the promoter (Kustu et al., 1989). These sigma-54-dependent transcriptional activators usually contain three domains: an amino-terminal regulatory domain, a carboxyl-terminal DNA binding domain, and a highly conserved ATP binding catalytic central domain (North et al., 1993). To determine if the *M. xanthus* chromosome might encode a sigma-54-dependent transcriptional activator, Kaufman and Nixon (1996) used degenerate oligonucleotides derived from the conserved central domain of these activators and PCR to amplify gene fragments from the *M. xanthus* chromosome. They identified 14 different fragments, suggesting that the *M. xanthus* genome encodes many sigma-54-dependent transcriptional activators. Among these activators are the previously identified PilR, which regulates *M. xanthus pilA* (the pilin structural gene) (Wu and Kaiser, 1995), and SasR, a putative activator of *4521* expression (Gorski and Kaiser, submitted for publication; Yang and Kaplan, unpublished data).

Related to the possibility that the *4521* promoter is a member of the sigma-54 family, a *M. xanthus* gene, *rpoN*, predicted to encode a sigma-54 homolog was identified and characterized (Keseler and Kaiser, 1995). This gene was identified by cross-hybridization to the *Caulobacter crescentus rpoN* gene. Interestingly, attempts to generate an *rpoN* null mutation were not successful, suggesting that this sigma-54 gene is essential for *M. xanthus* growth. This represents the first example of an alternative sigma factor that is essential for cell growth. Unfortunately, the inability to generate an *rpoN* null mutant makes it difficult to test the sigma-54-dependent expression of *4521* and other genes in this class.

The chromosomal *M. xanthus* DNA into which the Ω*4521* Tn*5lac* was inserted was cloned, and the nucleotide sequence of a 1.7-kb region has been determined. The sequence analysis revealed one open reading frame of 1,311 bp that codes for a predicted polypeptide that shares 24 to 33% identity with members of the serpin family of serine proteinase inhibitors (Gulati et al., unpublished data). Members of this family include human plasma proteins such as antithrombin and antitrypsin, and human tissue proteins such as plasminogen activator inhibitor (Potempa et al., 1994). To our knowledge this is the first bacterial member of the serpin family.

One model suggested by the *4521* sequence analysis is that the *M. xanthus* serpin homolog is a proteinase inhibitor that feed-back-inhibits a serine proteinase, thus decreasing the concentration of extracellular A signal. The pattern of A-signal activity reveals that extracellular A signal increases about 1 h after development is initiated and begins to decrease about 3 h later (Kuspa et al., 1986). This decrease in A-signal activity could be a result of a number of overlapping phenomena, such as a decrease in the release of proteinases, a reduction in the amount of protein substrate, and the inhibition of the proteinases. Interestingly, the Ω*4521* Tn*5lac* insertion does not have an obvious effect on growth or development, suggesting either that *4521* is not critical for these processes, or that other genes can serve similar functions.

GENETIC SUPPRESSION ANALYSIS IDENTIFIES A TWO-COMPONENT SYSTEM REGULATING *4521* EXPRESSION

Dissecting the circuit that connects extracellular A signal to its responsive genes is an important aspect of the analysis of the A-signaling system. The A-signaling system appeared to be an ideal system for genetic analysis; the phenotypic characteristics of mutants could be predicted, and a few of the components were known. In addition, the genetic tools for such an analysis in *M. xanthus* were available. It was possible to transfer genes between *M. xanthus* strains and between *E. coli* and *M. xanthus*, allowing mutations to be mapped and tests of complementation, dominance, and epistasis to be performed (Gill and Shimkets, 1993). A suppressor screen could therefore be used to identify A-signal transducers.

Early studies indicated that the control of *4521* expression by extracellular A signal is at the transcriptional level (Kaplan et al., 1991). This suggested that the A-signal transducers are regulators of *4521* transcription. A genetic screen was developed to identify these elements based on their ability to suppress the *asgB480* mutation. This scheme was expected to identify mutations that permitted *4521* expression in the absence of A signal. The mutant phenotypes could result from loss-of-function mutations of negative regulators or gain-of-function mutations of positive regulators.

Specifically, *asgB480* mutants containing the Tn*5lac* Ω*4521* fusion strain were UV mutagenized and plated on nutrient agar containing 5-bromo-4-chloro-3-indolyl-β-D-galactopyranoside (X-Gal) (Kaplan et al., 1991). Those mutants that expressed *4521* generated blue colonies among a background of tan parents that did not express *4521*. Six suppressor mutants (*sasB5, -7, -14, -15, -16, -17*) were isolated that bypass both the starvation and A-signal requirements for *4521* expression. In these mutants, *4521* is expressed during growth and development in the absence of A signal.

The six suppressor mutations map to the same locus, *sasB* (Kaplan et al., 1991), which encodes a number of regulatory proteins (Fig. 2). One of the suppressor mutations, *sasB7*, maps to the *sasS* gene. The sequence of the wild-type *sasS* gene predicts that it encodes a transmembrane histidine kinase sensor typical of the type found in two-component signal transduction systems (Yang and Kaplan, 1997). The N terminus of SasS appears to contain two transmembrane domains. These two transmembrane domains are similar to TM1 and TM2 of the *E. coli* chemotaxis receptors (MCPs) (25 to 40% and 20 to 25% identical) (Bolinger et al., 1984; Boyd et al., 1983; Krikos et al., 1983). In addition, the entire N-terminal input domain (amino acids 1 to 190) of SasS has limited homology (about 21% identity) to the N terminus of the *E. coli* MCPs (Bolinger et al., 1984; Boyd et al., 1983; Krikos et al., 1983). The C terminus of SasS contains all of the conserved residues typically found in sensor histidine protein kinases (Parkinson and Kofoid, 1992).

To study the function of *sasS* in response to A signal, null mutants of *sasS* were generated and characterized (Yang and Kaplan, 1997). In contrast to the effect of the *sasB7* point mutation, which allows for A-signal-independent *4521* expression, the *sasS* null mutations abolish *4521* expression. These data indicate that the wild-type *sasS* gene product functions as a positive regulator of *4521* expression and that *sasB7* is a gain-of-function allele. Further analysis of the *sasS* null mutants revealed that the mutants generate normal A-signal levels, but they do not respond to exogenous A signal. In addition, the *sasS* null mutants form defective fruiting bodies and sporulate less efficiently, indicating that the wild-type *sasS* gene product participates in *M. xanthus* development. Most important, SasS appears to be a key element in the transduc-

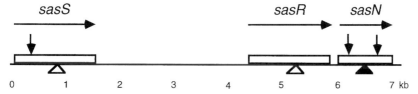

FIGURE 2 Physical map of the *sasB* locus. Genetic and molecular analysis of wild-type and mutated *sasB* alleles were used to develop a map of the *sasB* locus. The line represents the *M. xanthus* chromosome. The horizontal arrows indicate genetically determined individual transcription units and their direction of transcription. The vertical arrows indicate the location of point mutations that identified the region. The boxes indicate the position of open reading frames. The wedges indicate the phenotypes of null mutations in the open reading frames: open wedges represent null mutants that express the *4521* reporter gene at basal levels, and closed wedges represent null mutants that express the *4521* reporter gene at high levels during growth and development.

tion of starvation and extracellular A signal. It is predicted to function by controlling the phosphorylation level of a downstream response regulator that affects expression of genes such as *4521*.

Subsequent suppressor analysis has identified, 3 kb downstream of *sasS*, a putative cognate response regulator for SasS (Yang and Kaplan, unpublished data) (Fig. 2). This *sasR* gene is predicted to encode a sigma-54-dependent transcriptional activator based on its sequence similarity to other activators. This predicted protein falls into the subset of sigma-54-dependent activator proteins that have a response regulatory receiver domain at their amino terminus. This gene was among those identified by Kaufman and Nixon (1996) in their PCR amplification. The phenotype of the *sasR* null mutants indicates that SasR, like SasS, functions as a positive regulator of *4521* expression and is important for normal fruiting body formation and sporulation (Gorski and Kaiser, submitted for publication; Yang and Kaplan, unpublished data).

Having identified *sasS* and *sasR* and generated mutations with different phenotypes, it was possible to examine their relationship genetically. Epistasis analysis can order the function of gene products when two mutations that generate different phenotypes are placed into the same strain. The outcome phenotype represents the function that is farther downstream in a signal transduction pathway (Parkinson, 1995). Such an epistasis test was performed by placing an *sasS* gain-of-function mutation and an *sasR* null mutation together within the same strain. In this case, the *sasS* mutation causes high expression of the *4521* reporter gene during growth and development, and the *sasR* mutation abolishes *4521* expression. The outcome phenotype was identical to the *sasR* phenotype (abolished *4521* expression), suggesting that *sasR* functions downstream of *sasS*, which is typical of a cognate sensor and response regulator pair (Yang and Kaplan, unpublished data).

A simple mechanism of A-signal transduction based on SasS and SasR being a cognate

pair (Fig. 3) would predict that SasS is stably phosphorylated when extracellular A signal is above its minimum threshold concentration and would serve as a phosphoryl donor for SasR, which when phosphorylated would activate *4521* transcription. It is possible that SasS senses A signal by directly binding the A signal amino acids or through interactions with amino acid binding proteins. In either case, SasS would sense whether the concentration of A-signal amino acids in the periplasmic space was above 10 μM. Such a mechanism would be typical of the interactions between a ligand and its binding protein. This possibility is particularly intriguing when it is considered that A signal is a set of amino acids, and that the predicted SasS N-terminal input domain structure is similar to that of the MCP chemoreceptors, many of which directly sense amino acids.

SasN PREVENTS *4521* EXPRESSION DURING GROWTH

The five other original suppressor mutations (*sasB5*, *-14*, *-15*, *-16*, and *-17*) map to the *sasN* gene, which is located directly downstream of *sasR* (Xu et al., in press) (Fig. 2). The *sasN* gene product has no sequence similarities to known proteins. Its distinguishing characteristics are in its predicted N terminus and include a hydrophobic region and a leucine zipper motif. The phenotypes of the *sasN* null mutants are the same as those of the original point mutants. The null mutants bypass both the starvation and A-signal requirements for *4521* expression, which is reflected in very high expression of *4521* during growth and development. These data indicate that SasN functions as a critical negative regulator that prevents *4521* expression during growth. Cloning and sequencing of the mutant *sasN* genes have shown that two independent alleles were isolated. The *sasB16* mutation generates a nonsense codon at amino acid 47. Mutations *sasB5*, *-14*, *-15*, and *-17* are identical and cause a threonine-to-proline substitution at amino acid 280. In addition, the *sasN* null mutants form defective fruiting bodies and

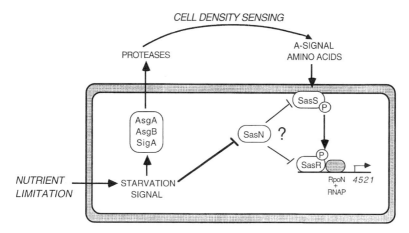

FIGURE 3 A model of the A-signal-response pathway. An individual *M. xanthus* cell is represented during early development to illustrate the relationship between the elements and the signaling molecules controlling *4521* gene expression. Arrows indicate positive biochemical or genetic functions, and lines ending in bars indicate negative biochemical or genetic functions. Genes enclosed in white shapes are currently under study in our laboratories. RpoN + RNAP represents *M. xanthus* RNA polymerase sigma-54 holoenzyme. The bent arrow indicates transcription of the *4521* gene. In this model, the SasN inhibitor, whose target is currently unknown, is inactivated in starving cells. By 1 to 2 h after starvation at high cell density, the extracellular A-signal concentration surpasses the activation threshold of 10 μM, thereby activating the autokinase activity of the SasS sensor. This phosphate is then transferred to the SasR activator, resulting in *4521* transcription.

sporulate at about 10% of the wild-type level, indicating that the wild-type SasN plays a role in normal fruiting body formation and sporulation.

Two models, in which SasN is a central element, suggest how the high level of extracellular amino acids present during growth are prevented from triggering *4521* expression (Fig. 3). In one model, SasN functions upstream of SasS and SasR, and SasN inhibits the activity of the SasS sensor during growth, resulting in basal *4521* expression. As a consequence of starvation, SasN is inactivated, allowing SasS to be stably phosphorylated when the A-signal concentration surpasses its threshold, leading to *4521* expression. In the second model, the SasN negative activity impacts the *4521* promoter more directly. If this second model is true, activated SasR must be able to induce *4521* transcription in the presence of active SasN, in order to explain the

high *4521* expression in strains containing the *sasS* gain-of-function mutation.

FUTURE DIRECTIONS

The numerous signaling pathways involved early in *M. xanthus* fruiting body development indicate that many aspects of the cells' environment are assessed when they are poised at the transition between growth and development. Our current knowledge suggests that what was initially viewed as a simple linear set of signaling pathways—starvation sensing, generation of A signal, and reception and response to A signal—is more likely to be an interconnected network. The challenge for the future will be to understand these complex interconnections.

The application of genetic methods to identify the components of these systems, followed by biochemical studies to characterize their mechanisms of action, is an approach that has proved to be successful in the past and will be continued into the future. Specifically, for the A-signal-generating pathway, a genetic suppression analysis is currently in use to identify genes whose products interact with the known *asg* gene products. Further understanding of the A-signal-response pathway requires biochemical approaches to examine the activities of the SasS, SasR, and SasN regulators. These experiments will address, among other things, the specificity for the subset of amino acids that comprise A signal and the concentration threshold for activity. Finally, a comprehensive analysis of A-signal generation and response requires that the scope of the investigation be broadened to include *asg*-dependent genes other than *4521*, some of which may be required for fruiting body formation.

ACKNOWLEDGMENTS

We gratefully acknowledge the support from the National Institute of General Medical Sciences (grant GM47265 to L.P. and GM47444 to H.B.K.) and the Army Research Office (grant 38554-LS to L.P.).

REFERENCES

Appleby, J. L., J. S. Parkinson, and R. B. Bourret. 1996. Signal transduction via the multistep phosphorelay: not necessarily a road less traveled. *Cell* **86**:845–848.

Bolinger, J., C. Park, S. Harayama, and G. L. Hazelbauer. 1984. Structure of the Trg protein: homologies with and differences from other sensory transducers of *Escherichia coli*. *Proc. Natl. Acad. Sci. USA* **81**:3287–3291.

Bowden, M. G., and H. B. Kaplan. 1996. The *Myxococcus xanthus* developmentally expressed *asgB*-dependent genes can be targets of the A signal-generating or A signal-responding pathway. *J. Bacteriol.* **178**:6628–6631.

Boyd, A., K. Kendall, and M. I. Simon. 1983. Structure of the serine chemoreceptor in *Escherichia coli*. *Nature* **301**:623–626.

Bramucci, M. G., B. D. Green, N. Ambulos, and P. Youngman. 1995. Identification of a *Bacillus subtilis spo0H* allele that is necessary for suppression of the sporulation-defective phenotype of a *spo0A* mutation. *J. Bacteriol.* **177**:1630–1633.

Burbulys, D., K. A. Trach, and J. A. Hoch. 1991. Initiation of sporulation in *B. subtilis* is controlled by a multicomponent phosphorelay. *Cell* **64**:545–552.

Davis, J. M., J. Mayor, and L. Plamann. 1995. A missense mutation in *rpoD* results in an A-signalling defect in *Myxococcus xanthus*. *Mol. Microbiol.* **18**:943–952.

Dombroski, A. J., W. A. Walter, J. M. T. Record, D. A. Siegele, and C. A. Gross. 1992. Polypeptides containing highly conserved regions of transcription initiation factor σ^{70} exhibit specificity of binding to promoter DNA. *Cell* **70**:501–512.

Downard, J. S., and D. R. Zusman. 1985. Differential expression of protein S genes during *Myxococcus xanthus* development. *J. Bacteriol.* **161**:1146–1155.

Dworkin, M. 1973. Cell-cell interactions in the myxobacteria. *Symp. Soc. Gen. Microbiol.* **23**:125–147.

Dworkin, M., and D. Kaiser. 1985. Cell interactions in myxobacterial growth and development. *Science* **230**:18–24.

Gardella, T., H. Moyle, and M. M. Susskind. 1989. A mutant *Escherichia coli* σ^{70} subunit of RNA polymerase with altered promoter specificity. *J. Mol. Biol.* **206**:579–590.

Gill, R. E., and L. J. Shimkets. 1993. Genetic approaches for analysis of myxobacterial behavior, p. 129–156. *In* M. Dworkin and D. Kaiser (ed.), *Myxobacteria II.* American Society for Microbiology, Washington, D.C.

Gorski, L., and D. Kaiser. Targeted mutagenesis of sigma-54 activator proteins in *Myxococcus xanthus*. *J. Bacteriol.*, in press.

Grossman, A. D. 1991. Integration of developmental signals and the initiation of sporulation in *B. subtilis*. *Cell* **65**:5–8.

Gulati, P., D. Xu, and H. B. Kaplan. 1995. Identification of the minimum regulatory region of a *Myxococcus xanthus* A-signal-dependent developmental gene. *J. Bacteriol.* **177**:4645–4651.

Gulati, P., J. Gibson, and H. B. Kaplan. Unpublished data.

Hagen, D. C., A. P. Bretscher, and D. Kaiser. 1978. Synergism between morphogenetic mutants of *Myxococcus xanthus*. *Dev. Biol.* **64**:284–296.

Hernandez, V. J., and M. Cashel. 1995. Changes in conserved region 3 of *Escherichia coli* σ^{70} mediate ppGpp-dependent functions *in vivo*. *J. Mol. Biol.* **252**:536–549.

Hu, J. C., and C. A. Gross. 1985. Mutations in the sigma subunit of *E. coli* RNA polymerase which affect positive control of transcription. *Mol. Gen. Genet.* **199**:7–13.

Ireton, K., D. Z. Rudner, K. J. Siranosian, and A. D. Grossman. 1993. Integration of multiple developmental signals in *Bacillus subtilis* through the Spo0A transcription factor. *Genes Dev.* **7**:283–294.

Ishihama, A. 1993. Protein-protein communication within the transcription apparatus. *J. Bacteriol.* **175**:2483–2489.

Kaiser, D. 1984. Regulation of multicellular development in myxobacteria, p. 197–218. *In* R. Losick and L. Shapiro (ed.), *Microbial Development*. Cold Spring Harbor Laboratory, Cold Spring Harbor, N.Y.

Kaplan, H. B., A. Kuspa, and D. Kaiser. 1991. Suppressors that permit A-signal-independent developmental gene expression in *Myxococcus xanthus*. *J. Bacteriol.* **173**:1460–1470.

Kaufman, R. I., and B. T. Nixon. 1996. Use of PCR to isolate genes encoding σ54-dependent activators from diverse bacteria. *J. Bacteriol.* **178**:3967–3970.

Keseler, I. M., and D. Kaiser. 1995. An early A-signal-dependent gene in *Myxococcus xanthus* has a σ^{54}-like promoter. *J. Bacteriol.* **177**:4638–4644.

Kolb, A., K. Igarashi, A. Ishihama, M. Lavigne, M. Buckle, and H. Buc. 1993. *E. coli* RNA polymerase, deleted in the C-terminal part of its a-subunit, interacts differently with the cAMP-CRP complex at the *lac*P1 and at the *gal*P1 promoter. *Nucleic Acids Res.* **21**:319–326.

Krikos, A., N. Mutoh, A. Boyd, and M. I. Simon. 1983. Sensory transducers of *E. coli* are composed of discrete structural and functional domains. *Cell* **33**:615–622.

Kroos, L., and D. Kaiser. 1984. Construction of Tn*5lac*, a transposon that fuses *lacZ* expression to exogenous promoters, and its introduction into *Myxococcus xanthus*. *Proc. Natl. Acad. Sci. USA* **81**:5816–5820.

Kroos, L., A. Kuspa, and D. Kaiser. 1986. A global analysis of developmentally regulated genes in *Myxococcus xanthus*. *Dev. Biol.* **117**:252–266.

Kuldell, N., and A. Hochschild. 1994. Amino acid substitutions in the −35 recognition motif of σ^{70} that results in defects in phage λ repressor-stimulated transcription. *J. Bacteriol.* **176**:2991–2998.

Kumar, A., B. Grimes, N. Fujita, K. Makino, R. A. Malloch, R. S. Hayward, and A. Ishihama. 1994. Role of the sigma 70 subunit of *Escherichia coli* RNA polymerase in transcriptional activation. *J. Mol. Biol.* **235**:405–413.

Kuner, J. M., and D. Kaiser. 1981. Introduction of transposon Tn*5* into *Myxococcus* for analysis of developmental and other nonselectable mutants. *Proc. Natl. Acad. Sci. USA* **78**:425–429.

Kuspa, A., and D. Kaiser. 1989. Genes required for developmental signalling in *Myxococcus xanthus*: three *asg* loci. *J. Bacteriol.* **171**:2762–2772.

Kuspa, A., L. Kroos, and D. Kaiser. 1986. Intercellular signaling is required for developmental gene expression in *Myxococcus xanthus*. *Dev. Biol.* **117**:267–276.

Kuspa, A., L. Plamann, and D. Kaiser. 1992a. A-signalling and the cell density requirement for *Myxococcus xanthus* development. *J. Bacteriol.* **174**:7360–7369.

Kuspa, A., L. Plamann, and D. Kaiser. 1992b. Identification of heat-stable A-factor from *Myxococcus xanthus*. *J. Bacteriol.* **174**:3319–3326.

Kustu, S., E. Santero, J. Keener, D. Popham, and D. Weiss. 1989. Expression of σ54 (*ntrA*)-dependent genes is probably united by a common mechanism. *Microbiol. Rev.* **53**:367–376.

LaRossa, R., J. Kuner, D. Hagen, C. Manoil, and D. Kaiser. 1983. Developmental cell interactions of *Myxococcus xanthus*: analysis of mutants. *J. Bacteriol.* **153**:1394–1404.

Li, M., H. Moyle, and M. M. Susskind. 1994. Target of the transcriptional activation function of phage λ cI protein. *Science* **263**:75–77.

Li, Y., and L. Plamann. 1996. Purification and phosphorylation of *Myxococcus xanthus* AsgA protein. *J. Bacteriol.* **178**:289–292.

Lonetto, M., M. Gribskov, and C. A. Gross. 1992. The sigma^{70} family: sequence conservation and evolutionary relationships. *J. Bacteriol.* **174**:3843–3849.

Makino, K., M. Amemura, S.-K. Kim, A. Nakata, and H. Shinagawa. 1993. Role of the σ^{70} subunit of RNA polymerase in transcriptional activation by activator protein PhoB in *Escherichia coli*. *Genes Dev.* **7**:149–160.

Margolis, P., A. Driks, and R. Losick. 1991. Establishment of cell type by compartmentalized activation of a transcription factor. *Science* **254**:562–565.

McVittie, A., F. Messik, and S. A. Zahler. 1962. Developmental biology of *Myxococcus*. *J. Bacteriol.* **84**:546–551.

North, A. K., K. E. Klose, K. M. Stedman, and S. Kustu. 1993. Prokaryotic enhancer-binding proteins reflect eukaryote-like modularity: the puzzle of nitrogen regulatory protein C. *J. Bacteriol.* **175**:4267–4273.

O'Connor, K. A., and D. R. Zusman. 1983. ColiphageP1-mediated transduction of cloned DNA from *Escherichia coli* to *Myxococcus xanthus*: use for complementation and recombinational analyses. *J. Bacteriol.* **155**:317–329.

Parkinson, J. S. 1993. Signal transduction schemes of bacteria. *Cell* **73**:857–871.

Parkinson, J. S. 1995. Genetic approaches for signaling pathways and proteins, p. 9–23. *In* J. A. Hoch and T. J. Silhavy (ed.), *Two-Component Sig-*

nal Transduction. American Society for Microbiology, Washington, D.C.

Parkinson, J. S., and E. C. Kofoid. 1992. Communication modules in bacterial signaling proteins. *Annu. Rev. Genet.* **26:**71–112.

Plamann, L., A. Kuspa, and D. Kaiser. 1992. Proteins that rescue A-signal-defective mutants of *Myxococcus xanthus. J. Bacteriol.* **174:**3311–3318.

Plamann, L., J. M. Davis, B. Cantwell, and J. Mayor. 1994. Evidence that *asgB* encodes a DNA-binding protein essential for growth and development of *Myxococcus xanthus. J. Bacteriol.* **176:** 2013–2020.

Plamann, L., Y. Li, B. Cantwell, and J. Mayor. 1995. The *Myxococcus xanthus asgA* gene encodes a novel signal transduction protein required for multicellular development. *J. Bacteriol.* **177:**2014–2020.

Potempa, J., E. Korzus, and J. Travis. 1994. The serpin superfamily of proteinase inhibitors: structure, function and regulation. *J. Biol. Chem.* **269:** 15957–15960.

Reichenbach, H. 1993. Biology of the myxobacteria: ecology and taxonomy, p. 13–62. *In* M. Dworkin and D. Kaiser (ed.), *Myxobacteria II.* American Society for Microbiology, Washington, D.C.

Shi, W., and H. Kaplan. Personal communication.

Shimkets, L. J., R. E. Gill, and D. Kaiser. 1983. Developmental cell interactions in *Myxococcus xan-* *thus* and the *spoC* locus. *Proc. Natl. Acad. Sci. USA* **80:**1406–1410.

Siegele, D. A., J. C. Hu, W. A. Walter, and C. A. Gross. 1989. Altered promoter recognition by mutant forms of the σ^{70} subunit of *Escherichia coli* RNA polymerase. *J. Mol. Biol.* **206:**591–603.

Singer, M., and D. Kaiser. 1995. Ectopic production of guanosine penta- and tetraphosphate can initiate early developmental gene expression in *Myxococcus xanthus. Genes Dev.* **9:**1633–1644.

Stock, J. B., A. J. Ninfa, and A. M. Stock. 1989. Protein phosphorylation and regulation of adaptive responses in bacteria. *Microbiol. Rev.* **53:**450–490.

Versalovic, J., T. Koeuth, R. Britton, K. Geszvain, and J. R. Lupski. 1993. Conservation and evolution of the *rpsU-dnaG-rpoD* macromolecular synthesis operon in bacteria. *Mol. Microbiol.* **8:**343–355.

Wu, S. S., and D. Kaiser. 1995. Genetic and functional evidence that type IV pili are required for social gliding motility in *Myxococcus xanthus. Mol. Microbiol.* **18:**547–558.

Xu, D., C. Yang, and H. B. Kaplan. *Myxococcus xanthus sasN* encodes a regulator that prevents developmental gene expression during growth. *J. Bacteriol.,* in press.

Yang, C., and H. B. Kaplan. 1997. *Myxococcus xanthus sasS* encodes a sensor histidine kinase required for early developmental gene expression. *J. Bacteriol.* **179:**7759–7767.

Yang, C., and H. B. Kaplan. Unpublished data.

CELL CONTACT-DEPENDENT
C SIGNALING IN *MYXOCOCCUS XANTHUS*

Lawrence J. Shimkets and Dale Kaiser

6

The myxobacteria possess a communal lifestyle that has fascinated biologists for over a century (Thaxter, 1892; Kaiser, 1993). The multicellular fruiting body has stood in defiance of a concept still found in collegiate biology textbooks that bacteria are solitary, unicellular creatures. It is now clear that cell-cell interactions are more than just an idiosyncrasy of the myxobacteria. This chapter will focus on one of the cell-cell signaling systems of *Myxococcus xanthus* but will not attempt to be a comprehensive review of myxobacterial physiology. There are a number of recent review articles and a book that describe the developmental biology, motility, physiology, and genetics of myxobacteria (Dworkin, 1996; Dworkin and Kaiser, 1993; Hartzell and Youderian, 1995; Kaiser, 1996; Kaplan and Plamann, 1996; Kim and Kaiser, 1992; Reichenbach and Dworkin, 1992; Shimkets, 1996, 1997a,b; Shimkets and Dworkin, 1997; chapter 5 of this volume).

When starving *M. xanthus* cells are placed on a solid surface at high cell density, the cells progress through a series of three partially overlapping morphological stages referred to as rippling, aggregation, and sporulation to yield a spore-filled fruiting body. Myxobacteria typically form traveling waves, which are also referred to as ripples, as a prelude to fruiting body development (Reichenbach, 1965; 1966; Shimkets and Kaiser, 1982a) (Fig. 1). Cells form a series of equidistant ridges that advance in a rhythmic manner to generate traveling waves like ripples on a water surface. The waves have a wavelength of 45 to 70 μm, depending on the type of media, a period of about 20 min, and a maximum velocity of 2.2 to 3.7 μm min^{-1}, which is the rate of cell movement (Shimkets and Kaiser, 1982a). Shortly after the traveling waves are visible, the cells begin the process of fruiting body development, and for a short time rippling and aggregation occur concurrently (Fig. 1). During aggregation, cells migrate to well-spaced foci and construct a fruiting body using another form of directed movement (Fig. 2). In *M. xanthus* the fruiting body is a raised mound of approximately 50,000 cells. Finally, the long, thin, rod-shaped vegetative cells within the fruiting body differentiate into spherical, dormant spores (Fig. 2).

McVittie and colleagues (1962) described a type of conditional developmental mutant of

Lawrence J. Shimkets, Department of Microbiology, 527 Biological Sciences Building, University of Georgia, Athens, GA 30602-2605.   *Dale Kaiser*, Biochemistry Department, Stanford University, Stanford, CA 94305.

Cell-Cell Signaling in Bacteria, Edited by Gary M. Dunny and Stephen C. Winans
©1999 American Society for Microbiology, Washington, D.C.

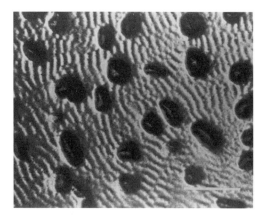

FIGURE 1 Traveling waves and fruiting bodies generated during the early stages of *M. xanthus* development. Traveling waves, or ripples, are the ridges of cells observed in the areas between the dark, immature fruiting bodies. Bar, 400 μm. Reprinted from Shimkets and Kaiser (1982a).

M. xanthus that was unable to form fruiting bodies alone but could form fruiting bodies when mixed with another developmental mutant. By making mixtures of each pair of mutants, they discovered that their six isolates fell into two extracellular complementation groups. They reasoned that one of the complementation groups might be unable to produce an essential chemoattractant necessary for fruiting body formation, but could respond to that attractant when it was presented by adjacent cells. Hagen et al. (1978) realized that any developmental cell-to-cell signal could be revealed by this method, and they isolated a more extensive set of mutants that have provided the framework for much of the research in this area. Today five extracellular complementation groups, A through E, are recognized (Kroos and Kaiser, 1987; Downard, 1993).

This chapter will focus on C signaling, which differs in several respects from the other signaling systems described in this book. First, it is a tactile signaling system that requires cell alignment during presentation of the stimulus. Second, it modulates cell reversal frequency to regulate directional movement of cells. Third,

expression of the C-signaling gene *csgA* is regulated by a feedback loop that gradually increases the CsgA concentration over the course of development. The gradual buildup of CsgA induces rippling, fruiting body formation, and sporulation in their proper temporal order at increasing CsgA threshold concentrations.

THE *csgA* GENE AND GENE PRODUCT

All mutations that render a cell unable to produce the C-signal map to a gene known as *csgA* (Shimkets et al., 1983; Shimkets and Asher, 1988), which encodes a 24.6-kDa gene product (Lee et al., 1995). CsgA purified from wild-type cells (Kim and Kaiser, 1990c,d) or expressed in *Escherichia coli* as a MalE-CsgA fusion (Lee et al., 1995) restores development when added exogenously to starving *csgA* mutant cells. In addition, anti-CsgA antibodies inhibit wild-type development (Shimkets and Rafiee, 1990). Together, these data provide clear experimental evidence that CsgA signals cells from the outside.

CsgA has striking amino acid identity with members of the short-chain alcohol dehydrogenase family (Lee and Shimkets, 1994; Baker, 1994). These enzymes use NAD(H) or NADP(H) to catalyze the interconversion of secondary alcohols and ketones or mediate decarboxylation (Persson et al., 1991; Neidle et al., 1992). The structure has been solved for many members of this family by X-ray crystallography, for *Streptomyces hydrogenans* $3\alpha/20\beta$-hydroxysteroid dehydrogenase (Ghosh et al., 1991, 1994), rat and human dihydropteridine reductase (Varughese et al., 1992, 1994; Su et al., 1993), human 17β-hydroxysteroid dehydrogenase-type 1 (Ghosh et al., 1995), *Mycobacterium tuberculosis* and plant enoyl-acyl carrier protein reductase (Dessen et al., 1995; Rafferty et al., 1995), and mouse carbonyl reductase (Tanaka et al., 1996). The amino acids lining the NAD(H) or NADP(H) binding pocket are known with some certainty since the protein crystals also contained crystallized coenzyme. From the structure a threonine res-

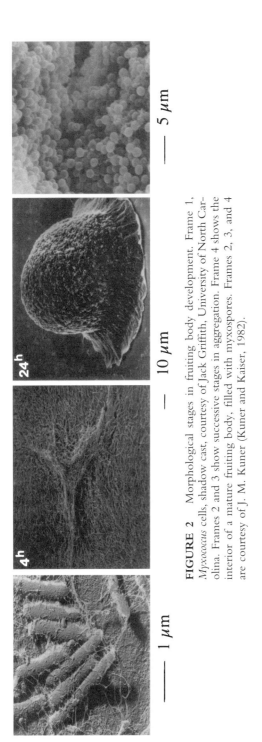

FIGURE 2 Morphological stages in fruiting body development. Frame 1, *Myxococcus* cells, shadow cast, courtesy of Jack Griffith, University of North Carolina. Frames 2 and 3 show successive stages in aggregation. Frame 4 shows the interior of a mature fruiting body, filled with myxospores. Frames 2, 3, and 4 are courtesy of J. M. Kuner (Kuner and Kaiser, 1982).

idue near the N terminus (T6 in CsgA) is anticipated to stabilize coenzyme binding through hydrogen bonding (Ghosh et al., 1994). A T6A amino acid substitution renders CsgA unable to restore development to *csgA* mutants and unable to bind radiolabeled NAD$^+$ in vitro (Lee et al., 1995). Deletion of the entire coenzyme fold also produces a protein that is unable to rescue development of *csgA* mutants (Shimkets and Rafiee, 1990). Furthermore, 10 mM NAD and NADP$^+$ stimulate rescue of *csgA* development, whereas NADH and NADPH delay development over 24 h (Lee et al., 1995). These results argue that coenzyme binding is important for CsgA activity.

Short-chain alcohol dehydrogenases contain a conserved $Sx_{12-14}YxxxK$ motif near the middle of the protein that serves a central role in chemical catalysis (Persson et al., 1991). At ambient cellular pH the ε-amino group of lysine is protonated and donates a hydrogen bond to the 2′ hydroxyl of the NAD(H) nicotinamide ribose to help orient it next to the substrate (Varughese et al., 1994). The proximity of lysine is thought to lower the pK of the phenolic group of tyrosine, allowing it to serve as a general acid. The serine side chain may provide a path for proton extraction from the reactive center. Conservative substitution of the CsgA active site residues, S135T and K155R, results in inactive *csgA* alleles (Lee et al., 1995). A protein containing the S135T substitution was unable to restore development to *csgA* mutants.

Further support for an enzymatic function for CsgA was uncovered with the discovery that two transposon insertions in the *socA* operon restore C signaling to *csgA* mutants (Lee and Shimkets, 1994). The *socA* operon consists of three genes. The first gene in the operon, *socA*, encodes a member of the short-chain alcohol dehydrogenase family. SocB is a membrane-spanning protein of unknown function. SocC is a negative regulator of the operon. One insertion that causes suppression is located between *socA* and *socB*; the other is located in *socC* (Lee and Shimkets, 1996).

These insertions result in a 30- to 100-fold increase in *socA* transcription. Lee and Shimkets (1996) proposed that suppression is due to overexpression of SocA, which substitutes for CsgA in C signaling. It has proved difficult to conclusively verify this model since *socA* mutations are lethal. SocA has strongest identity with 3-oxyacyl-reductase, which plays an essential role in fatty acid biosynthesis.

The model that CsgA is an extracellular enzyme has a strong foundation of support but will obviously require identification of the putative CsgA substrate. Short-chain alcohol dehydrogenases catalyze chemical reactions involving such a wide variety of substrates that it is not possible to deduce the actual substrate from the CsgA amino acid sequence alone. One might imagine that mutations that eliminate any substrates for CsgA would also confer the conditional developmental phenotype typical of *csgA*, yet *csgA* is the sole known member of this signaling class. If saturating mutagenesis has been performed, such a substrate may have an essential role in cell growth.

What is the source of coenzyme? Currently, no examples of secretory pathways for NAD(P)H are known. The coenzyme could be released by autolysis of developing cells (Wireman and Dworkin, 1975) or cosecreted with CsgA. An alternative possibility is that extracellular CsgA is internalized and the enzymatic function occurs cytoplasmically. This possibility is also without precedent in other *Proteobacteria*.

REGULATION OF *csgA* EXPRESSION

The *csgA* gene produces an approximately 800-nucleotide transcript that is flanked by divergent transcriptional units (Fig. 3). Downstream of *csgA* is the *fprA* gene, which encodes a 26.6-kDa protein (Shimkets, 1990). FprA has 43% amino acid identity with *E. coli* PdxH, which encodes a flavin mononucleotide (FMN)-dependent enzyme that oxidizes pyridoxine 5′-phosphate and pyridoxamine 5′-phosphate to pyridoxal 5′-phosphate, the active coenzyme derived from vitamin B_6,

with the production of H_2O_2 (Lam and Winkler, 1992). FprA overexpression in *E. coli* derepresses FMN biosynthesis, and purified FprA contains noncovalently bound FMN (Shimkets, 1990). Although FprA has not been directly demonstrated to catalyze the same chemical reaction as PdxH, an *fprA* knockout mutation renders *M. xanthus* vegetative cells inviable, as would be expected for a gene required for the biosynthesis of an essential coenzyme (Shimkets, 1990). Upstream of *csgA* is the *hemG* gene, which encodes protoporphyrinogen oxidase (Fig. 3). *M. xanthus* HemG produces a homodimer with subunit molecular mass of about 50 kDa (Dailey and Dailey, 1996). The purified protein contains a noncovalently bound flavin adenine dinucleotide but no detectable redox-active metal. HemG utilizes protoporphyrinogen IX, but not coproporphyrinogen III, as a substrate and produces 3 mol of H_2O_2 per mol of protoporphyrin. Attempts to disrupt *hemG* have been unsuccessful, suggesting that it is an essential gene (Shimkets, unpublished data).

Both the *fprA* and *hemG* transcriptional units overlap the *csgA* transcriptional unit, providing potential points for the regulation of *csgA* expression. The 3′ end of the *fprA* coding region is separated by just 8 bp from the 3′ end of the *csgA* coding region (Hagen and Shimkets, 1990). A Tn*5lac* insertion located 13 codons upstream from the *csgA* termination codon, *csgA278* (Lee et al., 1995), is oriented to detect transcription from *fprA* and demonstrates that the *fprA* transcription continues into the *csgA* coding region (Hagen and Shimkets, 1990). What effect, if any, this has on transcription of *csgA* has not been determined. Both genes are only weakly transcribed during growth and development, reducing the potential for collision of converging RNA polymerase molecules.

The *csgA* start codon is located 80 bp from the proposed *hemG* start codon (Dailey and Dailey, 1996), and the two regulatory regions overlap. The first base of the *csgA* transcript is also the first base of the translation start codon (Li et al., 1992; Lee et al., 1995). Upstream of

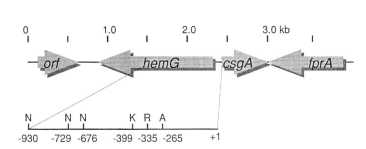

FIGURE 3 Gene arrangement in the vicinity of *csgA*. Arrows denote the direction of transcription. The *csgA* gene encodes a member of the short-chain alcohol dehydrogenase family (Lee et al., 1995). The *hemG* gene encodes protoporphyrinogen oxidase (Dailey and Dailey, 1996). The *frpA* gene is thought to encode pyridoxine 5'-phosphate oxidase (Lam and Winkler, 1992). Orf encodes an unknown function. The restriction map below is an enlargement of the *csgA* regulatory region upstream of *csgA*; this region overlaps most of the *hemG* gene. Abbreviations for restriction sites A, *Apa*I; R, *Eco*RI; K, *Kpn*I; N, *Nar*I.

the *csgA* transcriptional start is a σ^{70}-like promoter sequence that deviates from the *E. coli* consensus sequence in enough positions that one would suspect transcription to be dependent on activator proteins (Li et al., 1992).

The *csgA* gene is weakly expressed in vegetative cells and increases several-fold during development (Hagen and Shimkets, 1990). A *csgA-lacZ* transcriptional reporter system has been constructed by transposition of Tn*5lac* (Kroos and Kaiser, 1984) into the *csgA* gene (Shimkets and Asher, 1988). Readout from the reporter suggests a four- to sixfold increase in *csgA* expression from beginning to end of development, peaking about the time of sporulation (Hagen and Shimkets, 1990). The developmentally regulated expression of *csgA* appears to play a central role in fruiting body morphogenesis. An upstream regulatory region that controls *csgA* expression has been defined with a nested series of deletions (Li et al., 1992). Maximal *csgA* reporter expression, specifically >80% of the wild-type expression level, required about 400 bp upstream from the start site of transcription under conditions of extreme starvation and yielded the full allotment of fruiting bodies and spores. A much larger upstream region of between 729 and 930 bp was required when cells were placed on clone fruiting (CF) agar, which contains pyruvate as a carbon source and NH_4^+ as a

nitrogen source. In both cases the maximal level of expression decreased as the extent of deletion increased (Fig. 4). This suggests that multiple activator proteins may bind to this upstream regulatory region to regulate *csgA* transcription. Only a portion of the *hemG* gene is required for maximum expression, so the large size of the upstream regulatory region is not due to the production of a *cis*-acting regulatory protein (Fig. 3).

The *csgA* upstream regulatory region appears to serve as the focal point for the processing of information about the environment that regulates CsgA production. The CsgA concentration in turn acts as a developmental timer to induce each developmental stage in its proper temporal order. CsgA concentrations rise during development, reaching their maximum at the completion of aggregation (Kim and Kaiser, 1990c). Similarly, *csgA* expression increases slowly over the course of development, reaching its peak at the onset of sporulation (Hagen and Shimkets, 1990). Each of the morphological events observed with starving cells—rippling, aggregation, and sporulation—are induced at different levels of *csgA* expression (Li et al., 1992). Rippling requires less than 20% maximum *csgA* expression; however, this is insufficient to induce aggregation or sporulation (Fig. 4). Aggregation requires 30% of maximum expression,

FIGURE 4 The *csgA* upstream regulatory region acts as a developmental timer to induce, in sequence, traveling waves, aggregation, and sporulation. It times by sequentially raising the level of the *csgA* product because these three processes have different concentration thresholds. A strain whose *csgA* gene upstream regulatory region was trimmed to the points shown on the *x* axis was assayed for level of *csgA* expression (*y* axis) and developmental morphological stage (arrows) under conditions of severe starvation (closed circles) or partial and more gradual starvation (open circles). Under the two starvation regimens, traveling waves are observed at less than 20% maximal *csgA* expression, aggregation at 30%, and sporulation at 80%. A more extended upstream regulatory region is required under conditions of partial starvation. Abbreviations: A, *Apa*I; R, *Eco*RI; K, *Kpn*I; N, *Nar*I; S, *Sal*I.

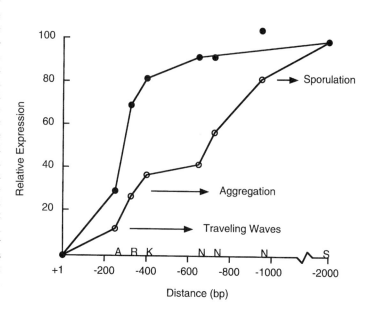

and sporulation requires about 80%. Parallel results are obtained by adding various quantities of CsgA (the protein) to rescue *csgA* mutants. Low levels (0.8 unit) induce aggregation and expression of an early CsgA-dependent developmental gene (Kim and Kaiser, 1991). One unit induces sporulation and late CsgA-dependent developmental gene expression. Three to four units inhibits development entirely (Kim and Kaiser, 1990d). The model to emerge from these studies is that CsgA triggers each developmental stage when it reaches the appropriate concentration threshold, thereby ensuring that these events occur in their correct temporal order.

The implication of this study is that there are multiple binding sites for one or more regulatory proteins in the *csgA* upstream regulatory region. Carbon and nitrogen levels may augment the production or binding of these proteins, since a much larger regulatory region is necessary to achieve maximum *csgA* expression and fruiting body development in the presence of small amounts of carbon and nitrogen sources. The nature of the regulators of *csgA* expression is under investigation. One of the regulators may be CsgA itself. *csgA* mutants exhibit reduced *csgA* expression during development (Kim and Kaiser, 1991; Li et al., 1992). Addition of CsgA restores wild-type levels of *csgA* expression, arguing that CsgA is involved in a positive feedback loop that regulates its own production (Kim and Kaiser, 1991). The *bsgA* gene also regulates *csgA* expression, since *bsgA* mutants exhibit about 50% of maximum *csgA* expression (Li et al., 1992). The *bsgA* gene encodes an intracellular ATP-dependent protease (Gill et al., 1993).

C-SIGNAL TRANSDUCTION PATHWAY

Despite uncertainties as to the chemical form of C-signal transmission from one cell to another, the cellular responses to C signaling are revealed by the *csgA* mutant phenotype. As mentioned above, C factor (the active product of the *csgA* gene) is required for at least four major responses during starvation-induced fruiting body morphogenesis: rippling, aggregation, sporulation, and expression of numerous genes, including full expression of the *csgA* gene. Although not unusual for some developmental signals, such as cyclic AMP in *Dictyostelium discoideum* and wg in *Drosophila melanogaster*, this set of responses to C factor includes modulation of cell movement for rippling and aggregation, as well as changes in the expression of a number of genes, thus raising the question of how a signal transduction pathway can be structured to provide such qualitatively different responses to the same signal molecule, as well as ordering those responses in time.

After 1 day of normal development, *M. xanthus* cells have aggregated and constructed hemispherical, steep-sided fruiting bodies, whereas a *csgA* mutant has formed loose, irregular aggregates that later, with accretion of more cells, become highly branched linear ridges. Transposon insertion mutants that began but failed to complete fruiting body aggregation were isolated, to mimic *csgA*-deficient mutants (Sogaard-Anderson et al., 1996). The new mutants were also required to be cell autonomous, not complementable for sporulation by admixed wild-type cells. This requirement helped to ensure they had alterations in the response pathway. Seven of eight mutants identified in this screen mapped to the same two genetic regions.

One set of transposons had inserted in three different genes of the well-studied *frz* gene region. These mutations removed only the cell movement responses to C signaling of rippling and aggregation; sporulation and *csgA* expression were normal. The other set of transposons had inserted in the *fruA* gene, a putative DNA binding protein with a helix-turn-helix motif (Ogawa et al., 1996). The *fruA* mutations (originally designated class II) removed both the cell movement responses and gene expression responses. Sporulation was highly deficient; *csgA* expression was low, and expression of other genes that depend on C

factor (C-dependent genes) was low or significantly reduced. The contrasting properties of these two sets of insertions immediately suggested that the C-signal transduction pathway was branched, one branch leading to the movement responses, the other branch to sporulation and to control of gene expression, including *csgA* itself. This branched pathway has been confirmed by epistasis tests with characterized *frz* mutations, a *csgA* null mutation, and a mutation in *fruA* (Sogaard-Anderson et al., 1996, 1997).

Developmental mutants that either fail to sporulate but aggregate normally, or fail to aggregate but sporulate normally, suggested that aggregation and sporulation can be uncoupled (Morrison and Zusman, 1979; Zusman, 1984). Yet, during fruiting body morphogenesis, aggregation precedes sporulation and spores form inside fruiting bodies. Sporulation in groups is advantageous for myxobacteria because subsequent germination reestablishes a swarm of contiguous cells that feed and grow more efficiently than dispersed cells (Rosenberg et al., 1977). Moreover, a macroscopic mass of spores would be more readily dispersed in the soil than a single spore (Bonner, 1982; Kaiser, 1986). A branched C-signal transduction pathway explains the normal coupling but also allows mutants that uncouple these processes.

The *frz* gene cluster is known to control the frequency of gliding reversal in growing cells (Blackhart and Zusman, 1985). The products of *frz* genes generate a phosphorylation cascade, and their amino acid sequences are similar to the che proteins of *E. coli* and *Salmonella* sp. Sensory activation of FrzCD protein, working through FrzA, induces the autophosphorylation of a His residue near the amino end of FrzE, which in turn transfers the P to an Asp residue near the C terminus of FrzE (Acuna et al., 1995). Phosphorylation of FrzE, in turn, modulates the frequency of reversal of gliding direction. Sensory input to FrzCD also induces the methylation of this protein by means of the *frzF*-encoded methyl transferase (McBride et al., 1993). Transposon insertions in the *frzCD*, *frzE*, and *frzF* genes were isolated in the screen for *csgA*-like aggregation-defective mutants. Because the *frzCD*, *frzE*, and *frzF* genes were identified in the mutant screen, and addition of C factor induces methylation of *frzCD* during development (Sogaard-Anderson and Kaiser, 1996), C factor relies on the *frz* cascade for aggregation and fruiting body morphogenesis.

Having first excluded the possibilities that C factor regulates *frz* gene transcription or translation, Sogaard-Anderson and Kaiser (1996) showed that exposing cells to C factor increases the ratio of methylated to nonmethylated FrzCD protein. FrzCD is the sequence homolog of the methyl-accepting chemotaxis proteins (MCPs) in *E. coli* and *Salmonella* sp., although the *M. xanthus* protein lacks the membrane-spanning and extracellular domains of other MCPs (McBride et al., 1993). McBride et al. (1992) had observed that FrzCD is methylated during development. Sogaard-Anderson and Kaiser (1996) showed that the pattern of FrzCD methylation changed during development in a regular way from the roughly equal parts of methylated and nonmethylated protein found during growth. During the first 6 h of development, as the cells recognized starvation and built small asymmetric aggregates, the protein first became fully nonmethylated, then at 9 and 12 h, the protein shifted to the fully methylated state as the cells formed symmetrical aggregates and accumulated many cells in them. These regular changes were lost in a *csgA* mutant, including the shift to full methylation at 9 and 12 h. Instead, most of the FrzCD protein remains nonmethylated at 9 h and about half at 12 h in the *csgA* mutant strain. At 18 h and beyond, FrzCD protein is lost in *csgA*+ cells, making it difficult to evaluate changes at these times. While these results do not exclude other possible stimuli for FrzCD methylation during development, they do implicate C factor. A major role of methylation in *E. coli* chemotaxis is to desensitize the MCP receptor, adjusting the chemotaxis phosphorylation cascade to perceive increases in attractant

concentration above the ambient level of attractant. In *M. xanthus* where FrzCD, lacking membrane-spanning and extracellular domains, appears to be a cytoplasmic protein, and where the only evidence for adaptation is to a nonphysiologic repellent, isoamyl alcohol (Shi and Zusman, 1994), the role of methylation could be different. Based on the correlation between methylation and reversal frequency in growing cells (McBride et al., 1992; Shi et al., 1993), Sogaard-Anderson and Kaiser (1996) suggested that C-signaling-induced methylation would decrease the frequency of reversal of gliding direction; however, direct behavioral assays remain to be carried out in C-factor-responding cells.

The C-signal transduction pathway as currently understood is represented in Fig. 5. The lower branch of the pathway includes the *dev* operon (Thony-Meyer and Kaiser, 1993). *devR* expression is C factor dependent (Kroos et al., 1986). Its placement in the lower branch ahead of sporulation gene expression is based on the observation that sporulation in *devRS* mutants is decreased to 0.1 to 1% that of wild-type cells (Kroos et al., 1990). *fruA* is epistatic to *dev* (Sogaard-Anderson et al., 1997), and *devR* mutants aggregate, although the aggregates remain translucent and less compact than wild-type (Kroos et al., 1990; Thony-Meyer and Kaiser, 1993).

The *lacZ* transcriptional fusions to *devR*, which were generated by transpositional inactivation of *devR* with Tn*5lac* Ω*4414* and Ω*4473*, are unusual among developmentally regulated fusions examined. At 9 h of development and long before these fusions would have reached their maximum levels of expression (Kroos et al., 1986), enzyme production in individual bacterial cells was examined by means of fluorescence-activated cell sorting and a fluorogenic β-galactosidase substrate (Russo-Marie et al., 1993). The cells sorted cleanly into two subpopulations: one expressed the maximum β-galactosidase value; the other expressed less than 5% of maximum. Cells of the highly expressing subpopulation, after storage in buffer on ice for approximately

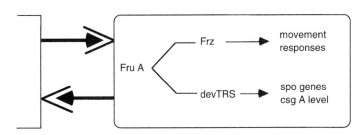

FIGURE 5 Model for the C-signal transduction pathway. Ends of two cells engaged in C signaling are represented. The large arrowheads symbolize C factor presented at the ends of both cells. These arrows engage C-factor sensors through the ends of the other cell. There is as yet no information as to whether the sensor is on the cell surface, in the periplasm, or in the cytoplasm. FruA may be the sensor, or it may receive a signal from the sensor. Lines and arrows indicate the direction of signal flow.

2 h, were reanalyzed. All cells continued to express high levels of β-galactosidase, none the low. Apparently, *devR* can be in either of two mutually exclusive stable regulatory states, high expressing or low expressing (Russo-Marie et al., 1993). Subsequent investigation showed that a product of the *dev* operon feeds back negatively on expression of the operon, which could account for the stability of the regulatory states (Thony-Meyer and Kaiser, 1993; Julien and Kaiser, 1997). Measurement of mRNA levels in *dev*[+] cells implies that the *dev* operon is usually expressed at a very low level. In addition to strong negative feedback and a switch-like quality, *dev* expression also depends strongly on cell density, such that the higher the cell density, the lower the level of *lacZ* expression from $\Omega 4414$ (Thony-Meyer and Kaiser, 1993). It is interesting that this multiply regulated, switch-like system precedes spore morphogenesis; perhaps *dev* is a trigger for spore initiation.

In addition to *dev*, expression of a series of developmental markers is partially or absolutely dependent on C signaling, and in general these markers initiate their expression after 6 h of development (Kaiser and Kroos, 1993). They include Tn*5lac* insertions $\Omega 4427$, $\Omega 4514$, $\Omega 4492$, $\Omega 4499$, $\Omega 4400$, $\Omega 4474$, $\Omega 4406$, $\Omega 4403$, $\Omega 4529$, $\Omega 4435$, $\Omega 4497$, $\Omega 4401$, $\Omega 4506$, $\Omega 4459$, alkaline and neutral phosphatases, and *mbhA* (Kaiser and Kroos, 1993). Where tested, all these markers depend on A-signaling as well.

No mutation in *csgA* has yet been found to alter expression of a gene expressed prior to 6 h of development apart from the *csgA* gene itself. During development, the expression of *csgA* increases, and that increase is C factor dependent (Hagen and Shimkets, 1990; Kim and Kaiser, 1990b). With the *csgA-lacZ* transcriptional fusion strain, LS269, which inactivates the *csgA* gene (Shimkets and Asher, 1988), addition of purified C factor or wild-type cells rescues *csgA* expression, as indicated above, and rescues aggregation and sporulation as well.

The C-factor-dependent increase in *csgA* expression helps to explain how the two Tn*5lac* reporters $\Omega 4499$ (expressed during aggregation) and $\Omega 4435$ (expressed during sporulation) can have different thresholds for their expression. It has been observed that 0.8 unit of C factor was sufficient for full expression of reporter $\Omega 4499$, whereas 1.0 unit of C factor was necessary for any expression of $\Omega 4435$ and sufficient for its full expression (Kim and Kaiser, 1991). The measurements were made after adding correspondingly different volumes of the same preparation of purified C factor.

Curiously, four components of *M. xanthus* peptidoglycan, N-acetyl glucosamine, N-acetyl muramic acid, D-alanine, and diaminopimelic acid, have two disparate effects: (i) they rescue the sporulation defect of a *csgA* mutant, and (ii) they induce rippling in a wild-type strain (Shimkets and Kaiser,

1982a,b). *M. xanthus* peptidoglycan contains diaminopimelate, not lysine, which is found in many gram-positive bacteria, and lysine cannot replace diaminopimelate in an active mixture. Thus, both effects seemed peptidoglycan specific. Moreover, any peptidoglycan not derived from *M. xanthus* had both activities provided that it contained these particular components (Shimkets and Kaiser, 1982a,b). A decade later, Li et al. (1992) showed that these same peptidoglycan components induce *csgA* gene expression. Since CsgA is required for rippling, the peptidoglycan components may induce rippling by simply activating *csgA* expression.

But what of the rescue of sporulation by peptidoglycan components, and why the component specificity? In brief, the idea is that peptidoglycan components, which are released upon damage to the *M. xanthus* cell wall, could serve to induce the part of the C-signal transduction pathway that leads to sporulation. Licking and Kaiser (unpublished data) obtained evidence that glycerol induction of sporulation lies downstream of devTRS in the sporulation branch of the C-signaling pathway shown in Fig. 5. Sporulation in *M. xanthus* has been linked to β-lactamase induction by O'Connor and Zusman (1997). They discovered that peptidoglycan components, as well as antibiotics that inhibit peptidoglycan synthesis, induce β-lactamase activity in *M. xanthus* and sporulation of nonstarved cells. In enteric bacteria like *E. coli*, peptidoglycan fragments are recycled and serve as regulators of a cell-wall sensing circuit that regulates β-lactamase production (Jacobs et al., 1997). Perhaps C signaling destabilizes the cell wall, triggering release of cell wall components and inducing sporulation.

END-TO-END C SIGNALING

Myxococcus cells are polarized; the individual cells are highly asymmetric rods with an axial ratio of 1:10 (0.5 × 7 μm) (Fig. 2). Moreover, the cells always glide along the direction of their long axis; in addition, they bear pili at one cell pole, having the type IV structure

(Wu and Kaiser, 1995). An early indication of end-to-end C signaling was that nonswarming mutants arrest fruiting body development at the same stage as *csgA* mutants (Kroos et al., 1988). Like those mutants, nonswarming mutants fail to ripple, aggregate, or sporulate. Developmental expression of C-dependent *lacZ* fusions in nonswarming strains is reduced or abolished in the same pattern and to the same degree as expression in *csgA* mutants. These observations suggest that nonswarming mutants are defective in C signaling.

Nonswarming mutant strains fail to complete C-dependent developmental events even when they are packed by centrifugation, in an attempt to compensate for their failure to aggregate. Wild-type concentrations of active C factor can be purified from nonswarming mutants (Kim and Kaiser, 1990b). Yet when nonswarming cells are mixed with *csgA* responder cells, the *csgA* cells do not complete development. Despite close proximity in these cell mixtures, nonswarming cells appear unable to present C factor to the motile *csgA* cells. Nevertheless, solubilized, purified C factor does restore nonswarming cell sporulation and gene expression to wild-type levels (Kim and Kaiser, 1990b). (Nonswarming cell aggregation is not restored because C factor does not rescue their swarming defect).

It was suggested that swarming movements were necessary to align cells such that they came into end-to-end contact with each other, and this hypothesis was tested by placing the nonswarming cells on a solid developmental surface of agar that had been scored in one dimension with fine sandpaper to form microscopic grooves (Kim and Kaiser, 1990a). Indeed, these grooves oriented the long axis of the nonswarming cells within them parallel to the axis of the groove and created end-to-end contacts between them. High levels of both C-dependent gene expression and sporulation were evident among these aligned nonswarming cells but were not detected in similarly aligned isogenic *csgA* mutant cells. These geometric constraints on cell-to-cell C-factor transmission are eliminated when solubilized, purified C factor is added to nonswarming cells (Kim and Kaiser, 1990b).

A second independent indication of end-to-end C signaling derived from close observations of the rippling waves (Fig. 1). Sager and Kaiser (1994) observed that when two sets of rippling waves collided, they failed to interfere with each other, despite the fact that the waves are dense heaps of cells and would be physically unable to pass through each other as they are perceived to do. Sager and Kaiser also observed that most of the cells within a wave crest are oriented with their long axes aligned in the direction of wave propagation, just as Reichenbach had observed for the migrating swarm front of *Stigmatella aurantiaca* (Reichenbach et al., 1975). Sager and Kaiser saw that the lack of interference would vanish as a paradox if two colliding circular wave fronts reflected from each other with precision when they met. Reflection, they proposed, could follow from a reversal response to C signaling. If individual pairs of cells collided end to end, then each reversed their direction of gliding, an entire wave front would reflect cell by cell. This would provide the precision necessary to account for the visual perception of interpenetration because individual waves would maintain their precollision form after collision. To make the reflection scheme work, it must be assumed that C signaling follows from end-to-end contact between cells; side-to-side or side-to-end will not work.

AGGREGATION AND INDUCTION OF SPORE FORMATION

Following A-signaling and starting between 8 and 9 h after starvation, the early asymmetric aggregates accrete many more cells (Fig. 2). Sogaard-Anderson and Kaiser (1996) proposed a model for this accretion that is based on the observed changes in FrzCD methylation, described above. The elementary event in this model is an aligned contact between cells followed by C-factor signaling. Initiated by cells protruding from an edge of an early aggregate that happen to be moving into the aggregate,

C-factor signaling would increase the level of FrzCD methylation, as demonstrated. Assuming that an increase in FrzCD methylation correlates with a decrease in the frequency of reversal of gliding direction, as observed in growing cells (McBride et al., 1992; Shi et al., 1993), the cells would form chains, each moving into the aggregation center (Sogaard-Anderson and Kaiser, 1996). Chains of cells are evident streaming directly into aggregates in the time-lapse film of Kuhlwein and Reichenbach (1968). According to this scheme, the *frz* mutants would be incapable of building a fruiting body, as observed (McBride et al., 1993), because being unable to reverse their gliding direction, they could not engage the end of a cell chain.

It should be noted that there is no necessary conflict between the proposed nonreversal response of 9- to 12-h cells to C signaling and the proposed reversal response of 1- to 6-h cells to C signaling that is needed for the precise reflection of colliding waves discussed above. FrzCD protein is fully nonmethylated in the 1- to 6-h cells, whereas it is fully methylated in the 9- to 12-h cells (Sogaard-Anderson and Kaiser, 1996), opening the way for a different cellular response to C signal that is related to methylation state. Moreover, the C-factor level is observed to rise sharply after 6 h (Kim and Kaiser, 1990b); its level might also contribute to the cell behavioral response.

Returning to the period of aggregation, the aggregates would grow larger as more and more cells entered, and they would become symmetric since the cells that can initiate chains protrude from all directions at the edges of early foci (Fig. 2; Kuner and Kaiser, 1982). The finished mound becomes dark and hemispherical, the darkness suggesting that the cells have achieved a very high density. Although often equated with sporulation, darkening starts before sporulation. The cells within a nascent fruiting body, despite their high density, remain in motion as indicated by continual changes in overall shape of the cell mass evident in the time-lapse moving pictures of Kuhlwein and Reichenbach (1968) and from the tracking of individual cells (Sager and Kaiser, 1993). The hemispherical nascent fruiting bodies actually have two cell-density domains: an outer domain of densely packed and ordered rod cells and an inner domain of less ordered rod cells having one-third the density of the outer shell. Tracks of individual cells within the outer domain of a fruiting body show them moving in circular bidirectional streams. Earlier, O'Connor and Zusman (1989) had described spiral arrangements of cells in fixed 7-h aggregates. Within the inner domain, cell movement ceases as the fruiting body matures. The counter-circulating outer-domain cells begin to differentiate from rods into spherical spores; however, these spores accumulate in the central domain. As more and more rods become spores, the inner domain enlarges at the expense of the outer domain, eventually to fill the entire fruiting body. Based on the evidence that the highest levels of C signaling are required to initiate spore differentiation (Li et al., 1992; Kim and Kaiser, 1991), and on their own studies of spatially differential gene expression within a fruiting body, Sager and Kaiser (1993) proposed a morphogenetic mechanism for the packing of spores in the central domain. C signaling in the outer domain would be enhanced by repeated end-to-end collisions between cells owing to the counter-rotation of chains of cells packed at very high density. The continued stream of pulses from this sort of biological supercollider would elevate C-factor expression to its highest possible level. Intense C signaling would initiate spore differentiation in the outer domain. Once initiated, as a cell became round and lost its motility, it would be passively transported from the outer to the inner domain by the movements of the undifferentiated rod cells, transported in the same way that boulders are passively carried by a stream and then are deposited on the inside bank of meanders of that stream. Prespores would thus accumulate in the inner domain, where they could mature. They could also pack closely as spheres because of the pressure of prespore

transport. In this way virtually all viable cells within a mound could become spores. Spores would be close-packed as is evident in Fig. 2, and they would fill the entire fruiting body since the volume of a spore is roughly the same as the cell it replaces (White, 1993).

REFERENCES

Acuna, G., W. Shi, K. Trudeau, and D. Zusman. 1995. The *cheA* and *cheY* domains of *Myxococcus xanthus* FrzE function independently in vitro as an autokinase and a phosphate acceptor. *FEBS Lett.* **358:**31–33.

Baker, M. 1994. *Myxococcus xanthus* C-factor, a morphogenic paracrine signal, is similar to *Escherichia coli* 3-oxoacyl-[acyl-carrier-protein] reductase and human 17β-hydroxysteroid dehydrogenase. *Biochem. J.* **301:**311–312.

Blackhart, B. D., and D. Zusman. 1985. The frizzy genes of *Myxococcus xanthus* control directional movement of gliding motility. *Proc. Natl. Acad. Sci. USA* **82:**8767–8770.

Bonner, J. T. 1982. Evolutionary strategies and developmental constraints in the cellular slime molds. *Am. Nat.* **119:**530–552.

Dailey, H. A., and T. A. Dailey. 1996. Protoporphyrinogen oxidase of *Myxococcus xanthus*: cloning, purification, and characterization of the cloned enzyme. *J. Biol. Chem.* **271:**8714–8718.

Dessen, A., A. Quemard, J. S. Blanchard, W. R. Jacobs, Jr., and J. C. Sacchettini. 1995. Crystal structure and function of the isoniazid target of *Mycobacterium tuberculosis*. *Science* **267:**1638–1641.

Downard, J. 1993. Identification of *esg*, a genetic locus involved in cell-cell signaling during *Myxococcus xanthus* development. *J. Bacteriol.* **175:**7762–7770.

Dworkin, M. 1996. Recent advances in the social and developmental biology of myxobacteria. *Microbiol. Rev.* **60:**70–102.

Dworkin, M., and D. Kaiser (ed.). 1993. *Myxobacteria II.* American Society for Microbiology, Washington, D.C.

Ghosh, D., C. M. Weeks, P. Grochulski, W. L. Duax, M. Erman, R. L. Rimsay, and J. C. Orr. 1991. Three-dimensional structure of 3α,20β-hydroxysteroid dehydrogenase: a member of the short-chain dehydrogenase family. *Proc. Natl. Acad. Sci. USA.* **88:**10064–10068.

Ghosh, D., Z. Wawrzak, C. M. Weeks, W. L. Duax, and M. Erman. 1994. The refined three-dimensional structure of 3α, 20β-hydroxysteroid dehydrogenase and possible roles of the residues conserved in short-chain dehydrogenases. *Structure* **2:**629–640.

Ghosh, D., V. Z. Pletnev, D.-W. Zhu, Z. Wawrkak, W. L. Pangborn, F. Labrie, and S.-X. Lin. 1995. Structure of human estrogenic 17β-hydroxysteroid dehydrogenase at 2.20 A resolution. *Structure* **3**:503–513.

Gill, R. E., M. Karlok, and D. Benton. 1993. *Myxococcus xanthus* encodes an ATP-dependent protease which is required for development gene transcription and intercellular signaling. *J. Bacteriol.* **175**:4538–4544.

Hagen, T. J., and L. J. Shimkets. 1990. Nucleotide sequence and transcriptional products of the *csg* locus of *Myxococcus xanthus*. *J. Bacteriol.* **172**:15–23.

Hagen, D. C., A. P. Bretscher, and D. Kaiser. 1978. Synergism between morphogenetic mutants of *Myxococcus xanthus*. *Dev. Biol.* **64**:284–296.

Hartzell, P. L., and P. Youderian. 1995. Genetics of gliding motility and development in *Myxococcus xanthus*. *Arch. Microbiol.* **164**:309–323.

Jacobs, C., J.-M. Frere, and S. Normark. 1997. Cytosolic intermediates for cell wall biosynthesis and degradation control inducible β-lactam resistance in Gram-negative bacteria. *Cell* **88**:823–832.

Julien, B., and D. Kaiser. 1997. Analysis of the *devTRS* operon from *M. xanthus*, p. 28. *In* H. Kaplan (ed.), *Abstr. 24th Meeting on the Biology of Myxobacteria*. New Braunfels, Tex.

Kaiser, D. 1986. Control of multicellular development: *Dictyostelium* and *Myxococcus*. *Annu. Rev. Genet.* **20**:539–566.

Kaiser, D. 1993. Roland Thaxter's legacy and the origins of multicellular development. *Genetics* **135**:249–254.

Kaiser, D. 1996. Bacteria also vote. *Science* **272**:1598–1599.

Kaiser, D., and L. Kroos. 1993. Intercellular signaling, p. 257–283. *In* M. Dworkin and D. Kaiser (ed.), *Myxobacteria II*. American Society for Microbiology, Washington, D.C.

Kaplan, H. B., and L. Plamann. 1996. A *Myxococcus xanthus* cell density sensing system required for multicellular development. *FEMS Microbiol. Lett.* **139**:89–95.

Kim, S. K., and D. Kaiser. 1990a. Cell alignment required in differentiation of *Myxococcus xanthus*. *Science* **249**:926–928.

Kim, S. K., and D. Kaiser. 1990b. Cell motility is required for the transmission of C-factor, an intercellular signal that coordinates fruiting body morphogenesis of *Myxococcus xanthus*. *Genes Dev.* **4**:896–905.

Kim, S. K., and D. Kaiser. 1990c. C-factor: A cell-cell signaling protein required for fruiting body morphogenesis of *M. xanthus*. *Cell* **61**:19–26.

Kim, S. K., and D. Kaiser. 1990d. Purification and properties of *Myxococcus xanthus* C-factor, an intercellular signaling protein. *Proc. Natl. Acad. Sci. USA* **87**:3635–3639.

Kim, S. K., and D. Kaiser. 1991. C-factor has distinct aggregation and sporulation thresholds during *Myxococcus* development. *J. Bacteriol.* **73**:1722–1728.

Kim, S. K., and D. Kaiser. 1992. Control of cell density and pattern by intercellular signaling in *Myxococcus* development. *Annu. Rev. Microbiol.* **46**:117–139.

Kroos, L., and D. Kaiser. 1984. Construction of Tn*5 lac*, a transposon that fuses *lacZ* expression to exogenous promoters, and its introduction in *Myxococcus xanthus*. *Proc. Natl. Acad. Sci. USA* **81**:5816–5820.

Kroos, L., and D. Kaiser. 1987. Expression of many developmentally regulated genes depends on a sequence of cell interactions. *Genes Dev.* **1**:840–854.

Kroos, L., A. Kuspa, and D. Kaiser. 1986. A global analysis of developmentally regulated genes in *Myxococcus xanthus*. *Dev. Biol.* **117**:252–266.

Kroos, L., P. Hartzell, K. Stephens, and D. Kaiser. 1988. A link between cell movement and gene expression argues that motility is required for cell-cell signalling during fruiting body development. *Genes Dev.* **2**:1677–1685.

Kroos, L., A. Kuspa, and D. Kaiser. 1990. Defects in fruiting body development caused by Tn*5lac* insertions in *Myxococcus xanthus*. *J. Bacteriol.* **172**:484–487.

Kuhlwein, H., and H. Reichenbach. 1968. Schwarmentwicklung und Morphogenese bei Myxobacterien. Film C 893/1965, Institut fur den Wissenschaftlichen Film, Gottingen, Germany.

Kuner, J. M., and D. Kaiser. 1982. Fruiting body morphogenesis in submerged cultures of *Myxococcus xanthus*. *J. Bacteriol.* **151**:458–461.

Lam, H.-M., and M. E. Winkler. 1992. Characterization of the complex *pdxH-tyrS* operon of *Escherichia coli* K12 and pleiotropic phenotypes caused by *pdxH* insertion mutations. *J. Bacteriol.* **174**:6033–6045.

Lee, B.-U., K. Lee, J. Mendez, and L. J. Shimkets. 1995. A tactile sensory system of *Myxococcus xanthus* involves an extracellular NAD(P)$^+$-containing protein. *Genes Dev.* **9**:2964–2973.

Lee, K., and L. J. Shimkets. 1994. Cloning and characterization of the *socA* locus which restores development to *Myxococcus xanthus* C-signaling mutants. *J. Bacteriol.* **176**:2200–2209.

Lee, K., and L. J. Shimkets. 1996. Suppression of a signaling defect during *Myxococcus xanthus* development. *J. Bacteriol.* **178**:977–984.

Li, S.-F., B.-U. Lee, and L. J. Shimkets. 1992. *csgA* expression entrains *Myxococcus xanthus* development. *Genes Dev.* **6**:401–410.

Licking, E., and D. Kaiser. Unpublished data.

McBride, M. J., T. Kohler, and D. R. Zusman. 1992. Methylation of FrzCD, a methyl-accepting taxis protein of *Myxococcus xanthus*, is correlated with factors affecting cell behavior. *J. Bacteriol.* **174:**4246–4257.

McBride, M. J., P. Hartzell, and D. R. Zusman. 1993. Motility and tactic behavior of *Myxococcus xanthus*, p. 285–305. *In* M. Dworkin and D. Kaiser (ed.), *Myxobacteria II.* American Society for Microbiology, Washington, D.C.

McVittie, A., F. Messik, and S. A. Zahler. 1962. Developmental biology of *Myxococcus. J. Bacteriol.* **84:**546–551.

Morrison, C., and D. Zusman. 1979. Mutants with temperature sensitive stage specific defects. *J. Bacteriol.* **140:**1036–1042.

Neidle, E., C. Hartnett, L. N. Ornston, A. Bairoch, M. Rekik, and S. Harayama. 1992. *Cis*-diol dehydrogenases encoded by the TOL pWW0 plasmid *xylL* gene and the *Acinetobacter calcoaceticus* chromosomal *benD* gene are members of the short-chain alcohol dehydrogenase superfamily. *Eur. J. Biochem.* **204:**113–120.

O'Connor, K. A., and D. R. Zusman. 1989. Patterns of cellular interactions during fruiting-body formation in *Myxococcus xanthus. J. Bacteriol.* **171:** 6013–6024.

O'Connor, K. A., and D. R. Zusman. 1997. Starvation-independent sporulation in *Myxococcus xanthus* involves the pathway for *β*-lactamase induction and provides a mechanism for competitive cell survival. *Mol. Microbiol.* **24:**839–850.

Ogawa, M., S. Fujitani, X. Mao, S. Inouye, and T. Komano. 1996. FruA, a putative transcription factor essential for the development of *Myxococcus xanthus. Mol. Microbiol.* **22:**757–767.

Persson, B., M. Krook, and H. Jornvall. 1991. Characteristics of short-chain alcohol dehydrogenases and related enzymes. *Eur. J. Biochem.* **200:** 537–543.

Rafferty, J. B., J. W. Simon, C. Baldock, P. J. Artymiuk, A. R. Stuitje, A. R. Slabas, and D. W. Rice. 1995. Common themes in redox chemistry emerge from the X-ray structure of oilseed rape (*Brassica napus*) enoyl acyl carrier protein reductase. *Structure* **3:**927–938.

Reichenbach, H. 1965. Rhythmische Vorgange bei der Schwarmentfaltung von Myxobacterien. *Ber. Deutsch. Bot. Ges.* **78:**102–105.

Reichenbach, H. 1966. *Myxococcus* spp. (Myxobacterales) Schwarmentwicklung und Bildung von Protocysten, p. 557–578. *In* G. Wolf (ed.), *Encyclopaedia Cinematographica.* Film E778/1965. Institut fur den Wissenschaftlichen Film, Gottingen, Germany.

Reichenbach, H., and M. Dworkin. 1992. The myxobacteria, p. 3416–3487. *In* H. G. Truper, A. Balows, M. Dworkin, W. Harder, and K.-H. Schleifer (ed.), *The Prokaryotes.* Springer-Verlag, New York.

Reichenbach, H., H. Galle, and H. Heunert. 1975. *Stigmatella aurantiaca* (Myxobacterales)— Schwarmentwicklung und Morphogenese. Film E2421, Institut fur den Wissenschaftlichen Film, Gottingen, Germany.

Rosenberg, E., K. Keller, and M. Dworkin. 1977. Cell-density dependent growth of *Myxococcus xanthus* on casein. *J. Bacteriol.* **120:**770–777.

Russo-Marie, F., M. Roederer, B. Sager, L. A. Herzenberg, and D. Kaiser. 1993. *β*-galactosidase activity in single differentiating bacterial cells. *Proc. Natl. Acad. Sci. USA* **90:**8194–8198.

Sager, B., and D. Kaiser. 1993. Spatial restriction of cellular differentiation. *Genes Dev.* **7:**1645–1653.

Sager, B., and D. Kaiser. 1994. Intercellular C-signaling and the traveling waves of *Myxococcus. Genes Dev.* **8:**2793–2804.

Shi, W., and D. R. Zusman. 1994. Sensory adaptation during negative chemotaxis in *Myxococcus xanthus. J. Bacteriol.* **176:**1517–1520.

Shi, W., T. Kohler, and D. R. Zusman. 1993. Chemotaxis plays a role in the social behaviour of *Myxococcus xanthus. Mol. Microbiol.* **9:**601–611.

Shimkets, L. J. 1990. The *Myxococcus xanthus* FprA protein causes increased flavin biosynthesis in *Escherichia coli. J. Bacteriol.* **172:**24–30.

Shimkets, L. J. 1996. *Myxococcus* coadhesion and role in the life cycle, p. 333–347. *In* M. Fletcher (ed.), *Bacterial Adhesion: Molecular and Ecological Diversity.* Wiley-Liss, New York.

Shimkets, L. J. 1997a. Structure and sizes of archaeal and bacterial genomes, p. 5–11. *In* F. J. deBruijn, J. R. Lupski, and G. Weinstock (ed.), *Bacterial Genomes: Physical Structure and Analysis.* Chapman & Hall, New York.

Shimkets, L. J. 1997b. Physical map of *Myxococcus xanthus*, p. 695–701. *In* F. J. deBruijn, J. R. Lupski, and G. Weinstock (ed.), *Bacterial Genomes: Physical Structure and Analysis.* Chapman & Hall, New York.

Shimkets, L. J. Unpublished data.

Shimkets, L. J., and S. J. Asher. 1988. Use of recombination techniques to examine the structure of the *csg* locus of *Myxococcus xanthus. Mol. Gen. Genet.* **211:**63–71.

Shimkets, L. J., and M. Dworkin. 1997. Myxobacterial multicellularity, p. 220–244. *In* J. Shapiro and M. Dworkin (ed.), *Bacteria as Multicellular Organisms.* Oxford University Press, New York.

Shimkets, L. J., and D. Kaiser. 1982a. Induction of coordinated movement of *Myxococcus xanthus* cells. *J. Bacteriol.* **152:**451–461.

Shimkets, L., and D. Kaiser. 1982b. Murein components rescue developmental sporulation of *Myxococcus xanthus. J. Bacteriol.* **152:**462–470.

Shimkets, L. J., and H. Rafiee. 1990. CsgA, an extracellular protein essential for *Myxococcus xanthus* development. *J. Bacteriol.* **172:**5299–5306.

Shimkets, L. J., R. E. Gill, and D. Kaiser. 1983. Developmental cell interactions in *Myxococcus xanthus* and the *spoC* locus. *Proc. Natl. Acad. Sci. USA* **80:**1406–1410.

Sogaard-Andersen, L., and D. Kaiser. 1996. C-factor, a cell surface-associated intercellular signaling protein, stimulates the cytoplasmic Frz signal transduction system in *Myxococcus xanthus. Proc. Natl. Acad. Sci. USA* **93:**2675–2679.

Sogaard-Andersen, L., F. J. Slack, H. Kimsey, and D. Kaiser. 1996. Intercellular C-signaling in *Myxococcus xanthus* involves a branched signal transduction pathway. *Genes Dev.* **10:**740–754.

Sogaard-Anderson, L., P. Soholt, E. Ellehauge, and A. Boysen. 1997. *fruA,* an essential locus in the C-signaling pathway, may serve to coordinate early developmental signals and C-factor signaling during fruiting body morphogenesis in *M. xanthus,* p. 27. *In* H. Kaplan (ed.), *Abstr. 24th Meeting on the Biology of Myxobacteria.* New Braunfels, Tex.

Su, Y., K. I. Varughese, N. H. Xuong, T. L. Bray, D. J. Roche, and J. M. Whiteley. 1993. The crystallographic structure of a human dihydropteridine reductase NADH binary complex expressed in *Escherichia coli* by a cDNA constructed from the rat homologue. *J. Biol. Chem.* **268:** 26836–26841.

Tanaka, N., T. Nanaka, M. Nakanishi, Y. Deyashiki, A. Hara, and Y. Mitsui. 1996. Crystal structure of the ternary complex of mouse lung carbonyl reductase at 1.8 A resolution: the structural origin of coenzyme specificity in the short-chain alcohol dehydrogenase/reductase family. *Structure* **4:**33–45.

Thaxter, R. 1892. Contributions from the Cryptogamic Laboratory of Harvard University. XVIII. On the Myxobacteriaceae, a new order of Schizomycetes. *Bot. Gaz.* **17:**389–406.

Thony-Meyer, L., and D. Kaiser. 1993. *devRS,* an autoregulated and essential genetic locus for fruiting body development in *Myxococcus xanthus. J. Bacteriol.* **175:**7450–7462.

Varughese, K. I., M. M. Skinner, J. M. Whiteley, D. A. Matthews, and N. H. Xuong. 1992. Crystal structure of rat liver dihydropteridine reductase. *Proc. Natl. Acad. Sci. USA* **89:**6080–6084.

Varughese, K. I., N. H. Xuong, P. M. Kiefer, D. A. Matthews, and J. M. Whiteley. 1994. Structural and mechanistic characteristics of dihydropterine reductase: a member of the Tyr-(Xaa)$_3$-Lys-containing family of reductases and dehydrogenases. *Proc. Natl. Acad. Sci. USA* **91:**5582–5586.

White, D. 1993. Myxospore and fruiting body morphogenesis, p. 307–332. *In* M. Dworkin and D. Kaiser (ed.), *Myxobacteria II.* American Society for Microbiology, Washington, D.C.

Wireman, J. W., and M. Dworkin. 1975. Morphogenesis and developmental interactions in myxobacteria. *Science* **189:**516–522.

Wu, S., and D. Kaiser. 1995. Genetic and functional evidence that Type IV pili are required for social gliding motility in *Myxococcus xanthus. Mol. Microbiol.* **18:**547–558.

Zusman, D. R. 1984. Developmental program of *Myxococcus xanthus,* p. 185–213. *In* E. Rosenberg (ed.), *Myxobacteria.* Springer, New York.

SYMBIOSES

QUORUM SENSING IN PLANT-ASSOCIATED BACTERIA

Leland S. Pierson III, Derek W. Wood, and Susanne Beck von Bodman

7

Numerous gram-negative bacteria produce one or more *N*-acyl-homoserine lactones (acyl-HSLs), which regulate the expression of a diverse range of bacterial traits that generally facilitate microbe-microbe or microbe-host interactions. These acyl-HSL regulatory systems were identified first in the marine symbiont *Vibrio fischeri* and were termed autoinduction (Eberhard, 1972; Nealson et al., 1970; Engebrecht et al., 1983). The term quorum sensing has been introduced to convey the cell density requirement for acyl-HSL-mediated gene regulation (Fuqua et al., 1994). The discovery that acyl-HSL signaling is necessary for carbapenem antibiotic production in *Erwinia carotovora* (Bainton et al., 1992) and Ti plasmid conjugation in *Agrobacterium tumefaciens* (Piper et al., 1993; Zhang et al., 1993) sparked general interest in quorum sensing as a form of regulation in plant-associated

bacteria. To date, a number of beneficial and pathogenic plant-associated bacteria have been shown to utilize acyl-HSL signals (Shaw et al., 1997; Fuqua et al., 1996; Swift et al., 1993, 1996). This chapter examines the role of acyl-HSL signals (plus a few non–acyl-HSL signals) in other pathogenic and beneficial plant-associated bacteria.

EVOLUTION OF PLANT-ASSOCIATED BACTERIA

Plant evolution occurred in the presence of bacteria, thus forging intimate associations between them. As a result of this coevolution, some bacteria have emerged to use plants as a source of nutrients, and in doing so cause injury, disease, and sometimes even death to the plant. These bacteria are generally referred to as pathogenic. Their ability to successfully invade specific hosts depended on the evolution of unique gene systems, termed virulence or pathogenicity genes, which are frequently activated in the presence of the plant host, and the products of which alter the plant in some way. In several cases the expression of virulence genes is regulated by acyl-HSL signals. For example, the regulation of Ti plasmid

Leland S. Pierson III and Derek W. Wood, Department of Plant Pathology, 204 Forbes Hall, University of Arizona, Tucson, AZ 85721-0036. *Susanne Beck von Bodman*, Department of Plant Science and Department of Molecular and Cell Biology, University of Connecticut, 1376 Storrs Road, U-67, Storrs, CT 06269-4067.

Cell-Cell Signaling in Bacteria, Edited by Gary M. Dunny and Stephen C. Winans
©1999 American Society for Microbiology, Washington, D.C.

conjugation in *A. tumefaciens*, the synthesis of plant tissue-macerating exoenzymes in *E. carotovora*, and the production of wilt-inducing exopolysaccharides in *Pantoea stewartii* subsp. *stewartii* all depend on the presence of inducing levels of acyl-HSL signals.

Plant-beneficial bacteria have evolved not only to profit from, but also to provide certain benefits for, a plant host. Examples of this group of bacteria, shown to rely on acyl-HSL signal-mediated gene regulation, include the symbiotic bacterium *Rhizobium leguminosarum* bv. *viciae*, which fixes atmospheric nitrogen for the use of the plant host, and the free-living, plant-associated bacterium *Pseudomonas aureofaciens*, which wards off plant pathogens through the production of antibiotics. In the latter case, the development of disease-suppressive soils is sufficient to protect a susceptible plant host from harm by a virulent pathogen.

EXAMPLES OF PLANT-ASSOCIATED BACTERIA THAT UTILIZE ACYL-HSL REGULATION

Pathogenic Bacteria

AGROBACTERIUM TUMEFACIENS
See chapter 8 for a complete discussion of the role of acyl-HSL signals in the tumor-inducing plant pathogenic bacterium *A. tumefaciens*.

ERWINIA CAROTOVORA
E. carotovora causes soft-rot in a variety of plant hosts (Barras et al., 1994). Strains of *E. carotovora* produce several extracellular enzymes, such as pectate lyase, pectin lyase, polygalacturonase, cellulase, and proteases, that are required for pathogenesis. The production of these exoenzymes is responsible for the extensive tissue maceration that is required for the pathogen to colonize and to propagate within the plant host. The regulation of extracellular enzyme production by strains of *E. carotovora* is extremely complex and is not completely understood (Fig. 1). Early evidence suggested that the production of these extracellular enzymes is regulated by N-3-oxohexanoyl-L-HSL (3-oxo-C6-HSL). Acyl-HSL synthases involved in 3-oxo-C6-HSL production were identified in three different strains of *E. carotovora* and named ExpI (Pirhonen et al., 1993), CarI (Swift et al., 1993), and HslI (Chatterjee et al., 1995). The *expI* gene was discovered in *E. carotovora* subsp. *carotovora* SCC3193. Mutants defective in *expI* have pleiotropic defects in the activation of exoenzyme expression (Pirhonen et al., 1993; Jones et al., 1993). Linked to the *expI* locus is a gene called *expR*, which encodes a LuxR homolog (Heikinheimo, 1994). Presumably, ExpI and ExpR constitute the quorum-sensing regulatory pair of *E. carotovora* subsp. *carotovora* SCC3193 and together control the coordinate, cell density-dependent expression of the tissue-macerating enzymes in this bacterium. However, the precise roles of ExpR and 3-oxo-C6-HSL in this regulation still need to be determined. For example, while mutations in *expI* block the expression of exoenzyme synthesis and mascerating capability, a mutation in *expR* had no effect on their expression. However, overexpression of *expR* resulted in reduced production of these enzymes (McGowan et al., 1995). Taken together, these genetic data suggest that ExpR may function as a repressor of extracellular enzyme production, possibly by titrating the 3-oxo-C6-HSL signal. Alternatively, ExpR may not play a role in extracellular enzyme production but may play some as yet undetermined role.

The CarI function was simultaneously discovered in *E. carotovora* GS101 governing the cell density-dependent expression of carbapenem antibiotic biosynthesis (Bainton et al., 1992; Chhabra et al., 1993). Carbapenems are broad-spectrum β-lactam antibiotics produced by numerous bacteria. Because carbapenem production is coordinated with the expression of extracellular mascerating enzymes, it is hypothesized that they function to reduce competition by other bacteria for the nutrients released due to tissue masceration (Salmond et al., 1995). CarI and ExpI are distinct proteins with only 70% identity; nevertheless, they

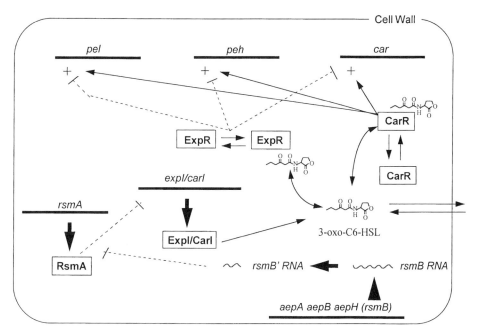

FIGURE 1 Model for the regulation of virulence factors in *E. carotovora* subsp. *carotovora*. Gene products are boxed. Virulence factors include pectate lyase (Pel), polygalacturonase (Peh), and carbapenem antibiotics (Car). + indicates a stimulation of transcription. | indicates a repression either of transcription or posttranscriptionally. Specific features and interactions are described in the text.

both catalyze the synthesis of 3-oxo-C6-HSL in their respective isolates of *E. carotovora*. The 3-oxo-C6-HSL produced by CarI serves as a coinducer for the CarR-specific transcriptional activation of the carbapenem antibiotic biosynthetic genes, and separately for the regulation of exoenzyme synthesis mediated by a factor encoded in the *rex* (regulation of exoenzymes) locus of strain GS101 (McGowan et al., 1995). This *rex*-encoded function is likely related to the ExpR regulator found in strain SCC3191. The *carI* gene is not physically linked to *carR*. Interestingly, CarR is specific for the regulation of carbapenem biosynthesis because CarR cannot complement *rex* mutations (McGowan et al., 1995). However, the expression of *carR* on a high-copy-number plasmid resulted in reduced exoenzyme synthesis. Two possible explanations could account for these results. First, overexpression of CarR from a multicopy plasmid could repress

exoenzyme expression. Second, additional copies of CarR may sequester the intracellular levels of 3-oxo-C6-HSL, thus limiting the 3-oxo-C6-HSL pool for Rex-mediated stimulation of expression of the exoenzyme genes (Swift et al., 1996).

Two additional regulatory systems, RsmA and Aep, may be linked to the ExpI/ExpR/CarR regulatory circuit in *E. carotovora*. RsmA is a homolog of CsrA, a protein found in *Escherichia coli* that functions to destabilize the mRNA transcripts of target gene systems, including glycogen accumulation, cell size, and cell surface properties (Romeo et al., 1993). CsrA has been shown to bind mRNA (Liu and Romeo, 1997). RsmA-mediated reduction of the mRNA levels from the *pel*, *peh* (also *hrpN*$_{Ecc}$), and *hslI* (the *luxI* homolog of strain 71) genes indicates that RsmA may function similarly in *E. carotovora* subsp. *carotovora* 71 (Y. Cui et al., 1995; Chatterjee et al., 1995). It was found that RsmA expression is controlled by RsmB (see below), and this regulation is dependent on the presence of the stationary-phase sigma factor RpoS. RsmA has been identified in a number of other *Erwinia* species where it controls the production of antibiotics, flagella, carotenoids, and extracellular polysaccharides (Mukherjee et al., 1996).

The AepA, AepB, and AepH (RsmB) proteins of *E. carotovora* are required for the production of extracellular enzymes in a number of *Erwinia* species, including *E. carotovora* (Liu et al., 1993; Murata et al., 1994). AepA and AepB activate the transcription of the genes encoding extracellular enzymes in response to specific plant-produced compounds. AepH (RsmB) appears to enhance the transcriptional activity of AepA/B. *rsmB* RNA is processed in vivo to yield *rsmB'* RNA, which counteracts the negative effect of RsmA (Chatterjee, personal communication). It was proposed that the Aep regulatory system senses the plant host environment and then exerts a negative effect on *rsmA* expression or RsmA function. Limiting the regulatory role of RsmA under these conditions may lead to enhanced expression of the acyl-HSL synthase, yielding high enough levels of 3-oxo-C6-HSL signal for the quorum sensing-mediated stimulation of extracellular enzyme synthesis and antibiotic production.

PANTOEA STEWARTII SUBSP. STEWARTII

P. stewartii subsp. *stewartii* (formerly *Erwinia stewartii*) is the causal agent of Stewart's wilt and leaf blight of corn. This is an insect-vectored disease whereby the corn flea beetle delivers the initial inoculum while foraging on corn. Stewart's wilt is characterized by the development of localized water-soaked lesions shortly after infection, followed by longitudinal chlorosis and necrosis as a result of systemic vascular infection, which eventually leads to the wilting and death of the plant. The major virulence factor produced by *P. stewartii* subsp. *stewartii* is an extracellular polysaccharide

(EPS) capsule (stewartan) that obstructs water flow in the xylem tissue in the infected plant (Braun, 1982). EPS production, and hence pathogenicity, is regulated by the LuxR homolog EsaR in conjunction with 3-oxo-C6-HSL produced by EsaI (Beck von Bodman and Farrand, 1995) (Fig. 2). The *esaR* and *esaI* genes are transcribed convergently and their open reading frames overlap by 31 bp. 3-oxo-C6-HSL is the major acyl-HSL species produced by *P. stewartii* subsp. *stewartii*, although minor levels of other acyl-HSL species can be detected, particularly when the bacterium is grown in minimal medium. The accumulation of acyl-HSL in *P. stewartii* subsp. *stewartii* is linear with growth, and *esaI* appears to be expressed constitutively. In contrast, *esaR* is negatively regulated by EsaR, as shown by the decreased expression of an *esaR::lacZ* reporter when functional EsaR is expressed from a recombinant plasmid (Beck von Bodman and Farrand, 1995). In addition, evidence suggests that EsaR binds to a region in the *esaR* promoter that contains the conserved *lux* box-like element (Beck von Bodman, unpublished results). This sequence spans and includes the -10 consensus sequence of the *esaR* promoter as opposed to its typical localization within the -35 consensus RNA polymerase interaction site of other acyl-HSL-regulated promoters (Fuqua et al., 1994). Gel shift assays with bacterial extracts of *E. coli* expressing *esaR* at high levels show that EsaR retards the mobility of a 61-bp *esaR* promoter fragment that contains this *lux* box element. Interestingly, EsaR-specific DNA binding occurs in the absence of any detectable acyl-HSL species, and the affinity of EsaR for this binding sequence decreases with the addition of increasing amounts of acyl-HSL (Beck von Bodman, unpublished results). EsaR also represses the expression of the EPS biosynthetic locus, *cps*, under 3-oxo-C6-HSL-limiting conditions, and derepression requires a threshold level of 3-oxo-C6-HSL. A chromosomal mutation in *esaI* blocked 3-oxo-C6-HSL production and completely abolished EPS. This mutant is also nonpathogenic. In contrast, a mutation in

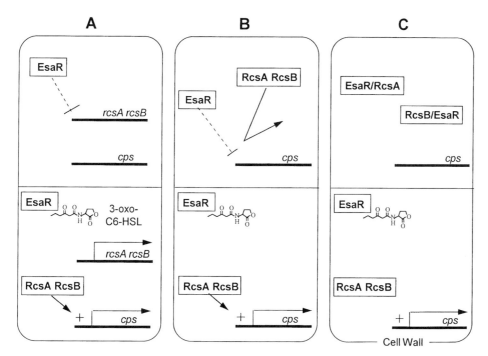

FIGURE 2 Three models for the regulation of EPS production in *P. stewartii* subsp. *stewartii*. Gene products are boxed. (A) EsaR may act as a transcriptional repressor of *rcsA* and *rcsB* in the absence of 3-oxo-C6-HSL. RcsA and RcsB are transcriptional activators essential for the expression of the *cps* gene cluster. In the presence of 3-oxo-C6-HSL, this repression is relieved, allowing expression of the RcsA/B transcriptional factors, which then can activate *cps* gene cluster transcription. (B) EsaR may serve as a transcriptional repressor of the *cps* genes directly. Binding of EsaR at one or multiple operator sites within the *cps* gene cluster may block access of RcsA/B in the absence of inducing levels of 3-oxo-C6-HSL. (C) Posttranslational model in which EsaR forms inactive heterodimers with RcsA and RcsB. Binding of 3-oxo-C6-HSL releases RcsA and RcsB. EsaR, RcsA, and RcsB proteins have structurally related helix-turn-helix domains (Beck von Bodman, unpublished results).

esaR alone, or in both *esaI* and *esaR*, yields mutants that are derepressed for EPS production. These strains appear mucoid even on nutrient agar medium lacking sugars, in contrast to the nonmucoid wild-type strain. Quantitative EPS measurements of cultures indicate that wild-type *P. stewartii* subsp. *stewartii* DC283 produces EPS in a cell density-dependent manner. Cultures grown in liquid case amino acid-peptone-glucose medium, which stimulates EPS production (Bradshaw-Rouse et al., 1981), are essentially repressed for EPS synthesis during early growth. Measurable EPS occurs only after the cell population reaches 2×10^8 CFU/ml. Such cultures contain ca. 1 μM 3-oxo-C6-HSL, which may represent the minimum threshold concentration for EsaR derepression. Interestingly, strains deficient for *esaR* are severely reduced in virulence and produce derepressed levels of EPS even at low cell densities, while the *esaI* mutant, which contains a functional *esaR* gene but lacks *esaI*, remains repressed for EPS production regardless of cell density. These observations suggest that EsaR functions as a repressor under acyl-HSL-limiting conditions (Beck von Bodman et al., 1998). The amino acid sequence of EsaR shows the same degree of structural conservation as LuxR and its homologs that function as transcriptional activators. Given these structural similarities, it is puzzling that EsaR maintains a DNA binding conformation in the absence of acyl-HSL, since it is believed that the LuxR homologs that function as transcriptional activators require acyl-HSL to promote DNA binding.

The precise molecular mechanism of how EsaR controls the expression of the *cps* gene cluster, encoding the EPS biosynthetic functions, in *P. stewartii* subsp. *stewartii* is an interesting question in itself. *P. stewartii* subsp. *stewartii* contains two-component signal transduction systems homologous to the Rcs-type two-component signal transduction system of colanic acid synthesis in *E. coli* (Torres-Cabassa et al., 1987). It appears that in *P. stewartii* subsp. *stewartii*, but perhaps not in *E. coli*, the expression or activity of these transcriptional activators is further governed by the quorum-sensing regulatory process mediated by EsaR. Fig. 2 provides three possible models for the action of EsaR in *P. stewartii* subsp. *stewartii*.

RALSTONIA SOLANACEARUM

R. solanacearum causes vascular wilt in many plants primarily via the production of exopolysaccharide fraction I (EPS I) (Buddenhagen, 1960; Hayward, 1995; Schell, 1996). Production of EPS I is regulated by a complex sensory transduction network involving the central regulatory gene *phcA,* which is required for EPS I production (Brumbley and Denny, 1990; Brumbley et al., 1993) (Fig. 3). *phcA* is autogenously regulated and encodes a LysR-type transcriptional activator that is required for the expression of a second regulatory gene *xpsR* (Hwang et al., 1995). Transcription of *phcA* and the activity of PhcA is negatively modulated by the *phcBSR* operon. *phcB* is required for the production of the volatile signal 3-hydroxypalmitic acid methyl ester (3-OH-PAME) (Clough et al., 1994), while *phcSR* apparently encodes a two-component regulatory system that functions to negatively affect *phcA* transcription, thus resulting in lower amounts of EPS I production (Clough et al., 1997). The current model suggests that PhcS and PhcR function together to negatively regulate PhcA transcription and activity. This negative regulation is reversed apparently by the presence of 3-OH-PAME, which may interfere with the ability of PhcS/PhcR to affect *phcA* or PhcA function negatively. Thus, 3-OH-PAME serves as a regulator of virulence in *R. solanacearum* (Flavier, 1997).

Highly concentrated ethyl acetate culture supernatants of *R. solanacearum* AW1 were shown to contain two acyl-HSL signals with high-pressure liquid chromatography retention times similar to those of C6-HSL and C8-HSL (Flavier et al., 1997). Three loci were identified that apparently are involved in the regulation and production of acyl-HSL. A 4.1-kb fragment was cloned from one locus that contains the genes *solI/solR*, which belong to the LuxI/LuxR family. Inactivation of *solI* had no effect on the expression of virulence genes in culture or on the ability of *R. solanacearum* to cause wilt in tomatoes. A gene *aidA* adjacent to *solI/solR* was identified that was dependent on acyl-HSL production for expression. However, no similarity between AidA and other proteins within the databases has been found. Expression of *solI* and *solR* required PhcA.

Non-Acyl-HSL Diffusible Signals in Plant-Associated Bacteria

XANTHOMONAS CAMPESTRIS

Although no acyl-HSL-mediated regulatory system has been identified in *X. campestris* strain 8004, an unidentified diffusible signal has been found to be correlated with virulence factor production. *Xanthomonas campestris* pv. *campestris* is a major pathogen worldwide causing black rot in cruciferous plants (Onsando, 1992). It produces several virulence factors, including proteinases, pectinases, and an endoglucanase. A 27-kb region of the genome contains at least seven genes (*rpfA-rpfG*) involved in the regulation of these virulence factors. Mutations in the *rpf* region result in decreased virulence on susceptible hosts, but

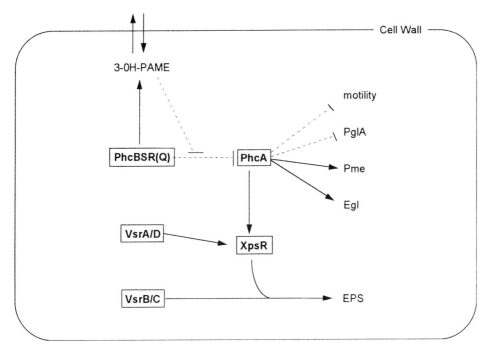

FIGURE 3 Model for the regulation of virulence in *R. solanacearum*. Gene products are boxed. Virulence factors include pectin methyl esterase (Pme), endoglucanase (Egl), and exopolysaccharide fraction I (EPS). Factors believed important during nonpathogenic growth include motility and endopolygalacturonase A (PglA). 3-OH-PAME represents the extracellular factor 3-hydroxypalmitic acid methyl ester, believed to be produced by PhcB. Specific interactions are described in the text. This figure was modified from Clough et al. (1997).

unlike *hrp* mutations, do not affect elicitation of the hypersensitive response on incompatible hosts (Arlat et al., 1991). *rpfF* and *rpfB* are involved in the production of these exocellular enzymes via a diffusible substance called DSF (Barber et al., 1997). Protease production in *rpfF* mutants is restored to normal levels by the presence of other strains of *X. campestris*. However, no acyl-HSLs have been identified yet in *X. campestris,* and the addition of DSF to exponentially growing cultures of *X. campestris* does not result in early activation of protease production. Therefore, DSF is not believed to function as a classic autoinducer. The predicted amino acid sequences of RpfF and RpfB have no similarity to the LuxI/LuxR family of quorum-sensing regulators. RpfF shares similarity to enoyl-CoA hydratase of *Rhizobium meliloti* (Margolin et al., 1995),

while RpfB shares similarity to the long-chain fatty acyl ligase of *E. coli* (Fulda et al., 1994). DSF accumulates in early stationary phase but subsequently declines. One hypothesis regarding the possible structure of DSF is that upon entering stationary phase, fatty acid precursor pools are diverted through RbfF and then RbfB into a fatty acyl-CoA derivative that is next conjugated to another unknown factor and secreted (Barber et al., 1997). Slater et al. (1997) used a endoglucanase::*uidA* (β-glucuronidase) fusion as a reporter of DSF activity to identify novel transposon mutants in strain 8004 that hyperexpress DSF.

X. campestris pv. *campestris* B-24 also produces a diffusible signal called DF involved in the production of the yellow membrane-bound brominated aryl-polyene pigment xanthomonadin and EPS production (Poplawsky and Chun, 1997). Analysis of an 18.6-kb region involved in xanthomonadin biosynthesis identified seven transcriptional units (*pigA* through *pigG*). *pigB* mutants were defective in the production of xanthomonadins, EPS, and DF, and were complemented for xanthomonadin and EPS production by supernatants from 33 of 37 strains of *X. campestris* pathovars and *Xanthomonas translucens* and *Xanthomonas vesicatoria*. *pigB* mutants were also deficient in epiphytic survival (Poplawsky and Chun, 1995). DF is active at low concentrations, and preliminary analysis indicates that it may be a butyrolactone derivative (J. Cui et al., 1995; Chun et al., 1998). Preliminary evidence suggests that DSF and DF are different signals that independently regulate exocellular enzyme and xanthomonadin production, respectively, while both appear to regulate EPS production (Poplawsky, personal communication; Daniels, personal communication).

Beneficial Bacteria

RHIZOBIUM LEGUMINOSARUM

R. leguminosarum bv. *viciae* is a gram-negative bacterium that is capable of forming a symbiotic relationship with certain leguminous plants such as peas and vetch (van Rhijn and Vanderleyden, 1995). During infection by *R. leguminosarum*, it attaches to root hairs resulting in root hair curling and subsequent internalization of the bacterium. The bacterium moves through an infection thread into the root cortex in which it differentiates into bacteroids within structures termed nodules where atmospheric nitrogen fixation occurs. Strains of *R. leguminosarum* bv. *viciae* that are able to form nodules contain many of the genes required for host range, nodulation, and nitrogen fixation on a large Sym plasmid, pRL1JI (Hirsch, 1979). A gene *rhiR* encoding the LuxR homolog RhiR has been identified on pRL1JI which controls the expression of the adjacent *rhiABC* (Gray et al., 1996; Schripsema et al., 1996). There appear to be numerous acyl-HSLs produced by *R. leguminosarum,* and they are probably produced by more than one "I" gene. Gray et al. (1996) indicated that there are two acyl-HSL signals that activate *rhiABC* expression in conjunction with RhiR. Crockford et al. (1995) have reported also that catalase activity in *R. leguminosarum* is regulated in a manner consistent with acyl-HSL-mediated control (Crockford et al., 1995). However, the specific signal utilized in this regulation has not been identified.

The function of the *rhiABC* genes in *R. leguminosarum* is unclear. The *rhiA* gene is expressed by free-living bacteria in the rhizosphere but not by bacteroids (Cubo et al., 1992; Dibb et al., 1984). Mutations in RhiA and RhiR do not affect nodulation unless they occur in bacteria that already harbor a mutation in the *nodFEL* operon involved in specifying host range (Cubo et al., 1992; van Rhijn and Vanderleyden, 1995). Such double or triple mutants are unable to form nodules. These observations have led to speculation that *rhiABC* genes play some role in microbe-host interactions within the rhizosphere.

N-3-hydroxy-7-*cis*-tetradecanoyl-L-homoserine lactone (Δ7-3hydroxy-C14-HSL) was the first acyl-HSL reported as being produced by *R. leguminosarum* and was originally believed to be the cognate signal for RhiR (Gray et al., 1996). It is now known that Δ7-3hydroxy-C14-HSL is produced by CinI (Lithgow and Downie, personal communication). It was also initially hypothesized that Δ7-3hydroxy-C14-HSL was the same compound as the *Rhizobium* bacteriocin *small*, which inhibited the growth of several *R. leguminosarum* biovars in conjunction with the *sbs* gene(s) located on the Sym plasmid (Hirsch, 1979; Wijffelman et al., 1983). Bacteriocin *small* is only made by conjugal recipients of *Rhizobium* sp. and is believed to act by causing conjugal donor cells of *Rhizobium* sp. to cease cell division and most probably induce the expression of conjugal transfer genes. One clue that Δ7-3hydroxy-C14-HSL and *small* are not identical was that in contrast to *small*, Δ7-3hydroxy-C14-HSL production does not require the Sym plasmid and in fact is completely repressed by the *rps* gene(s) that resides on the Sym plasmid (Gray et al., 1996; Wijffelman et al., 1983). However, further work is required before the roles of Δ7-3hydroxy-C14-HSL and *small* are fully understood.

Rosemeyer et al. (1998) demonstrated that *Rhizobium etli* (formerly *R. leguminosarum* bv. *phaseoli*), a nitrogen fixer on common bean, produces at least seven different acyl-HSL signals. One of them appears to act similarly to bacteriocin *small*. Two genes, *raiI* and *raiR*, were identified. Mutants defective in *raiI* produce only three of the acyl-HSL signals while *raiR* mutants produce four different acyl-HSL signals.

PSEUDOMONAS AUREOFACIENS

P. aureofaciens 30-84 was isolated originally from wheat roots taken from a wheat field in which take-all disease caused by the ascomycete fungus *Gaeumannomyces graminis* var. *tritici* had been naturally suppressed. Strain 30-84 can be used as a seed treatment to protect wheat from take-all disease (reviewed by Weller, 1988; Pierson and Pierson, 1996). The ability of this strain to reduce the severity of take-all is due to the production of the phenazine antibiotics phenazine-1-carboxylic acid, 2-hydroxy-phenazine-1-carboxylic acid, and 2-hydroxy-phenazine (Pierson and Thomashow, 1992). Phenazine production is also responsible in part for the rhizosphere competence of *P. aureofaciens*. Loss of phenazine production resulted in a 10^4-fold decrease in the persistence of strain 30-84 on wheat roots in natural soils in the presence of other microorganisms (Mazzola et al., 1992). In contrast, the loss of phenazine production had no effect on the persistence of strain 30-84 in steam-sterilized soils, which lack competing populations.

The *phzFABCD* operon is responsible for the biosynthesis of the phenazine antibiotics by strain 30-84 (Pierson et al., 1995). Expression of *phzFABCD* is regulated at one level by PhzI and PhzR, members of the LuxI/LuxR family of quorum-sensing regulators (Pierson et al., 1994, 1996; Wood and Pier-

son, 1996) (Fig. 4). PhzI encodes an acyl-HSL synthase that is responsible for the production of hexanoyl-homoserine lactone (C6-HSL), which is required for the expression of *phzFABCD* (Wood et al., 1997). At a second level, the LemA/GacA two-component regulatory system (Parkinson, 1993) is required for the production of the C6-HSL signal, presumably by controlling the expression of *phzI* (Pierson et al., 1996).

ACYL-HSL REQUIRED FOR GENE EXPRESSION IN SITU

Although numerous acyl-HSL regulatory systems exist in nature, evidence for their direct role in regulating gene expression in situ is limited. The first biochemical evidence that acyl-HSL signaling occurred in situ was provided by Boettcher and Ruby (1995), who showed that *N*-(3-oxohexanoyl) homoserine lactone (3-oxo-C6-HSL), the cognate *V. fischeri* autoinducer, could be extracted directly from the light organs of *Euprymna* species at concentrations sufficient to induce bioluminescence in vitro. Genetic evidence for the role of C6-HSL in expression of the *phzFABCD* operon in situ on wheat roots by *P. aureofaciens* 30-84 was shown by the use of isogenic strains (Wood et al., 1997). A *phzB:inaZ* reporter that expresses ice nucleation in place of the phenazine antibiotics was constructed along with a *phzI⁻ phzB:inaZ* reporter that expresses ice nucleation only in the presence of exogenous acyl-HSL signal. The *phzI⁻ phzB:inaZ* reporter expressed ice nucleation activity at a 1,000-fold lower level on wheat roots as compared to the *phzI⁺* strain. Introduction of a *phzI⁺* strain in combination with the *phzI⁻ phzB:inaZ* reporter restored ice nucleation activity to wild-type levels when introduced at a 50:50 ratio. In addition, restoration of ice nucleation activity in the PhzI⁻ reporter was proportional to the relative amount of the PhzI⁺ donor at various ratios. Recent work utilizing a library of 700 bacterial isolates from wheat roots taken from diverse geographic areas indicates that approximately 8% restored phenazine gene expres-

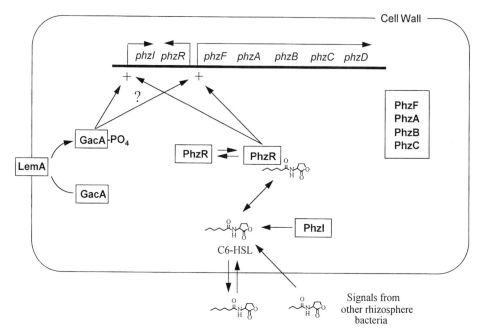

FIGURE 4 Model for the regulation of phenazine antibiotic production in *P. aureofaciens*. Gene products are boxed. PhzF, PhzA, PhzB, and PhzC represent the phenazine antibiotic biosynthetic complex. Signals from other rhizobacteria represent the potential role of cross-communication in the rhizosphere community. Specific interactions are described in the text.

sion in a *phzI⁻ phzB:inaZ* strain of 30-84 to wild-type levels in situ, suggesting that acyl-HSL-mediated communication among unrelated bacteria may be a ubiquitous phenomenon (Pierson et al., in press).

BIOLOGICAL ROLES FOR ACYL-HSL SIGNALING

Clearly, many different plant-associated bacteria utilize diffusible acyl-HSL signals to regulate the expression of various genes that encode products important for the success of the bacterium in its interactions with other bacteria or with its host plants. However, exact roles for these diffusible signals in plant-associated bacteria in nature have yet to be clearly elucidated. One common theme that is emerging is their potential use to delay the expression of factors that could elicit host defensive responses until the bacterial population has accumulated to sufficient levels to overcome these responses. Such a rationale has

been proposed in regard to the role of 3-oxo-C6-HSL-mediated control of both extracellular enzymes and carbapenem antibiotics in pathogenesis by *E. carotovora* (Pirhonen et al., 1993; Swift et al., 1996). During colonization by a few bacterial cells, insufficient 3-oxo-C6-HSL is present, resulting in the lack of extracellular enzyme production, which may help the bacterium avoid elicitation of plant defense responses. However, once the bacterial population reaches a critical density, the coordinated production of extracellular enzymes by the entire population may allow the pathogen to overwhelm host defenses. The concurrent production of carbapenem antibiotics would presumably allow *E. carotovora* to inhibit the growth of other microorganisms, which would compete for nutrients released by the activity of the extracellular enzymes. The linkage of these two apparently unrelated systems by a single acyl-HSL signal molecule would therefore ensure that *E. carotovora* could effectively avoid both host plant defenses and competition by other microorganisms.

A similar hypothesis has been proposed for the vascular wilt pathogen *P. stewartii* subsp. *stewartii*. Upon introduction into the plant by the insect vector, the pathogen must create an environment suitable for growth without eliciting the host defenses. It is generally thought that the development of localized water-soaking lesions shortly after infection is a requisite for successful in planta establishment. The bacteria cause a disruption of the plant cell membrane, which results in the leakage of cellular fluids to generate a favorable environment for the pathogen without invoking a defense response by the plant. The water-soaking phenotype derives in part from the activity of the functions encoded by the *hrp* (formerly called *wts*) locus of *P. stewartii* subsp. *stewartii* (Coplin et al., 1992). The *hrp* cluster of *P. stewartii* subsp. *stewartii* is related to the well-characterized *hrp* system of *Erwinia amylovora*. Unlike *E. amylovora*, *P. stewartii* subsp. *stewartii* does not elicit a natural hypersensitive response, presumably because of the inefficient expression of *hrpN*, which encodes a harpin elicitor. In addition, there is preliminary genetic evidence that EsaR has a subtle repressive effect on the expression of certain *hrp*-encoded genes (Coplin and Beck von Bodman, unpublished data). Perhaps *P. stewartii* subsp. *stewartii* has managed to fine-tune the selective expression of *hrp*-encoded functions to favor those required for cell wall damage and water-soaking, while keeping functions that may elicit plant defense responses in check. One envisions that the development of water-soaking lesions follows a direct interaction between pathogen and plant cell. It is reasonable, therefore, to assume that the synthesis of an EPS capsule during the early stages of infection might interfere with a direct contact between pathogen and plant cell. In addition, the synthesis of EPS is a highly energy-consuming activity and might well restrict the growth of the pathogen as it attempts to establish itself in planta. The delayed systemic infection of *P. stewartii* subsp.

stewartii requires that the bacterium migrate from the point of infection to the vascular system of the plant. Perhaps this migration is triggered and even facilitated by the production of the capsular EPS once the population has reached a specific cell density. Alternatively, EPS synthesis may occur primarily in the xylem tissue, where it may serve to capture plant vascular fluids to generate a plentiful nutritional environment. Taken together, these considerations suggest that a primary purpose of quorum sensing in *P. stewartii* subsp. *stewartii* may be to delay the onset of capsular EPS synthesis until the infecting population has had a chance to grow and multiply in planta, and not before the presence of an EPS capsule serves a beneficial role.

A rationale for acyl-HSL-mediated regulation in plant-beneficial bacteria is not as obvious. In regard to the nitrogen-fixing symbiont *R. leguminosarum* bv. *viciae*, the bacteriocin *small*, which inhibits the growth of closely related *R. leguminosarum* biovars, is structurally an acyl-HSL (Schripsema et al., 1996). Thus, this acyl-HSL signal may be important in competition for nodule occupancy between *Rhizobium* strains in the plant rhizosphere. Additionally, the observation that lack of $\Delta7$-3hydroxy-C14-HSL results in reduced nodulation efficiency in *nodFEL*-defective strains suggests that it may play a role in the interaction between *R. leguminosarum* bv. *viciae* and the host (Cubo et al., 1992; Dibb et al., 1984).

In the beneficial plant-associated bacterium *P. aureofaciens*, the role of C6-HSL in regulating phenazine antibiotic production is even more intriguing. Classically, antibiotic production has been assumed to be utilized as a method of establishing a niche in competition with other microorganisms. However, the demonstration that phenazine production occurs only after sufficient acyl-HSL signal has accumulated would suggest that phenazines play an important role after initial colonization and growth has occurred. One possibility is that phenazines are utilized to maintain the bacterial population in a previously established site, or that they may play an as yet unrecognized competitive or physiological role for the producer strain. In addition, since in many cases acyl-HSLs regulate more than one function, it is possible that the PhzI/PhzR system regulates some as yet unknown factor that plays an important role in the life cycle of the bacterium.

SUMMARY AND FUTURE DIRECTIONS

Since the original identification of acyl-HSL-mediated gene regulation in *V. fischeri*, much progress has been made in understanding this extremely important form of gene regulation at the molecular level. Through the work primarily on *V. fischeri* and *Vibrio harveyi*, and more recently *Pseudomonas aeruginosa* and *A. tumefaciens*, we have gained fundamental information regarding how the diffusible signals may be synthesized and how they may interact with the LuxR protein, resulting in changes in gene expression patterns. However, much basic work remains to be done. This need for continued research is further underscored by the recent identification of analogous regulatory systems in a wide range of plant-associated and other bacteria. Since bacteria appear to utilize this regulatory scheme in different ways, it is essential that many of the basic hypotheses worked out in other systems be confirmed in these new systems. It is also becoming increasingly obvious that acyl-HSL regulatory schemes are intimately linked to other regulatory cascades in numerous intricate ways (Pierson et al., 1997, 1998). Thus, continued research on many systems will be required to fully understand how bacteria sense and respond to their biotic environment and to place this knowledge within an overall ecological framework.

ACKNOWLEDGMENTS

We thank Arun Chatterjee, Mike Daniels, Tim Denny, Allan Downie, Kendall Gray, Al Poplawsky, and Simon Swift, who provided us with information prior to publication and/or for reviewing parts of the manuscript prior to submission. In addition, we thank E. Pierson and M. Hawes for critical review of the manuscript. Some of the work presented from the lead author's lab was supported by NSF grant No. 9514074.

REFERENCES

Arlat, M., C. L. Gough, C. E. Barber, C. Boucher, and M. J. Daniels. 1991. *Xanthomonas campestris* contains a cluster of *hrp* genes related to the larger *hrp* cluster of *Pseudomonas solanacearum*. *Mol. Plant-Microbe Interact.* **8**:593–601.

Bainton, N. J., P. Stead, S. R. Chhabra, B. W. Bycroft, G. P. C. Salmond, G. S. A. B. Stewart, and P. Williams. 1992. *N*-(3-Oxohexanoyl)-L-homoserine lactone regulates carbapenem antibiotic production in *Erwinia carotovora*. *Biochem. J.* **288**:997–1004.

Barber, C. E., J. L. Tang, J. X. Feng, M. Q. Pan, T. J. G. Wilson, H. Slater, J. M. Dow, P. Williams, and M. J. Daniels. 1997. A novel regulatory system required for pathogenicity of *Xanthomonas campestris* is mediated by a small diffusible signal molecule. *Mol. Microbiol.* **24**:555–566.

Barras, F., F. van Gijsegem, and A. K. Chatterjee. 1994. Extracellular enzymes and pathogenesis of soft-rot *Erwinia*. *Annu. Rev. Phytopathol.* **32**:201–234.

Beck von Bodman, S. Unpublished results.

Beck von Bodman, S., and S. K. Farrand. 1995. Capsular polysaccharide biosynthesis and pathogenicity in *Erwinia stewartii* require induction by a *N*-acyl-homoserine lactone autoinducer. *J. Bacteriol.* **177**:5000–5008.

Beck von Bodman, S., D. R. Majerczak, and D. L. Coplin. A negative regulator mediates quorum-sensing control of exopolysaccharide production in *Pantoea stewartii* subsp. *stewartii*. *Proc. Natl. Acad. Sci. USA* **95**:7687–7692.

Boettcher, K. J., and E. G. Ruby. 1995. Detection and quantification of *Vibrio fischeri* autoinducer from symbiotic squid light organs. *J. Bacteriol.* **177**:1053–1059.

Bradshaw-Rouse, J. J., M. A. Whatley, D. L. Coplin, A. Woods, L. Sequeira, and A. Kelman. 1981. Agglutination of *Erwinia stewartii* strains with a corn agglutinin: correlation with extracellular polysaccharide production and pathogenicity. *Appl. Environ. Microbiol.* **42**:344–350.

Braun, E. J. 1982. Ultrastructural investigation of resistant and susceptible maize inbreds infected with *Erwinia stewartii*. *Phytopathology* **72**:159–166.

Brumbley, S. M., and T. P. Denny. 1990. Cloning of *phcA* from wild-type *Pseudomonas solanacearum*, a gene that when mutated alters expression of multiple traits that contribute to virulence. *J. Bacteriol.* **172**:5677–5685.

Brumbley, S. M., B. F. Carney, and T. P. Denny. 1993. Phenotype conversion in *Pseudomonas solanacearum* due to spontaneous inactivation of PhcA, a putative LysR transcriptional activator. *J. Bacteriol.* **175:**5477–5487.

Buddenhagen, I. W. 1960. Strains of *Pseudomonas solanacearum* in indigenous hosts in banana plantations in Costa Rica, and their relationship to bacterial wilt of bananas. *Phytopathology* **50:**660–664.

Chatterjee, A. Personal communication.

Chatterjee, A., Y. Cui, Y. Liu, C. K. Dumenyo, and A. K. Chatterjee. 1995. Inactivation of *rsmA* leads to overproduction of extracellular pectinases, cellulases, and proteases in *Erwinia carotovora* subsp. *carotovora* in the absence of the starvation/cell density-sensing signal, N-(3-oxohexanoyl)-L-homoserine lactone. *Appl. Environ. Microbiol.* **61:** 1959–1967.

Chhabra, S. R., P. Stead, N. J. Bainton, G. P. C. Salmond, G. S. A. B. Stewart, P. Williams, and B. W. Bycroft. 1993. Autoregulation of carbapenem biosynthesis in *Erwinia carotovora* ATCC 39048 by analogues of N-3-(oxohexanoyl)-L-homoserine lactone. *J. Antibiot.* **46:**441–454.

Chun, W., J. Cui, and A. Poplawsky. 1998. Purification, characterization and biological role of a pheromone produced by *Xanthomonas campestris* pv. *campestris. Physiol. Mol. Plant Pathol.* **51:**1–14.

Clough, S. J., M. A. Schell, and T. P. Denny. 1994. Evidence for involvement of a volatile extracellular factor in *Pseudomonas solanacearum* virulence gene expression. *Mol. Plant-Microbe Interact.* **7:**621–630.

Clough, S. J., K.-E. Lee, M. A. Schell, and T. P. Denny. 1997. A two-component system in *Ralstonia* (*Pseudomonas*) *solanacearum* modulates production of *phcA*-regulated virulence factors in response to 3-hydroxypalmitic acid methyl ester. *J. Bacteriol.* **179:**3639–3648.

Coplin, D. L. and S. Beck von Bodman. Unpublished data.

Coplin, D. L., R. D. Frederick, D. R. Majerczak, and L. D. Tuttle. 1992. Characterization of a gene cluster that specifies pathogenicity in *Erwinia stewartii. Mol. Plant-Microbe Interact.* **5:**81–88.

Crockford, A. J., G. A. Davis, and H. D. Williams. 1995. Evidence for cell-density-dependent regulation of catalase activity in *Rhizobium leguminosarum* bv. *phaseoli. Microbiology* **141:**843–851.

Cubo, M. T., A. Economou, G. Murphy, A. W. B. Johnston, and J. A. Downie. 1992. Molecular characterization and regulation of the rhizosphere-expressed genes *rhiABCR* that can influence nodulation by *Rhizobium leguminosarum* biovar *viciae. J. Bacteriol.* **174:**4026–4035.

Cui, J., J. K. Fellman, A. R. Poplawsky, and W. Chun. 1995. Partial purification of a putative pheromone from *Xanthomonas campestris* pv. *campestris. Phytopathology* **85:**1148.

Cui, Y., A. Chatterjee, Y. Liu, C. K. Dumenyo, and A. K. Chatterjee. 1995. Identification of a global repressor gene, *rsmA*, of *Erwinia carotovora* subsp. *carotovora* that controls extracellular enzyme N-(3-oxohexanoyl)-L-homoserine lactones, and pathogenicity in soft-rotting *Erwinia* spp. *J. Bacteriol.* **177:**5108–5115.

Daniels, M. Personal communication.

Dibb, N. J., J. A. Downie, and N. J. Brewin. 1984. Identification of a rhizosphere protein encoded by the symbiotic plasmid of *Rhizobium leguminosarum. J. Bacteriol.* **158:**621–627.

Eberhard, A. 1972. Inhibition and activation of bacterial luciferase synthesis. *J. Bacteriol.* **109:**1101–1105.

Engebrecht, J., K. Nealson, and M. Silverman. 1983. Bacterial bioluminescence: isolation and genetic analysis of functions from *Vibrio fischeri. Cell* **32:**773–781.

Flavier, A. B. 1997. *Endogenous Signal Molecules and Gene Regulation in Ralstonia solanacearum.* Ph.D. thesis. University of Georgia, Athens.

Flavier, A. B., S. J. Clough, M. A. Schell, and T. P. Denny. 1997. Identification of 3-hydroxypalmitic acid methyl ester as a novel autoregulator controlling virulence in *Ralstonia solanacearum. Mol. Microbiol.* **26:**251–259.

Fulda, M., E. Heinz, and E. F. P. Wolter. 1994. The *fadD* gene of *E. coli* K-12 is located close to *rnd* at 39-6 min of the chromosomal map and is a new member of the AMP-binding protein family. *Mol. Gen. Genet.* **242:**241–249.

Fuqua, W. C., S. C. Winans, and E. P. Greenberg. 1994. Quorum sensing in bacteria: the LuxR-LuxI family of cell density-responsive transcriptional regulators. *J. Bacteriol.* **176:**269–275.

Fuqua, C., S. C. Winans, and E. P. Greenberg. 1996. Census and consensus in bacterial ecosystems: The LuxR-LuxI family of quorum sensing transcriptional regulators. *Annu. Rev. Microbiol.* **50:** 727–751.

Gray, K. M. 1996. Identification of multiple autoinducer signals in *Rhizobium leguminosarum.* Presented as poster X15 at the 8th International Congress on Molecular Plant-Microbe Interactions, Knoxville, Tenn.

Gray, K. M., J. P. Pearson, J. A. Downie, B. E. A. Boboye, and E. P. Greenberg. 1996. Cell-to-cell signaling in the symbiotic nitrogen fixing bacterium *Rhizobium leguminosarum*: autoinduction of a stationary phase and rhizosphere-expressed genes. *J. Bacteriol.* **178:**372–376.

Hayward, A. C. 1995. *Pseudomonas solanacearum*, p. 139–151. *In Pathogenesis and Host Specificity in Plant*

Diseases: Histopathological, Biochemical, Genetic and Molecular Basis. Elsevier Science Publishing, Inc., New York.

Heikinheimo, R. 1994. Ph.D. thesis. Swedish University of Agricultural Sciences, Uppsala, Sweden.

Hirsch, P. R. 1979. Plasmid-determined bacteriocin production by *Rhizobium leguminosarum. J. Gen. Microbiol.* **113**:219–228.

Hwang, I. Y., D. M. Cook, and S. K. Farrand. 1995. A new regulatory element modulates homoserine lactone mediated autoinduction of Ti plasmid conjugal transfer. *J. Bacteriol.* **177**:449–458.

Jones, S., B. Yu, N. J. Bainton, M. Birdsall, B. W. Bycroft, S. R. Chhabra, A. J. R. Cox, P. Golby, P. J. Reeves, S. Stephens, M. K. Winson, G. P. C. Salmond, G. S. A. B. Stewart, and P. Williams. 1993. The *lux* autoinducer regulates the production of exoenzyme virulence determinates in *Erwinia carotovora* and *Pseudomonas aeruginosa. EMBO J.* **12**:2477–2482.

Lithgow, J., and A. Downie. Personal communication.

Liu, M. Y., and T. Romeo. 1997. The global regulator CsrA of *Escherichia coli* is a specific mRNA-binding protein. *J. Bacteriol.* **179**:4639–4642.

Liu, Y., H. Murata, A. Chatterjee, and A. K. Chatterjee. 1993. Characterization of a novel regulatory gene *aepA* that controls extracellular enzyme production in the phytopathogenic bacterium *Erwinia carotovora* subsp. *carotovora. Mol. Plant-Microbe Interact.* **6**:299–308.

Margolin, W., D. Bramhill, and S. R. Long. 1995. The *dnaA* gene of *Rhizobium meliloti* lies within an unusual gene arrangement. *J. Bacteriol.* **177**:2892–2900.

Mazzola, M., R. J. Cook, L. S. Thomashow, D. M. Weller, and L. S. Pierson III. 1992. Contribution of phenazine antibiotic biosynthesis to the ecological competence of fluorescent pseudomonads in soil habitats. *Appl. Environ. Microbiol.* **58**:2616–2624.

McGowan, S., M. Sebaihia, S. Jones, S. Yu, N. Bainton, P. F. Chan, B. W. Bycroft, G. S. A. B. Stewart, G. P. C. Salmond, and P. Williams. 1995. Carbapenem antibiotic production in *Erwinia carotovora* is regulated by CarR, a homologue of the LuxR transcriptional activator. *Microbiology* **141**:541–550.

Mukherjee, A., Y. Cui, Y. Liu, C. K. Dumenyo, and A. K. Chatterjee. 1996. Global regulation in *Erwinia* species by *Erwinia carotovora rsmA*, a homologue of *Escherichia coli csrA* repression of secondary metabolites, pathogenicity and hypersensitive reaction. *Microbiology* **142**:427–434.

Murata, H., A. Chatterjee, Y. Liu, and A. K. Chatterjee. 1994. Regulation of the production of extracellular pectinase, cellulase, and protease in the soft rot bacterium *Erwinia carotovora* subsp. *carotovora*: evidence that *aepH* of *E. carotovora* subsp. *carotovora* strain 71 activates gene expression in *E. carotovora* subsp. *carotovora, E. carotovora* subsp. *atroseptica*, and *Escherichia coli. Appl. Environ. Microbiol.* **60**:3150–3159.

Nealson, K. H., T. Platt, and J. W. Hastings. 1970. Cellular control of the synthesis and activity of the bacterial luminescent system. *J. Bacteriol.* **104**:313–322.

Onsando, J. M. 1992. *Black Rot of Crucifers*. Prentice-Hall, Englewood Cliffs, N.J.

Parkinson, J. S. 1993. Signal transduction schemes of bacteria. *Cell* **73**:857–871.

Pierson, E. A., D. W. Wood, J. A. Cannon, F. M. Blachere, and L. S. Pierson III. 1998. Interpopulation signaling via *N*-acyl-homoserine lactones among bacteria in the wheat rhizosphere. *Mol. Plant-Microbe Interact.*, vol. 11, in press.

Pierson, L. S., III, and E. A. Pierson. 1996. Phenazine antibiotic production in *Pseudomonas aureofaciens*: role in rhizosphere ecology and pathogen suppression. *FEMS Microbiol. Lett.* **136**:101–108.

Pierson, L. S., III, and L. S. Thomashow. 1992. Cloning and heterologous expression of the phenazine biosynthetic locus form *Pseudomonas aureofaciens* 30-84. *Mol. Plant-Microbe Interact.* **5**:330–339.

Pierson, L. S., III, V. D. Keppenne, and D. W. Wood. 1994. Phenazine antibiotic biosynthesis in *Pseudomonas aureofaciens* 30-84 is regulated by PhzR in response to cell density. *J. Bacteriol.* **176**:3966–3974.

Pierson, L. S., III, T. Gaffney, S. Lam, and F. Gong. 1995. Molecular analysis of genes encoding phenazine biosynthesis in the biological control bacterium *Pseudomonas aureofaciens* 30-84. *FEMS Microbiol. Lett.* **134**:299–307.

Pierson, L. S., III, D. W. Wood, and S. T. Chancey. 1996. Phenazine antibiotic biosynthesis in the biological control bacterium *Pseudomonas aureofaciens* 30-84 is regulated at multiple levels, p. 463–468. *In* G. Stacey, B. Mullin, and P. M. Gresshoff (ed.), *Biology of Plant-Microbe Interactions.* IS-MPMI Press, St. Paul, Minn.

Pierson, L. S., III, E. A. Pierson, D. W. Wood, S. T. Chancey, and D. E. Harvey. 1997. Recent advances in the genetic regulation of the activity of plant growth-promoting rhizobacteria, p. 94–101. *In* O. Ogoshi, K. Kobayashi, Y. Homma, F. Kodano, N. Kondo, and S. Akino (ed.), *Plant Growth-Promoting Rhizobacteria: Present Status and Future Prospects.* OECD Press, Sapporo, Japan.

Pierson, L. S., III, D. W. Wood, E. A. Pierson, and S. T. Chancey. 1998. *N*-acyl-homoserine lactone-mediated gene regulation in biological control by fluorescent pseudomonads: current knowledge and future work. *Eur. J. Plant Pathol.* **104**:1–9.

Piper, K. R., S. Beck von Bodman, and S. K. Farrand. 1993. Conjugation factor of *Agrobacterium tumefaciens* regulates Ti plasmid transfer by autoinduction. *Nature* **362**:448–450.

Pirhonen, M., D. Flego, R. Heikinheimo, and E. T. Palva. 1993. A small diffusible signal molecule is responsible for the global control of virulence and exoenzyme production in the plant pathogen *Erwinia carotovora. EMBO J.* **12**:2467–2476.

Poplawsky, A. R. Personal communication.

Poplawsky, A. R., and W. Chun. 1995. A *Xanthomonas campestris* pv. *campestris* mutant negative for production of a diffusible signal is also impaired in epiphytic development. *Phytopathology* **85**:1148.

Poplawsky, A. R., and W. Chun. 1997. *pigB* determines a diffusible factor needed for extracellular polysaccharide slime and xanthomonadin production in *Xanthomonas campestris* pv. *campestris. J. Bacteriol.* **179**:439–444.

Romeo, T., M. Gong, M. Y. Liu, and A.-M. Brun-Zinkernagel. 1993. Identification and molecular characterization of *csrA*, a pleiotrophic gene from *Escherichia coli* that affects glycogen biosynthesis, gluconeogenesis, cell size, and surface properties. *J. Bacteriol.* **175**:4744–4755.

Rosemeyer, V., J. Michiels, C. Verreth, and J. Vanderleyden. 1998. *luxI-* and *luxR-*homologous genes of *Rhizobium etli* CNPAF512 contribute to synthesis of autoinducer molecules and nodulation of *Phaseolus vulgaris. J. Bacteriol.* **180**:815–821.

Salmond, G. P. C., B. W. Bycroft, G. S. A. B. Stewart, and P. Williams. 1995. The bacterial "enigma": cracking the code of cell-cell communication. *Mol. Microbiol.* **16**:615–624.

Schell, M. A. 1996. To be or not to be: how *Pseudomonas solanacearum* decides whether or not to express virulence genes. *Eur. J. Plant Pathol.* **102**:459–469.

Schripsema, J., K. E. E. de Rudder, T. B. van Vliet, P. P. Lankhorst, E. de Vroom, J. W. Kijne, and A. A. N. van Brussel. 1996. Bacteriocin *small* of *Rhizobium leguminosarum* belongs to the class of *N*-acyl-L-homoserine lactone molecules, known as autoinducers and as quorum sensing co-transcription factors. *J. Bacteriol.* **178**:366–371.

Shaw, P. D., G. Ping, S. L. Daly, C. Cha, J. E. Cronan, Jr., K. L. Rinehart, and S. K. Farrand. 1997. Detecting and characterizing *N*-acyl-homoserine lactone signal molecules by thin-layer chromatography. *Proc. Natl. Acad. Sci. USA* **94**:6036–6041.

Slater, H., C. Barber, A. Alvarez-Morales, M. Dow, and M. Daniels. 1997. The regulation of pathogenicity gene expression in *Xanthomonas campestris* mediated by a small diffusible signal molecule. *In* Proceedings of the Conference on Molecular Genetics of Bacteria and Phage, Madison, Wis.

Swift, S., M. K. Winson, P. F. Chan, N. J. Bainton, M. Birstall, P. J. Reeves, D. E. D. Rees, S. R. Chhabra, P. J. Hill, B. W. Bycroft, G. P. C. Salmond, P. Williams, and G. S. A. B. Stewart. 1993. A novel strategy for the isolation of luxI homologues, evidence for the widespread distribution of a LuxR, LuxI superfamily in enteric bacteria. *Mol. Microbiol.* **10**:511–520.

Swift, S., G. S. A. B. Stewart, and P. Williams. 1996. The inner workings of a quorum sensing signal generator. *Trends Microbiol.* **4**:463–465.

Torres-Cabassa, A., S. Gottesman, R. D. Frederick, P. J. Dolph, and D. L. Coplin. 1987. Control of extracellular polysaccharide synthesis in *Erwinia stewartii* and *Escherichia coli* K-12: a common regulatory function. *J. Bacteriol.* **169**:4525–4531.

van Rhijn, P., and J. Vanderleyden. 1995. The *Rhizobium*-plant symbiosis. *Microbiol. Rev.* **59**:124–142.

Weller, D. M. 1988. Biological control of soilborne plant pathogens in the rhizosphere with bacteria. *Annu. Rev. Phytopathol.* **26**:379–407.

Wijffelman, C. A., E. Pees, A. A. N. van Brussel, and P. J. J. Hooykaas. 1983. Repression of small bacteriocin excretion in *Rhizobium leguminosarum* and *Rhizobium trifolii* by transmissible plasmids. *Mol. Gen. Genet.* **192**:171–176.

Wood, D. W., and L. S. Pierson III. 1996. The *phzI* gene of *Pseudomonas aureofaciens* 30-84 is responsible for the production of a diffusible signal required for phenazine antibiotic production. *Gene* **168**:49–53.

Wood, D. W., F. Gong, M. M. Daykin, P. Williams, and L. S. Pierson III. 1997. *N*-acyl-homoserine lactone-mediated regulation of phenazine gene expression by *Pseudomonas aureofaciens* 30-84 in the wheat rhizosphere. *J. Bacteriol.* **179**:7663–7670.

Zhang, L., P. J. Murphy, A. Kerr, and M. E. Tate. 1993. *Agrobacterium* conjugation and gene regulation by *N*-acyl-L-homoserine lactones. *Nature* **362**:446–448.

CELL DENSITY-DEPENDENT GENE EXPRESSION BY *AGROBACTERIUM TUMEFACIENS* DURING COLONIZATION OF CROWN GALL TUMORS

Stephen C. Winans, Jun Zhu, and Margret I. Moré

8

Agrobacterium tumefaciens is widely known for its ability to transfer oncogenic DNA from its tumor-inducing (Ti) plasmid to the nuclei of infected plant cells, resulting in the formation of crown gall neoplasms. It is perhaps less widely appreciated that the DNA transfer system that mediates this trans-kingdom genetic exchange is only one of two DNA transfer systems found on these megaplasmids. In addition to the approximately 20 *vir* genes required for plant transformation, Ti plasmids contain a complete set of *tra* genes that are responsible for the interbacterial conjugal transfer of the entire Ti plasmid (Alt-Mörbe et al., 1996; Farrand et al., 1996). These two genetic exchange systems, though homologous, are functionally quite distinct (Cook et al., 1997) (Fig. 1). Conjugal transfer of Ti plasmids was the subject of another review (Farrand, 1993).

Conjugation played a central role in establishing the importance of the Ti plasmid in tumorigenesis. It was demonstrated in 1971 that the ability to incite tumorigenesis was readily and permanently lost when culturing

agrobacteria at high temperatures, suggesting that this property might be encoded epigenetically (Hamilton and Fall, 1971). Conclusive proof that tumorigenesis required a plasmid or lysogenic bacteriophage came when Kerr and colleagues demonstrated that the tumorigenesis phenotype could be transferred to a nontumorigenic strain by superinfection of a tumor that contained virulent bacteria (Kerr, 1969, 1971). Physical evidence for a tumor-inducing plasmid was obtained shortly thereafter (van Larebeke et al., 1974). Subsequently, a portion of this plasmid (T-DNA) was found in the genome of axenic plant tumors (Chilton et al., 1977).

As described above, early conjugation experiments were carried out by superinfecting a crown gall tumor that contained conjugal donor bacteria with a conjugal recipient. Conjugation ex planta was virtually undetectable, suggesting that crown gall tumors provided a factor essential for transfer. About the same time, it was demonstrated that crown gall tumors release compounds called opines that serve as nutrient sources for the colonizing bacteria (see Dessaux et al., 1992, for a review of these compounds). Opines are imported into the bacterial cells and catabolized using specific permeases and catabolic enzymes, both generally encoded on the Ti plas-

Stephen C. Winans, Jun Zhu, and Margret I. Moré, Section of Microbiology, Cornell University, Ithaca, NY 14853.

Cell-Cell Signaling in Bacteria, Edited by Gary M. Dunny and Stephen C. Winans
©1999 American Society for Microbiology, Washington, D.C.

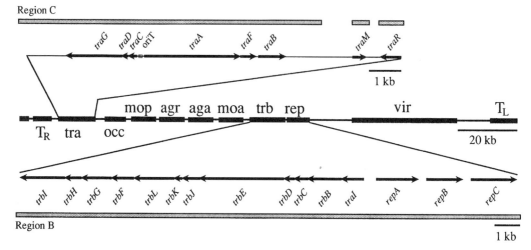

FIGURE 1 Genetic map of the *tra* and *trb* clusters of octopine-type Ti plasmids. *traR*, *traI*, and *traM* are described in the text. The remaining *tra* genes are thought to be required for conjugal DNA processing, while *trb* genes are thought to encode a conjugal pore. Other Ti plasmid loci include T_L and T_R (DNA fragments that are transferred to plant nuclei), *occ* (octopine catabolism), *mop* (mannopine catabolism), *agr* (agropine catabolism), *aga* (agropinic acid catabolism), *moa* (mannopinic acid catabolism), *rep* (vegetative replication), and *vir* (required for T_L-DNA and T_R-DNA transfer).

mid. Kerr and colleagues therefore attempted to detect ex planta conjugation in the presence of opines and found that Ti plasmid-containing strains reproducibly transferred this plasmid ex planta at relatively high efficiencies under these conditions (Kerr et al., 1977; see also Genetello et al., 1977). For octopine-type Ti plasmids such as pTi15955, pTiA6, pTiR10, and pTiAch5, conjugation was stimulated by octopine (the product of a reductive condensation between arginine and pyruvate), while conjugation of nopaline-type Ti plasmids such as pTiC58 required agrocinopines A or B (two related sugar phosphodiester opines).

When Ti plasmid conjugation was first described, it was speculated that the same proteins might be required for T-DNA transfer to plant nuclei. However, subsequent studies showed that the two transfer systems are functionally independent (Cook et al., 1997). DNA sequence analyses indicates that both *vir* and *tra* systems are chimeric. Most components of the *tra* system resemble IncP-type *tra* genes, while one *tra* gene resembles an

RSF1010 *mob* gene at its 5′ half and resembles an IncF *tra* gene at its 3′ half. Likewise, all 11 genes in the *virB* operon resemble IncN *tra* genes, while two genes in the *virD* operon resemble IncP-type *tra* genes (Alt-Mörbe et al., 1996; Farrand et al., 1996; Pohlman et al., 1994). In all cases, sequence similarities between *tra* genes and *vir* genes are comparatively weak. Ti plasmid *tra* genes also closely resemble a cluster of *tra* genes of the pSym plasmid of *Rhizobium* strain NGR234 (Freiberg et al., 1997).

DISCOVERY OF N-ACYL-HOMOSERINE LACTONES AS CONJUGAL PHEROMONES

In 1991, Kerr and colleagues made the surprising discovery that *A. tumefaciens* strains containing octopine-type Ti plasmids produced a diffusible compound that stimulated Ti plasmid conjugation when cultured in the presence of octopine (Zhang and Kerr, 1991). Experiments were carried out comparing two classes of Ti plasmids. One class, denoted transfer-efficient (Tra[e]) plasmids, transferred at high levels in the presence of low concentrations of octopine. The other class, denoted transfer-inefficient (Tra^ie) plasmids, transferred only after extended culturing in the presence of high concentrations of octopine, and even then, transferred their Ti plasmids at 10- to 100-fold lower efficiencies than Tra^e strains. Strains containing either class of plasmid, when cultured in the presence of octopine, produced a low-molecular-weight compound, originally called conjugation factor (CF), that dramatically stimulated the transfer of Tra^ie plasmids. Production of this factor could be stimulated by octopine or by other imino acids, including nopaline, mannityl opines, proline, and several other compounds. Conjugal donors treated with CF still required octopine for transfer; therefore, CF was described as potentiating conjugation in the presence of octopine. CF was active only if provided to conjugal donors, since exposing conjugal recipients to CF did not elevate conjugation efficiency. It is not yet understood why some Ti plasmids transfer more efficiently than others.

Purification and chemical analysis of CF required a bioassay that was simpler and more reproducible than measuring conjugation efficiencies. Such a bioassay was made available by Farrand and colleagues, who discovered that CF could increase the transcription of a *tra-lacZ* transcriptional fusion (Piper et al., 1993). Armed with this assay, Kerr, Tate, and colleagues purified CF from a strain containing an octopine-type Ti plasmid (Zhang et al., 1993). Two related bioactive compounds were described, 3-oxo-C8-homoserine lactone (3-oxo-C8-HSL) and 3-oxo-C6-HSL. Two additional compounds were also described, one having a molecular mass identical to that of 3-oxo-C6-HSL, and another having a molecular mass and other properties identical to an acyclic methyl ester of 3-oxo-C8-HSL. Strains containing a conjugation-constitutive nopaline-type Ti plasmid were subsequently shown also to produce 3-oxo-C8-HSL, although production of related compounds was not described (Hwang et al., 1994). The structural similarity between CF and the autoinducers of *Vibrio fischeri* and *Erwinia caratovora* led to speculation that the production and recognition of the conjugal pheromone might be similar to that of other autoinducers.

IDENTIFICATION OF TraR, TraI, AND TraM IN A NOPALINE-TYPE Ti PLASMID

In the course of transposon mutagenesis of a region required for conjugal transfer, a Tn5 insertion was identified that derepressed conjugation (Beck von Bodman et al., 1989). The constitutive conjugation phenotype of this Tn5 insertion was dominant in merodiploid strains, indicating that the transposon caused a gain of function rather than a loss of function. Sequence analysis indicated that this Tn5 was located 238 nucleotides upstream of a gene denoted *traR*, which showed sequence similarity to *luxR* of *V. fischeri* (Piper et al., 1993). This and other genetic analysis suggested that an outward reading promoter in the transposon was probably responsible for overexpression of *traR*. An *A. tumefaciens* Ti plasmidless strain containing a plasmid that expresses TraR and a second plasmid with a *tra-lacZ* fusion expressed elevated levels of β-galactosidase in the presence of exogenous crude 3-oxo-C8-HSL, suggesting the TraR was a receptor for this autoinducer. Exogenous 3-oxo-C6-HSL also caused induction of this fusion, though not as efficiently as 3-oxo-C8-HSL (Piper et al., 1993).

The fact that *traR* resembled *luxR* and 3-oxo-C8-HSL resembled the *V. fischeri* autoinducer suggested that a *luxI*-type gene probably directed production of the pheromone, and genetic analysis suggested that this gene was located on the Ti plasmid. Such a gene, denoted *traI*, was identified unlinked from *traR* (Hwang et al., 1994). A plasmid containing *traI* synthesized 3-oxo-C8-HSL, although at low levels. When a compatible plasmid expressing TraR was introduced into this strain, the resulting strain produced higher levels of this autoinducer, suggesting that TraR positively regulated *traI*. This was confirmed by constructing a *traI-lacZ* fusion and demonstrating that β-galactosidase production required expression of TraR and exogenous autoinducer. This indicates that, analogous to many other quorum-sensing systems, *traI* is positively regulated by TraR, leading to positive autoregulation of pheromone production. Genetic evidence indicated that *traI* was the first gene in an operon of genes required for conjugation (Hwang et al., 1994).

It was subsequently shown that quorum-dependent expression of *tra* genes involves at least one additional gene, denoted *traM* (Hwang et al., 1995). Expression of *traM* on a multicopy plasmid abolished *tra* gene expression in a strain expressing wild-type levels of TraR. Providing high levels of 3-oxo-C8-HSL did not stimulate *tra* gene expression. However, *tra* gene expression was restored by overexpressing TraR, suggesting that TraM and TraR may interact stoichiometrically. A null mutation in *traM* dramatically increased conjugation and *tra* gene expression. Expression of *traM* was positively regulated by TraR and 3-oxo-C8-HSL, indicating that this quorum-sensing regulon induces its own antagonist and suggesting that TraM may be acting as a governor to limit expression of this regulon.

THE *traR*, *traI*, AND *traM* GENES OF OCTOPINE-TYPE Ti PLASMID

The discovery of quorum-dependent *tra* gene expression proceeded in our laboratory contemporaneously with the discoveries described above, but by a completely different route. In an effort to identify genes that are induced by the opine octopine, several thousand random Tn*5gusA7* insertions were screened for octopine-inducible patterns of gene expression, and several dozen strongly induced fusions were obtained. Most of these insertions were later found to lie in genes that are required for the complete catabolism of octopine (Cho et al., 1996). Since it was known that octopine stimulates Ti plasmid conjugation, all inducible insertions were screened for a defect in conjugation. Surprisingly, none of the insertions was defective, but even more surprisingly, one insertion showed dramatic elevation of conjugation (Fuqua and Winans, 1994). We learned later that this insertion disrupted *traM*, explaining the elevated conjugation efficiency of this mutant (Fuqua et al., 1995).

The insertion in *traM* ultimately led to the discovery of the *traR* gene of this plasmid, in part due to the fact that these genes are adjacent to one another. It was first shown that a multicopy plasmid containing the *traM*::Tn*5gusA7* insertion and flanking DNA elevated both conjugation frequencies and *tra* gene expression. Subcloning this plasmid ultimately indicated that neither increase was due to the *traM*::Tn*5gusA7* insertion. Rather, both increases were due to multicopy expression of the adjacent *traR* gene. The octopine-type *traR* gene was sequenced and found to be 81% identical to that of the nopaline-type Ti plasmid (Fuqua and Winans, 1994).

In the same study, overlapping cosmid clones were screened for a gene that directed the production of 3-oxo-C8-HSL. Several such cosmids were found, and *traI* was ultimately mapped to a position directly upstream of a cluster of *trb* genes (Fuqua and Winans, 1994). The *traI* genes of the two Ti plasmids are 88% identical and are located in identical positions with respect to *trb* genes and *rep* genes, which direct the vegetative replication of the two Ti plasmids. A *traI*::Tn*5gusA7* insertion was deficient in production of 3-oxo-C8-HSL and was also deficient in conjugation. A multicopy plasmid containing *traI* restored 3-oxo-C8-HSL production but did not restore conjugation, indicating that *traI* is the first gene in the *trb* operon (Fuqua and Winans, 1994). TraI was subsequently purified as a His$_6$-tagged fusion protein, and its activities were reconstituted in a purified biochemical assay (Moré et al., 1996). The substrates in this reaction were demonstrated to be 3-oxo-C8-acyl carrier protein and S-adenosyl-methionine (see chapter 14 of this volume).

Although the discovery of *traM* preceded that of *traR* or *traI*, additional experiments were required to understand the hyperconjugal phenotype of the *traM*::Tn*5gusA7* mutation. First, the mutation was found to cause a loss of function, since providing a wild-type copy in *trans* reduced conjugation to wild-type levels (Fuqua et al., 1995). Second, although this fusion was originally identified as octopine-inducible, octopine was found to act only indirectly, and TraR and 3-oxo-octanoyl-HSL are the direct inducers, as is true of the *traM* gene of the nopaline-type Ti plasmid. Third, like the TraM protein of the nopaline-type Ti plasmid, the decrease in conjugation caused by TraM overexpression could be countered by TraR overexpression, but unlike the nopaline system, the decrease could also be countered by providing high levels of exogenous 3-oxo-C8-HSL (Fuqua et al., 1995). A model of quorum-dependent expression of the *tra* regulon is shown in Fig. 2.

CONTROL OF *traR* EXPRESSION BY OPINES

The discovery of the TraR-TraI-TraM regulatory system helps to explain older data showing that opines are required for efficient

conjugation (see above). The *traR* gene of octopine-type Ti plasmids is located at the 3' end of a 14-kb operon that is induced by octopine and positively regulated by the OccR transcriptional regulator (Fuqua and Winans, 1996b) (Fig. 3A). This fully accounts for the octopine requirement, since strains that artificially overexpress *traR* do not require octopine or OccR for efficient conjugation. This *traR* gene is also expressed from a second promoter that lies midway within this operon, just upstream of the *msh* gene. This promoter is positively regulated by TraR and 3-oxo-C8-HSL, creating a positive autoregulatory loop (Fuqua and Winans, 1996b).

Some years ago, it was discovered that sulfur-containing amino acids strongly inhibit conjugation of octopine-type Ti plasmids (Hooykaas et al., 1979). We have found that these amino acids also inhibit *tra* gene expression in strains that express *traR* from its native promoter. However, inhibition was not detected in a strain in which *traR* is expressed from an artificial promoter, suggesting that these amino acids act in wild-type strains by decreasing expression of *traR*. This was con-

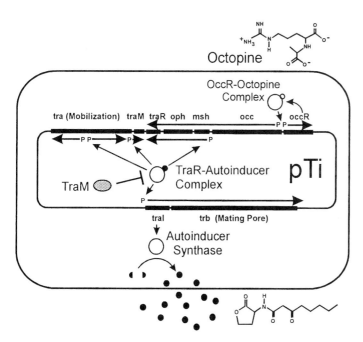

FIGURE 2 A model describing quorum-dependent regulation of octopine-type Ti plasmid *tra* and *trb* genes. Octopine-OccR complexes elevate transcription of *traR*. TraR-autoinducer complexes cause a further elevation in *traR* and *traI* expression, as well as inducing other *tra* and *trb* genes, including *traM*, which encodes a TraR antagonist.

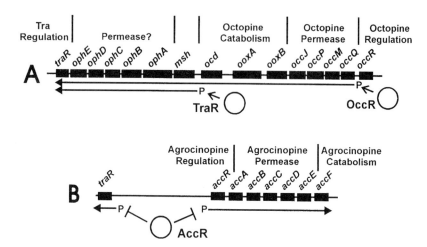

FIGURE 3 Opine-responsive expression of the *traR* genes of octopine-type and nopaline-type Ti plasmids. (A) Positive regulation of the octopine-type *traR* gene by OccR and by TraR. (B) Negative regulation of the nopaline-type *traR* gene by AccR.

firmed by showing that cysteine and methionine block the expression of a *traR-lacZ* fusion introduced into the Ti plasmid (Zhu and Winans, unpublished data).

The *traR* gene of the nopaline-type Ti plasmid pTiC58, like that of octopine-type Ti plasmids, is induced in response to opines. The so-called conjugal opine for this plasmid is not nopaline, but rather agrocinopines A and B. Either of these related opines induces a set of agrocinopine uptake and catabolic genes and the set of *tra* genes, and a null mutation in the *accR* repressor causes both regulons to be expressed constitutively (Beck von Bodman et al., 1992). In contrast, overexpression of *traR* causes constitutive *tra* gene expression but does not affect regulation of agrocinopine catabolic genes. The most probable explanation is that AccR directly represses *traR* and that agrocinopines A and B antagonize this repression (Fig. 3B), although direct evidence of this has not been reported. Assuming that this model is correct, it would appear that control of *traR* expression by opines has evolved independently at least twice, since the AccR repressor is not homologous to the OccR activator, and the

genes upstream of each *traR* gene are not homologous. Positive autoregulation of *traR* by the TraR protein has not been reported in the nopaline-type Ti plasmid.

cis-ACTING SITES FOUND IN TraR-DEPENDENT PROMOTERS

Five TraR-dependent promoters have been identified in the octopine-type Ti plasmid, those expressing the *traC*, *traA*, *traI*, *traM*, and *msh* genes, and the start site of each promoter was identified by primer extension or RNase protection assays (Fuqua and Winans, 1996a,b; Fuqua, unpublished data). The *traC*, *traA*, and *traI* promoters contain sequences (centered at approximately −42 nucleotides) denoted *tra* boxes that resemble each other and also resemble several putative binding sites for LuxR of *V. fischeri* or LasR of *Pseudomonas aeruginosa* (Fig. 4A). These sites show a marked dyad symmetry, suggesting that they could provide binding sites for a TraR dimer. The *traC* and *traA* promoters are of special interest, since these promoters are arranged divergently and share a single *tra* box (Fig. 4B). The position of this *tra* box suggests that RNA polymerase bound to *traC* could contact one monomer of

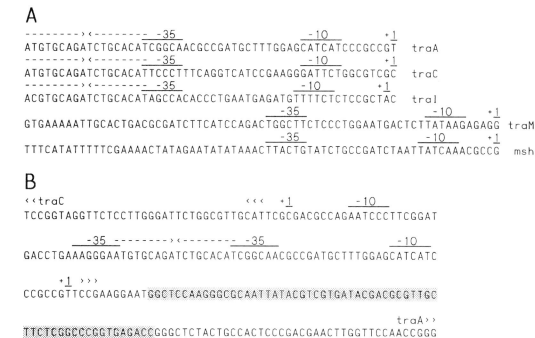

FIGURE 4 Putative TraR binding sites upstream of five TraR-regulated promoters. (A) Alignment of the *traA*, *traC*, *traI*, *traM*, and *msh* promoters of the octopine-type Ti plasmid. Putative TraR binding sites are shown as dashed arrows. (B) Sequence of divergent *traC* and *traA* promoters, which share a single *tra* box. A region similar to the *oriT* of the nopaline-type Ti plasmid is indicated in gray, directly downstream from *traA*.

a TraR dimer, while a second RNA polymerase bound to *traA* could make identical contacts to the other monomer of the same dimer. If so, these contacts would probably be made sequentially rather than simultaneously. The origin of conjugal transfer (*oriT*) is located immediately downstream of the *traA* promoter (Fig. 4B). *oriT* almost certainly is a binding site for the TraA protein, which is homologous to an *oriT*-specific endonuclease of plasmid RSF1010. If so, it seems plausible that TraA might negatively autoregulate its own transcription, as has been reported in another conjugation system (Moncalian et al., 1997). Two TraR-dependent promoters, those upstream of *traM* and *msh*, do not have canonical *tra* boxes. This observation may help to explain why the *msh* promoter is expressed at extremely low levels compared to all the other TraR-dependent promoters.

A DOMINANT DEFECTIVE *traR*-LIKE GENE FOUND ON THE OCTOPINE-TYPE Ti PLASMID

Octopine-type Ti plasmids cause tumors that synthesize no fewer than eight opines, four of the octopine class (octopine, octopinic acid, lysopine, and histopine) and four of the mannopine class (mannopine, agropine, mannopinic acid, and agropinic acid). In the course of analyzing the mannopine catabolic region of one such plasmid (pTi15955), Farrand and colleagues discovered that this plasmid contains a *traR*-like gene (Oger et al., 1998). This gene would encode a truncated protein that contains the amino-terminal autoinducer binding domain (which corresponds to a region of LuxR that is thought to mediate dimerization) but lacks the DNA binding domain, owing to a frameshift mutation between these domains. Sequence inspection suggested that this gene, denoted *trlR*, might be coregulated with the mannopine permease, which is induced by mannopine. This was confirmed by constructing a *trlR-lacZ* fusion and demonstrating mannopine-inducible production of β-galactosidase (Oger et al., 1998).

In an earlier study in the laboratory of E. P. Greenberg, truncated LuxR proteins lacking their carboxyl-terminal domains were shown not only to be defective, but to have a dominant defective phenotype, that is, to be able to interfere with the functions of the full-length protein (Choi and Greenberg, 1992; see chapter 15 of this volume). This was taken as evidence that the amino-terminal domain mediates LuxR multimerization, such that overexpression of the amino-terminal domain results in accumulation of inactive heteromultimers and a decrease in active LuxR homodimers. Using this reasoning, it was postulated that TrlR might interfere with TraR in *tra* gene activation. It was shown that mannopine inhibited conjugation and *tra* gene expression in a wild-type strain, but did not affect either in a strain having an insertion mutation that was polar upon *trlR*. As predicted, artificial overproduction of TrlR inhibited conjugation and *tra* gene expression. TrlR did not affect TraR-independent expression of the *traR* gene. However, TraR was reported to activate expression of *trlR*. Finally, it was found that most but not all Ti plasmids contain more than one *traR*-like gene. In a separate study, two copies of *traR* were described in a Ti plasmid of *Agrobacterium vitis* (Fournier et al., 1994).

In a parallel study, it was found that pTiR10, pTiA6NC, and pTiB6 all contain genes identical to *trlR* (denoted *traS* in that study), indicating that this gene is not a laboratory artifact but rather is widely disseminated (Zhu and Winans, 1998). This study reached generally similar conclusions to the study described above, although TraR-dependent expression of *trlR* was not detected in pTiR10, possibly reflecting a difference between these plasmids. The frameshift mutation in *trlR* was corrected by site-directed mutagenesis, and the resulting gene encoded a fully functional protein. Expression of *trlR* was strongly inhibited by favored catabolites, including succinate, glutamine, and tryptone, and these catabolites thereby restored *tra* gene expression, pheromone production, and conjugation (Zhu and Winans, 1998). It was speculated that this could help to explain the adaptive significance of the *trlR* gene. Since TrlR would not accumulate when the bacteria have ample sources of energy and carbon, this could ensure that the energetically expensive process of conjugation occurs only when donor bacteria are satiated for carbon and energy. Figure 5 summarizes a model for the activities of this protein.

Directly downstream of *trlR* is a gene whose product resembles the output module of methyl-accepting chemoreceptors. Expression of similar fragments of methyl-accepting chemoreceptors strongly impairs chemotaxis and, as predicted, artificial expression of this gene in both *A. tumefaciens* and *Escherichia coli* strongly impaired chemotaxis in swarm plates (Zhu and Winans, 1998).

BIODETECTION OF *N*-ACYL-HSL BY *A. TUMEFACIENS*

Farrand and colleagues constructed a Ti plasmidless strain of *A. tumefaciens* that expresses TraR and contains a *traG-lacZ* fusion, and tested the production of β-galactosidase in response to different autoinducers (Shaw et al., 1997). While this strain detected 3-oxo-C8-HSL more sensitively than other autoinducers, it detected a surprisingly wide range of compounds, including compounds with acyl chains greater than four carbons, and compounds with 3-oxo or 3-hydroxyl substituents, as well as compounds that were unsubstituted at this position. In this study, an elegant reverse-phase thin-layer chromatography system was described that involved the separation of autoinducers followed by detection using the reporter strain immobilized in agar that contained the chromogenic β-galactosidase substrate 5-bromo-4-chloro-3-indolyl-β-D-galactopyranoside (X-Gal). Autoinducers cause the formation of blue spots on

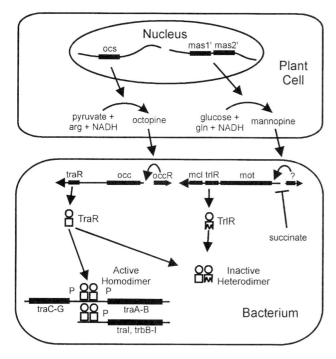

FIGURE 5 Antagonism of TraR function by a dominant defective TraR-like protein. *trlR* synthesizes a truncated protein that closely resembles TraR but that lacks a DNA binding domain. TrlR is able to antagonize TraR function, probably by forming inactive heterodimers. Transcription of *trlR* is induced by mannopine and strongly inhibited by favored catabolites, including succinate, glutamate, and tryptone. *mot* encodes an ABC-type mannopine uptake system, while *mcl* encodes a defective methyl-accepting chemoreceptor-like protein, whose expression inhibits chemotaxis.

these thin-layer chromatography plates. A similar biodetection system has been described using the native pigment produced by *Chromobacter violaceum* (see chapter 19 of this volume).

In a separate study, it was found that an *A. tumefaciens* strain expressing wild-type levels of TraR showed an extremely narrow autoinducer specificity (Zhu et al., 1998). This strain was far more sensitive to 3-oxo-C8-HSL than to any other compound and was detectably stimulated only by compounds that were extremely similar to 3-oxo-C8-HSL. Many other autoinducer analogs showed strongly antagonistic effects. In contrast, a congenic strain that overexpressed TraR was stimulated by an extremely broad spectrum of autoinducers, comparable to the data of Shaw et al. (1997). Furthermore, in this strain none of the autoinducer analogs was a potent antagonist. It was concluded that TraR overproduction drastically broadens its substrate specificity, enhancing its utility in biodetection of autoinducers (McLean et al., 1997; Stickler et al.,

submitted). These data also suggest that autoinducer antagonists may be much easier to identify than previously thought, which could have important clinical implications in fighting pathogens that use quorum-sensing proteins to regulate pathogenesis genes.

PERSPECTIVES AND FUTURE STUDIES

Although a great deal has been learned about this quorum-sensing system, clearly much remains to be done. For example, the mechanism of action of TraM remains poorly understood. Further studies are needed to determine whether TraM acts catalytically or stoichiometrically, and also to determine whether TraM-mediated inhibition is responsive to some environmental stimulus. Although the only TraM homolog currently described is found in a conjugation system in a *Rhizobium* plasmid (Freiberg et al., 1997), it is possible that TraM homologs will be described in other regulatory systems.

Biochemical studies of LuxR-type proteins are currently in their infancy. An important challenge, therefore, will be to reconstitute the activities of a full-length LuxR-type protein in a purified system. If this can be done, it should be possible to understand precisely how autoinducers convert these proteins from inactive forms to active forms. Questions about how these proteins interact with DNA and with RNA polymerase could also be addressed.

Quorum sensors of some other bacteria have been shown to regulate a wide variety of target genes. A third question for future studies therefore concerns the number of genes regulated by TraR. We have used Tn5*gusA7* to screen for genes that are induced by 3-oxo-C8-HSL and have so far identified eight inducible insertions (unpublished data). Of these, seven insertions are located in previously described *tra* or *trb* genes, while one insertion is located on an entirely different replicon. This gene resembles the response regulator component of two-component proteins. It is not known whether expression of this gene requires TraR. Identifying additional quorum-inducible genes may help to provide new insights about how *A. tumefaciens* colonizes plant hosts.

One of the remaining mysteries about the regulation of Ti plasmid *tra* genes is its adaptive value. It is easy to understand why conjugating only in the presence of opines would benefit a conjugal recipient, since a Ti plasmid is required to catabolize opines. However, this does not explain the benefit to the conjugal donor or to the Ti plasmid itself. The conjugal donor should benefit by not transferring the Ti plasmid under these conditions, since conjugation will lead to increased numbers of bacteria that can compete for opines. The Ti plasmid should benefit from horizontal transfer under all conditions. Perhaps the Ti plasmid balances the benefit it derives from horizontal dissemination against the energetic costs that transfer imposes upon the donor, and transfers only under conditions in which the donor can tolerate these costs.

The adaptive significance of quorum-dependent transfer provides an even more perplexing question. Since both *traI* and *traR* are found on the Ti plasmid, conjugal donors both release and perceive this conjugal pheromone and, as expected, high numbers of conjugal donors dramatically increase the conjugation efficiency of each donor (Fuqua and Winans, 1996a). Why do conjugal donors take a census of other conjugal donors rather than of recipients? One possibility is that conjugal donors conjugate with each other. However, Ti plasmid *tra* genes include a gene, *trbK*, that is homologous to a gene of IncP-type plasmids that is required for entry exclusion. Any donor that is synthesizing autoinducers should be expressing *trbK* (since this gene is in the same operon as *traI*) and should therefore be a poor conjugal recipient. A second possibility is that different Ti plasmids may compete for available recipients, but if so, why would each donor signal its presence to competitors? Perhaps the physical act of conjugation requires multiple donors per recipient, such that solitary donors will transfer their plasmids inefficiently. However, this has not been reported of other conjugal systems. One final possibility is that environments containing a quorum of conjugal donors may be likely also to contain large numbers of other bacteria that could serve as conjugal recipients.

REFERENCES

Alt-Mörbe, J., J. L. Stryker, C. Fuqua, S. K. Farrand, and S. C. Winans. 1996. The conjugal transfer system of *A. tumefaciens* octopine-type Ti plasmids is closely related to the transfer system of an IncP plasmid and distantly related to Ti plasmid *vir* genes. *J. Bacteriol.* **178:**4248–4257.

Beck von Bodman, S., J. E. McCutchan, and S. K. Farrand. 1989. Characterization of conjugal transfer functions of *Agrobacterium tumefaciens*. *J. Bacteriol.* **171:**5281–5289.

Beck von Bodman, S., G. T. Hayman, and S. K. Farrand. 1992. Opine catabolism and conjugal transfer of the nopaline Ti plasmid pTiC58 are coordinately regulated by a single repressor. *Proc. Natl. Acad. Sci. USA* **89:**643–647.

Chilton, M.-D., M. H. Drummond, D. J. Merlo, D. Sciaky, A. L. Montoya, M. P. Gordon, and E. W. Nester. 1977. Stable incorporation of plasmid DNA into higher plant cells: the molecular basis of crown gall-tumorigenesis. *Cell* **11:** 263–271.

Cho, K., C. Fuqua, B. S. Martin, and S. C. Winans. 1996. Identification of *Agrobacterium tumefaciens* genes that direct the complete catabolism of octopine. *J. Bacteriol.* **178:**1872–1880.

Choi, S. H., and E. P. Greenberg. 1992. Genetic evidence for multimerization of LuxR, the transcriptional activator of *Vibrio fischeri* luminescence. *Mol. Mar. Biol. Biotechnol.* **1:**408–413.

Cook, D. M., P.-L. Li, F. Ruchaud, S. Padden, and S. K. Farrand. 1997. Ti plasmid conjugation is independent of *vir*: reconstitution of the *tra* functions from pTiC58 as a binary system. *J. Bacteriol.* **179:**1291–1297.

Dessaux, Y., A. Petit, and J. Tempe. 1992. Opines in *Agrobacterium* biology, p. 109–136. *In* D. P. S. Verma (ed.), *Molecular Signals in Plant-Microbe Interactions*. CRC Press, Boca Raton, Fla.

Farrand, S. K. 1993. Conjugal transfer of *Agrobacterium* plasmids. *In* D. B. Clewell (ed.), *Bacterial Conjugation*. Plenum Press, New York.

Farrand, S. K., I. Hwang, and D. M. Cook. 1996. The *tra* region of the nopaline-type Ti plasmid is a chimera with elements related to the transfer systems of RSF1010, RP4, and F. *J. Bacteriol.* **178:**4233–4247.

Fournier, P., P. de Ruffray, and L. Otten. 1994. Natural instability of *Agrobacterium vitis* Ti plasmid due to unusual duplication of a 2.3-kb DNA fragment. *Mol. Plant-Microbe Interact.* **7:**164–172.

Freiberg, C., R. Fellay, A. Bairoch, W. J. Broughton, A. Rosenthal, and X. Perret. 1997. Molecular basis of symbiosis between *Rhizobium* and legumes. *Nature* **387:**394–401.

Fuqua, C. Unpublished data.

Fuqua, C., and S. C. Winans. 1996a. Conserved *cis*-acting promoter elements are required for density-dependent transcription of *Agrobacterium tumefaciens* conjugal transfer genes. *J. Bacteriol.* **178:** 435–440.

Fuqua, C., and S. C. Winans. 1996b. Localization of OccR-activated and TraR-activated promoters that express two ABC-type permeases and the *traR* gene of Ti plasmid pTiR10. *Mol. Microbiol.* **20:** 1199–1210.

Fuqua, C., M. Burbea, and S. C. Winans. 1995. Activity of the *Agrobacterium* Ti plasmid conjugal transfer regulator TraR is inhibited by the product of the *traM* gene. *J. Bacteriol.* **177:**1367–1373.

Fuqua, W. C., and S. C. Winans. 1994. A LuxR-LuxI type regulatory system activates *Agrobacterium* Ti plasmid conjugal transfer in the presence of a plant tumor metabolite. *J. Bacteriol.* **176:**2796–2806.

Genetello, C., N. van Larebeke, M. Holsters, A. De Picker, M. van Montagu, and J. Schell.

1977. Ti plasmids of *Agrobacterium* as conjugative plasmids. *Nature* **265:**561–563.

Hamilton, R. H., and M. Z. Fall. 1971. The loss of tumor-initiating ability in *Agrobacterium tumefaciens* by incubation at high temperatures. *Experientia* **27:**229–230.

Hooykaas, P. J. J., C. Roobol, and R. A. Schilperoort. 1979. Regulation of the transfer of Ti plasmids of *Agrobacterium tumefaciens*. *J. Gen. Microbiol.* **110:**99–109.

Hwang, I., L. Pei-Li, L. Zhang, K. R. Piper, D. M. Cook, M. E. Tate, and S. K. Farrand. 1994. TraI, a LuxI homolog, in responsible for production of conjugation factor, the Ti plasmid N-acylhomoserine lactone autoinducer. *Proc. Natl. Acad. Sci. USA* **91:**4639–4643.

Hwang, I., D. M. Cook, and S. K. Farrand. 1995. A new regulatory element modulates homoserine lactone-mediated autoinduction of Ti plasmid conjugal transfer. *J. Bacteriol.* **177:**449–458.

Kerr, A. 1969. Transfer of virulence between isolate of *Agrobacterium*. *Nature* **223:**1175–1176.

Kerr, A. 1971. Acquisition of virulence by nonpathogenic isolates of *Agrobacterium radiobacter*. *Physiol. Plant Pathol.* **1:**241–246.

Kerr, A, P. Manigault, and J. Tempe. 1977. Transfer of virulence in vivo and in vitro in *Agrobacterium*. *Nature* **265:**560–561.

McLean, R. J. C., M. Whiteley, D. J. Stickler, and W. C. Fuqua. 1997. Evidence of autoinducer activity in naturally-occurring biofilms. *FEMS Microbiol. Lett.* **154:**259–263.

Moncalian, G., G. Grandoso, M. Llosa, and F. de la Cruz. 1997. *oriT*-processing and regulatory roles of TrwA protein in plasmid R388 conjugation. *J. Mol. Biol.* **270:**188–200.

Moré, M. I., D. L. Finger, J. L. Stryker, C. Fuqua, A. Eberhard, and S. C. Winans. 1996. Enzymatic synthesis of a quorum-sensing autoinducer through use of defined substrates. *Science* **272:**1655–1658.

Oger, P., K.-S. Kim, R. L. Sackett, K. R. Piper, and S. K. Farrand. 1998. Octopine-type Ti plasmids code for a mannopine-inducible dominant-negative allele of *traR*, the quorum-sensing activator that regulates Ti plasmid conjugal transfer. *Mol. Microbiol.* **27:**277–288.

Piper, K. R., S. Beck von Bodman, and S. K. Farrand. 1993. Conjugation factor of *Agrobacterium tumefaciens* regulates Ti plasmid transfer by autoinduction. *Nature* **362:**448–450.

Pohlman, R. F., H. D. Genetti, and S. C. Winans. 1994. Common ancestry between IncN conjugal transfer genes and macromolecular export systems of plant and animal pathogens. *Mol. Microbiol.* **14:**655–668.

Shaw, P. D., G. Ping, S. L. Daly, C. Cha, J. E. Cronan, Jr., K. L. Rinehart, and S. K. Farrand. 1997. Detecting and characterizing N-acyl-homoserine lactone signal molecules by thin-layer chromatography. *Proc. Natl. Acad. Sci. USA* **94:** 6036–6041.

Stickler, D. J., N. S. Morris, R. J. C. McLean, and C. Fuqua. 1998. Biofilms on indwelling urethral catheters produce quorum-sensing signal molecules in situ and in vitro. *Appl. Environ. Microbiol.* **64:**3486–3490.

van Larebeke, N., G. Engler, M. Holsters, S. van den Elsacker, I. Zaenen, R. A. Schilperoort, and J. Schell. 1974. Large plasmid in *Agrobacterium tumefaciens* essential for crown gall-inducing ability. *Nature* **252:**169–170.

Zhang, L, and A. Kerr. 1991. A diffusible compound can enhance conjugal transfer of the Ti plasmid in *Agrobacterium tumefaciens. J. Bacteriol.* **173:**1867–1872.

Zhang, L., P. J. Murphy, A. Kerr, and M. E. Tate. 1993. *Agrobacterium* conjugation and gene regulation by N-acyl-L-homoserine lactones. *Nature* **362:**446–448.

Zhu, J., J. Beaber, M. I. Moré, C. Fuqua, A. Eberhard, and S. C. Winans. 1998. Analogs of the autoinducer 3-oxooctanoyl-homoserine lactone strongly inhibit activity of the TraR protein of *Agrobacterium tumefaciens. J. Bacteriol.* **180:**5398–5405.

Zhu, J., and S. C. Winans. 1998. Activity of the quorum-sensing regulator TraR of *Agrobacterium tumefaciens* is inhibited by a truncated, dominant defective TraR-like protein. *Mol. Microbiol.* **27:** 289–297.

Zhu, J., and S. C. Winans. Unpublished data.

REGULATION OF PATHOGENICITY
IN *STAPHYLOCOCCUS AUREUS*
BY A PEPTIDE-BASED
DENSITY-SENSING SYSTEM

Richard P. Novick

9

Staphylococcus aureus is an extraordinarily versatile pathogen, causing a wide spectrum of infections ranging from small skin abscesses to life-threatening endocarditis, pneumonia, sepsis, and toxinoses such as toxic shock and food poisoning. As is typical of gram-positive bacteria, *S. aureus* depends for its pathogenicity largely on a set of extracellular proteins, and, accordingly, staphylococcal pathogenicity is multifactorial. In most of the staphylococcal disease conditions, with the exception of the toxinoses, no single factor is responsible, and it has generally been difficult to define precisely the role of any given factor. This point is addressed at some length in a review by Projan and Novick (1997). Thus, the standard list of pathogenicity factors (Table 1) is simply an inclusive list of all known exoproteins, some of which are obviously involved in pathogenicity while the role of others is far from clear. The list includes secreted proteins, such as cytotoxins, immunotoxins, and enzymes, and surface proteins, such as protein A, fibrinogen binding protein, and fibronectin binding protein, which are assumed collectively to enable

the organism to adhere to tissue components, to resist phagocytosis and other host defenses, and to attack the local cellular and structural elements of diverse tissues and organs (Arvidson, 1983; Projan and Novick, 1997).

The pathogenicity factors listed in Table 1 are expressed in vitro by *S. aureus* according to a carefully orchestrated temporal program. This in vitro expression program, schematized in Fig. 1, is assumed to be relevant to the development of an infection in vivo, which requires that the various proteins be produced at appropriate times and in appropriate amounts. According to this scheme, expression of these factors begins very early in the exponential phase, at which time the synthesis of protein A (PrA), coagulase (Coa), fibronectin binding proteins (FnbpA and FnbpB), and doubtless other surface proteins is rapidly induced. Later in the exponential phase, in one group of strains, including our standard strain, RN6390B, synthesis of these proteins is sharply down-regulated at the transcriptional level. In a second group of strains, particularly those causing toxic shock syndrome, the synthesis of many surface proteins continues throughout exponential growth, resulting in much higher levels of these proteins than are seen with the first group (Kreiswirth et al., unpublished data). During the postexponential

Richard P. Novick, Skirball Institute of Biomolecular Medicine, New York University Medical Center, 540 First Avenue, New York, NY 10016.

Cell-Cell Signaling in Bacteria, Edited by Gary M. Dunny and Stephen C. Winans
©1999 American Society for Microbiology, Washington, D.C.

TABLE 1 Virulence factors in *S. aureus*

Factor	Pathogenic activity[a]	Regulation agr	Regulation sar	Regulation sae
Superantigens				
Enterotoxins				
A	Food poisoning, TSS	−		
B	Food poisoning, TSS	+		
C	Food poisoning, TSS	+		
D	Food poisoning, TSS	+		
E	Food poisoning, TSS	+		
Toxic shock toxins				
TSST-1	TSS	+	+	
TSST-O	None (?)	+		
Exfoliative toxins				
ETA	Scalded skin syndrome	+		
ETB	Scalded skin syndrome	+		
Cytotoxins				
α-Hemolysin	Hemolysis, necrosis	+	+	+
β-Hemolysin	Hemolysis, necrosis	+	+	+
δ-Hemolysin	Weak hemolysis	+	+	
γ-Hemolysin	Hemolysis, necrosis	+		
Leukocidin	Leukolysis	+		
Enzymes				
Proteases	Spread, nutrition	+		
Nucleases	Spread, nutrition	−		
Lipases	Spread, nutrition	+		
Hyaluronidase	Spread, nutrition	−		
Esterases (e.g., FAME)	Inactivation of toxic fatty acids	+		
Surface factors				
Protein A	Antiphagocytosis	+	+	+
Coagulase	?	+		+
Clumping factors	?	+		+
Fibronectin binding protein	Adhesion	+		

[a] TSS, toxic shock syndrome.

phase, in both groups, transcription of a large set of genes encoding extracellular pathogenicity factors, including cytotoxins, hemolysin, superantigens, and tissue-degrading enzymes, is up-regulated. Synthesis of these factors continues for a relatively short time and is then switched off as cells enter stationary phase (Arvidson, personal communication). It is not known whether specific regulatory factors are responsible for this downshift. Similar in vitro temporal programs have been described for other bacterial pathogens, e.g., *Bordetella pertussis* (Huh and Weiss, 1991; Scarlato et al., 1991).

Several regulatory elements are involved in this temporal program in vitro, of which the best known and probably most important is *agr* (accessory gene regulation). The importance of the *agr* system is supported by in vivo studies showing that *agr* mutants are greatly attenuated in virulence in each of several animal models tested, including mammary infections (Foster et al., 1990), arthritis (Abdelinour et al., 1993), and subcutaneous abscesses in mice (Barg et al., 1992) and endocarditis in rabbits (Cheung et al., 1994). Results such as these obviously support the inference that the above list of products of *agr*-up-regulated

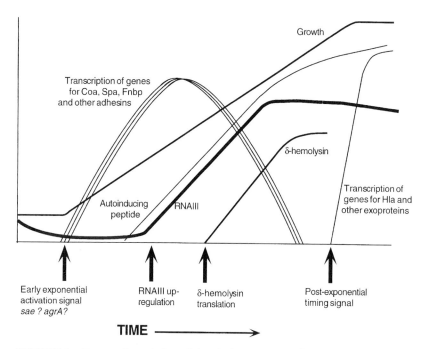

FIGURE 1 Temporal regulation of the virulence response in *S. aureus*. Regulatory events in the virulence response are shown in relation to a standard in vitro growth curve. At the beginning of exponential phase, two or more signals activate synthesis of protein A, fibronectin binding protein, and other surface proteins and of coagulase. During exponential phase, the autoinducing peptide accumulates, eventually reaching a concentration sufficient to activate synthesis of RNAIII, which immediately down-regulates transcription of the surface protein and coagulase genes. δ-Hemolysin, encoded by RNAIII, is not translated until approximately 1 h later. As the cells enter postexponential phase, RNAIII plus an unknown timing signal combine to up-regulate transcription of the hemolysins, exotoxins, and other secreted proteins.

genes includes many that are frank pathogenicity factors; but the reciprocal down-regulation of surface proteins by *agr* is also relevant: although *agr* mutants are greatly attenuated, they are not avirulent; *agr spa* mutants, however, are entirely avirulent in the mouse mammary infection model (Foster et al., 1990), which is particularly noteworthy because protein A is overproduced in *agr* mutants and may compensate in some unknown way for the absence of the other factors.

A few other pleiotropic regulatory genes affecting exoprotein production and virulence have been identified (Cheung et al., 1995; Giraudo et al., 1994a), and several environmental factors, such as subinhibitory antibiotics (Gemmel and Shibl, 1976), increased osmolarity, short-chain alcohols (Foster, personal communication; Arvidson, personal communication), and a specific ester (glycerol monolaurate) (Projan et al., 1994; Schlievert et al., 1992), have been observed to affect the expression of many of these same virulence genes.

We note also that the products listed in Table 1 are all accessory proteins, not needed for normal growth and cell division, that many are encoded by accessory genetic elements such as plasmids, transposons, temperate phages, and pathogenicity islands, and that this is true of bacterial exoproteins in general.

THE *agr* SYSTEM

The idea that staphylococcal virulence factors are coordinately regulated was originally conceived on the basis of mutations that jointly affected their expression (Bjorklind and Arvidson, 1980; Yoshikawa et al., 1974) and was confirmed by the isolation of a chromosomal Tn551 insertion that had this phenotype (Mallonee et al., 1982; Recsei et al., 1986). Cloning of the chromosomal sequences surrounding this insertion (Lofdahl et al., 1988; Peng et al., 1988) led to the identification of a global regulatory locus, agr, that reciprocally regulates the level of expression of surface proteins and secreted proteins. agr is a complex locus that consists of two divergent transcription units, driven by promoters P2 and P3 (Fig. 2). The P2 operon contains four genes, agrB, agrD, agrC, and agrA, all of which are required for transcriptional activation of the agr system (Novick et al., 1995). The primary function of this four-gene unit is to activate the two major agr promoters, P2 and P3 (Kornblum et al., 1990; Lofdahl et al., 1988; Peng et al., 1988), while the actual effector of agr-dependent exoprotein gene regulation is the P3 transcript RNAIII (Janzon and Arvidson, 1990; Kornblum et al., 1990; Novick et al., 1993), which acts primarily at the level of transcription. agr is conserved throughout the staphylococci with interesting variations, especially in the B-D-C region, that have helped to illuminate the structure and function of the system. In Fig. 3 is presented a comparison of the agrB, C, D sequences from several S. aureus strains and from Staphylococcus lugdunensis (Ji et al., 1997; Vandenesch et al., 1993).

The P2 Operon

Examination of the agr sequence suggested that AgrC and AgrA represent the two components of a classical bacterial signal transduction system (Nixon et al., 1986), of which AgrC would correspond to the receptor and AgrA to the response regulator (Kornblum et al., 1990). The predicted transmembrane nature of AgrC suggested that there might be a soluble activator, and, indeed, such an activator was readily demonstrated in culture supernatants (Balaban and Novick, 1995a). Subsequent studies showed that this activator is produced by agr$^+$ but not by agr-null strains,

and purification and sequencing revealed that it is an octapeptide processed from within the 46-amino-acid AgrD product (Ji et al., 1995) and that AgrB is required for processing (Ji et al., 1997). Since the activator is encoded within the locus that it activates, it is an autoinducer and is referred to as the agr autoinducing peptide (AIP). Note that the P2 operon shows two levels of autocatalysis since it encodes its own autoactive response regulator, AgrA, as well as the activator for its own signal receptor, the AIP.

The AIP

A commercially prepared synthetic peptide corresponding to the AIP from RN6390B was dimeric owing to an intermolecular disulfide and had no activity. Monomerization by reduction of the S—S bond, however, did not cause activation of the material, indicating that the dimerization was not responsible for the observed lack of activity and suggesting that the native material was posttranslationally modified in some other way. Mass spectroscopy showed that unlike the synthetic peptide, the native one was largely monomeric and that its molecular mass was about 18 u smaller than that predicted by its sequence (Ji et al., 1995). These results suggested that there was an internal anhydride bond, probably involving the cysteine thiol group. Consistent with this was inactivation by hydroxylamine (which would cleave a thioester) but not by iodoacetic acid (which would react with a free thiol group but not with a thioester) (Ji et al., 1997). Based on the likelihood that any thioester would probably be introduced during processing, we predicted that the acceptor group would be the C-terminal carboxyl. In collaboration with Muir, we have developed a solid-phase method for the synthesis of peptides with an internal thioester ring (Fig. 4). We have prepared two of these to date, the original one described above and a variant produced by S. aureus 502A (Ji et al., 1997) (Fig. 3), and have found that each has the same activity as the native peptide, confirming the structure (Muir et al., unpublished data). It is

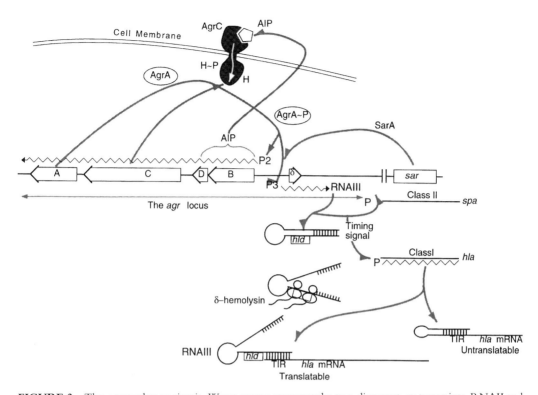

FIGURE 2 The *agr* regulatory circuit. Wavy arrows represent the two divergent *agr* transcripts, RNAII and RNAIII, generated from promoters P2 and P3, respectively. RNAII determines four gene products, AgrA, AgrB, AgrC, and AgrD. AgrA and AgrC are the response regulator and receptor-histidine protein kinase components of a classical signal transduction pathway, and AgrB and AgrD combine to generate an autoinducing peptide (AIP), which is the activating ligand for AgrC. Activated AgrA, plus an accessory transcription factor, SarA, up-regulates both P2 and P3. The latter transcript, RNAIII, is the effector of *agr*-specific regulation and also determines δ-hemolysin. The AIP is shown binding to the extracellular domain of AgrC and inducing a signal (white arrow) that causes autophosphorylation of the conserved histidine in the cytoplasmic domain. This phosphate is transferred to AgrA, giving AgrA~P, which, in conjunction with SarA, activates P2 and P3, setting in motion the autocatalytic *agr* activation circuit. RNAIII immediately blocks transcription of class II exoprotein genes (*spa* and other surface protein genes) and is predicted to fold into an untranslatable configuration. An unknown cellular factor then causes unfolding and permits translation of *hld* and, at the same time, permits the interaction with *hla* mRNA that allows translation of α-hemolysin. In conjunction with the timing signal (see Fig. 6), RNAIII up-regulates transcription of the class I exoprotein genes. It is not clear whether the putative unfolding of RNAIII is required for transcriptional activation of these genes.

proposed that the unique thioester ring structure is a key feature of this ligand-receptor interaction, and studies are in progress to determine its significance.

It is noted that peptides are responsible for the autoinduction of bacteriocin and lantibiotic syntheses by lactobacilli and other gram-positive bacteria (Klein et al., 1993; Nes et al., 1995), and for competence autoinduction in *Streptococcus pneumoniae* (Håvarstein et

al., 1995) and *Bacillus subtilis* (Solomon et al., 1996). All of these serve as ligands for receptors in typical signal transduction pathways; however, none contains an internal thioester ring, nor has any other naturally occurring peptide with such a ring been identified. It is also notable that all autoinducers in gram-negative bacteria are *N*-acyl homoserine lactones, rather than peptides, and, with the exception of the *V. harveyi* autoinducer (Cao

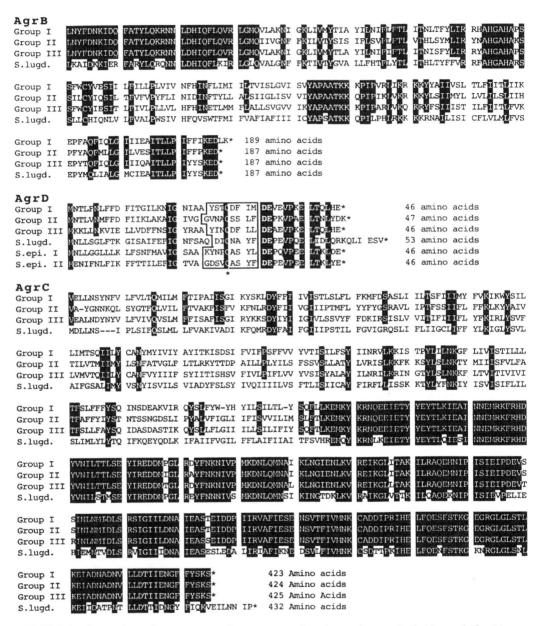

FIGURE 3 Comparison of *agr* sequences. Sequences are aligned according to similarities, and identities are shaded. Sequences were determined in the Novick and Arvidson laboratories, with the exception of the two *Staphylococcus epidermidis agr*Ds, which were determined by Van Wamel et al. (in press). S. epi., *S. epidermidis*; S lugd., *S. lugdunensis*.

et al., 1995), these act intracellularly rather than via signal transduction.

AgrB is a 26-kDa protein that is required for the production of the AIP. Strains express-ing both AgrB and AgrD produce the peptide, but those with mutations or deletions in either do not (Ji et al., 1995). AgrB was predicted by computer and confirmed by PhoA fusions

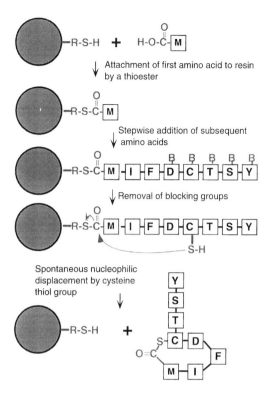

FIGURE 4 Solid-phase synthesis of cyclic thioester-containing octapeptide.

(Ji and Novick, unpublished data) to be a transmembrane protein, and it is proposed that it catalyzes processing and secretion of the AIP in a concerted manner. We have prepared a series of subclones containing either *agrB* or *agrD* or both from the strains with divergent *agr* sequences shown in Fig. 3 and have found that AgrB will process a closely related AgrD but not a distantly related one, indicating that it determines the specificity of processing (Fig. 5) (Ji et al., 1997). We have prepared a His-epitope-tagged derivative of AgrD and have been able to demonstrate that this product is processed in two steps and that AgrB is absolutely required. One step would be a simple peptide cleavage to release the N-terminal amino acid, and the second is likely to involve displacement of the C-terminal peptide bond by the cysteine thiol to generate the cyclic thioester. We have not identified any other

gene product involved in activator production, and we predict that AgrB catalyzes both the processing and secretion of the peptide. This would represent an entirely novel type of protein. AgrB, however, lacks detectable sequence similarity to the combined ATP transporter-cysteine proteases involved in processing and secreting the autoinducing peptides of lactobacilli (Nes et al., 1996).

AgrC is a 46-kDa protein that was predicted by computer and proved by the construction and analysis of PhoA fusions (Lina et al., 1997; Manoil and Beckwith, 1985) to have an N-terminal domain containing six or seven transmembrane helices and a C-terminal cytoplasmic domain containing the conserved histidine common to bacterial signal receptors. The presence of multiple transmembrane helices is typical of signal receptors in gram-positive bacteria (Axelsson and Holck, 1995; Ba–Thein et al., 1996; Diep et al., 1994; Håvarstein et al., 1996; Huhne et al., 1996; Quadri et al., 1997; Solomon et al., 1995), which are activated by peptides, in contrast to those of gram–negative bacteria, which generally have one transmembrane helix, rarely two, and are activated by a variety of signals other than peptides. The significance of this difference remains to be determined. AgrC has been convincingly demonstrated to be the signal receptor for the system: cells expressing AgrC quantitatively remove the AIP from a culture supernatant, whereas cells not expressing AgrC have no effect (Ji et al., 1995); AgrC is phosphorylated in vitro as well as in vivo, probably on its conserved histidine, in response to the AIP (Lina et al., 1997). A fusion between the maltose binding protein of *Escherichia coli* and AgrC at position 139 (near the beginning of the last predicted extracellular loop) retained the AIP-inducible autophosphorylation activity of the native receptor, suggesting that the AIP binding site is in this loop (Lina et al., 1997).

AgrA is a 34-kDa cytoplasmic protein (Peng et al., 1988) with the general features of bacterial response regulators. It is presumably activated by transphosphorylation of its con-

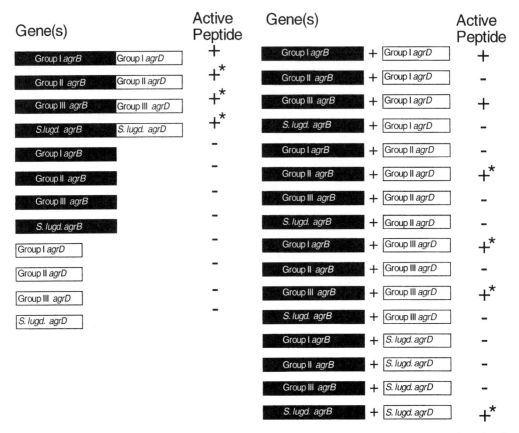

FIGURE 5 Processing specificity. The four *agrB* and *agrD* determinants were cloned separately under β-lactamase promoter control and introduced into strain RN7667 (*agr*-null plus pI524). Culture supernatants were then tested for the presence of activation/inhibition activity. A plus (+) sign indicates that the strain activated RN6390B (group I). A starred plus (+*) sign indicates inhibition of RN6390B. A minus (−) sign indicates that no activation/inhibition activity was detectable. As can be seen, the *agrD* and *agrB* clones alone were inactive; the matching pairs were active whether the genes were on the same plasmid (top four lines) or on separate plasmids, and the combinations of groups I and III were active in both directions. All other combinations were inactive.

served aspartate residue by AgrC∼P. Although the primary function of AgrA is activation of the two major *agr* promoters (Lofdahl et al., 1988), purified AgrA does not detectably bind to the intergenic region between the P2 and P3 transcription initiation sites (Morfeldt et al., 1996), either because only the activated form binds or because some other factor such as SarA is jointly required (Cheung and Projan, 1994). Suggestive evidence has recently been obtained that AgrA may participate in the activation of *coa* transcription (Lebeau et al., 1994).

At this point, we note that the organization of the P2 operon is precisely parallel to that of the *comAP* operon, which autoinduces the state of competence in *B. subtilis* (Magnuson et al., 1994). Autoinducing circuits of this type, in which the autoinducer accumulates during growth until a specific threshold concentration is attained, at which point the system is switched on, are considered to represent cell density-sensing mechanisms and thus to represent a means of intercell communication in a population of bacteria (Fuqua et al., 1996; Kleerebezem et al., 1997). This type of density sensing could be highly useful in the context of an abscess, where the up-regulated genes would be activated when the organisms become crowded and nutritionally deprived. This is parallel in an interesting way to the activation of *V. fischeri* bioluminescence (Eberhard, 1972; Fuqua, 1994 and Winans, 1994), which is the prototypical density-sensing autoinduction system. Here, the bioluminescence response, which occurs in the light organs of various deep-sea fishes and other animals, does not occur until the bacteria have become sufficiently dense to generate the amount of light required by the host organism.

The P3 Operon

The P3 operon encodes one translated product, the 26-amino-acid δ-hemolysin peptide, which has been shown by mutations and deletions to have no role in the regulatory function of the *agr* locus (Janzon and Arvidson, 1990; Novick et al., 1993). It also contains two small potentially translatable reading frames, which have also been shown by mutation and deletion to have no regulatory function. It is therefore clear that the 514 P3 transcript RNAIII is itself the effector of *agr*-specific target gene regulation (Novick et al., 1993). It is certainly likely that RNAIII analogs encoded by other staphylococci have a similar regulatory role; this has yet to be demonstrated.

The primary regulatory function of RNAIII is at the level of transcription, as shown by a series of transcriptional fusions of target gene promoters to the staphylococcal β-lactamase structural gene as a reporter (Novick et al., 1993). The mechanism by which RNAIII regulates transcription has yet to be determined. It is very unlikely that RNAIII interacts directly with target gene DNAs; more likely is an interaction involving one or more regulatory proteins. Two simple possibilities are that RNAIII regulates the translation of proteins that are themselves regulators of target gene transcription, and that RNAIII interacts directly with regulatory pro-

teins and affects their function by inducing allosteric shifts, as is the case with classical apo-regulators. It is also possible that RNAIII regulates the stability of target gene mRNAs. We have obtained indirect evidence that rules out the third possibility in a series of experiments examining the effects of inhibitors of protein synthesis on exoprotein gene transcription. In these experiments, we observed a strong increase in *hla* transcription and a similarly strong inhibition of *spa* transcription immediately following addition of the antibiotic to a growing culture (unpublished data). These effects were equally strong in the absence or in the presence of RNAIII and were specific for the respective promoters. They are consistent with effects of RNAIII on the translation or on the function of labile regulatory proteins. We note that regulatory RNAs interacting directly with regulatory proteins have been described recently in several systems (Sledjeski et al., 1996).

In a separate series of experiments (Fig. 6), we have demonstrated that a second signal, in addition to RNAIII, is necessary for the post-exponential up-regulation of exoprotein expression but is not involved in the down-regulation of surface protein genes, such as *spa*. In these experiments, using a pRN5548 clone in which *rnaIII* transcription is driven by the β-lactamase promoter, we found that exoprotein gene transcription is not up-regulated until the postexponential phase, even when RNAIII transcription is induced as long as 6 h earlier (Vandenesch et al., 1991). *Spa* transcription, however, is always stopped immediately upon RNAIII induction. This timing signal, as well as RNAIII regulation, was overridden by the antibiotics, suggesting that the two signals converge on a labile transcriptional regulatory protein or proteins.

Translational Regulation

RNAIII contains a series of nucleotides in its 5' region complementary to the leader region of *hla* and another series of nucleotides in its 3' region complementary to the 5' region of *spa*. The existence of these complementary se-

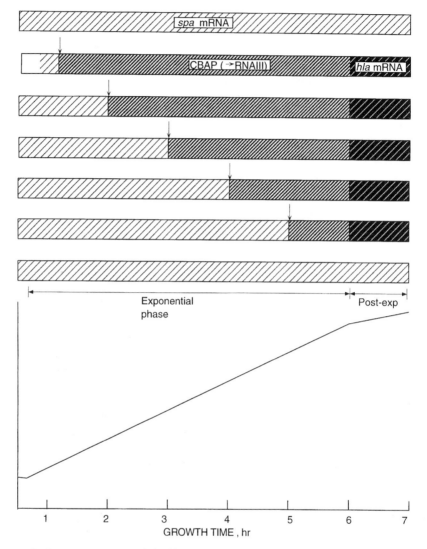

FIGURE 6 Timing signal for *hla* mRNA induction by RNAIII. Arrows represent time of addition of the β-lactamase inducer CBAP to a growing culture of RN6911 (*agr*-null) containing plasmid pRN6848, a pC194-based vector with *rnaIII* cloned behind P~bla~. Light shading represents *spa* transcription, medium shading represents CBAP-induced RNAIII synthesis and the inhibition of *spa* transcription; dark shading represents *hla* transcription. Adapted from Vandenesch et al. (1991).

quences suggested that RNAIII might regulate translation of these proteins as well as transcription of their genes. This was confirmed by deletion analysis of RNAIII: deletion of the 5′ end, containing the sequences complementary to the α-hemolysin mRNA leader, abolished translation of α-hemolysin without affecting transcription of the *hla* gene (Novick et al., 1993), and deletion of the 3′ end, containing the sequences complementary to the *spa* mRNA leader, eliminated RNAIII-specific inhibition of PrA translation (unpublished data). The requirement of the 5′ end of RNAIII for α-hemolysin translation suggested that the *hla* mRNA leader must have a secondary structure that precludes translation and that pairing with the complementary sequences in RNAIII must prevent this inhibitory secondary structure from forming so as to permit translation. This hypothesis has been confirmed by Arvidson and coworkers, who demonstrated the formation of a specific complex between RNAIII and the *hla* leader that preempted the formation of an inhibitory pairing in the *hla* leader that blocked the translation initiation signals (Morfeldt et al., 1995). In the predicted secondary structure of RNAIII, however, the sequences that would unblock *hla* translation are paired and would be unavailable for complementary pairing with the *hla* leader. This pairing would also be expected to block δ-hemolysin translation, and so we suspected that if RNAIII assumed this configuration, it would have to be unfolded to permit δ-hemolysin translation as well as to interact with the *hla* leader. We have observed that δ-hemolysin is not translated until about 1 h after the synthesis of RNAIII, consistent with a requirement for a posttranslational change in its conformation (Balaban and Novick, 1995b). Moreover, deletion of the 3′ half of RNAIII, which would eliminate the inhibitory intramolecular pairing, eliminated this delay. Presumably, such a change in conformation would also be required for the activation of α-hemolysin translation by RNAIII. The agent of this putative posttranslational change in RNAIII conformation, however, is not known. One possibility is translation of the 18 translatable codons 5′ to *hld*, the δ-hemolysin coding sequence. It has recently been observed that a small H-NS analog, StpA, in *E. coli* affects RNA secondary structure and is synthesized for a short period during the exponential phase of growth (Zhang et al., 1995). Such a protein could be responsible for modifying RNAIII conformation so as to permit its translation. In Fig. 2 is a scheme modeling this rather complex regulatory circuit.

THE AUTOINDUCTION-DENSITY-SENSING PARADIGM

Bacterial autoinduction pathways represent a cell-cell communication modality (Fuqua et al., 1996) since the autoinducer accumulates extracellularly and increases in concentration as a function of population density. It is widely accepted that genes regulated by autoinducers are not expressed until a critical population density is reached, at which point, but not earlier, their expression becomes adaptive. The critical threshold population density would then coincide with a critical autoinducer concentration. This line of reasoning is based on the kinetics of autoinducer synthesis by symbiotic *V. fischeri* (Eberhard, 1972), in which a dramatic activation threshold was readily demonstrated in flask cultures and used to support a very compelling rationale for population density control of light production: only at a relatively high population density could the organisms produce enough light to satisfy the needs of its host. However, the activation of bioluminescence has a very clearly linear dose-response curve (Nealson, 1977), as does the *agr* response to the staphylococcal AIP (unpublished data). The threshold here would be the very low autoinducer concentration, which is just sufficient to initiate activation but far below that required for the maximal response. It is suggested, therefore, that there is no need to postulate any critical threshold for these responses because an upwardly accelerating response is inherent in any autoinduction mechanism. This can occur quite rapidly and may thus give the appearance of a critical threshold; indeed, the type of data that can be generated would not easily distinguish between these two possibilities. It would seem that the actual response kinetics will reflect the rate of synthesis of the autoinducer, the molar activity of the autoinducer, and the kinetics of the responding gene expression system. The population density at which rapid activation occurs is likely to vary between wide limits among various organisms and may not necessarily be determined by explicit biological needs. For example, the initiation of sporulation in *B. subtilis* involves the activation of one or more density sensors and consequently is very inefficient at subcritical cell densities. Since sporulation is primarily a protective response to nutritional deprivation or other unfavorable conditions, its density dependence would seem to be counteradaptive.

In the case of *S. aureus*, on the one hand, it could be argued that the density-dependent activation of pathogenicity determinants in the well-developed abscess pocket is clearly adaptive. The bacteria have been walled off by the host defense system, are crowded, and are under nutritional stress. The rapidly achieved high-level expression of secreted pathogenicity factors could enable the bacteria to express these factors while there is still time, enabling them to break out of the pocket and disseminate to initiate new foci of infection. This rationale is not easily applicable to non-localized infections, however, and in vivo studies are greatly needed for any real understanding of the role of autoinducing peptides and of density sensing and other regulatory functions in the disease process. Moreover, although *agr* activation occurs early, during mid-exponential phase, in some strains, including the standard strain, RN6390B, used for most of our studies (Kornblum et al., 1990), it occurs late, during the postexponential phase, in many others (unpublished data). The consequences of late *agr* activation are quite significant: synthesis of RNAIII-repressed proteins, such as PrA, continues at a high level into the postexponential phase, whereas activation of some of the postexponentially expressed genes may be very weak or absent, especially if translation, as well as transcription, is *agr* regulated, as is the case with α-hemolysin (Morfeldt et al., 1995; Novick et al., 1993). The biological adaptivity of this interstrain variation is not obvious; it may be merely a reflection of differences in parameters, such as the basal rate of AIP synthesis or the strength of the AIP-receptor interaction. It will be interesting to determine whether the timing of *agr* expression in *S. aureus* is correlated with differences in the type of infection caused by different strains.

BACTERIAL INTERFERENCE CAUSED BY THE STAPHYLOCOCCAL AUTOINDUCTION SYSTEM

A survey of various staphylococci, including diverse *S. aureus* strains and representatives of other staphylococcal species, revealed, remarkably, that the AIPs from some strains could inhibit the *agr* response in other strains. On the basis of this activity, *S. aureus* strains could be divided into three major groups such that within any one group, each strain produced a peptide that could activate the *agr* response in the others, whereas between groups the autoinducers were mutually inhibitory (Ji et al., 1997).

Genetic analysis (Fig. 5) revealed that the inhibitory and activating activities of any strain were always encoded by a single gene, *agrD*, suggesting that a single peptide was involved in each case. This was confirmed by purification of the autoinducer from several strains; in each case, a single peptide species was obtained that activated the *agr* response by the strain that produced it and others in its group and inhibited the response in strains of the other two groups. Finally, the above-mentioned thioester-containing synthetic autoinducers had the same activating and inhibiting activities as the culture supernatants containing the native peptides (Muir et al., unpublished data).

Two observations suggest that the peptide-determined autoinduction specificity may be biologically relevant. First, *S. aureus* 502A was used in the late 1960s to block umbilical stump colonization of newborn infants by the then-rampant 80/81 complex and was also used successfully to treat furunculosis by displacing the causative strains (Shinefield et al., 1971). In our hands, 80/81 strains usually belong to autoinducer group I, whereas 502A belongs to group II and its autoinducer is a highly potent inhibitor of *agr* activation in group I and III strains. A possible implication is that the 502A peptide was responsible for the observed interference; if so, one might infer that this interference involves inhibition of the expression of *agr*-regulated factors. An obvious problem with this hypothesis is that *agr* down-regulates most or all of the known surface proteins, many of which are adhesins and are generally regarded as colonization factors. However, the actual factors that are responsible for interference are not known, nor has it been demonstrated that the known adhesins correspond to colonization factors in vivo.

A second observation is that the vast majority of menstrual toxic shock strains of *S. aureus* belong to autoinducer group III. These strains have many other features in common, including phage type (Altemeier et al., 1982), electrotype (Musser et al., 1990), and weak expression of most extracellular toxic products (Schlievert et al., 1982), suggesting that they may represent a clone. Group III strains are also overrepresented among non-toxic shock vaginal isolates (Musser, personal communication), suggesting that group III may represent a predilection for the human vaginal mucosa. Whether any such predilection involves *agr*-regulated factors has yet to be determined. Again, this possibility may not be consistent with the *agr* down-regulation of adhesins.

SEQUENCE VARIATION IN THE *agr* LOCUS

As shown in Fig. 3, in which the AgrBs, AgrDs, and AgrCs from representative strains are aligned, the N-terminal third of AgrB and the C-terminal half of AgrC are highly conserved; between these conserved regions is a highly variable region encompassing the C-terminal two-thirds of AgrB, all of AgrD, and the N-terminal half of AgrC (Ji et al., 1997). Note that the dividing line between the variable and conserved regions is very sharp, consistent with a cassette-switching or other site-specific recombinational exchange mechanism, and that this sequence organization is conserved in the two non-*aureus* staphylococci thus far analyzed. We note that a remarkable evolutionary mechanism must underlie this divergence, since the three genes must have diverged in concert so as to maintain the specificity of their processing and activities. This mechanism could involve some sort of hypervariability-generating mechanism in which large numbers of random mutations occur, with selection for the functional combinations.

OTHER REGULATORY GENES AND FACTORS

Although it was once supposed that *agr* was the primary global regulator of virulence factor synthesis in *S. aureus*, studies from a number of laboratories have revealed a variety of other regulatory elements, indicating that the situation is much more complex than originally envisioned. The temporal program of exoprotein gene expression shown in Fig. 1 implies the existence of several regulatory elements in addition to *agr*. Fragmentary observations point to four of these: (i) *sae* is required for the expression of *coa* and may thus correspond to an early-exponential-phase signal; (ii) AgrA may also be involved in *spa* up-regulation at this stage, although it is clearly not required; (iii) an RNA structure modifier, such as the H-NS-like protein StpA (Zhang et al., 1995), is likely to act on RNAIII during mid-exponential phase; (iv) an independent timing signal is necessary for the postexponential up-regulation of exoprotein expression (Fig. 6). There are also a number of environmental factors that act independently of the known regulatory genes. For example, glycerol monolaurate has been found to block the expression of most exoprotein genes independently of *agr* (Projan et al., 1994) and of two different signal transduction-mediated induction pathways, β-lactamase (Projan et al., 1994) and VanA induction (Ruzin and Novick, 1998); NaCl, ethanol and certain antibiotics also inhibit the synthesis of certain exoproteins independently of *agr* (Foster, personal communication; Arvidson, personal communication; Gemmel and Shibl, 1976). Additionally, recent studies have identified a number of genes that are expressed in vivo but not in vitro and are therefore postulated to have a role in virulence. None of these latter genes is regulated by any of the known virulence regulators that have been identified thus far.

In Table 2 are summarized the known genes and environmental factors influencing the expression of staphylococcal virulence factors. Included are some of the genes expressed in vivo. Of particular interest is the *sae* mu-

TABLE 2 Genetic and environmental factors affecting exoprotein expression

Gene or factor	Description	Effect	Reference
sarA	Encodes a regulatory protein of 14 kDa that serves as an accessory transcription factor.	Greatly enhances expression of *agr* P2 and P3 promoters; may also regulate surface protein gene expression.	Cheung et al., 1996; Cheung and Projan, 1994
sae	Transposon-identified gene	Transposon insertion blocks the expression of *coa* as well as that of several *agr*-up-regulated genes, independently of *agr*	Giraudo et al., 1994b, 1995
			Cheung et al., 1995
Glycerol mono-laurate at subinhibitory concentrations	Fatty acid ester probably acting as a mild surfactant	Blocks the expression of most exoprotein genes, including those down-regulated by *agr*, probably by interfering with signal transduction	Projan et al., 1994
Chloramphenicol, tetracycline, clindamycin, at subinhibitory concentrations	Inhibitors of protein synthesis	Block the expression of *hla*, tst, and other *agr*-up-regulated accessory genes	Gemmel and Shibl, 1976; Schlievert and Kelly, 1984
Limitation of tryptophan (or other essential amino acids [?])		Blocks the expression of *hla*, tst, and other *agr*-up-regulated accessory genes without affecting growth	Leboeuf-Trudeau et al., 1969; Sharma and Haque, 1973
1 M NaCl		Blocks the expression of *hla*, *tst*, and other *agr*-up-regulated accessory genes	Foster, personal communication; Arvidson, personal communication
2% Ethanol		Induces the expression of *tst*	Foster, personal communication; Arvidson, personal communication

tation, whose effects are independent of *agr* and which seems to interfere with a regulatory paradigm rather different from that of *agr*: a different set of exoprotein genes is blocked, including *coa*, and genes that are reciprocally regulated by *agr* do not show this reciprocity. Mutants such as this indicate clearly that we still have a great deal to learn about virulence gene regulation in *S. aureus*.

The effects of environmental factors, many of which inhibit exoprotein synthesis at concentrations that do not affect growth, and which act independently of *agr*, suggest that the cell has a specific regulatory modality that selectively switches off expression of accessory genes, such as virulence factors and other exoproteins, when the organism is stressed by nutritional deprivation or inhibitors. The nature of such a switch is unknown. Since 1 M NaCl is rather harmless for staphylococci, it may be that its effects are the result of an osmoregulatory cascade. Clearly, the identification of the regulatory modalities that are responsible for these effects is a major goal in the understanding of bacterial biology.

Although these regulatory modalities have been identified on the basis of their effects on expression of the factors listed in Table 1, it is notable that studies designed to identify genes expressed in vivo but not in vitro have generally not come up with any of the factors on this list, owing to the fact that the latter are all produced in vitro; moreover, the factors that have thus far been identified by in vivo expression would not have been predicted a priori to be important for pathogenicity, nor has any one of them yet been demonstrated to be a pathogenicity factor. It may be true, nevertheless, that in vivo conditions, possibly including specific signals, serve to activate a set of genes that are explicitly necessary for growth in vivo and, therefore, for the causation of disease. Thus, present studies focusing on in vitro analysis of the regulatory modalities affecting the list of intuitive pathogenicity factors must be regarded as seriously incomplete.

REFERENCES

Abdelinour, A., S. Arvidson, T. Bremell, C. Ryden, and A. Tarkowski. 1993. The accessory gene regulator (*agr*) controls *Staphylococcus aureus* virulence in a murine arthritis model. *Infect. Immun.* **61:**3879–3885.

Altemeier, W. A., S. A. Lewis, P. M. Schlievert, M. S. Bergdoll, H. S. Bjornson, J. L. Staneck, and B. A. Crass. 1982. *Staphylococcus aureus* associated with toxic shock syndrome. *Ann. Intern. Med.* **96:**978–982.

Arvidson, S. Personal communication.

Arvidson, S. O. 1983. Extracellular enzymes from *Staphylococcus aureus*, p. 745–808. *In Staphylococci and Staphylococcal Infections*, vol. 2. Academic Press Ltd., London.

Axelsson, L., and A. Holck. 1995. The genes involved in production of and immunity to sakacin A, a bacteriocin from *Lactobacillus sake* Lb706. *J. Bacteriol.* **177:**2125–2137.

Balaban, N., and R. P. Novick. 1995a. Autocrine regulation of toxin synthesis by *Staphylococcus aureus*. *Proc. Natl. Acad. Sci. USA* **92:**1619–1623.

Balaban, N., and R. P. Novick. 1995b. Translation of RNAIII, the *Staphylococcus aureus agr* regulatory RNA molecule, can be activated by a 3'-end deletion. *FEMS Microbiol Lett.* **133:**155–161.

Barg, N., C. Bunce, L. Wheeler, G. Reed, and J. Musser. 1992. Murine model of cutaneous infection with gram-positive cocci. *Infect. Immun.* **60:**2636–2640.

Ba-Thein, W., M. Lyristis, K. Ohtani, I. T. Nisbet, H. Hayashi, J. I. Rood, and T. Shimizu. 1996. The *virR/virS* locus regulates the transcription of genes encoding extracellular toxin production in *Clostridium perfringens*. *J. Bacteriol.* **178:**2514–2520.

Bjorklind, A., and S. Arvidson. 1980. Mutants of *Staphylococcus aureus* affected in the regulation of exoprotein synthesis. *FEMS Microbiol. Lett.* **7:**203–206.

Cao, J. G., Z. Y. Wei, and E. A. Meighen. 1995. The lux autoinducer-receptor interaction in *Vibrio harveyi* binding. *Biochem. J.* **312:**439–444.

Cheung, A. L., and S. J. Projan. 1994. Cloning and sequencing of *sarA* of *Staphylococcus aureus*, a gene required for the expression of *agr*. *J. Bacteriol.* **176:**4168–4172.

Cheung, A. L., K. J. Eberhardt, E. Chung, M. R. Yeaman, P. M. Sullam, M. Ramos, and A. S. Bayer. 1994. Diminished virulence of sar⁻/agr⁻ mutant of *Staphylococcus aureus* in the rabbit model of endocarditis. *J. Clin. Invest.* **94:**1815–1822.

Cheung, A. L., C. Wolz, M. R. Yeaman, and A. S. Bayer. 1995. Insertional inactivation of a chromosomal locus that modulates expression of potential virulence determinants in *Staphylococcus aureus*. *J. Bacteriol.* **177:**3220–3226.

Cheung, A. L., J. H. Heinrichs, and M. G. Bayer. 1996. Characterization of the *sar* locus and its interaction with *agr* in *Staphylococcus aureus*. *J. Bacteriol.* **178:**418–423.

Diep, D. B., L. S. Håvarstein, J. Nissen-Meyer, and I. F. Nes. 1994. The gene encoding plantaricin A, a bacteriocin from *Lactobacillus plantarum* C11, is located on the same transcription unit as *agr*-like regulatory system. *Appl. Environ. Microbiol.* **60:**160–166.

Eberhard, A. 1972. Inhibition and activation of bacterial luciferase synthesis. *J. Bacteriol.* **109:**1101–1105.

Foster, S. Personal communication.

Foster, T. J., M. O'Reilly, P. Phonimdaeng, J. Cooney, A. H. Patel, and A. J. Bramley. 1990. Genetic studies of virulence factors of *Staphylococcus aureus*. Properties of coagulase and gamma-toxin, alpha-toxin, beta-toxin and protein A in the pathogenesis of *S. aureus* infections, p. 403–420. *In* R. P. Novick (ed.), *Molecular Biology of the Staphylococci*. VCH Publishers, New York.

Fuqua, W. C., and S. C. Winans. 1994. A LuxR-LuxI type regulatory system activates *Agrobacterium* Ti plasmid conjugal transfer in the presence of a plant tumor metabolite. *J. Bacteriol.* **176:**2796–2806.

Fuqua, C., S. C. Winans, and E. P. Greenberg. 1996. Census and consensus in bacterial ecosystems: the LuxR-LuxI family of quorum-sensing transcriptional regulators. *Annu. Rev. Microbiol.* **50:**727–751.

Gemmel, C. G., and A. M. A. Shibl. 1976. The control of toxin and enzyme biosynthesis in staphylococci by antibiotics, p. 657–664. *In* J. Jeljasze-wicz (ed.), *Staphylococci and Staphylococcal Diseases.* Gustav Fischer Verlag, Stuttgart.

Giraudo, A., G. Martinez, A. Calzolari, and R. Nagel. 1994a. Characterization of a Tn*925*-induced mutant of *Staphylococcus aureus* altered in exoprotein production. *J. Basic Microbiol.* **34:**317–322.

Giraudo, A., C. Raspanti, A. Calzolari, and R. Nagel. 1994b. Characterization of a Tn*551*-mutant of *Staphylococcus aureus* defective in the production of several exoproteins. *Can. J. Microbiol.* **8:**677–681.

Giraudo, A., H. Rampone, A. Calzolari, and R. Nagel. 1995. Phenotypic characterization and virulence of a *sae agr* mutant of *Staphylococcus aureus*. *Can. J. Microbiol.* **42:**20–123.

Håvarstein, L. S., G. Coomaraswamy, and D. A. Morrison. 1995. An unmodified hepta-decapeptide pheromone induces competence for genetic transformation in streptococcus pneumoniae. *Proc. Natl. Acad. Sci. USA* **92:**11140–11144.

Håvarstein, L. S., P. Gaustad, I. F. Nes, and D. A. Morrison. 1996. Identification of the streptococcal competence-pheromone receptor. *Mol. Microbiol.* **21:**863–869.

Huh, Y. J., and A. A. Weiss. 1991. A 23-kilodalton protein, distinct from BvgA, expressed by virulent *Bordetella pertussis* binds to the promoter region of *vir*-regulated toxin genes. *Infect. Immun.* **59:**2389–2395.

Huhne, K., L. Axelsson, A. Holck, and L. Krockel. 1996. Analysis of the sakacin P gene cluster from *Lactobacillus sake* Lb674 and its expression in sakacin-negative *Lb. sake* strains. *Microbiology* **142:**1437–1448.

Janzon, L. and S. Arvidson. 1990. The role of the delta-lysin gene (*hld*) in the regulation of virulence genes by the accessory gene regulator (*agr*) in *Staphylococcus aureus*. *EMBO J.* **9:**1391–1399.

Ji, G., and R. P. Novick. Unpublished data.

Ji, G., R. Beavis, and R. Novick. 1995. Cell density control of staphylococcal virulence mediated by an octapeptide pheromone. *Proc. Natl. Acad. Sci. USA* **92:**12055–12059.

Ji, G., R. Beavis, and R. P. Novick. 1997. Bacterial interference caused by autoinducing peptide variants. *Science* **276:**2027–2030.

Kleerebezem, M., L. E. Quadri, O. P. Kuipers, and W. M. de Vos. 1997. Quorum sensing by peptide pheromones and two-component signal-transduction systems in Gram-positive bacteria. *Mol. Microbiol.* **24:**895–904.

Klein, C., C. Kaletta, and K. D. Entian. 1993. Biosynthesis of the lantibiotic subtilin is regulated by a histidine kinase/response regulator system. *Appl. Environ. Microbiol.* **59:**296–303.

Kornblum, J., B. Kreiswirth, S. J. Projan, H. Ross, and R. P. Novick. 1990. Agr: a polycistronic locus regulating exoprotein synthesis in *Staphylococcus aureus*, p. 373–402. *In* R. P. Novick (ed.), *Molecular Biology of the Staphylococci*. VCH Publishers, New York.

Kreiswirth, B., H. Ross, and R. P. Novick. Unpublished data.

Lebeau, C., F. Vandenesch, T. Greeland, R. P. Novick, and J. Etienne. 1994. Coagulase expression in *Staphylococcus aureus* is positively and negatively modulated by an *agr*-dependent mechanism. *J. Bacteriol.* **176:**5534–5536.

Leboeuf-Trudeau, T., J. de Repentigny, R. M. Frenette, and S. Sonea. 1969. Tryptophan metabolism and toxin formation in *S. aureus* Wood 46 strain. *Can. J. Microbiol.* **15:**1–7.

Lina, G., S. Jarraud, G. Ji, T. Greenland, A. Pedraza, J. Etienne, R. P. Novick, and F. Vandenesch. 1998. Transmembrane topology and histidine protein kinase activity of AgrC, the *agr* signal receptor in *Staphylococcus aureus*. *Mol. Microbiol.* **28:**655–662.

Lofdahl, S., E. Morfeldt, L. Janzon, and S. Arvidson. 1988. Cloning of a chromosomal locus (*exp*) which regulates the expression of several exoprotein genes in *Staphylococcus aureus*. *Mol. Gen. Genet.* **211:**435–440.

Magnuson, R., J. Solomon, and A. D. Grossman. 1994. Biochemical and genetic characterization of a competence pheromone from B. subtilis. *Cell* **77:**207–216.

Mallonee, D. H., B. A. Glatz, and P. Pattee. 1982. Chromosomal mapping of a gene affecting enterotoxin A production in *Staphylococcus aureus*. *Appl. Environ. Microbiol.* **43:**397–402.

Manoil, C. and J. Beckwith. 1985. Tn phoA: a transposon probe for protein export signals. *Proc. Natl. Acad. Sci. USA* **82:**8129–8133.

Morfeldt, E., D. Taylor, A. von Gabain, and S. Arvidson. 1995. Activation of alpha-toxin translation in *Staphylococcus aureus* by the trans-encoded antisense RNA, RNAIII. *EMBO J.* **14:**4569–4577.

Morfeldt, E., K. Tegmark, and S. Arvidson. 1996. Transcriptional control of the *agr*-dependent virulence gene regulator. *Mol. Microbiol.* **21:**1227–1237.

Muir, T., G. Ji, and R. P. Novick. Unpublished data.

Musser, J. Personal communication.

Musser, J. M., P. M. Schlievert, A. W. Chow, P. Ewan, B. N. Kreiswirth, V. T. Rosdahl, A. S. Naidu, W. Witte, and R. K. Selander. 1990. A single clone of *Staphylococcus aureus* causes the majority of cases of toxic shock syndrome. *Proc. Natl. Acad. Sci. USA* **87:**225–229.

Nealson, K. H. 1977. Autoinduction of bacterial luciferase. Occurrence, mechanism and significance. *Arch. Microbiol.* **112**:73–79.

Nes, I. F., L. S. Håvarstein, and H. Holo. 1995. Genetics of non-lantibiotic bacteriocins. *Dev. Biol. Stand.* **85**:645–651.

Nes, I. F., D. B. Diep, L. S. Håvarstein, M. B. Brurberg, V. Eijsink, and H. Holo. 1996. Biosynthesis of bacteriocins in lactic acid bacteria. *Antonie van Leeuwenhoek* **70**:113–128.

Nixon, B. T., C. W. Ronson, and R. M. Ausubel. 1986. Two-component regulatory systems responsive to environmental stimuli share strongly conserved domains with the nitrogen assimilation regulatory genes ntrB and ntrC. *Proc. Natl. Acad. Sci. USA* **83**:7850–7854.

Novick, R . P., H. F. Ross, S. J. Projan, J. Kornblum, B. Kreiswirth, and S. Moghazeh. 1993. Synthesis of staphylococcal virulence factors is controlled by a regulatory RNA molecule. *EMBO J.* **12**:3967–3975.

Novick, R. P., S. Projan, J. Kornblum, H. Ross, B. Kreiswirth, and S. Moghazeh. 1995. The *agr* P-2 operon: an autocatalytic sensory transduction system in *Staphylococcus aureus*. *Mol. Gen. Genet.* **248**:446–458.

Peng, H.-L., R. P. Novick, B. Kreiswirth, J. Kornblum, and P. Schlievert. 1988. Cloning, characterization and sequencing of an accessory gene regulator (*agr*) in *Staphylococcus aureus*. *J. Bacteriol.* **179**:4365–4372.

Projan, S., and R. Novick. 1997. The molecular basis of virulence, p. 55–81. *In* G. Archer and K. Crossley (ed.), *Staphylococci in Human Disease*. Churchill Livingstone, New York.

Projan, S. J., S. Brown-Skrobot, P. Schlievert, S. L. Moghazeh, and R. P. Novick. 1994. Glycerol monolaurate inhibits the production of β-lactamase, toxic shock syndrome toxin-1 and other staphylococcal exoproteins by interfering with signal transduction. *J. Bacteriol.* **176**:4204–4209.

Quadri, L. E., M. Kleerebezem, O. P. Kuipers, W. M. de Vos, K. L. Roy, J. C. Vederas, and M. E. Stiles. 1997. Characterization of a locus from *Carnobacterium piscicola* LV17B involved in bacteriocin production and immunity: evidence for global inducer-mediated transcriptional regulation. *J. Bacteriol.* **179**:6163–6171.

Recsei, P., B. Kreiswirth, M. O'Reilly, P. Schlievert, A. Gruss, and R. Novick. 1986. Regulation of exoprotein gene expression by *agr*. *Mol. Gen. Genet.* **202**:58–61.

Ruzin, A., and R. Novick. 1998. Glycerol monolaurate inhibits induction of vancomycin resistance in *Enterococcus faecalis*. *J. Bacteriol.* **180**:182–185.

Scarlato, V., B. Arico, A. Prugnola, and R. Rappuoli. 1991. Sequential activation and environmental regulation of virulence genes in *Bordetella pertussis*. *EMBO J.* **10**:3971–3975.

Schlievert, P., M. Osterholm, J. Kelly, and R. Nishimura. 1982. Toxin and enzyme characterization of *Staphylococcus aureus* isolates from patients with and without toxic-shock syndrome. *Ann. Intern. Med.* **96**:937–940.

Schlievert, P. M., and J. A. Kelly. 1984. Clindamycin-induced suppression of toxic-shock syndrome-associated exotoxin production. *J. Infect. Dis.* **149**:471.

Schlievert, P. M., J. R. Deringer, M. H. Kim, S. J. Projan, and R. P. Novick. 1992. Effect of glycerol monolaurate on bacterial growth and toxin function. *Antimicrob. Agents Chemother.* **36**:626–631.

Sharma, B., and R. Haque. 1973. Effect of tryptophan analogues on synthesis of staphylococcal beta-hemolysin. *J. Gen. Microbiol.* **77**:221–224.

Shinefield, H. R., J. C. Ribble, and M. Boris. 1971. Bacterial interference between strains of *Staphylococcus aureus*, 1960-1971. *Am. J. Dis. Child.* **121**:148–152.

Sledjeski, D. D., A. Gupta, and S. Gottesman. 1996. The small RNA, DsrA, is essential for the low temperature expression of RpoS during exponential growth in *Escherichia coli*. *EMBO J.* **15**:3993–4000.

Solomon, J., R. Magnuson, A. Sruvastavam, and A. Grossman. 1995. Convergent sensing pathways mediate response to two extracellular competence factors in *Bacillus subtilis*. *Genes Dev.* **9**:547–558.

Solomon, J., B. Lazazzera, and A. Grossman. 1996. Purification and characterization of an extracellular peptide factor that affects two different developmental pathways in *Bacillus subtilis*. *Genes Dev.* **10**:2014–2024.

Vandenesch, F., J. Kornblum, and R. Novick. 1991. A temporal signal, independent of *agr*, is required for *hla* but not *spa* transcription in *Staphylococcus aureus*. *J. Bacteriol.* **173**:6313–6320.

Vandenesch, F., S. Projan, B. Kreiswirth, J. Etienne, and R. P. Novick. 1993. Agr-related sequences in *Staphylococcus lugdunensis*. *FEMS Microbiol. Lett.* **111**:115–122.

Van Wamel, W. J. B., A. G. A. Welten, J. Verhoef, and A. C. Fluit. 1997. Diversity in the accessory gene regulator (*agr*) locus of *Staphylococcus epidermidis*, abstr. B-33, p. 34. *Abstr. 97th Gen. Meet. Am. Soc. Microbiol. 1997*, American Society for Microbiology, Washington, D.C.

Yoshikawa, M., F. Matsuda, M. Naka, E. Murofushi, and Y. Tsunematsu. 1974. Pleiotropic alteration of activities of several toxins and enzymes in mutants of *Staphylococcus aureus*. *J. Bacteriol.* **119:**117–122.

Zhang, A., V. Derbyshire, J. L. Salvo, and M. Belfort. 1995. *Escherichia coli* protein StpA stimulates self-splicing by promoting RNA assembly in vitro. *RNA* **1:**783–793.

QUORUM SENSING IN
PSEUDOMONAS AERUGINOSA

Everett C. Pesci and Barbara H. Iglewski

10

Pseudomonas aeruginosa is a highly adaptive, nonsporeforming, gram-negative bacterium. This ubiquitous microbe can grow in almost any moist environment at temperatures between 10 and 43°C (Joklik et al. 1992; Brock et al., 1994). Its ability to grow in a multitude of settings indicates that it must constantly monitor its surroundings for changes that require an adaptive response. One particular method used by *P. aeruginosa* to monitor and respond to its environment is the phenomenon of quorum sensing. This chapter will focus on research conducted over the last several years that shows *P. aeruginosa* uses at least two complete, separate quorum-sensing systems to control the expression of multiple genes in response to its own cell density.

DISCOVERY OF THE *P. AERUGINOSA las* QUORUM-SENSING SYSTEM

The first hint that *P. aeruginosa* was controlling gene expression in a cell density-dependent manner came with the report by Whooley et

al. (1983) that exoprotease production occurred in a "growth-associated" manner. The mechanism responsible for this control was uncovered when Gambello and Iglewski (1991) complemented an elastase-deficient *P. aeruginosa* strain. They showed that this elastase-negative strain contained the *lasB* gene, which encodes elastase, but did not produce a *lasB* mRNA. The subsequent search for a positive regulator that would complement the elastase phenotype of this strain led to the discovery of the *lasR* gene (Gambello and Iglewski, 1991).

The *lasR* gene encodes a 239-amino-acid homolog of the prototypical quorum-sensing transcriptional activator LuxR of *Vibrio fischeri* (Gambello and Iglewski, 1991; Fuqua et al., 1996). The LasR protein, like other LuxR homologs, has two highly conserved regions. One is the putative autoinducer-binding region (36% identity to the LuxR autoinducer-binding region), which constitutes the amino two-thirds of the protein, and the other is the putative DNA-binding region (53% identity to the LuxR DNA-binding region) which consists of a helix-turn-helix motif located in the carboxyl third of the protein (Gambello and Iglewski, 1991). The *lasR* gene is transcribed as a monocistronic operon (Gambello and Iglewski, 1991), which is directly fol-

Everett C. Pesci, Department of Microbiology and Immunology, School of Medicine, East Carolina University, Greenville, NC 27858.   *Barbara H. Iglewski*, Department of Microbiology and Immunology, Box 672, Strong Memorial Hospital, University of Rochester, Rochester, NY 14642.

Cell-Cell Signaling in Bacteria, Edited by Gary M. Dunny and Stephen C. Winans
©1999 American Society for Microbiology, Washington, D.C.

lowed on the *P. aeruginosa* chromosome by the *lasI* gene (Passador et al., 1993).

The *lasI* gene encodes LasI, a 201-amino-acid homolog (35% identity, 56% similarity) of the prototypical *V. fischeri* LuxI autoinducer synthase protein (Passador et al., 1993; Fuqua et al., 1996). LasI was shown to direct the synthesis of *N*-(3-oxododecanoyl)-L-homoserine lactone (3-oxo-C12-HSL), which is the autoinducer molecule of the *las* quorum-sensing system (Pearson et al., 1994). 3-oxo-C12-HSL is a typical autoinducer in that it consists of a homoserine lactone ring with an acyl side chain (see chapter 14 of this volume). The discovery of this quorum-sensing system controlling virulence factor production in a human pathogen such as *P. aeruginosa* was very exciting, and with this excitement came a flurry of research designed to better understand the mechanisms behind this regulation.

PROPERTIES OF LasR AND 3-OXO-C12-HSL

After the components of the *las* quorum-sensing system were defined, some of their properties were determined. Deletion analysis of LasR suggested that the 3-oxo-C12-HSL interaction site was within amino acids 3 to 155, and that transcriptional activation by LasR required only the carboxyl-terminal portion of LasR from amino acids 160 to 239 (Passador et al., unpublished data). The ability of LasR to bind to the *lasB* promoter region was demonstrated (You et al., 1996), and it has been shown that 3-oxo-C12-HSL will specifically bind to cells expressing LasR (Passador et al., 1996). Passador et al. (1996) also showed that the length of the acyl side chain was the most important factor that affected the ability of 3-oxo-C12-HSL analogs to activate or bind to LasR. As acyl side-chain length decreased from the normal 12-carbon length of 3-oxo-C12-HSL, the ability to activate or bind to LasR decreased, indicating that the LasR/3-oxo-C12-HSL interaction was one of high specificity. To study specificity between quorum-sensing systems from different species, Gray et al. (1994) performed inter-changeability studies in which different combinations of the *las* and *lux* quorum-sensing systems were reconstituted in *Escherichia coli*. They found that these two systems were specific concerning the autoinducer required to activate LasR or LuxR (3-oxo-C12-HSL or 3-oxo-C6-HSL, respectively), and that some cross reading of promoters could occur (LasR-3-oxo-C12-HSL could effectively activate the *lux* promoter and LuxR-3-oxo-C6-HSL could partially activate the *lasB* promoter).

GENES CONTROLLED BY THE *las* SYSTEM

Together, LasR and 3-oxo-C12-HSL have proved to control a number of *P. aeruginosa* virulence genes (Table 1). After LasR's role in elastase production was discovered, it was shown that this protein was also required for the production of two other *P. aeruginosa* proteases, LasA protease and alkaline protease (Toder et al., 1991; Gambello et al., 1993). Gambello et al. (1993) also found that LasR was required for full expression of exotoxin A. While these studies showed that LasR was involved in the control of the above-mentioned virulence factors, the role of 3-oxo-C12-HSL had not yet been determined. Passador et al. (1993) showed that an unidentified factor (3-oxo-C12-HSL) in the supernatant of *P. aeruginosa* cultures was capable of activating *lasB* expression through LasR. Pearson et al. (1994) then dissected the *las* system to show that LasR and 3-oxo-C12-HSL were required and sufficient to activate *lasB* in *E. coli*. Finally, the last obvious target for LasR control was *lasI*. Seed et al. (1995) showed that LasR and 3-oxo-C12-HSL controlled the transcription of the monocistronic *lasI* operon, indicating that this autoinducer synthase gene was controlled by a positive feedback loop. This suggested that when a threshold concentration of 3-oxo-C12-HSL was reached, the LasR-3-oxo-C12-HSL complex could form and induce the production of more 3-oxo-C12-HSL. Seed et al. (1995) also discovered that 10-fold more 3-oxo-C12-HSL was re-

TABLE 1 *P. aeruginosa* quorum-sensing controlled genes

las system		rhl system	
Gene	Product or function	Gene or product	Product or function
lasI	Putative 3-oxo-C12-HSL synthase	*RhlI*	Putative, C4-HSL synthase
lasB	Elastase	*rhlAB*	Rhamnosyltransferase
rsaL	*lasI* repressor	*lasB*	Elastase
lasA	*LasA* protease	*rpoS*	Stationary-phase sigma factor
apr[a]	Alkaline protease	*apr*[a]	Alkaline protease
toxA	Exotoxin A	Pyocyanin	Phenazine antibiotic
xcpP, R	General secretory pathway proteins		
rhlR	RhlR transcriptional activator protein		
lasR	LasR transcriptional activator protein		

[a] Latifi et al. (1995) presented data that suggested *apr* is controlled by the *rhl* system. The effect of *lasR* on *apr* seen by Gambello et al. (1993) may have been indirect because of the ability of LasR-3-oxo-C12-HSL to activate *rhlR*.

quired for LasR to activate *lasB* than the amount needed to activate *lasI*, indicating that a hierarchy existed with respect to the order of gene activation by LasR-3-oxo-C12-HSL. In this hierarchy, *lasI* could be activated before *lasB*, causing the 3-oxo-C12-HSL concentration to increase so that *lasB* and other LasR-3-oxo-C12-HSL-controlled genes could be activated. Finally, the truly global role of the *las* quorum-sensing system was strengthened with the report of Chapon-Herve et al. (1997) that demonstrated LasR-3-oxo-C12-HSL activated the *xcpP* and *xcpR* genes, which encode proteins of the *P. aeruginosa* general secretory pathway.

CHARACTERIZATION OF THE *P. AERUGINOSA rhl* QUORUM-SENSING SYSTEM

It became apparent that *P. aeruginosa* contained another quorum-sensing system when Ochsner et al. (1994a) discovered a second *P. aeruginosa* LuxR homolog. The gene that encoded this homolog was named *rhlR* for its ability to regulate the production of rhamnolipid, which acts as a biosurfactant/hemolysin. RhlR is 31% identical to LasR and 23% identical to LuxR and contains a typical helix-turn-helix motif in its carboxyl region (Ochsner et al., 1994a). Shortly after this discovery, Pearson et al. (1995) reported that *P.*

aeruginosa produced a second autoinducer molecule, *N*-butyryl-L-homoserine lactone (C4-HSL), that affected *lasB* expression. The components of the *rhl* quorum-sensing system fell into place when Ochsner and Reiser (1995) reported that the *rhlI* gene, which encodes a LuxI homolog, was located directly downstream from *rhlR*. The *rhlI* gene was subsequently shown to direct the synthesis of C4-HSL (Winson et al., 1995).

The *rhl* system has not been studied to the extent that the *las* system has, but it appears to be a typical quorum-sensing system. Pearson et al. (1997) showed that C4-HSL binds specifically to cells expressing RhlR, but unlike the binding of 3-oxo-C12-HSL to LasR-expressing cells (Passador et al., 1996), this binding is significantly enhanced by overexpression of the chaperone GroESL. While the significance of the involvement of GroESL in this indirect binding assay is unknown, binding of C4-HSL to RhlR-expressing cells suggests that the RhlR-C4-HSL complex is responsible for activating genes controlled by the *rhl* system.

The genetic organization of the *rhl* system is relatively similar to that of the *las* system in that *rhlI* is located directly downstream from *rhlR*. However, unlike the *las* system, one of the primary targets of the *rhl* system, *rhlAB*, is located directly upstream from *rhlR*. The *rhlA* and *rhlB* genes, which encode for a rhamnosyltransferase required for rhamnolipid production (Ochsner et al., 1994b), were shown to be cotranscribed on a single mRNA (Pearson et al., 1997) while *rhlR* and *rhlI* are transcribed from their own promoters (Pearson et al., 1997; Brint and Ohman, 1995).

GENES CONTROLLED BY THE *rhl* QUORUM-SENSING SYSTEM

Ochsner et al. (1994a) found that *rhlR* was not only involved in rhamnolipid production, but that it was also important for full elastase activity. The involvement of the *rhl* system in elastase production was reported by others (Brint and Ohman, 1995; Latifi et al., 1995; Ochsner and Reiser, 1995; Pearson et al., 1995; Winson et al., 1995). It was also shown that RhlR and RhlI were required for full production of LasA protease, pyocyanin, and alkaline protease (Brint and Ohman, 1995; Latifi et al., 1995), indicating that the *rhl* system affects multiple virulence products (Table 1). Winson et al. (1995) showed that expression of *rhlI* was positively controlled by RhlR and C4-HSL to create a typical positive autoregulation loop in which RhlR-C4-HSL activated the production of more C4-HSL. Pearson et al. (1997) then reconstituted the *rhl* system in *E. coli* to develop a C4-HSL bioassay that proved RhlR and C4-HSL were required and sufficient to activate *rhlA* transcription. Finally, Latifi et al. (1996) showed that RhlR-C4-HSL could activate *rpoS*, which encodes a stationary-phase sigma factor involved in the regulation of numerous stress-response genes.

INTRACELLULAR COMMUNICATION BETWEEN QUORUM-SENSING SYSTEMS

With the realization that *P. aeruginosa* contained two separate, complete quorum-sensing systems came the potential for communication between these systems. Pearson et al. (1997) showed that the components of the *las* and *rhl* systems were not compatible in that C4-HSL does not activate LasR nor does 3-oxo-C12-HSL activate RhlR in *E. coli*. They also found

that LasR-3-oxo-C12-HSL preferentially activated *lasB* over *rhlA* and that RhlR-C4-HSL preferentially activated *rhlA* over *lasB*. These results indicated that the "R proteins" and autoinducers of the *las* and *rhl* systems were not interchangeable and that there was relatively high specificity with regard to the quorum sensing-controlled promoters that each system could activate. However, it was apparent that these two systems were not completely independent of one another. The first indication that these systems actually were communicating came when Pearson et al. (1995) presented data that suggested LasR may control the production of C4-HSL. The link between the *las* and *rhl* system was finally elucidated when Latifi et al. (1996) showed that LasR-3-oxo-C12-HSL activated the transcription of *rhlR*. This implied that there existed a hierarchy of *P. aeruginosa* quorum sensing in which the *las* system lies upstream of the *rhl* system in a signaling cascade (Latifi et al., 1996). The transcriptional control of *rhlR* expression by the *las* system was confirmed by Pesci et al. (1997), who also found that the *las* system controlled RhlR at yet another level. They showed that 3-oxo-C12-HSL could block C4-HSL from binding to cells expressing RhlR. This result was unexpected because it has been shown that C4-HSL will not block 3-oxo-C12-HSL from binding to cells expressing LasR, and that 3-oxo-C12-HSL will not activate RhlR (Passador et al., 1996; Pearson et al., 1997). This blocking was shown to have a biological effect as 3-oxo-C12-HSL significantly inhibited the ability of RhlR and C4-HSL to activate *rhlA* (Pesci et al., 1997). These data suggested that the *las* system controlled RhlR at the posttranslational level. This meant that the posttranslational control of RhlR by 3-oxo-C12-HSL could occur before *rhlI* was induced, when the concentration of 3-oxo-C12-HSL would be higher than the concentration of C4-HSL. The excess 3-oxo-C12-HSL could block the association of RhlR and C4-HSL until enough RhlR and/or C4-HSL were present to overcome the blocking effect. This would provide *P. aeru-*

ginosa with two mechanisms to ensure that the *rhl* system is activated after the *las* system, implying that only the *las* system directly "senses a quorum." It has been speculated that perhaps multiple levels of RhlR control are required because this protein affects a critical gene(s) (Pesci and Iglewski, 1997). The demonstration that *rpoS* is positively regulated by RhlR–C4-HSL adds greater importance to RhlR regulation and could explain the multiple control levels that exist (Latifi et al., 1996).

REGULATION OF *lasR*

The addition of *rhlR* to the growing list of *lasR*-controlled genes means that *lasR* is at the top of the *P. aeruginosa* quorum-sensing hierarchy. When one considers this, the regulation of *lasR* is obviously of great importance to *P. aeruginosa* quorum sensing. Pesci et al. (1997) found that in *P. aeruginosa*, *lasR* was expressed at a basal level of transcription until becoming activated in the last half of log phase growth. This result was in contrast to that of Latifi et al. (1996) who reported that *lasR* is expressed constitutively in *P. aeruginosa*. Both studies appear to be carefully documented so the reason for the conflict of data is not apparent. Pesci et al. (1997) demonstrated that in *P. aeruginosa*, *lasR* is partially expressed in the absence of 3-oxo-C12-HSL, and mildly up-regulated by 3-oxo-C12-HSL. In similar studies, Latifi et al. (1996) reported that LasR negatively autoregulates *lasR* transcription in *E. coli*. Some of the confusion about *lasR* regulation may have been answered with a report by Albus et al. (1997) on the control of *lasR*. They showed that *lasR* transcription could begin at two different locations that are 30 bp apart. The control of *lasR* by two different promoters that potentially could activate transcription under different conditions might be the reason for the puzzling data obtained during the study of *lasR* regulation. Albus et al. (1997) also showed that *lasR* transcription is positively regulated through a cyclic AMP receptor protein consensus sequence centered 77.5 and 47.5 bp upstream from the two respective *lasR*

transcriptional start sites. Vfr, which is the *P. aeruginosa* homolog of cyclic AMP receptor protein, was required for *lasR* expression and was able to bind to the *lasR* promoter region, indicating that it was directly affecting *lasR* regulation (Albus et al., 1997). Finally, the discovery of a potential *lux* box sequence (discussed below) in the *lasR* promoter region adds another possible mode of regulation for this gene (Albus et al., 1997). When all data are considered, the regulation of *lasR* is obviously multifactorial and will require much future research before being completely understood.

PROMOTERS OF *P. AERUGINOSA* QUORUM SENSING-CONTROLLED GENES

Genes that are controlled by quorum sensing often have a *lux* box consensus sequence close to, or overlapping with, their promoter (Fuqua et al., 1996). Rust et al. (1996) showed that the promoter region of *lasB* contained two such operator sequences that were involved in quorum sensing-mediated regulation of *lasB*. One operator sequence, OP1, was centered 42.5 bp upstream from the transcriptional start site, where it could easily overlap with the −35 region of the *lasB* promoter. The other operator sequence, OP2, was centered 102 bp upstream from the start of transcription. The fact that *lasB* expression is controlled by two quorum-sensing operator sites could be the reason that both the *las* and *rhl* systems affect *lasB* transcription. Further experiments will be required to determine if each operator is affected by a different quorum-sensing system. A number of other *P. aeruginosa* quorum sensing-controlled genes have since been shown to contain a *lasB*-type operator sequence in their promoter region (Fig. 1). Of these putative operators, the sequences upstream from *rhlI*, *rhlA*, *lasA*, *lasI*, *rhlR*, and *lasR* match 17, 14, 11, 11, 10, and 8 nucleotides, respectively, of 20 nucleotides with the *lasB* OP1 sequence. Such sequence similarities allow for interesting speculations; however, whether they have any bearing on

FIGURE 1 Alignment of conserved regions in *P. aeruginosa* quorum sensing-controlled promoters. Putative operators are compared to the OP1 operator of *lasB*, with boxes indicating conserved nucleotides. For each gene except *rhlI* and *rhlR*, the distance between the 3′ end of each potential operator and the start of transcription is indicated on the right. For *rhlI* and *rhlR*, the distance between the 3′ end of the potential operator and the translational start site is given. Transcriptional start sites were mapped by the following: *lasB*, Rust et al., 1996; *rhlA*, Pearson et al., 1997; *lasI*, Seed et al., 1995; *lasA*, Freck-O'Donnell and Darzins, 1993; *lasR*, Albus et al., 1997. The conserved sequence in the *rhlI* promoter region was identified by Latifi et al. (1995). The *rhlR* promoter region was mapped to within 80 bp upstream from the *rhlR* start codon by deletion analysis (Ochsner et al., 1994a).

Gene	Operator Sequence	Distance
lasB OP1	ACCTGCCAGTTCTGGCAGGT	32
rhlI	CCCTACCAGATCTGGCAGGT	135
lasB OP2	ACCTGCTT-TTCTGCTAGCT	93
rhlA	TCCTGTGAAATCTGGCAGTT	32
lasI	ATTTGCTAGTTAT-AAAATT	22
lasA	TACTGCAAAAGCTGATAGTT	34
rhlR	TTTTGCCGTATCGGCAAGGC	71
lasR	AGGACGGGTATCGTACTAGGT	26

the ability of a particular quorum-sensing system to activate a particular gene remains to be determined.

QUORUM SENSING AND *P. AERUGINOSA* VIRULENCE

P. aeruginosa is an opportunistic pathogen that primarily infects immunocompromised individuals, but it is also a serious problem for cystic fibrosis patients who are generally not considered to be immunocompromised (Joklik et al., 1992). These properties combined with the ubiquitous nature of this pathogen cause it to be a major source of nosocomial infections (Joklik et al., 1992). The ability of *P. aeruginosa* to also infect plants and insects (Lory, 1990; Rahme et al., 1995) underscores the capacity of this organism to adapt to changes in its environment and to produce a variety of virulence factors. As discussed above, a number of *P. aeruginosa* virulence factors, including elastase, LasA protease, alkaline protease, exotoxin A, and rhamnolipid, as well as critical genes such as *xcpP*, *xcpR*, and *rpoS* are controlled by the *las* and/or *rhl* quorum-sensing systems. The reason that *P. aeruginosa* controls many of its virulence factors through quorum sensing is an often discussed topic, with one of the more common speculative

answers being that perhaps quorum sensing allows the coordinated timing of virulence factor expression so their production occurs when sufficient cell numbers are present to overwhelm host defenses. With so many important genes under the quorum-sensing umbrella, it was not surprising to learn that Tang et al. (1996) showed a *P. aeruginosa lasR* mutant was significantly less virulent than the wild-type strain in a neonatal mouse model of infection. This indicated that *lasR* was required for full virulence of *P. aeruginosa*. Another interesting observation that links quorum sensing to virulence in a different way was that 3-oxo-C12-HSL stimulated respiratory epithelial cells to produce interleukin-8, a potent activator of the inflammatory response (DiMango et al., 1995). This suggests 3-oxo-C12-HSL acts directly as a virulence factor in addition to its role as an indirect virulence factor as part of the LasR-3-oxo-C12-HSL activation team.

FUTURE DIRECTIONS

The study of *P. aeruginosa* quorum sensing continues to generate interesting results. Our laboratory has discovered that a negative regulator of *lasI* exists. This regulator has been named RsaL and is encoded by a gene located

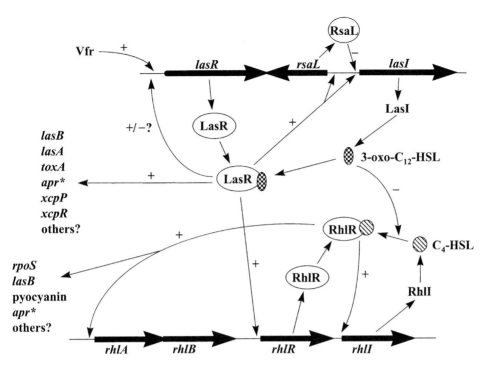

FIGURE 2 Model of quorum sensing in *P. aeruginosa*. LasR, RhlR, RsaL, 3-oxo-C12-HSL, and C4-HSL are symbolized by circles. Plus (+) or minus (−) symbols indicate transcriptional activation or repression of the gene(s) at the end of an arrow, respectively. The minus symbol by the arrow between 3-oxo-C12-HSL and C4-HSL indicates blocking of the association between C4-HSL and RhlR by 3-oxo-C12-HSL. The effect of LasR-3-oxo-C12-HSL on *lasR* transcription is unclear and is indicated by the "+/−?" symbol. The quorum-sensing hierarchy begins with basal production of LasR and 3-oxo-C12-HSL. At a high culture density, *lasR* is induced by Vfr, and 3-oxo-C12-HSL reaches a threshold concentration where it binds to LasR. LasR-3-oxo-C12-HSL can then activate genes controlled by the *las* system. Speculatively, the autoinduction of *lasI* could then cause 3-oxo-C12-HSL concentration to be in excess of C4-HSL concentration. This could allow 3-oxo-C12-HSL to block the binding of C4-HSL to RhlR until enough RhlR and/or C4-HSL are produced to overcome the blocking effect of 3-oxo-C12-HSL. Once RhlR associates with C4-HSL, autoinduction of *rhlI* occurs, and the remainder of the RhlR-C4-HSL-controlled genes are activated. ★, See the footnote in Table 1 for an explanation of *apr* regulation.

between, and in the opposite orientation of, *lasR* and *lasI* (deKievit et al., submitted). This obviously indicates that the regulation of *las* quorum sensing is even more complex than it originally appeared. We have also found that *P. aeruginosa* can generate spontaneous mutants that partially suppress mutations in *lasR*, suggesting that a back-up virulence factor activation system may exist (Van Delden et al., 1998). Taken together, these findings indicate that *P. aeruginosa* quorum sensing is more complex than it appears; further research is re-

quired to fully understand this sensitive form of cell-to-cell communication.

CONCLUSION

To help simplify *P. aeruginosa* quorum sensing, we have included a model that summarizes our current level of understanding of this complex subject (Fig. 2). This model is based on the generally accepted model of quorum sensing (Fuqua et al., 1996) in which a diffusible autoinducer is produced at a basal level at low cell densities. Autoinducer concentration then increases with cell density until a threshold concentration is reached, when it binds to a LuxR-type protein. The protein-autoinducer complex then activates the autoinducer synthase gene and other genes that it controls. In *P. aeruginosa*, the quorum-sensing hierarchical chain appears to begin with the activation of *lasR* by Vfr and the buildup of 3-oxo-C12-HSL concentration at higher cell densities. When 3-oxo-C12-HSL builds to a threshold level, it binds to LasR, and the LasR-3-oxo-C12-HSL complex induces the expression of several genes, including *lasI* and *rhlR*. We speculate that 3-oxo-C12-HSL could then block C4-HSL from binding to RhlR until enough RhlR and/or C4-HSL are present to overcome the 3-oxo-C12-HSL blocking effect. When the RhlR-C4-HSL complex forms, *rhlI* and the other *rhl*-controlled genes would subsequently be induced. In such a model, *P. aeruginosa* could delay the induction of *rhl*-controlled genes to a later time in the growth cycle when certain genes (*rpoS*, for example) may be needed.

The *las* and *rhl* quorum-sensing systems of *P. aeruginosa* provide this pathogen with an elegant mechanism to temporally control the production of numerous important genes. While a great amount of information has been learned about these systems, there is obviously a lot that remains unknown. As we learn more about this widespread form of cell-to-cell communication, it will add to our understanding of how intrinsic chemical signals allow bacteria to coordinate a specific response, which will hopefully lead to practical applications that will improve our ability to treat *P. aeruginosa* infection and other gram-negative bacterial infections.

ACKNOWLEDGMENTS

E. C. Pesci was supported by Research Fellowship Grant PESCI96FO from the Cystic Fibrosis Foundation. This work was also supported by National Institutes of Health research grant R01A133713-04. We thank L. Passador, T. deKievit, J. P. Pearson, C. Van Delden, J. Raymond, R. Smith, and C. S. Pesci for help in manuscript preparation and thoughtful insight.

REFERENCES

Albus, A. M., E. C. Pesci, L. J. Runyen-Janecky, S. E. H. West, and B. H. Iglewski. 1997. Vfr controls quorum sensing in *Pseudomonas aeruginosa. J. Bacteriol.* **179:**3928–3935.

Brint, J. M., and D. E. Ohman. 1995. Synthesis of multiple exoproducts in *Pseudomonas aeruginosa* is under the control of RhlR-RhlI, another set of regulators in strain PAO1 with homology to the autoinducer-responsive LuxR-LuxI family. *J. Bacteriol.* **177:**7155–7163.

Brock, T. D., M. T. Madigan, J. M. Martinko, and J. Parker. 1994. Pseudomonads, p. 774–777. *In Biology of Microorganisms*, 7th ed. Prentice-Hall, Englewood Cliffs, N.J.

Chapon-Herve, V., M. Akrim, A. Latifi, P. Williams, A. Lazdunski, and M. Bally. 1997. Regulation of the *xcp* secretion pathway by multiple quorum-sensing modulons in *Pseudomonas aeruginosa. Mol. Microbiol.* **24:**1169–1178.

deKievit, T., P. C. Seed, L. Passador, and B. H. Iglewski. RsaL regulates cell-to-cell communication and expression of global virulence determinants *in Pseudomonas aeruginosa* by inhibiting production of the autoinducer molecule 3-oxo-C12-HSL. Submitted for publication.

DiMango, E., H. J. Zar, R. Bryan, and A. Prince. 1995. Diverse *Pseudomonas aeruginosa* gene products stimulate respiratory epithelial cells to produce interleukin-8. *J. Clin. Invest.* **96:**2204–2210.

Freck-O'Donnell, L. C., and A. Darzins. 1993. *Pseudomonas aeruginosa lasA* gene: determination of the transcription start point and analysis of the promoter/regulatory region. *Gene* **129:**113–117.

Fuqua, W. C., S. C. Winans, and E. P. Greenberg. 1996. Census and consensus in bacterial ecosystems: the LuxR-LuxI family of quorum-sensing transcriptional regulators. *Annu. Rev. Microbiol.* **50:**727–751.

Gambello, M. J., and B. H. Iglewski. 1991. Cloning and characterization of the *Pseudomonas aeruginosa lasR* gene: a transcriptional activator of elastase expression. *J. Bacteriol.* **173:**3000–3009.

Gambello, M. J., S. Kaye, and B. H. Iglewski. 1993. LasR of *Pseudomonas aeruginosa* is a transcriptional activator of the alkaline protease gene (*apr*) and an enhancer of exotoxin A expression. *Infect. Immun.* **61:**1180–1184.

Gray, K. M., L. Passador, B. H. Iglewski, and E. P. Greenberg. 1994. Interchangeability and specificity of components from the quorum-sensing regulatory systems of *Vibrio fischeri* and *Pseudomonas aeruginosa. J. Bacteriol.* **176:**3076–3080.

Joklik, W. K., H. P. Willett, D. B. Amos, and C. M. Wilfert. 1992. *Pseudomonas,* p. 576–583.

In Zinsser's Microbiology, 20th ed. Appleton & Lange, Stamford, Conn.

Latifi, A., M. K. Winson, M. Foglino, B. W. Bycroft, G. S. A. B. Stewart, A. Lazdunski, and P. Williams. 1995. Multiple homologues of LuxR and LuxI control expression of virulence determinants and secondary metabolites through quorum sensing in *Pseudomonas aeruginosa* PAO1. *Mol. Microbiol.* **17:**333–343.

Latifi, A., M. Foglino, K. Tanaka, P. Williams, and A. Lazdunski. 1996. A hierarchical quorum-sensing cascade in *Pseudomonas aeruginosa* links the transcriptional activators LasR and RhlR to expression of the stationary-phase sigma factor RpoS. *Mol. Microbiol.* **21:**1137–1146.

Lory, S. 1990. *Pseudomonas* and other nonfermenting bacilli, p. 595–600. *In* B. D. Davis, R. Dulbecco, H. N. Eisen, and H. S. Ginsberg (ed.), *Microbiology.* J. P. Lippincott, Philadelphia.

Ochsner, U. A., and J. Reiser. 1995. Autoinducer-mediated regulation of rhamnolipid biosurfactant synthesis in *Pseudomonas aeruginosa. Proc. Natl. Acad. Sci. USA* **92:**6424–6428.

Ochsner, U. A., A. K. Koch, A. Fiechter, and J. Reiser. 1994a. Isolation and characterization of a regulatory gene affecting rhamnolipid biosurfactant synthesis in *Pseudomonas aeruginosa. J. Bacteriol.* **176:**2044–2054.

Ochsner, U. A., A. K. Koch, A. Fiechter, and J. Reiser. 1994b. Isolation, characterization, and expression in *Escherichia coli* of the *Pseudomonas aeruginosa rhlAB* genes encoding a rhamnosyltransferase involved in rhamnolipid biosurfactant synthesis. *J. Biol. Chem.* **269:**19787–19795.

Passador, L., J. M. Cook, M. J. Gambello, L. Rust, and B. H. Iglewski. 1993. Expression of *Pseudomonas aeruginosa* virulence genes requires cell-to-cell communication. *Science* **260:**1127–1130.

Passador, L., K. D. Tucker, K. R. Guertin, M. P. Journet, A. S. Kende, and B. H. Iglewski. 1996. Functional analysis of the *Pseudomonas aeruginosa* autoinducer PAI. *J. Bacteriol.* **178:**5995–6000.

Passador, L., K. D. Tucker, W. H. Kuhnert, and B. H. Iglewski. Unpublished data.

Pearson, J. P., K. M. Gray, L. Passador, K. D. Tucker, A. Eberhard, B. H. Iglewski, and E. P. Greenberg. 1994. Structure of the autoinducer required for expression of *Pseudomonas aeruginosa* virulence genes. *Proc. Natl. Acad. Sci. USA* **91:**197–201.

Pearson, J. P., L. Passador, B. H. Iglewski, and E. P. Greenberg. 1995. A second *N*-acylhomoserine lactone signal produced by *Pseudomonas aeruginosa. Proc. Natl. Acad. Sci. USA* **92:**1490–1494.

Pearson, J. P., E. C. Pesci, and B. H. Iglewski. 1997. Roles of *Pseudomonas aeruginosa las* and *rhl* quorum sensing systems in control of elastase and rhamnolipid biosynthesis genes. *J. Bacteriol.* **179:**5756–5767.

Pesci, E. C., and B. H. Iglewski. 1997. The chain of command in *Pseudomonas* quorum sensing. *Trends Microbiol.* **5:**132–135.

Pesci, E. C., J. P. Pearson, P. C. Seed, and B. H. Iglewski. 1997. Regulation of *las* and *rhl* quorum sensing in *Pseudomonas aeruginosa. J. Bacteriol.* **179:**3127–3132.

Rahme, L. G., E. J. Stevens, S. F. Wolfort, J. Shao, R. G. Tompkins, and F. M. Ausubel. 1995. Common virulence factors for bacterial pathogenicity in plants and animals. *Science* **268:**1899–1902.

Rust, L., E. C. Pesci, and B. H. Iglewski. 1996. Analysis of the *Pseudomonas aeruginosa* elastase (*lasB*) regulatory region. *J. Bacteriol.* **178:**1134–1140.

Seed, P. C., L. Passador, and B. H. Iglewski. 1995. Activation of the *Pseudomonas aeruginosa lasI* gene by LasR and the *Pseudomonas* autoinducer PAI: an autoinduction regulatory hierarchy. *J. Bacteriol.* **177:**654–659.

Tang, H. B., E. DiMango, R. Bryan, M. Gambello, B. H. Iglewski, J. B. Goldberg, and A. Prince. 1996. Contribution of specific *Pseudomonas aeruginosa* virulence factors to pathogenesis of pneumonia in a neonatal mouse model of infection. *Infect. Immun.* **64:**37–43.

Toder, D. S., M. J. Gambello, and B. H. Iglewski. 1991. *Pseudomonas aeruginosa* LasA; a second elastase gene under transcriptional control of *lasR. Mol. Microbiol.* **5:**2003–2010.

Van Delden, C., E. C. Pesci, J. P. Pearson, and B. H. Iglewski. 1998. Starvation selection restores elastase and rhamnolipid production in a *Pseudomonas aeruginosa* quorum-sensing mutant. *Infect. Immun.* **66:**4499–4502.

Whooley, M. A., J. A. O'Callaghan, and A. J. McLoughlin. 1983. Effect of substrate on the regulation of exoprotease production by *Pseudomonas aeruginosa* ATCC 10145. *J. Gen. Microbiol.* **129:**981–988.

Winson, M. K., M. Camara, A. Latifi, M. Foglino, S. R. Chhabra, M. Daykin, M. Bally, V. Chapon, G. P. C. Salmond, B. W. Bycroft, A. Lazdunski, G. S. A. B. Stewart, and P. Williams. 1995. Multiple *N*-acyl-L-homoserine lactone signal molecules regulate production of virulence determinants and secondary metabolites in *Pseudomonas aeruginosa. Proc. Natl. Acad. Sci. USA* **92:**9427–9431.

You, Z., J. Fukushima, T. Ishiwata, B. Chang, M. Kurata, S. Kawamoto, P. Williams, and K. Okuda. 1996. Purification and characterization of LasR as a DNA-binding protein. *FEMS Microbiol. Lett.* **142:**301–307.

PRODUCTION OF ANTIMICROBIAL COMPOUNDS

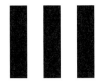

THE LANTIBIOTICS NISIN AND SUBTILIN ACT AS EXTRACELLULAR REGULATORS OF THEIR OWN BIOSYNTHESIS

Michiel Kleerebezem, Willem M. de Vos, and Oscar P. Kuipers

11

NISIN

Many gram-positive bacteria secrete polypeptides that show antimicrobial activity against competitive microorganisms. A large group of small antimicrobial peptides (AMPs) that have a wide host range has been described; these AMPs are usually cysteine-rich and hydrophobic (Klaenhammer, 1993). Especially heat-stable peptides secreted by lactic acid bacteria have been studied extensively because of their potential use as natural preservatives in food applications. Two major families of these small AMPs are generally recognized: (i) type I AMPs, or lantibiotics that are heat-stable peptides subjected to a high degree of posttranslational modification prior to secretion (Schnell et al., 1988; Jung, 1991; de Vos et al., 1995a), and (ii) type II AMPs, or heat-stable peptides that include unmodified but secreted peptides (Klaenhammer, 1993; Nes et al., 1996).

Lantibiotics constitute an uncommon family of biologically active peptides that are thought to exert their antimicrobial activity by formation of transient pores in the bacterial cytoplasmic membranes (Sahl, 1991). They contain uncommon, unsaturated residues like dehydroalanine and dehydrobutyrine and typical intramolecular thioether bridges called (β-methyl) lanthionines, which are formed by the addition of a cysteine-derived sulfhydryl group to the double bond of a dehydrated residue. These thioether bridges determine the characteristic polycyclic structure of lantibiotics, and their position within the polypeptides has led to the subdivision of this family of AMPs into linear (group A) and circular (group B) lantibiotics (Jung, 1991). The precursors of the lantibiotic molecules are ribosomally synthesized and subsequently modified in a series of posttranslational, enzymatic modifications. The unusual structural properties of this family of AMPs have led to considerable interest in their biosynthetic pathways and structure-function relationships. A number of group A lantibiotics have been characterized that are produced by a variety of gram-positive bacteria, such as *Enterococcus faecalis*, *Staphylococcus epidermidis*, *Lactococcus lactis*, *Lactobacillus sake*, *Bacillus subtilis*, *Carnobacterium piscicola*, *Streptococcus pyogenes*, *Streptococcus salivarius*, and *Streptococcus mutans*. The most extensively studied lantibiotic is produced by *Lactococcus lactis* and is called nisin, which is derived from N inhibitory substance (Mattick and Hirsch,

Michiel Kleerebezem, Willem M. de Vos, and Oscar P. Kuipers, Microbial Ingredients Section, NIZO Food Research, Kernhemseweg 2, 6710 BA Ede, The Netherlands.

Cell-Cell Signaling in Bacteria, Edited by Gary M. Dunny and Stephen C. Winans
©1999 American Society for Microbiology, Washington, D.C.

1944). The interest in this lantibiotic molecule was greatly stimulated by the broad spectrum of its antimicrobial activity and its increasing application as a natural preservative in the food industry.

The primary structure of nisin A was the first lantibiotic structure to be elucidated; it is a 34-amino-acid 3,353-Da peptide that contains four methyl-lanthionine and one lanthionine cyclic structures, and one dehydroalanine (Dha) and two dehydrobutyrine residues (Fig. 1A). In addition, a natural variant of nisin A (called nisin Z) has been identified, which differs only by the exchange His27→ Asn; except for this exchange, it has a structure that is identical to that of nisin A (Mulders et al., 1991). The structural genes encoding nisin A (Buchman et al., 1988) and nisin Z (Mulders et al., 1991) have been characterized,

and producers of both nisin species are widely distributed (de Vos et al., 1993). The predicted 57-residue pre-nisin peptide contains a 23-residue N-terminal extension called leader peptide that is absent in the mature nisin molecule (de Vos et al., 1991). This leader peptide is not subjected to the posttranslational modification reactions that occur within the mature domain (van der Meer et al. 1993). Furthermore, introduction of specific mutations in this leader peptide domain resulted in a loss of nisin modification and/or secretion (van der Meer et al., 1993, 1994), strongly suggesting a specific role of the leader peptide in directing the posttranslational modification and targeting processes. The characterization of a number of nisin mutants (and mutants of other lantibiotics) generated by protein engineering yielded valuable information on re-

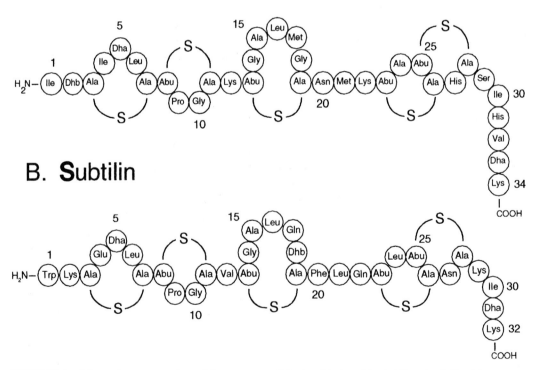

FIGURE 1 Schematic representation of the structure of the lantibiotics nisin A (A) and subtilin (B).

quirements for production, modification, antimicrobial activity, and the general structure-function relation of these molecules (Kuipers et al., 1996).

In this overview we describe the genetic and biochemical characterization of the nisin biosynthetic pathway. We will address, in more detail, the relatively recent finding that nisin acts as the signal molecule that induces its own biosynthesis and subsequent research based on this finding, e.g., lantibiotic-sensor interaction. In addition, with respect to this latter aspect, we will compare the nisin data with observations of the regulation of the biosynthetic pathway of another, structurally related lantibiotic called subtilin (Fig. 1B; see below), which is produced by *B. subtilis*.

THE NISIN GENE CLUSTER

In most strains of *Lactococcus lactis*, nisin production is encoded by large conjugative transposons, and the nucleotide sequences of 11 genes involved in the complex nisin biosynthesis, which are organized in a gene cluster, *nisABTCIPRKFEG*, have been determined (Fig. 2A) (Buchman et al., 1988; Kaletta and Entian, 1989; Engelke et al., 1992, 1994; van

der Meer et al., 1993; Kuipers et al., 1993; Siegers and Entian, 1995; de Vos et al., 1995b). Besides the structural *nisA* gene, the cluster contains genes encoding proteins involved in the intracellular posttranslational modification reactions (*nisB* and *nisC*, based on homology to genes found exclusively in other lantibiotic gene clusters) (Siezen et al., 1996). These modification enzymes form a membrane-localized complex with the putative transport protein of the ABC translocator family (encoded by *nisT*), which is probably involved in the secretion of the modified prenisin molecule (Engelke et al., 1992; Kuipers et al., 1993; Siegers et al., 1996). During or shortly after secretion of the pre-nisin molecule, the N-terminal leader peptide is proteolytically removed by the extracellular subtilisin-like protease encoded by *nisP* (van der Meer et al., 1993; Siezen et al., 1995). Two systems that are involved in immunity to nisin of the producing cell are encoded within the gene cluster. The first consists of a lipoprotein encoded by *nisI* that contributes to immunity to the producing cell via an unknown mechanism (Kuipers et al., 1993), and the second consists of the ABC exporter sys-

A. *Lactococcus lactis*; nisin gene cluster

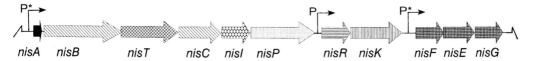

nisA   nisB   nisT   nisC nisI nisP   nisR nisK   nisF nisE nisG

B. *Bacillus subtilis*; subtilin gene cluster

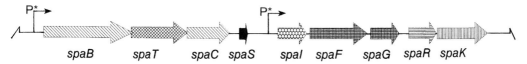

spaB   spaT   spaC spaS   spaI spaF   spaG spaR spaK

FIGURE 2 Organization of nisin (A) and subtilin (B) biosynthetic gene clusters. The structural lantibiotic genes are indicated in black, the sensor proteins are indicated by vertical hatching, and the response regulators are indicated by horizontal hatching. Genes that play a role in the following processes are also indicated: precursor modification (tilted hatching), export (tilted squares), leader processing (stippled), or immunity (horizontal squares and hexagonals). Known promoters are indicated by P* for regulated promoters and P for nonregulated promoters. For further information regarding the organization and individual functions of the gene products, the reader is referred to Siezen et al. (1996).

tem encoded by *nisFEG* that possibly exerts its immunity effects through active nisin extrusion from the cell (or cell membrane) (Siegers and Entian, 1995). Finally, the proteins encoded by *nisR* (response regulator) and *nisK* (sensor) have been shown to be involved in the regulation of nisin biosynthesis (van der Meer et al., 1993; Kuipers et al., 1993, 1995; Engelke et al., 1994; de Ruyter et al., 1996a,b; Kleerebezem et al., 1997a) and belong to the family of bacterial two-component signal transduction systems. Two-component regulatory systems form a widespread and important mechanism of signal transduction in bacteria, playing a key role in many of the changes in cellular physiology that result from changes in the environment. These systems use phosphorylation as a means to transfer information (Parkinson and Kofoid, 1992; Stock et al., 1989).

The production of nisin by *Lactococcus lactis* is regulated in a growth phase-dependent manner; production starts during mid-logarithmic phase and increases to reach a maximum level at the end of the logarithmic growth phase and the beginning of the stationary phase (Buchman et al., 1988; de Ruyter et al., 1996b). Although it was clear that the products encoded by the *nisR* and *nisK* genes were essential for regulation of nisin biosynthesis, until recently the nature of the environmental stimulus that activates the regulatory pathway involved in nisin biosynthesis remained unclear. This elementary question was resolved by introduction of a 4-bp deletion in the structural *nisA* gene (Δ*nisA*), in a *Lactococcus lactis* strain that normally produces nisin. The deletion resulted not only in the expected loss of nisin production but also in a complete abolition of transcription of the Δ*nisA* gene (Kuipers et al., 1993). An important finding leading to the insight that AMPs may have both antimicrobial and signal activity was the observation that transcription of Δ*nisA* could be restored by the addition of subinhibitory amounts of nisin to the culture medium (Kuipers et al., 1995). Experiments using engineered nisin variants or

nisin fragments indicated that the N-terminal residues of the fully matured nisin molecule have a more important role in its signal function than do the more C-terminally located residues (Kuipers et al., 1995; Dodd et al., 1996), although charge differences in the C terminus also affect induction capacity (van Kraaij et al., 1997). Interestingly, several nisin variants appeared much more (or much less) potent as inducer peptides whereas their antimicrobial activity was not proportionally increased (or decreased) compared to wild-type nisin, indicating that these activities are not necessarily correlated (Kuipers et al., 1995). Furthermore, disruption of *nisR* and/or *nisK* in a Δ*nisA* background abolishes activation of Δ*nisA* transcription by nisin (Kuipers et al., 1995; Kleerebezem et al., manuscript in preparation), indicating that the two-component regulatory system composed of NisR and NisK is essential for nisin-mediated signal transduction.

In conclusion, these data indicate that, besides its activity as AMP, nisin also acts as a secreted peptide pheromone that induces its own biosynthesis by triggering the NisRK signal transduction system in a quorum sensing-like manner (Kleerebezem et al., 1997b; Dunny and Leonard, 1997). A schematic representation of the factors involved in the complex biosynthetic pathway of nisin and its role as peptide pheromone is depicted in Fig. 3.

nis PROMOTERS AND THE DEVELOPMENT OF A NISIN-INDUCIBLE GENE EXPRESSION SYSTEM

Three promoters within the nisin gene cluster were identified by primer extension and transcriptional fusion to the *Escherichia coli* promoterless β-glucuronidase gene (*gusA*) in the promoter-probe vector pNZ273 (Kuipers et al., 1995; de Ruyter et al., 1996a; Platteeuw et al., 1994). Transcription from two of these promoters, preceding the *nisA* and the *nisF* gene, was shown to be regulated by nisin-mediated, NisRK-dependent signal transduction. When the *nisA* or *nisF* promoter-

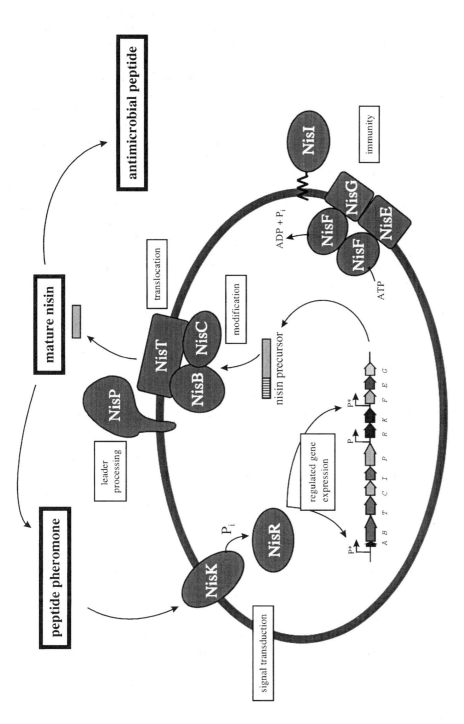

FIGURE 3 Model for nisin biosynthesis and regulation. Primarily, nisin acts as an antimicrobial peptide, and producing cells are protected against this activity of nisin by NisI and NisFEG. A second role of nisin in this model is as a peptide pheromone that is sensed by NisK, which triggers the signal transduction cascade, including phosphotransfer to NisR and subsequent activation of transcription of the nisin gene cluster. The *nisA* mRNA is ribosomally translated and subsequently modified (NisB and NisC) and secreted (NisT), followed by leader processing (NisP), finally resulting in the mature nisin molecule.

gusA fusion constructs were present in the Δ*nisA Lactococcus lactis* background, no β-glucuronidase activity could be detected unless nisin was added to the culture medium. The level of transcription of both promoters appeared directly dependent on the concentration of extracellularly added nisin, indicating a linear dose-response relationship between the inducer (nisin) concentration and the resulting transcriptional level (de Ruyter et al., 1996a,b). A typical nisin dose-response measurement, obtained when the *nisA* promoter-fusion construct is introduced in the Δ*nisA* background, is shown in Fig. 4A. When the same plasmid is introduced in either a Δ*nisA*/Δ*nisR* or a Δ*nisA*/Δ*nisK* strain, the observed nisin-dependent β-glucuronidase response is completely absent (Fig. 4A), which substantiates the *nisRK* dependence of the nisin-induced transcription activity of this promoter (similar results were obtained with the *nisF* promoter fusion; not shown). Furthermore, a transferable dual plasmid system was developed that relies on the presence of two compatible broad-host-range plasmids and allows the functional implementation of the nisin controlled expression system in lactic acid bacteria, including *Lactobacillus helveticus* and *Leuconostoc lactis* (Kleerebezem et al., 1997a) and other gram-positive bacteria (Eichenbaum et al., 1998). One of these plasmids encodes the *nisRK* genes (Kleerebezem et al., 1997a), whereas the second plasmid is a *nisA* promoter vector (see below) (de Ruyter et al., 1996b). Taken together, these data clearly show that NisR and NisK are the only protein components required for nisin-mediated signal transduction. The third promoter, preceding *nisR*, was shown to direct nisin-independent *gusA* expression in a nisin transposon- and plasmid-free strain of *Lactococcus lactis*, indicating that this promoter constitutively drives the expression of the *nisRK* genes (de Ruyter et al., 1996a).

A terminator sequence directly downstream of the structural *nisA* gene probably plays an important role in balancing the relative expression levels of the components in the *nisABTCIP* operon by allowing *nisA* promoter-driven transcription of the downstream genes (*nisB-P*) by limited readthrough (van der Meer, 1993; Kuipers et al., 1995; de Ruyter et al., 1996a). The role of the *nisFEG* gene products in nisin immunity of the producing cell provides a rationale for the observed nisin-mediated regulation of the *nisF* promoter, providing higher immunity levels when the cells meet higher nisin concentrations. The constitutive character of the *nisR* promoter is probably required for a rapid response of the cell in the presence of nisin, including a rapid build-up of immunity. Nevertheless, since no clear terminator sequence can be identified downstream of *nisP* the possibility is raised that by read-through of the *nisA* promoter the level of *nisRK* encompassing mRNA is raised in a nisin dependent manner, which is supported by the detection of a mRNA molecule encompassing *nisABTCIPRK* (Ra et al., 1996).

Stimulated by the broad interest in the development of food-grade microorganisms for the controlled production of desirable metabolites, enzymes, and other proteins for the food industry, the regulatory characteristics of nisin biosynthesis were exploited to develop a nisin-controlled expression system (NICE system) for *Lactococcus lactis*. Several *nisA* promoter vectors were constructed that allow either transcriptional or translational fusion of the gene encoding the desired protein or enzyme (de Ruyter et al., 1996b). These *nisA* promoter vectors appeared suitable for tightly controlled expression of both homologous and heterologous genes in *Lactococcus lactis*, achieving gene expression in a dynamic range of more than 1,000-fold, depending on the amount of nisin added. This led ultimately to very high protein yields (up to 60% of total cell protein) (de Ruyter et al., 1996b; Kuipers et al., 1997). In addition, a fully food-grade nisin-induced expression system was obtained by combination of the *nisA* promoter fusion with the food-grade selection marker *lacF*, in a *lacF*-deficient lactococcal strain that has the *nisRK* genes integrated on the chromosome

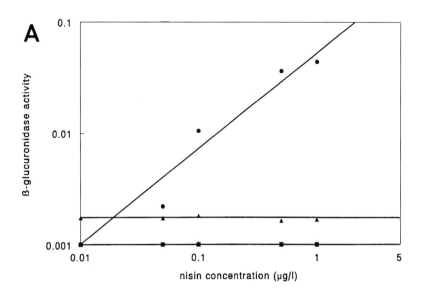

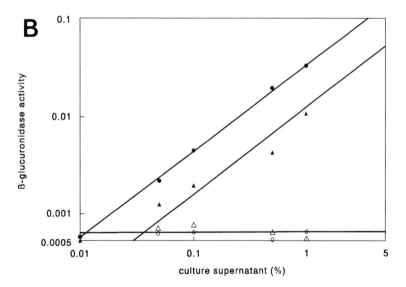

FIGURE 4 (A) Levels of β-glucuronidase activity in cells of *Lactococcus lactis* harboring pNZ8008 (*nisA* promoter-*gusA* fusion plasmid) as a function of the concentration of the extracellularly added nisin. Three different derivatives of strain NZ9700 (nisin-producing strain) were tested that contain deletions in one or more of the nisin genes: NZ9800 (filled circles; Δ*nisA*) (Kuipers et al., 1993), NZ9805 (filled squares; Δ*nisA*, Δ*nisK*) (Kleerebezem et al., manuscript in preparation), and NZ9808 (filled triangles; Δ*nisA*, Δ*nisR*) (Kleerebezem et al., manuscript in preparation). (B) *spaB* or *spaI* promoter-driven *gusA* expression in a derivative of *B. subtilis* ATCC 6633 (subtilin-producing strain) that does not produce subtilin owing to a disruption in the *spaB* gene. The *spaB* mutant cells contained either a *spaB* (circles) or *spaI* (triangles) promoter-*gusA* fusion plasmid. Promoter activities (β-glucuronidase activities) were analyzed as a function of the concentration of the extracellularly added culture supernatant of either the subtilin-producing strain ATCC 6633 (filled symbols) or its *spaB* derivative (open symbols).

(*pepN::nisRK*) (Platteeuw et al., 1996; de Ruyter et al., 1996b, 1997a). Furthermore, the tight regulation includes an essentially undetectable production of the protein of interest in the uninduced state, which allows the cloning and controlled expression of lethal proteins. The latter possibility was exploited in construction of food-grade lactococcal strains that express the lytic genes *lytA* and *lytH* (derived from lactococcal bacteriophage ΦUS3) upon induction with nisin. This allows controlled release of intracellular enzymes efficiently at an early stage of cheese production, which could result in accelerated cheese ripening (de Ruyter et al., 1997). The many desirable characteristics of the nisin-inducible gene expression system have initiated the development of a transferable dual-plasmid system that allows fully functional implementation of this gene expression system in other gram-positive bacteria, including several lactic acid bacteria (see above), thus widening the scope of potential use of this NICE system in the (food) industry (Kleerebezem et al., 1997a; Eichenbaum et al., 1998).

Alignment of the *nisA* and *nisF* promoters revealed that both regulated promoter sequences contain a −10 region that differs by one nucleotide from the TATAAT consensus −10 sequence (*nisA*: TACAAT; *nisF*: TATACT) (Fig. 5). However, instead of a standard −35 region, these promoters contain a pentanucleotide (TCTGA) direct repeat (PDR) separated by six nucleotides, positioned from −39 to −24 relative to the transcription start site (Fig. 5). This finding suggests that this PDR is important in the promoter function of these DNA elements and could possibly represent the binding position of the response regulator NisR. To investigate this possibility, several mutations were introduced into the *nisA* promoter region of a *nisA* promoter-*gusA* fusion construct, and the activity level and nisin inducibility of the resulting promoters were analyzed in a nisin transposon- and plasmid-free *Lactococcus lactis* strain (MG1363) and a strain containing the *nisRK* genes integrated in its chromosome (NZ3900) (de Ruyter et al., manuscript in

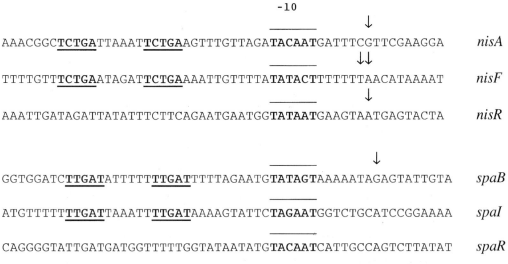

FIGURE 5 Comparison of the *nisA*, *nisR*, *nisF* and *spaB*, *spaI*, *spaR* promoters. Arrows indicate the transcription start sites mapped by primer extension (de Ruyter et al., 1996a; Chung and Hansen, 1992). For promoter sequences where no transcription start site is indicated, this site has not been mapped. The (putative) −10 regions are indicated in bold and by a line above the sequence. The pentanucleotide direct repeat (PDR) elements, which probably play an essential role in the regulatory features observed for these promoters, are indicated in bold and by underlining of the sequence.

preparation). Variation of the relative spacing between the two pentanucleotide elements and/or the −10 region (by introduction of a 3-bp deletion) resulted in all cases in a complete loss of promoter activity and inducibility (de Ruyter et al., in preparation). Moreover, by single-base-pair substitutions within the promoter sequence, several nucleotides could be identified that are essential for promoter activity and/or inducibility (Fig. 6) (de Ruyter et al., manuscript in preparation). Finally, conversion of the *nisA* promoter into a promoter element that exhibits constitutive transcription activity and loss of inducibility at least requires introduction of a consensus −35 region (TTGACA) with consensus spacing of 17 nucleotides relative to the −10 region (de Ruyter et al., manuscript in preparation).

Taken together, the results support an essential role of the PDR elements in the tight control of the *nisA* promoter activity. Preliminary data indicate that NisR can bind specifically to the *nisA* promoter in vitro, which results in a DNA-protein complex that exhibits a reduced mobility in polyacrylamide gels relative to the unbound *nisA*-promoter fragment (de Ruyter et al., manuscript in preparation). Quantitative analysis of NisR binding efficiency to mutant promoter elements in these gel retardation studies could resolve the specific requirements (intact PDR element?) for NisR-*nisA*-promoter binding. These experiments combined with DNA footprinting analysis of the NisR binding to the *nisA* promoter are currently being performed in our laboratory and will contribute to a better understanding of the regulatory mechanism underlying this tightly controlled gene expression system.

THE SUBTILIN GENE CLUSTER AND ITS REGULATION

Subtilin is a lantibiotic that is structurally closely related to nisin (Fig. 1B) and is produced by strains of *B. subtilis*. Subtilin biosynthesis appears to be regulated in a growth phase-dependent manner that strongly resembles that of nisin, including a start of production during the mid- to end-logarithmic-growth phase and reaching a maximum production level during early stationary phase (Gutowski-Eckel et al., 1994). The genes required for subtilin biosynthesis (*spa* genes) share a high degree of similarity with the genes in the nisin biosynthetic gene cluster (for a review see Siezen et al., 1996) (Fig. 2B). Interestingly, besides *spa* genes involved in pre-subtilin (encoded by *spaS*) modification (*spaB* and *spaC*), translocation (*spaT*) and immunity (*spaI*, *spaF*, and *spaG*), the subtilin gene cluster also contains *nisRK* homologs, designated *spaR* and *spaK*, that have been shown to be essential for subtilin biosynthesis (Klein et al., 1993). However, the subtilin biosynthetic gene cluster seems to lack a *nisP* homolog, and so far it remains unknown which component is involved in proteolytic removal of the subtilin leader peptide. Nevertheless, these findings strongly suggest that subtilin biosynthesis and regulation follow a scheme similar to that postulated for nisin (Fig. 3).

nisA promoter mutation analysis

AAACGGC**TCTGA**TTAAAT**TCTGA**AGTTTGTTAGA**TACAAT**GATTTCGTTCGAAGGA

FIGURE 6 Sequence of the wild-type *nisA* promoter element. Nucleotides that were identified by mutation analysis as essential for the nisin-inducible activation of transcription from this promoter are indicated by black triangles. Mutations of nucleotides indicated by the open triangles did not affect the promoter function, and mutation position with an intermediate effect on promoter function (i.e., still inducible, but with reduced levels of transcription at a given nisin concentration) is indicated by a gray triangle (I).

A single promoter, preceding the *spaB* gene, has been identified within the subtilin gene cluster (Chung and Hansen, 1992; Gutowski-Eckel et al., 1994). Interestingly, this promoter has similarities to vegetative sigma A promoters of *B. subtilis* (Gutowski-Eckel et al., 1994; Moran, 1990), which direct transcription of regulons expressed during stationary phase and endospore formation (Moran, 1990). However, alignment of the *spaB* promoter with the *nisA* and *nisF* promoters clearly reveals the presence of a PDR element within the *spaB* promoter with similar spacing to its −10 region, as compared to the *nisA* and *nisF* promoters (Kleerebezem et al., manuscript in preparation). Furthermore, a putative promoter sequence containing the same PDR was found upstream of *spaI* (Kleerebezem et al., 1997c). These findings suggest strongly that the *spaB* and putative *spaI* promoters are subject to regulatory modes similar to those described for the *nisA* and *nisF* promoters. To investigate this, both the *spaB* and *spaI* promoters were amplified from the *B. subtilis* chromosomal DNA and cloned in a *gusA* promoter-probe vector similar to pNZ273 (the only difference is replacement of chloramphenicol resistance marker by a tetracycline resistance marker Kleerebezem et al., manuscript in preparation). These *spaB* and *spaI* promoter constructs were transformed to a chloramphenicol-resistant *spaB* disruption strain that has a subtilin-negative phenotype (Klein et al., 1992). In these transformants, β-glucuronidase activity could be induced only with culture supernatant of the subtilin-producing *B. subtilis* strain, in contrast to the lack of induction using culture supernatant from its *spaB* derivative (Fig. 4B) (Kleerebezem et al., manuscript in preparation).

These data clearly show that the *spaB* and *spaI* promoters are both regulated promoters that can be induced by the addition of subinhibitory amounts of subtilin to the growth medium. These findings strongly resemble those obtained in similar experiments using the *nisA* and *nisF* promoters of *Lactococcus lactis* (Fig. 4A), which justifies the conclusion that the biosynthesis of both these lantibiotics is autoregulated. Although a definite involvement of the *spaRK* gene products cannot be concluded from these experiments, the expression of *spaB* has been shown to depend on *spaRK* (Gutowski-Eckel et al., 1994), indicating that the regulation of subtilin biosynthesis appears to follow the nisin model. In analogy to the nisin system, the *spaRK* genes are probably transcribed from their own promoter (Fig. 5) but, owing to the absence of a clear terminator sequence downstream of *spaG*, they could be transcribed at elevated levels by *spaI* promoter readthrough following subtilin induction. The location of the structural gene *spaS* within the subtilin gene cluster is essentially different from the location of *nisA* in the nisin gene cluster. However, balanced express of *spaS* and *spaBTC* is probably achieved by the activity of a putative subtilin-inducible promoter upstream of *spaS*, in which the PDR sequences are almost completely conserved (1 bp variation relative to the *spaB* and *spaI* promoters) (not shown; Kleerebezem et al., in preparation). Therefore, it seems very likely that the mechanism of regulation of subtilin biosynthesis by *B. subtilis* is very similar to the regulatory mechanism involved in nisin biosynthesis, including the peptide pheromone function of the mature subtilin molecule, the role of *spaRK* in the subtilin-mediated signal transduction, and the thereby activated *spaB*, *spaS*, and *spaI* promoter activity.

TWO-COMPONENT REGULATORY SYSTEMS INVOLVED IN REGULATION OF NISIN AND SUBTILIN BIOSYNTHESIS

Sequence analysis of NisK and SpaK shows that both proteins have two predicted hydrophobic transmembrane helices in their N-terminal "input" domain (Engelke et al., 1994; Klein et al., 1993) that enclose a hydrophilic, presumably extracellular domain of approximately 110 amino acid residues. In the C-terminal "transmitter" domain of the NisK protein, both conserved histidine kinase subdomains, containing the overall conserved histidine residue (Parkinson and Kofoid, 1992), are present. (A schematic topology model of the NisK protein is depicted in Fig. 7.) In contrast, the SpaK sequence (Klein et al., 1993) appears to be too short and lacks the last of the two conserved subdomains. However, this could be due to a sequencing error resulting in a frame shift in the published sequence (Klein et al., 1993), especially since another frame encodes a further 72 amino acids completely containing the missing conserved subdomain. This sequencing error was confirmed by sequence analysis of three independently obtained PCR products that were obtained by using two primers, one overlapping with the start codon and one completely downstream of the "extended" *spaK* gene and *B. subtilis* ATCC 6633 (subtilin producer) chromosomal DNA as a template.

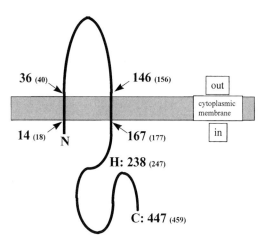

FIGURE 7 Membrane topology model of the NisK protein. Transmembrane helices were predicted on the basis of hydrophobicity plots of the protein sequences, and membrane topology could be predicted on the basis of the positive inside rules described by von Heijne (1992). Indicated are the positions of the membrane spanning regions and the conserved histidine residue (H:) that is the target for the autophosphorylation reaction upon ligand binding of the sensor protein. Similarly, the model for SpaK can be predicted, and the positions of SpaK membrane spanning regions and the histidine residue are indicated in parentheses.

This indicates that the full-length SpaK protein is 72 amino acid residues longer than the previously published protein (Klein et al., 1993) and contains all essential domains (Kleerebezem et al., manuscript in preparation). High identity is found between the sensor proteins NisK and SpaK (26.4%) that both belong to the EnvZ-like subgroup of known sensor proteins (Siezen et al., 1996). They share significant homology with, for example, the EnvZ sensor protein of *E. coli* and the PhoR sensor protein from *B. subtilis*. The latter similarity, in contrast to the former, is limited to the C-terminal "transmitter" domain. Likewise, remarkably high similarity is found between the response regulator proteins NisR and SpaR (41.3%) (Engelke et al., 1994), which both appear to belong to the OmpR-like subgroup of response regulators (Siezen et al., 1996), sharing significant homology with, for example, OmpR of *E. coli* and PhoP of *B. subtilis*.

The general model for sensor protein functioning predicts an interaction of the inducing ligand and the sensor molecule within the membrane-associated domain of the sensor protein (Parkinson and Kofoid, 1992). This model implies that the inducer molecule nisin interacts with the N-terminal domain of NisK, which, according to the topology model (Fig. 7), is enclosed between residues 14 and 167. Likewise, the inducer subtilin is supposed to interact with SpaK residues enclosed between residues 18 and 177 (Fig. 7). To validate this assumption, chimeric *spaK-nisK* genes were created that encode hybrid sensor proteins in which parts of the NisK input domain are replaced by the SpaK input domain (Fig. 7). These chimeric genes were expressed from a plasmid, in a strain that harbors a compatible *nisA* promoter-*gusA* fusion plasmid and has the *nisR* gene integrated in its chromosome (*pepN::nisR*). This experimental setup allowed quantitative analysis of the signal transduction efficiency by these hybrid two-component systems after induction with either nisin or subtilin. The results obtained in these experiments indicated that a functional chimeric sensor protein (that is able to activate *nisA* promoter activity with or without induction) is obtained only when (minimally) the entire input domain is derived from one of the original sensors (e.g., residues 1 to 167 in the NisK topology model or residues 1 to 177 in the SpaK topology model; Fig. 7). All hybrid sensors containing fusion points within these regions were unable to activate *nisA* promoter activity under any condition (Kleerebezem et al., manuscript in preparation), indicating that an intact input domain is the minimal requirement for input-domain functionality. Fusion of the complete input domain of SpaK (at least residues 1 to 177) to the transmitter domain of NisK resulted in functional hybrid sensor proteins that, via NisR, modulate the *nisA* promoter activity only when subtilin is present in the culture medium. Two types of these functional SpaK-NisK hybrids were obtained. One type exhibited "normal" sensor properties, i.e., no *nisA* promoter activity in the absence of subtilin and a linear dose-response correlation between promoter activity and the concentration of subtilin added to the culture medium. The second type exhibited "inverted" sensor properties, i.e., high *nisA* promoter activity in the absence of subtilin and a subtilin concentration-dependent reduction of this promoter activity (Kleerebezem et al., manuscript in preparation). The resulting sensory behavior of these fusion proteins (normal or inverted) appeared to depend on the position at which the two domains of the original sensors were joined in the hybrid protein (Kleerebezem et al., manuscript in preparation). This observation could generate insight into the mechanism of transmembrane signal transduction by this family of proteins.

These data strongly suggest that the subdomain of these sensors required for inducer interaction is located in the N-terminal domain and includes at least both transmembrane domains and the enclosed extracellular domain (SpaK, 1-177; NisK, 1-167; Fig. 7) (Kleerebezem et al., manuscript in preparation), which is in analogy with the model proposed for these sensor-inducer interactions (Park-

inson and Kofoid, 1992). Moreover, it is clear that these input domains contain all the information required to trigger the signal-transducing activity of the C-terminal transmitter domain, and that these input domains are interchangeable.

CONCLUDING REMARKS

From the data presented here, it is clear that an autoregulatory circuit is involved in the regulation of biosynthesis of both nisin and subtilin by *Lactococcus lactis* and *B. subtilis*, respectively. Therefore, besides their function as AMPs, these lantibiotics both exhibit a peptide pheromone function that plays an essential role in the regulation of their biosynthesis. The finding that the production of these lantibiotics is regulated in a growth phase-dependent manner (Gutowski-Eckel et al., 1994; de Ruyter et al., 1996b) indicates that these autoregulatory systems can be qualified as peptide-pheromone-induced quorum-sensing systems, which appears to be a common theme in growth phase-dependent regulation in gram-positive bacteria (Nes et al., 1996; Kleerebezem et al., 1997b; Dunny and Leonard, 1997).

A third gene cluster from *S. epidermidis*, involved in the biosynthesis of the lantibiotic epidermin, contains a gene encoding a regulatory protein, *epiQ* (Schnell et al., 1992), that shares significant homology with response regulator proteins but lacks a gene encoding the corresponding sensor protein. Since the homology between EpiQ and either NisR or SpaR is relatively low, EpiQ appears to belong to a different family of response regulators. Remarkably, the highest similarity of EpiQ and other response regulators is limited to the C-terminal domain of EpiQ, whereas normally the most conserved regions within response regulators are located in their N-terminal domain (Stock et al., 1989). Nevertheless, it was shown that EpiQ regulates epidermin production by binding to an inverted repeat that is located just upstream of the −35 region of the *epiA* (encoding the epidermin precursor) promoter (Peschel et al., 1993). Although no experimental data are available, on the basis of what has been found for nisin and subtilin autoregulation, one could speculate that regulation of epidermin biosynthesis is dependent on fully modified epidermin via a partially unknown induction system involving EpiQ and possibly a chromosomally encoded sensor protein. Alternatively, based on the lack of homology of EpiQ and NisR or SpaR, it could be that epidermin biosynthesis is regulated via a completely different mechanism that is triggered by an environmental factor (not epidermin) that has not been identified so far.

The observation that an AMP can also act as signal molecule involved in induction of its own biosynthesis is not unique to the lantibiotics nisin and subtilin. It has recently been shown that class II AMP CB2, which is produced by *C. piscicola*, is involved in the regulation of its own biosynthesis (Saucier et al., 1995; Quadri et al., 1996, 1997, submitted). Remarkably, the production of CB2 can also be triggered by another peptide (CbnS) that shares specific characteristics with the type II AMPs but lacks antimicrobial activity (Quadri et al., 1997, submitted; see also chapter 12 of this volume). Similar to this system found in *C. piscicola*, it has been shown that biosynthesis of several class II AMPs by different lactobacilli is also induced by a secreted, AMP-like (but lacking AMP activity) peptide (see chapter 12 of this volume). However, in contrast to the system found in *C. piscicola*, the class II AMPs produced by these lactobacilli do not function as inducers of their own biosynthesis, and their biosynthesis appears to be strictly dependent on the AMP-like inducer peptide.

In conclusion, several secreted antimicrobial peptides produced by gram-positive bacteria, belonging to either the family of lantibiotics or class II AMPs, appear to have a second function as peptide pheromones involved in the regulation of their own biosynthesis. In contrast, the regulation of production of several other AMPs belonging to the class II family of AMPs appears to depend on a specific autoregulated peptide that shares certain characteristics with class II AMPs but lacks any antimicrobial activity.

Clearly, it would make biological sense to control the concerted production of antimicrobial peptides in response to high cell density. This mechanism of regulation ensures that the concentration of the AMP in the environment rapidly reaches a level sufficient to kill the competitors of the producing cells. Moreover, by massive production of the AMP at higher cell densities, the producer cell does not allow target organisms to initiate or develop a defense reaction and thus ensures the effectiveness of the AMP. Alternatively, one could reason that producing cells are protected from ineffective high-level production of the AMP when growth conditions are such that the inducing agent (which in the case of nisin and subtilin is the AMP itself) rapidly diffuses away from the cellular environment, thereby preventing accumulation of this inducing agent to the level required to trigger the concerted production of the AMP.

ACKNOWLEDGMENTS

Part of the work presented here was supported by the European Community Biotechnology Program (Contracts BIOT-CT94-3055 and BIOT-CT96-0498). We thank Geert Ellen, Roland Siezen, and Jeroen Hugenholtz for critically reading this manuscript and Pascalle de Ruyter for making data available before publication. We are also grateful to Karl-Dieter Entian and Stefan Borchert of the Frankfurt University for generously supplying the *spaB* disruption mutant of *B. subtilis* ATCC 6633, which plays an essential role in a number of the experiments described.

REFERENCES

Buchman, G. W., S. Banerjee, and J. N. Hansen. 1988. Structure, expression, and evolution of a gene encoding the precursor of nisin, a small protein antibiotic. *J. Biol. Chem.* **263:**16260–16266.

Chung, Y. J., and N. J. Hansen. 1992. Determination of the sequence of *spaE* and identification of a promoter in the subtilin (*spa*) operon of *Bacillus subtilis. J. Bacteriol.* **174:**6699–6702.

de Ruyter, P. G. G. A., O. P. Kuipers, M. M. Beerthuyzen, I. van Alen-Boerrigter, and W. M. de Vos. 1996a. Functional analysis of promoters in the nisin gene cluster of *Lactococcus lactis. J. Bacteriol.* **178:**3434–3439.

de Ruyter, P. G. G. A., O. P. Kuipers, and W. M. de Vos. 1996b. Controlled gene expression systems for *Lactococcus lactis* with the food-grade inducer nisin. *Appl. Environ. Microbiol.* **62:** 3662–3667.

de Ruyter, P. G. G. A., O. P. Kuipers, and W. M. de Vos. 1997. Food-grade controlled lysis of *Lactococcus lactis* for accelerated cheese ripening. *Nat. Biotechnol.* **15:**976–979.

de Ruyter, P. G. G. A., M. Kleerebezem, O. P. Kuipers, and W. M. de Vos. Manuscript in preparation.

de Vos, W. M., G. Jung, and H.-G. Sahl. 1991. Appendix: Definitions and nomenclature of lantibiotics, p. 457–464. *In* J. Jung and H.-G. Sahl (ed.), *Nisin and Novel Lantibiotics: Proceedings of the First International Workshop on Lantibiotics.* Escom Publishers, Leiden, The Netherlands.

de Vos, W. M., J. W. M. Mulders, R. J. Siezen, J. Hugenholtz, and O. P. Kuipers. 1993. Properties of nisin Z and distribution of its gene, *nisZ,* in *Lactococcus lactis. Appl. Environ. Microbiol.* **59:** 213–218.

de Vos, W. M., O. P. Kuipers, J. R. van der Meer, and R. J. Siezen. 1995a. Maturation pathway of nisin and other lantibiotics: posttranslationally modified antimicrobial peptides exported by gram-positive bacteria. *Mol. Microbiol.* **17:**427–437.

de Vos, W. M., M. M. Beerthuyzen, E. J. Luesink, and O. P. Kuipers. 1995b. Genetics of the nisin operon and the sucrose-nisin conjugative transposon Tn*5276. Dev. Biol. Stand.* **85:**617-625.

Dodd, H. M., N. Horn, W. C. Chan, C. J. Giffard, B. W. Bycroft, G. C. K. Roberts, and M. J. Gasson. 1996. Molecular analysis of the regulation of nisin immunity. *Microbiology* **142:** 2385–2392.

Dunny, G. M., and B. A. B. Leonard. 1997. Cell-cell communication in Gram-positive bacteria. *Annu. Rev. Microbiol.* **51:**527–564.

Eichenbaum, Z., M. J. Federle, D. Marra, W. M. de Vos, O. P. Kuipers, M. Kleerebezem, and J. R. Scott. 1998. Use of the lactococcal *nisA* promoter to regulate gene expression in gram-positive bacteria: comparison of induction level and promoter strength. *Appl. Environ. Microbiol.* **64:**2763–2769.

Engelke, G., Z. Gutowski-Eckel, M. Hammelman, and K.-D. Entian. 1992. Biosynthesis of the lantibiotic nisin: genomic organisation and membrane localization of the NisB protein. *Appl. Environ. Microbiol.* **58:**3730–3743.

Engelke, G., Z. Gutowski-Eckel, P. Kiesau, K. Siegers, M. Hammelman, and K.-D. Entian. 1994. Regulation of nisin biosynthesis and immunity in *Lactococcus lactis* 6F3. *Appl. Environ. Microbiol.* **60:**814–825.

Gutowski-Eckel, Z., C. Klein, K. Siegers, K. Bohm, M. Hammelmann, and K.-D. Entian.

1994. Growth phase-dependent regulation and membrane localization of SpaB, a protein involved in biosynthesis of the lantibiotic subtilin. *Appl. Environ. Microbiol.* **60:**1–11.

Jung, G. 1991. Lantibiotics: a survey, p. 1–34. *In* J. Jung and H.-G. Sahl (ed.), *Nisin and Novel Lantibiotics: Proceedings of the First International Workshop on Lantibiotics.* Escom Publishers, Leiden, The Netherlands.

Kaletta, C., and K.-D. Entian. 1989. Nisin, a peptide antibiotic: cloning and sequencing of the *nisA* gene and posttranslational processing of its peptide product. *J. Bacteriol.* **171:**1597–1601.

Klaenhammer, T. R. 1993. Genetics of bacteriocins produced by lactic acid bacteria. *FEMS Microbiol. Rev.* **12:**39–86.

Kleerebezem, M., M. M. Beerthuyzen, E. E. Vaughan, W. M. de Vos, and O. P. Kuipers. 1997a. Controlled gene expression systems for lactic acid bacteria: transferable nisin inducible expression cassettes for *Lactococcus, Leuconostoc* and *Lactobacillus* spp. *Appl. Environ. Microbiol.* **63:**4581–4584.

Kleerebezem, M., L. E. N. Quadri, O. P. Kuipers, and W. M. de Vos 1997b. Quorum sensing by peptide pheromones and two-component signal-transduction systems in gram-positive bacteria. *Mol. Microbiol.* **24:**895–904.

Kleerebezem, M., W. M. de Vos, and O. P. Kuipers. Manuscript in preparation.

Klein, C., C. Kaletta, N. Schnell, and K.-D. Entian. 1992. Analysis of genes involved in biosynthesis of the lantibiotic subtilin. *Appl. Environ. Microbiol.* **58:**132–142.

Klein, C., C. Kaletta, and K.-D. Entian. 1993. Biosynthesis of the lantibiotic subtilin is regulated by a histidine kinase/response regulator system. *Appl. Environ. Microbiol.* **59:**296–303.

Kuipers, O. P., M. M. Beerthuizen, R. J. Siezen, and W. M. de Vos. 1993. Characterization of the nisin gene cluster *nisABTCIPR* of *Lactococcus lactis,* requirement of expression of the *nisA* and *nisI* genes for development of immunity. *Eur. J. Biochem.* **216:**281–291.

Kuipers, O. P., M. M. Beerthuyzen, P. G. G. A. de Ruyter, E. J. Luesink, and W. M. de Vos. 1995. Autoregulation of nisin biosynthesis in *Lactococcus lactis* by signal transduction. *J. Biol. Chem.* **270:**27299–27304.

Kuipers, O. P., G. Bierbaum, B. Ottenwälder, H. M. Dodd, N. Horn, J. Metzger, T. Kupke, V. Gnau, R. Bongers, P. van den Bogaard, H. Kosters, H. S. Rollema, W. M. de Vos, R. J. Siezen, G. Jung, F. Götz, H.-G. Sahl, and M. J. Gasson. 1996. Protein engineering of lantibiotics. *Antonie van Leeuwenhoek* **69:**161–170.

Kuipers, O. P., P. G. G. A. de Ruyter, M. Kleer-ebezem, and W. M. de Vos. 1997. Controlled overproduction of proteins by lactic acid bacteria. *Trends Biotechnol.* 15:135–140.

Mattick, A. T. R., and A. Hirsch. 1944. A powerful inhibitory substance produced by group N streptococci. *Nature* **154**:551.

Moran, C. P., Jr. 1990. Expression of sigma A and sigma H regulons during stationary phase and endospore formation, p. 287–294. *In* M. M. Zukowski, A. T. Ganesam, and J. A. Hoch (ed.), *Genetics and Biotechnology of Bacilli,* vol. 3. Academic Press, San Diego, Calif.

Mulders, J. W. M., I. J. Boerigter, H. S. Rollema, R. J. Siezen, and W. M. de Vos. 1991. Identification and characterization of the lantibiotic nisin Z, a natural nisin variant. *Eur. J. Biochem.* **201**:581–584.

Nes, I. F., D. B. Diep, L. S. Håvarstein, M. B. Brurberg, V. Eijsink, and H. Holo. 1996. Biosynthesis of bacteriocins in lactic acid bacteria. *Antonie van Leeuwenhoek* **70**:17–32.

Parkinson, J. S., and E. C. Kofoid. 1992. Communication modules in bacterial signalling proteins. *Annu. Rev. Genet.* **26**:71–112.

Peschel, A., J. Augustin, T. Kupke, S. Stevanovic, and F. Götz. 1993. Regulation of epidermin biosynthetic genes by *epiQ. Mol. Microbiol.* **9**:31–39.

Platteeuw, C., G. Simons, and W. M. de Vos. 1994. Use of the *Escherichia coli* β-glucuronidase (*gusA*) gene as a reporter gene for analyzing promoters in lactic acid bacteria. *Appl. Environ. Microbiol.* **60**:587–593.

Platteeuw, C., I. van Alen-Boerrigter, S. van Schalkwijk, and W. M. de Vos. 1996. Foodgrade cloning and expression system for *Lactococcus lactis. Appl. Environ. Microbiol.* **62**:1008–1013.

Quadri, L. E. N., L. Z. Yan, M. E. Stiles, and J. C. Vederas. 1996. Effect of amino acid substitutions on the activity of carnobacteriocin B2: overproduction of the antimicrobial peptide, its engineered variants and its precursor in *Escherichia coli. J. Biol. Chem.* **272**:3384–3388.

Quadri, L. E. N., M. Kleerebezem, O. P. Kuipers, W. M. de Vos, K. L. Roy, J. C. Vederas, and M. E. Stiles. 1997a. Characterization of a locus from *Carnobacterium piscicola* LV17B involved in bacteriocin production and immunity: evidence for global inducer-mediated transcription regulation. *J. Bacteriol.* **179**:6163–6171.

Quadri, L. E. N., M. Kleerebezem, O. P. Kuipers, W. M. de Vos, and M. E. Stiles. Two-component signal-transduction cascade from *Carnobacterium piscicola* LV17B: two signaling peptides and one sensor-transmitter. Submitted for publication.

Ra, S. R., M. Qiao, T. Immonen, I. Pujana, and P. E. J. Saris. 1996. Genes responsible for nisin synthesis, regulation and immunity form a regulon of two operons and are induced by nisin in *Lactococcus lactis* N8. *Microbiology* **142**:1281–1288.

Sahl, H.-G. 1991. Pore formation in bacterial membranes by cationic lantibiotics, p. 347–359. *In* J. Jung and H.-G. Sahl (ed.), *Nisin and Novel Lantibiotics: Proceedings of the First International Workshop on Lantibiotics.* Escom Publishers, Leiden, The Netherlands.

Saucier, L., A. Poon, and M. E. Stiles. 1995. Induction of bacteriocin in *Carnobacterium piscicola* LV17. *J. Appl. Bacteriol.* **78**:684–690.

Schnell, N., K.-D. Entian, U. Schneider, F. Götz, F. Zähner, R. Kellner, and G. Jung. 1988. Prepeptide sequence of epidermin, a ribosomally synthesized antibiotic with four sulphide-rings. *Nature* **333**:276–278.

Schnell, N., G. Engelke, J. Augustin, R. Rosenstein, V. Ungermann, F. Götz, and K.-D. Entian. 1992. Analysis of genes involved in the biosynthesis of lantibiotic epidermin. *Eur. J. Biochem.* **204**:57–68.

Siegers, K., and K.-D. Entian. 1995. Genes involved in immunity to the lantibiotic nisin produced by *Lactococcus lactis. Appl. Environ. Microbiol.* **61**:1082–1089.

Siegers, K., S. Heinzmann, and K.-D. Entian. 1996. Biosynthesis of the lantibiotic nisin: posttranslational modification of its prepeptide occurs at a multimeric membrane-associated lanthionine synthetase complex. *J. Biol. Chem.* **271**:12294–12301.

Siezen, R. J., H. S. Rollema, O. P. Kuipers, and W. M. de Vos. 1995. Homology modelling of the *Lactococcus lactis* leader peptidase NisP and its interaction with the precursor of the lantibiotic nisin. *Protein Eng.* **8**:117–125.

Siezen, R. J., O. P. Kuipers, and W. M. de Vos. 1996. Comparison of lantibiotic gene clusters and encoded proteins. *Antonie van Leeuwenhoek* **69**:171–184.

Stock, J. B., A. J. Ninfa, and A. N. Stock. 1989. Protein phosphorylation and regulation of adaptive responses in bacteria. *Microbiol. Rev.* **53**:450–490.

van der Meer, J. R., J. Polman, M. M. Beerthuizen, R. J. Siezen, O. P. Kuipers, and W. M. de Vos. 1993. Characterization of the *Lactococcus lactis* nisin A operon genes *nisP,* encoding a subtilisin-like serine protease involved in precursor processing, and *nisR,* encoding a regulatory protein involved in nisin biosynthesis. *J. Bacteriol.* **175**:2578–2588.

van der Meer, J. R., H. S. Rollema, R. J. Siezen, M. M. Beerthuyzen, O. P. Kuipers, and W. M. de Vos. 1994. Influence of amino acid substitutions in the nisin leader peptide on biosynthesis and secretion of nisin by *Lactococcus lactis. J. Biol. Chem.* **269:**3555–3562.

van Kraaij, C., E. Breukink, H. S. Rollema, R. J. Siezen, R. A. Demel, B. de Kruijff, and O. P. Kuipers. 1997. Influence of charge differences in the C-terminal part of nisin on antimicrobial activity and signaling capacity. *Eur. J. Biochem.* **247:**114–120.

von Heijne, G. 1992. Membrane protein structure prediction. Hydrophobicity analysis and the positive-inside rule. *J. Mol. Biol.* **225:**487–494.

REGULATION OF GROUP II PEPTIDE BACTERIOCIN SYNTHESIS BY QUORUM-SENSING MECHANISMS

Ingolf F. Nes and Vincent G. H. Eijsink

12

Prokaryotic and eukaryotic cells produce and secrete a variety of chemical substances that inhibit the growth of bacteria. Among these substances are the so-called bacteriocins, which are produced by bacteria and which are of proteinaceous nature. Bacteriocins constitute a large and diverse group of ribosomally synthesized antimicrobial proteins or peptides that may or may not undergo posttranslational modification. The term bacteriocins was originally defined quite specifically to refer to the colicin type of antimicrobial proteins produced by *Escherichia coli*. Distinguishing features of the colicin type of inhibitory proteins include their relatively high molecular weights and their narrow spectra of inhibitory activity (Pattus et al., 1990; Konisky, 1982; Pugsley, 1984). Nowadays, the term bacteriocin is used to define a much broader group of antimicrobial compounds. One well-studied group of bacteriocins are the antimicrobial peptides that are produced by lactic acid bacteria (LAB) (Klaenhammer, 1993; Jack et al., 1995; Nes et al., 1996).

In recent years LAB have been the focus of research because of their industrial application in the production of food and feed and because they are considered probiotic organisms. LAB are used in the production of most dairy products, such as cheese, yogurt, and butter, as well as in the production of a large variety of fermented meat and vegetable products. Furthermore, LAB are believed to play an important role in protecting mammals from microbial infections by being part of the normal and healthy microbial flora of the oral cavity, intestine, colon, vagina, and other body cavities, where they supposedly prevent colonization of harmful microorganisms.

The most important industrial property of LAB is their ability to preserve food and feed through a fermentation process that requires the presence of fermentable carbohydrates and that results in the production of lactic acid and a drop in pH. It has been known for a long time that LAB preserve food in one additional way, namely, by producing proteinaceous compounds that prevent growth of food-spoiling and pathogenic bacteria. During the last 10 years or so, it has been shown that these proteinaceous compounds can be classified as bacteriocins, but their chemical composition and biology can be both complex and diverse. These bacteriocins are found in many gram-

Ingolf F. Nes and Vincent G. H. Eijsink, Laboratory of Microbial Gene Technology, Department of Biotechnological Sciences, Agricultural University of Norway, P.O. Box 5051, N-1432 Ås, Norway.

Cell-Cell Signaling in Bacteria, Edited by Gary M. Dunny and Stephen C. Winans
©1999 American Society for Microbiology, Washington, D.C.

positive bacteria as well as in *E. coli* (Nes et al., 1996).

THE PEPTIDE BACTERIOCINS

In contrast to the original colicins, which are proteins of normal sizes, the bacteriocins found in LAB and other gram-positive bacteria are much smaller. They usually consist of only 30 to 60 amino acids and are therefore often referred to as peptide bacteriocins (Klaenhammer, 1993; Nes et al. 1996).

Two major classes of peptide bacteriocins have been identified in gram-positive bacteria: class I and II. Both classes contain small, ribosomally synthesized peptides that are cationic (pI usually between 9 and 11), hydrophobic or amphiphilic, and act by inducing leakage through the cell membrane of their target cells. The antimicrobial spectrum of these bacteriocins varies from a small group of very closely related microorganisms to a broader group of microorganisms that includes more distantly related species such as *Staphylococcus aureus*, *Listeria monocytogenes*, *Bacillus cereus*, and various clostridia. (Casaus et al., 1997; Cintas et al., 1997). Most bacteriocins seem to be primarily directed against closely related species that may be expected to compete for the same environment as the producing organisms, and the bacteriocin producers may therefore provide a selective advantage (Dykes, 1995).

Class I bacteriocins are the lantibiotics (Sahl et al., 1995); their most prominent member, nisin, is discussed elsewhere in this volume (see chapter 11). The lantibiotics are unique in that they contain modified amino acid residues, including lanthionine, methyllanthionine, didehydroalanine, and didehydrobutyrine. The lantibiotics will not be discussed further in this chapter.

Class II bacteriocins do not usually contain posttranslationally modified amino acids as found in lantibiotics. While the lantibiotics have been found exclusively in gram-positive bacteria, class II bacteriocins have also been found in *E. coli* (colicin V and microcin 24) (Fath et al., 1994; Håvarstein et al., 1994,

O'Brian et al., 1996). It must be emphasized that, although class II bacteriocins have been found most frequently in LAB, they are also found in several other gram-positive bacteria.

Initial genetic studies of bacteriocin production led to the general view that synthesis and secretion of class II bacteriocins requires the products of only four genes (Nes et al., 1995). These genes are the structural gene of the prebacteriocin, an immunity gene, and two genes encoding the secretion machinery, consisting of an ABC transporter and an accessory protein. The structural gene encodes a prebacteriocin containing an N-terminal leader sequence, which in almost all cases is of the so-called double-glycine type. This type of leader peptide is typical for bacteriocins and bacteriocin-like peptides and is characterized by certain conserved sequence elements as well as the presence of two glycine residues directly in front of the cleavage site (Håvarstein et al., 1994; see also below). The dedicated ABC transporter has been shown to contain an N-terminal proteolytic domain of approximately 170 residues that serves to cleave off the double-glycine leader peptide during or directly after transport over the cell membrane (Håvarstein et al., 1995b). Successful export of the bacteriocin also depends on the accessory protein, although the precise function of this protein is not known. In the absence of the accessory protein, the ABC transporter only cleaves off the double-glycine leader peptide without fulfilling the externalization of the bacteriocin (Venema et al., 1995). It has been shown that a few class II bacteriocins are transported out of the cell by the more general *sec*-dependent secretory system. In these cases the bacteriocin precursors contain typical *sec*-type leader peptides (Leer et al., 1995; Worobo et al., 1995; Cintas et al. 1997).

The presence of the leader peptide of the prebacteriocin most probably keeps the antimicrobial peptide in an inactive state as long as it is inside the producing cell. The immunity protein protects the producing cell against its own, mature bacteriocin coming from outside.

For a long time it has been known that bacteriocin production can sometimes be an unstable trait. This observation was most often explained by plasmid loss, since many bacteriocins are encoded by plasmid-borne genes (Klaenhammer, 1993). Alternatively, transposition of endogenous insertion elements into genes required for bacteriocin production has been shown to cause loss of the Bac$^+$ phenotype (Skaugen and Nes, 1994). The general observation that bacteriocin production in LAB is highly dependent on growth phase and growth conditions suggested, however, that more subtle regulatory mechanisms may, at least in part, cause the instability of the Bac$^+$ phenotype in some cases. Subtle regulatory mechanisms for bacteriocin production are known for some colicins. Plasmid-encoded colicin E1 is not expressed under ordinary growth conditions but is synthesized after treatment with DNA-damaging agents. This activation of ColE1 genes is considered to be an SOS response through the RecA inactivation of LexA repressor blocking the *colE1* transcription (Ebina et al., 1983). But positive regulation of expression of the colicin E1 gene by cyclic AMP and cyclic AMP receptor protein has also been reported (Shirabe et al., 1985).

In recent years it has been shown that, indeed, strict regulatory mechanisms exist for the regulation of the production of class II bacteriocins. These mechanisms are based on a secreted peptide pheromone that permits quorum sensing and thus permits cell density-dependent regulation of bacteriocin production.

THE THREE-COMPONENT REGULATORY SYSTEM

The Inducing Peptide (IP)— the Pheromone

During studies with the bacteriocin producer *Lactobacillus plantarum* C11, it was noticed that in some culture media the Bac$^+$ phenotype

was lost upon extensive dilution of producing cultures (Diep et al., 1995). The only way to retrieve the Bac$^+$ phenotype was by the addition of small amounts of spent culture medium from a Bac$^+$ culture. Initial studies indicated that the inducing component in spent culture medium was heat resistant and of proteinaceous nature, suggesting it to be a peptide. The inducing peptide (IP) was purified and found to be the 26-residue peptide termed plantaricin A (hereafter referred to as IP-C11) (Diep et al., 1995). The 26-mer as well as N-terminally truncated variants of 23, 22, and 15 residues long were synthesized, and the longest three of these peptides were shown to have inducing activity. At the time, IP-C11 was erroneously considered to be a bacteriocin, but it was later shown that the bacteriocin activity found in IP-C11 preparations is due to small amounts of contaminating "real" bacteriocins (Nes et al., 1996; Anderssen et al., 1998).

Bacteriocin production in *Lactobacillus sake* LTH673 also turned out to depend on a peptide pheromone, in a way very similar to that in *L. plantarum* C11. The pheromone (IP-673) was isolated from the culture medium of Bac$^+$ cultures and found to be a 19-mer with no significant bacteriocin activity but high inducing activity (Eijsink et al., 1996). A synthetic variant of IP-673 had full induction activity. As in *L. plantarum*, induction of Bac$^-$ cells with the IP resulted in the induction of production of both bacteriocins and the IP itself. Thus, the IP seemed to regulate bacteriocin production by an autoinduction mechanism that permitted rapid and explosive production of bacteriocins.

The genes for IP-C11 and IP-673 are both cotranscribed with genes encoding a two-component regulatory system (Diep et al., 1994, 1996; Hühne et al., 1996; Brurberg et al., 1997; see below for further details). On the basis of analogy to this particular genetic organization and/or by induction studies similar to the ones described above, peptide pheromones have now been discovered in other bacteriocin-producing LAB. Production of sakacin A by *L. sake* Lb706 is regulated by a 23-residue IP (IP-706) (Diep et al., manuscript in preparation); production of carnobacteriocins by *Carnobacterium piscicola* LV17 is, at least in part, regulated by a 24-residue IP (IP-LV17) (Saucier et al., 1995, 1997; Quadri et al., 1997); and bacteriocin production in *Enterococcus faecium* CTC492 is regulated by a 25-residue IP (IP-492) (Nilsen et al., 1998). The peptide pheromones are depicted in Fig. 1A.

Interestingly, the recent studies of the production of peptides (IPs or bacteriocins) by LAB, as well as concomitant genetic studies have revealed that many LAB produce more than one bacteriocin. Thus, the induction of gene expression that is accomplished by the

IPs results in the production of a battery of antimicrobial peptides (Diep et al., 1996; Brurberg et al., 1997; Quadri et al., 1994, 1997; Aymerich et al., 1996; Casaus et al., 1997; Anderssen et al., 1998; Eijsink et al., 1998). These peptides often have complementary antimicrobial activities, thus providing the producing bacterium with expanded competitiveness.

There are striking structural similarities between IPs and bacteriocins. Both types of peptides are strongly cationic, partly amphiphilic, and they adopt a partly helical structure in membrane-mimicking solvents and liposomes (Eijsink et al., 1996; Hauge et al., 1998, in press). Furthermore, they are produced as pre-

A	*Leader*	*IP*
pre IP-C11	*mkiqikgmkqlsnkemqkivgg*	**SSAYSLQMGATAIKQVKKLFKKWGW**
pre IP-706	*mklnyiekkqltnkqlkliigg*	**TNRNYGKPNKDIGTCIWSGFRHC**
pre IP-673	*mifkklsekelqkingg*	**MAGNSSNFIHKIKQIKTHR**
pre IP-LV17	*mkiktitkkqliqikgg*	**SKNSQIGKSTSSISKCVFSFFKKC**
pre IP-492	*meeknrlnakqcsdqelkkikgg*	**AGTKPQGKPASNLVECVFSLFKKCN**
Concensus leader :	*m.....................ls--el--i-gg*	

FIGURE 1 Inducing peptides (A) and the genetic organization of *bac* gene clusters (B). (A) The precursors of the (putative) inducing peptides from *L. plantarum* C11 (IP-C11), *L. sake* Lb706 (IP-706), *L. sake* LTH673 (IP-673), *C. piscicola* LV17 (IP-LV17), and *E. faecium* CTC492 (IP-492) are shown. The leader peptides resemble the typical double-glycine type of leader peptides that are normally found in bacteriocin precursors (see Håvarstein et al. [1994] for comparison). (B) Genetic organization of genes under control of the inducing peptide–histidine kinase–response regulator (IP-HPK-RR) system in *L. sake* LTH673 (Hühne et al., 1996; Brurberg et al., 1997), *L. plantarum* C11 (Diep et al., 1996), *L. sake* Lb706 (Axelsson and Holck, 1995), and *C. piscicola* LV17 (Quadri et al., 1997; Saucier et al., 1997). Below the arrows, open reading frames are denoted by the code used in the original publications. An asterisk indicates that the open reading frame has not been sequenced completely. Above the arrows, the genes are denoted by their (putative) functions: Im, immunity; B, bacteriocin; HPK, histidine kinase; RR, response regulator; IP, inducing peptide. Open arrows indicate open reading frames encoding proteins with unknown functions. P→ indicates promoters that are (putatively) regulated by the IP-HPK-RR system (see Fig. 3 for more details). For detailed analyses of the transcription of the shown operons, see the cited references. The *cbn* genes shown are from two different loci (one chromosomal and one on a plasmid; the vertical line in the figure indicates the separation) but are regulated by the same three-component (IP-HPK-RR) system. The IS*1163* element inserted into the *sap* cluster contains two open reading frames (*orf2* and *orf3*) that encode proteins with sequence similarities to transposases (Axelsson and Holck, 1995).

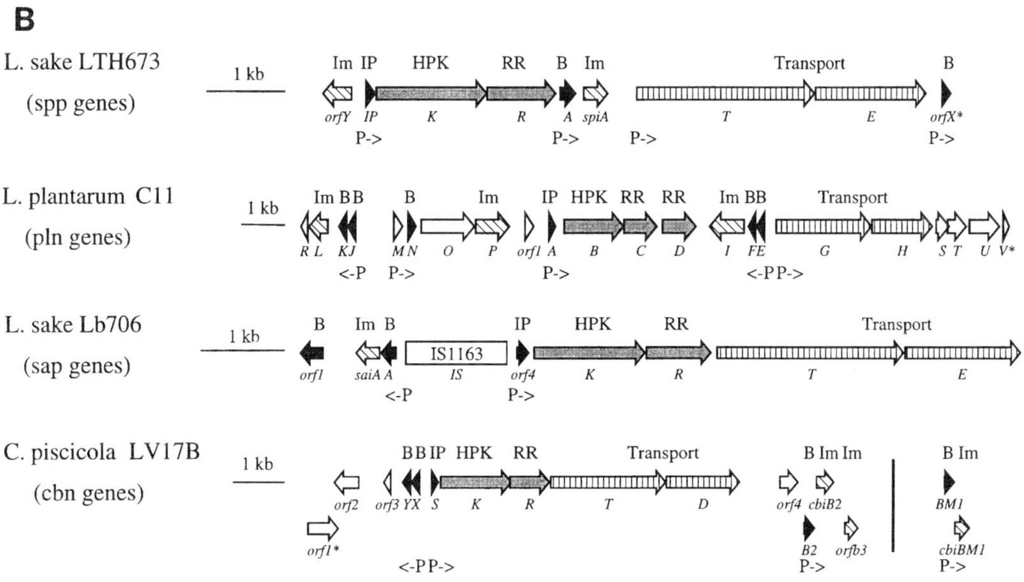

B

FIGURE 1 *(Continued)*

peptides containing a typical "bacteriocin-like" N-terminal double-glycine leader peptide (Fig. 1A), suggesting that IPs and bacteriocins are processed and secreted by the same ABC transporter system. The main difference between IPs and bacteriocins is their length. Whereas the smallest bacteriocins are more than 30 residues long, the largest induction factor known so far, IP-C11, consists of 26 amino acids. Shorter peptides can function as IPs, as illustrated by the natural 19-mer IP-673 as well as by the effectiveness of truncated forms of IP-C11 (Diep et al., 1995).

There is significant homology between some of the five known IPs depicted in Fig. 1A. Sequence analysis revealed that IP-492 and IP-LV17 peptides from *E. faecium* CTC492 and *C. piscicola* LV17, respectively, share 44% identity. IP-673 also shares significant homology with the above-mentioned IPs (Nilsen et al., 1998). The IPs are highly strain specific (Brurberg et al., 1997), and it will be of interest to see which structural characteristics of the peptide pheromone determine this specificity. All IPs act at very low concentrations, usually considerably lower than the bac-

teriocin concentrations needed for obtaining antimicrobial activity. Studies with *L. plantarum* C11 and *L. sake* LTH673 have shown that the concentration of IP-C11 and IP-673, respectively, needed for induction of bacteriocin production is in the order of 0.1 nM (Diep et al., 1995; Eijsink et al., 1996). Interestingly, it has been reported that only femtomolar concentrations of IP-492 are needed to induce bacteriocin production in Bac$^-$ cells of *E. faecium* CTC492 (Nilsen et al., 1998). These apparent differences in inducing power between the pheromones were also reflected in the production levels in inducible cells. The amount of pheromone produced by the enterocin-producing *E. faecium* CTC492 was much lower than the amounts of pheromones found in Bac$^+$ cultures of *L. sake* LTH673 and *L. plantarum* C11 (Nilsen et al., 1998).

Signal Transduction; the Histidine Kinase and the Response Regulator

Signal transduction, which is part of the bacterial response to environmental changes, normally involves two proteins with different functions that together make up a so-called two-component regulatory system (e.g., Stock et al., 1989, 1995): a histidine protein kinase (HPK) serving as an environmental sensor monitoring specific signals, and a response regulator that is activated by its cognate HPK to trigger the adaptive response of the host. Such two-component regulatory systems are found in most bacteria, but their number is variable. It has been estimated that enteric bacteria harbor some 50 different two-component systems mediating the response to a variety of chemical and physical signals (Stock et al., 1995).

All IPs (Fig. 1A) are cotranscribed with two genes encoding such a two-component regulatory system (Fig. 1B) (Diep et al., 1995; Axelsson and Holck, 1995; Brurberg et al., 1997; Quadri et al., 1997). Computer-assisted analysis of the HPKs of the class II bacteriocins identified the domains and sequence motifs that are commonly found within this superfamily of HPK, in particular, the well-conserved sequence that generally extends for approximately 200 residues in the C-terminal domain and corresponds to an ATP binding and histidine phosphorylation catalyzing domain. The N-terminal halves of most HPKs contain stretches of hydrophobic amino acids that presumably form transmembrane alpha-helices. Whereas two putative transmembrane regions are most commonly found in HPKs, the IP-associated HPKs, including *agr*-HPK, seem to have six to eight membrane-spanning stretches (Fig. 2) (Diep et al., 1994; Håvarstein et al., 1996).

The IPs described in this study resemble the competence peptide pheromones of streptococci (chapter 2 of this volume; Håvarstein et al., 1995a). Furthermore, the HPKs known to be involved in transducing the streptococcal competence signal (Håvarstein et al., 1996) seem to have the same topology as the IP-associated HPKs of bacteriocins (Fig. 2). Studies on closely related streptococcal systems indicated that the strain specificity determining part of the HPK competence-pheromone receptor is located in the N-terminal end of the HPK (Håvarstein et al., 1996). Because of

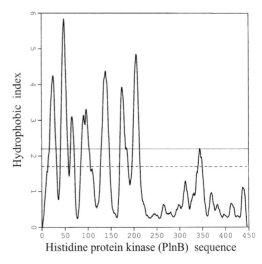

FIGURE 2 Prediction of transmembrane helices in HPK of the plantaricin system (encoded by the *plnB* gene from *L. plantarum* C11). The calculations and plotting were performed by using the Dense Alignment Surface method introduced by Cserzo et al. (1997). The solid line (———) indicates a strict cutoff, and the dashed line (- - - - -) indicates a loose cutoff value for the prediction of transmembrane helices in the protein.

the strong similarity between the regulatory systems for LAB bacteriocins and the *Streptococcus* competence system, it is likely that the strain- and IP-specific part of the bacteriocin HPK receptor is also located in the most N-terminal part of the protein. In the bacteriocins, however, sequence variability is not as exclusively limited to the first putative transmembrane helices as in *Streptococcus*. Instead, variability seems to involve the complete N-terminal, putatively membrane-spanning domain. In this context it is interesting to note that the HPKs belonging to IP-492 and IP-LV17 (44% identity between the pheromones) have N-terminal parts that show the most homology among the bacteriocin HPKs.

Recently Brurberg et al. (1997) have described experiments that confirm the hitherto putative role of the product of *sppK* (an HPK whose gene is cotranscribed with the gene encoding IP-673; see Fig. 1B) as the receptor for IP-673 (see below).

Since the IPs are known to adopt a partly alpha-helical structure in the presence of membrane-mimicking solvents and liposomes (Eijsink et al., 1996; Hauge et al., 1998) one could envisage that interaction of IP with the HPK involves a previous and/or concomitant interaction with the cell membrane.

It should be noted that the *agr* operon of *S. aureus,* which regulates its virulence phenotype, contains a two-component regulatory system of which *agrC* encodes an HPK with significant homology to bacteriocin HPKs. The similarity includes a predicted topology with six to eight transmembrane helices (Janzon et al., 1989; Ji et al., 1995, 1997). Of particular interest is the observation that production of the LAB bacteriocin plantaricin S involves a nonregular three-component regulatory system. In this case, the putative IP gene does not encode a double-glycine leader prepeptide, but rather a prepeptide that is probably processed differently and that could be more related to the pheromone in the *agr* system. (Ji et al., 1995; Stephens et al., manuscript in preparation).

The last step in transducing the signal to its target, the regulated gene, is the activation of a DNA binding protein that has been named the response regulator (RR). Among the regulatory systems associated with the synthesis of bacteriocins, only the plantaricin system from *L. plantarum* C11 contains two consecutive response regulators (PlnC and PlnD) following the pheromone gene (IP-C11, *plnA*) and the HPK (*plnB*) gene (Fig. 1B). These two response regulators are closely related, sharing almost 60% sequence identity. All other systems have only one response regulator (Fig. 1B). It has been shown that the products of *plnC* and *plnD* indeed bind to DNA regions just upstream of the transcriptional start sites of various *pln* operons (Risøen et al., 1998).

In conclusion, the IPs are cotranscribed with the genes encoding the proteins needed to transduce the inducing signal. As explained in detail below, induction by an IP results in the expression of all genes needed for the production of bacteriocins, the IP itself (autoinduction), and immunity proteins.

IP-Induced Transcription of *bac* Genes

Applying the generally accepted model for signal transduction through two-component systems (Stock et al., 1989; Parkinson, 1993), binding of the IP to its HPK receptor would result in autophosphorylation of this protein. Subsequently, the phosphate group is transferred to the RR, which then would mediate the response, in this case activation of gene expression. So far, it seems that the phosphorylated RR activates the transcription of *bac* genes directly by binding to specific elements in their promoters.

Transcription studies of Bac$^-$ and Bac$^+$ cells of *L. plantarum* C11 (Diep et al., 1995, 1996), *L. sake* LTH673 (Eijsink et al., 1996; Brurberg et al., 1997), and *C. piscicola* LV17 (Saucier et al., 1997; Quadri et al., 1997) have shown that mRNA of *bac* genes is difficult to detect in Bac$^-$ cells. Upon induction of such cells with the appropriate IP, a number of transcripts appear within 15 min. The rather extensive studies with *L. plantarum* C11 and *L. sake* LTH673 have shown that induction results in transcription of what are generally considered to be all genes necessary for the production of bacteriocins. The studies with *L. plantarum* C11 (Diep et al., 1996) disclosed that the IP-based regulatory mechanism affects the transcription of at least 22 genes, which encode for a variety of bacteriocins, IP-C11 itself, immunity proteins, a two-component signal transduction system, transport proteins, and a few proteins for which the function is not known (Fig. 1B).

Studies of the organization of various *bac* gene clusters as well as of their transcription show that there is variation in the way *bac* genes are organized and, therefore, in the size and composition of the different transcripts that are produced upon induction (Fig. 1B) (Axelsson and Holck, 1995; Hühne et al., 1996; Diep et al., 1996, manuscript in preparation; Brurberg et al., 1997; Quadri et al., 1997). Despite these differences, transcription of the *bac* operons depends in a similar way on an IP-activated three-component regulatory system in all cases that have been experimentally assessed so far. It is interesting to note that the IP-HPK-RR genes are in some cases cotranscribed with the two genes that encode the transport proteins (Fig. 1B) (*L. sake* Lb706 and *C. piscicola* LV17). These transport proteins are essential for activation and externalization of the pheromone, and it is therefore not surprising that their transcription may be coupled directly to the transcription of the IP gene.

As already mentioned, the IP pheromones share physicochemical properties with the bacteriocins. IP-673, which induces the production of sakacin P and other bacteriocins in *L. sake* LTH673, has no relevant bacteriocin activity (Eijsink et al., 1996); the only role of this peptide thus seems to be induction of *bac* gene expression. The somewhat longer IP-C11, however, does have modest bacteriocin activity in addition to its inducing potential (Anderssen et al., 1998). This modest activity, which possibly is of limited biological relevance, should not be confused with the high activity that has erroneously been ascribed to this peptide in one of the earlier studies (Nissen-Meyer et al., 1993). In the carnobacteriocin producer *C. piscicola* LV17, the situation may be more complex. In this case only one IP-HPK-RR gene cluster has been discovered, but bacteriocin production seems to be induced both by the IP and by one of the bacteriocins that are produced (Saucier et al., 1995; Quadri et al., 1996, 1997). This phenomenon is not fully understood, although it is tempting to speculate about some analogy to the regulation of nisin (see below, and chapter 11 of this volume).

Except for the induction of carnobacteriocins, the existing data do provide sufficient information to deduce a consistent and generally applicable definition of the roles of bacteriocins and IPs as bacteriocidal and/or regulating agents. The various studies referred to above make clear that striking similarities exist between the various peptides and indicate that

IPs and bacteriocins have a common ancestor, either a bacteriocin or a regulatory peptide.

The studies discussed above convincingly showed that transcription of *bac* genes is regulated by an autoinduction mechanism that is based on the use of an unmodified short-peptide cell-signaling factor (the IP). Similar regulatory schemes seem to be present in other gram-positive bacteria (e.g., the regulation of competence in *B. subtilis* [Magnusson et al., 1994] and of virulence in *S. aureus* [Ji et al., 1995]). The regulation of *bac* genes and the regulation of competence genes in *Streptococcus pneumoniae* are rather unique in that a simple, unmodified peptide functions as pheromone (see chapter 2 of this volume).

THE REGULATED PROMOTERS

Above, we described how *bac* genes are organized in transcriptional units in a way that varies from strain to strain. These transcriptional units are preceded by typical promoter elements that are recognizable by the presence of conserved direct and inverted repeats in the -40 to -80 region (Fig. 3). Such characteristic promoter elements have been detected in all *bac* gene clusters for which expression has been shown or suggested to be regulated by an IP (Diep et al., 1996; Brurberg et al., 1997; Quadri et al., 1997). The promoter elements do contain a normal -10 region, but their -35 regions are poorly conserved and hardly recognizable as such.

The characteristic elements in the -80 to -40 promoter region contain a direct repeat of approximately 10 residues with a spacing of approximately two turns of DNA. This is typical for regulated promoters, and one may envisage that the activated RR binds as a dimer to the -80 to -40 region. It has been shown that the RR indeed binds to a DNA fragment containing the direct repeat (Risøen et al., 1998), but details of the stoichiometry and precise location of binding are not yet available.

Promoter elements resembling the elements shown in Fig. 3 and those discussed above are also present in front of several bac-

Gene		Boxed promoter element	Downstream sequence
sppIP	TTC	**TTAACGTTAAT**CCGAAAAAAAC**TAACGTTAATA**	TTAAAAAATAAGATCCGCTTGTGAATTATGTATAAATTGATTA
sppA	ATA	**TTAACGTTTAA**CCGATAAAGTTGA**ACGTTAATA**	TTTTTTTGCGCAGAAATGGTAAATTGAAGCATAATAGTCTT
sppT	GCA	**TTAACGTTAAT**TTTGATAAACGTA**ACGTTAATG**	GATAATCATCCTGTTTACAAATAGTGTATGACATAATTAAGTA
orfX	ATA	**TTAGCGTTAA**CAGTTAAATTAATA**CGTTAATA**	ATTTTTTTGTCTTTAAATAGGGATTTGAAGCATAATGGTGTT
plnA	TGA	**TTCACGTTAA**AATTTAAAAAAAT**GTACGTTAATA**	GAAATAATTCCTCCGTACTTCAAAAACACATTATCCTAAAAGCG
plnJ	CAA	GTT**ACGTTAAA**TCGATTAAATAGTA**GCGATAACA**	AATTTAAAATAATTTTTTTTAAATTGTAGCGTATATTAATAAGTG
plnM	TAT	TGT**ACGATAA**TATCTAAAAATATT**ACGTTTATA**	AAAATATCGTACAATGATTCGACTAAGCGGTATATTAAAAGCG
plnE	TAT	**TTGACGTTAA**GAGAACGTTTTTTTT**ACTTTTATA**	ATTTTTTCAACAATCTGGTAAAAAAATAAATTAAACTAAATTTG
agrD	ATT	TAA**CAGTTAA**GTATTTATTCCT**ACAGTTA**GGC	AATATAATGATAAAAGATTGTACTAAATCGTATAATGACAGTG

FIGURE 3   Promoters regulated by the IP-HPK-RR system. Shown are the promoter elements found in front of *spp* genes from *L. sake* LTH673 (the upper four) (Brurberg et al., 1997) and *pln* genes from *L. plantarum* C11 (the next four) (Diep et al., 1996). See Fig. 1B for gene names and the location of the promoters. These promoter elements resemble regulated promoter elements found in the *agr* system in *S. aureus* (*agrD* promoter) (Janzon et al., 1989). The boxed part contains directly repeated sequences in the −40 to −80 region, which are underlined; within this box, residues present in at least six of the nine sequences shown are printed in bold. To the right of the box, Shine-Dalgarno sequences are underlined; experimentally determined transcriptional start sites are printed in bold (at the right end of the sequences).

teriocin structural genes from strains for which no regulatory mechanisms or the presence of an IP-HPK-RR gene cluster have been described or suggested. This clearly suggests that the production of many bacteriocins is regulated by the mechanisms discussed here. Existing data do indicate, however, that variations on the theme do occur: some IPs seem to be cotranscribed with more than one RR (*L. plantarum* C11; see above and Diep et al., 1994), whereas in the case of enterocin-producing *E. faecium* T136, the spacing between the two direct repeats in the regulatory promoter in front of the three-component regulatory operon seems to be three rather than two DNA turns (Nilsen et al., 1998).

The functionality of two of the regulatory promoters involved in production of sakacin P by *L. sake* LTH673 has been investigated in detail (Brurberg et al., 1997). The promoters preceding the IP-673 gene (denoted *sppIP* in Fig. 3) and the sakacin P structural gene (denoted *sppA* in Fig. 3) were cloned in front of a promoterless chloramphenicol acetyltransferase (CAT) gene in the promoter screening vector pGKV210 (Van der Vossen et al., 1987); the resulting constructs were called pMV10 and pMV11, respectively. The constructs were transformed to *L. plantarum* C11. Because the regulated promoter elements found in *L. sake* LTH673 and *L. plantarum* C11 are highly similar (Fig. 3), it was anticipated that the regulation and transcription machinery of the latter strain would recognize the promoters cloned in pMV10 and pMV11. This was indeed the case. Upon induction (with IP-C11), Bac⁻ cultures of *L. plantarum* C11 harboring pMV10 or pMV11 not only produced bacteriocins, but also produced CAT (Table 1).

As shown in Table 1, induced strains harboring pMV10 or pMV11 had approximately similar CAT activities. This correlates with the observation that IP-673 and sakacin P are produced in approximately equal amounts in the wild-type producer strain (Eijsink et al., 1996). No CAT activity could be detected in an uninduced strain harboring pMV11, indi-

cating that the promoter in front of the sakacin P structural gene is strictly regulated. Taking into account the detection limit of the assay, the conclusion is that induction results in an increase in activity of at least 435 times for the sakacin P structural gene promoter.

Interestingly, some CAT activity could be detected in uninduced cells harboring pMV10 (the IP-673 promoter), and the increase in activity upon induction was in this case limited to ~124 times. The apparent leakiness of this promoter is in agreement with the notion that under natural conditions, IP (and the two-component regulatory system encoded in the same operon) accumulates slowly as a consequence of low constitutive production. Such slow accumulation would then eventually result in triggering of the autoinduction loop (see below). However, as pointed out by Brurberg et al. (1997), the putative regulated promoter was cloned, including ~100 bp upstream. This small upstream region may contain a hitherto unidentified weak but constitutive promoter.

Clearly, further studies of transcription and promoter activity are necessary to unravel the details of the autoinduction mechanism described here. It should be noted in this respect that, for the autoinduction/quorum-sensing mechanism to function (see below), low constitutive production of a series of genes seems to be necessary (resulting in low production of the IP precursor, the two-component regulatory system, and the two transport-processing proteins).

The studies with pMV10 and pMV11 clearly show that the promoters from *L. sake* LTH673 are recognized by the transcription machinery of *L. plantarum* C11. This phenomenon was also exploited in experiments aimed at identification of the receptor of IP-673 (Brurberg et al., 1997). A DNA fragment containing the genes *sppIP-sppK-sppR-sppA-spiA* (encoding the three-component regulatory system, sakacin P, and an immunity protein) was introduced in *L. plantarum* C11. The resulting strain (*L. plantarum* C11/pMV40) could, in contrast to wild-type *L. plantarum*

TABLE 1 Chloramphenicol acetyl transferase (CAT) activity in transformed *L. plantarum* C11 strains[a]

Plasmid	Promoter in front of CAT gene	CAT activity (U/mg protein)		
		−IP–C11	+IP–C11	Increase upon induction
pGKV210	None	<0.2[b]	<0.2	
pMV10	*sppIP*	0.9 ± 0.3	112 ± 9	124-fold
pMV11	*sppA*	<0.2	87 ± 10	>435-fold

[a] Cells were harvested in the early stationary phase (A_{600} of ~2.5). At this stage of growth, bacteriocin levels in the culture supernatant of *L. plantarum* C11 are known to be at their maximum.

[b] The detection limit of the assay was 0.2 U/mg.

Source: Brurberg et al. (1997).

C11 as well as several control transformants, be induced not only by IP-C11 but also by IP-673. In other words, bacteriocin production in *L. plantarum* C11/pMV40 could be induced by both IP-C11 and IP-673. Regardless of the peptide used for induction, the induced cells produced sakacin P, IP-673, IP-C11, and the various plantaricins normally produced by Bac[+] *L. plantarum* C11 (Brurberg et al., 1997).

The latter experiment strongly supports the notion that *sppK* (encoding the HPK) encodes the receptor for IP-673. It also illustrates once more that each of the promoters from the two strains can "cross talk" with the regulatory machineries of the other. It will be interesting to study at what level the regulatory machinery is exchangeable. One may wonder what exactly happens upon inducing *L. plantarum* C11/pMV40 with IP-673. Possibly, IP-673 switches on the sakacin P signal transduction system (encoded by *sppK* and *sppR*), which then acts on all regulated promoters (foreign as well as host). Alternatively, induction and activation of the HPK encoded by *sppK* may result in phosphorylation of both the foreign and the host response regulators, which then each act on their cognate promoters.

QUORUM SENSING AND ENVIRONMENTAL SIGNALS

The results discussed above clearly show that bacteriocin production is, at least in some cases, controlled by an autoinduction mechanism that involves a secreted peptide pheromone that functions as a sensor for cell density. Within such a regulatory mechanism, one would expect low constitutive transcription of genes required for production of the IP. At a certain point accumulated IP would then trigger the autoinduction loop, resulting in a rapid and drastic increase in transcription of the genes involved in bacteriocin production. Cell density-dependent regulation of bacteriocin production makes sense from an ecological point of view since it would ensure that defense against related bacteria is reinforced when the competition for resources is getting tougher (Dykes, 1995). Regulation by a secreted peptide, meaning communication between cells, would help to synchronize bacteriocin production by all cells of one particular type of LAB in a complex microbial ecosystem.

In the study of the genetic determinant of sakacin A production by *L. sake* Lb706, Axelsson and Holck (1995) found a two-component system closely associated with the gene encoding a well-characterized bacteriocin. At that time, the biological significance of this finding was not recognized, and further details of the (putative) regulatory mechanism were not being unraveled because the production of sakacin A was apparently constitutive under the experimental conditions used. The breakthrough came when it was found that *L. plantarum* C11 lost its Bac⁺ phenotype under some conditions (extreme dilution) and that the phenotype can be restored only by the addition of an IP (Diep et al., 1995). In other words, under the conditions used in the laboratory, growing Bac⁻ cells were not able to restore a sufficient level of IP and they remained Bac⁻. This easy loss of the Bac phenotype has now been observed in several strains, but it is by no means a general phenomenon among bacteriocin-producing LAB. Many bacteriocin-producing LAB, including some in which genetic analyses indicate the presence of a regulatory mechanism, have a constitutive bacteriocin production and no "switching on" of bacteriocin production has been observed.

Growing Bac⁻ cells lose their inducibility quite early (at an A_{600} of ~0.5 in the case of *L. sake* LTH673 [Eijsink et al., 1996]), that is, at a stage during growth where the accumulated amounts of bacteriocin are usually still below the detection limits of bacteriocin assays. Thus, a fully functional autoinduction mechanism, which would, for example, result in triggering of the autoinduction loop at $A_{600} = 0.3$, is relatively difficult to detect in the laboratory by phenotypic studies. Sufficient production of IP under all conditions tested so far, in combination with triggering of bacteriocin production at low cell densities, may be the reason why the regulatory mechanisms described above have not been detected in many bacteriocin-producing LAB.

At present, it is not clear why some strains do and others do not lose the Bac phenotype upon extreme dilutions. This may simply be a consequence of natural variation in the basal levels of IP production under various conditions. In this respect, it is interesting to wonder to what extent the laboratory strains used nowadays to produce and study bacteriocins differ from the equivalent bacteriocin producer in its natural niche. The variation in behavior between bacteriocin-producing LAB suggests that environmental factors may play an important role in regulation of bacteriocin production (e.g., see Nes et al., 1996, for a discussion). Evidence supporting this notion is accumulating.

It was observed that changes in growth conditions could affect the frequency of spontaneous bacteriocin production under otherwise nonproducing conditions (Diep, 1996). In a more thorough study of the regulation of bacteriocin production in *L. sake* LTH673, Brurberg et al. (1997) set out to find growth conditions under which Bac⁻ cells would spontaneously become Bac⁺. In other words, attempts were made to show the functionality of the (putative) quorum-sensing system. Figure 4 shows that spontaneous restoration of the Bac⁺ phenotype could indeed be achieved by a change in growth conditions. Whereas plating cells from a Bac⁻ culture on MRS agar plates under aerobic conditions resulted in only Bac⁻ colonies (Fig. 4A), growing the bacteria on the same plates under anaerobic conditions resulted in a considerable number of Bac⁺ colonies (Fig. 4C). Interestingly, the Bac⁺ colonies were not randomly spread over the plates but, instead, clustered in a few groups (Fig. 4C). This nicely illustrates that the cells "talk" to each other. By changing to anaerobic conditions and another growth medium (BSM, Tichcazek et al., 1992), it was possible to obtain 100% Bac⁺ cells (Fig. 4D). These experiments clearly show the functionality of the proposed quorum-sensing mechanism, although many questions about the effects of environmental factors on the regulation of bacteriocin production remain unanswered.

In a reciprocal approach, Diep et al. (manuscript in preparation) attempted to make the apparently highly stable sakacin A producer *L. sake* Lb706 lose its Bac⁺ phenotype. In contrast to *L. sake* LTH673 and *L. plantarum* C11, this strain, like many other bacteriocin-producing LAB, does not lose the Bac phenotype upon extreme dilution. Still, genetic studies have shown that bacteriocin production must somehow be regulated (Axelsson and Holck, 1995). These studies showed that genes encoding a (putative) IP-706 as well as genes encoding the two-component regula-

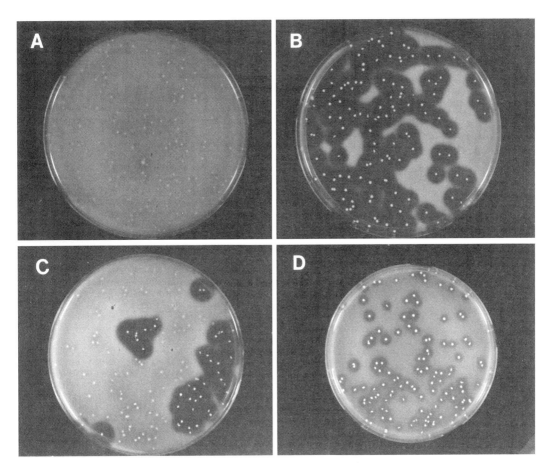

FIGURE 4 Loss and recovery of the Bac phenotype in *L. sake* LTH673 under various conditions. (A) Plating of a Bac⁻ culture of *L. sake* LTH673 on MRS medium under aerobic conditions; growth of indicator cells (present in an overlay) is not inhibited. (B) Same as in panel A, but IP-673 was added to the plate (50 ng/ml). (C) Plating of a Bac⁻ culture on MRS medium under anaerobic conditions. (D) Plating of a Bac⁻ culture on BSM medium under anaerobic conditions. Reprinted from Brurberg et al. (1997) with permission.

tory system (Fig. 1B) are each essential for bacteriocin production. It turned out to be difficult to make the induction system operate. Sakacin A was produced constitutively when incubation temperatures up to ~ 34°C were used, and the Bac⁺ phenotype was retained upon extreme dilution. At temperatures above ~35°C, the cells did not produce significant amounts of sakacin A even in the presence of synthetic pheromone (IP-706). However, in the small temperature range between ~34°C and ~35°C, it was possible to make the Bac⁺ phenotype disappear by extreme dilution and

full bacteriocin production could be regained by adding synthetic IP-706 (Diep et al., manuscript in preparation). By returning the growth temperature to 30°C, the bacteriocin production resumed without the supplement of IP-706. The same temperature-dependent regulation was found to take place in *Lactobacillus curvatus* LTH1174, which produces curvacin A (Tichaczek et al., 1992), a bacteriocin identical to sakacin A706 (Diep et al., manuscript in preparation). The observation of this remarkable phenomenon in two different bacteria indicates that it does have biological significance and that it is not merely due to some spontaneous mutation in the regulatory genes obtained during work with the two different isolates.

These experiments demonstrate the complex interaction that may exist between the environment and the regulatory system that controls bacteriocin production. Presently, it is not known how environmental effectors interact with the regulatory circuit of bacteriocin production. Changes in growth conditions may affect the activity of the promoter(s) that control(s) the low constitutive production of the pheromone. It remains to be seen whether this concerns a variation in "leakiness" of one or more of the regulated promoters or in the activity of a weak constitutive promoter. An alternative and attractive hypothesis to explain environmental effects on bacteriocin production is that the activation could be caused by an alternative two-component system. It is well recognized that two-component signal transduction systems can interact to process multiple environmental signals. During this process other two-component regulatory systems can be put into action, and cross talk may result from an activated HPK phosphorylating noncognate response regulators. The concept of cross talk was introduced to describe such nonspecific interactions with response regulators (Wanner, 1992).

Although the effects of growth conditions on the functionality of the mechanism for bacteriocin regulation are not well understood, it is without doubt that the quorum-sensing system described above is an essential part of it. For this, there is overwhelming genetic evidence (Axelsson and Holck, 1995; Diep et al., 1994, 1996, manuscript in preparation; Hühne et al., 1996; Brurberg et al., 1997; Quadri et al., 1997) as well as evidence derived from studies with synthetic peptide pheromones (Diep et al., 1995; Eijsink et al., 1996; Brurberg et al., 1997).

PERSPECTIVES

An overview of the regulation of the biosynthesis of class II bacteriocins is shown in Fig.

5. It is not known how common this type of regulation is within the groups of small bacteriocins, but it is found in both lantibiotics (chapter 11 of this volume) and class II bacteriocins. The regulatory cycle is apparently affected by growth conditions (e.g., environmental factors such as growth media, temperature, redox potential), which makes it difficult to identify this type of regulation just by studying the bacteriocin synthesis itself. The signal transduction regulatory circuit has been recognized through DNA sequence analysis and subsequent identification of genes encoding putative HPKs and RRs. In the case of class II bacteriocins, the genes constituting the regulatory systems are organized in an operon structure. The first gene in this operon encodes a bacteriocin-like induction peptide or, more correctly named, peptide pheromone. Another prevalent feature of the regulatory operon is the presence of the two direct DNA repeats in the promoters required for bacteriocin production. By searching for such regulatory elements in the DNA sequences of bacterial genomes, it should be possible to obtain information about how common these signal transduction systems are in the regulation of bacteriocin production in bacteria. One may speculate whether the search for new antimicrobial peptides may have been constrained because screening of potential producers was conducted under conditions in which *bac* genes were not expressed.

Among lantibiotics the transcriptional regulation of nisin has been studied in most detail. The nisin regulation resembles the regulation of class II bacteriocins in many ways, although no separate IP factor is required. The nisin molecule has a dual role: it exhibits antimicrobial activity and it serves as pheromone inducing the transcription of the genes required for its own biosynthesis (described in chapter 11).

It has been suggested that carnobacteriocin B2 (CbnB2) can also induce its own production in a Bac⁻ culture (Saucier et al., 1995, 1997; Quadri et al., 1997), and one could wonder if the regulation of carnobacteriocin

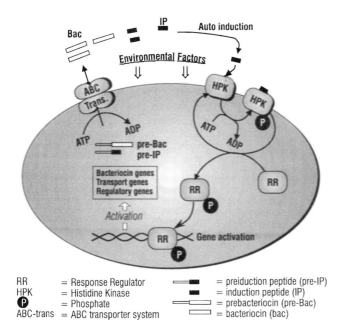

RR = Response Regulator
HPK = Histidine Kinase
Ⓟ = Phosphate
ABC-trans = ABC transporter system

━━■ = preiduction peptide (pre-IP)
■ = induction peptide (IP)
━━□ = prebacteriocin (pre-Bac)
□ = bacteriocin (bac)

FIGURE 5 An overview of bacteriocin production and regulation thereof.

production is more like that of nisin. However, the Cbn system has been shown to have its own three-component regulatory system, and it is difficult to envisage how autoregulation by CbnB2 is taking place. All the IPs have been shown to be very specific to their HPK-RR, and any cross induction by using different IPs has not been reported. It is therefore very unlikely that CbnB2 is able to act as a signaling molecule by directly interacting with the sensor domain of CbnK (HPK) (see Fig. 1B). On the other hand, it has been shown that environmental factors (i.e., growth conditions), which also may be defined as stress, are able to somehow increase the spontaneous production of bacteriocins in Bac⁻ strains without the initial supply of IP (Brurberg et al., 1997). One might therefore speculate that induction of carnobacteriocin synthesis by sublethal doses of CbnB2 is due to a kind of nonspecific induction, similar to what has been observed concerning spontaneous induction of sakacin P and plantaricin synthesis. How the environment interacts with the regulation of bacteriocins is not understood. But in some way the environmental

factors probably trigger the enhanced expression of the genes, leading to increased amounts of IP. The promoters of the various regulated operon sequences are not identical, and it has been shown that the promoter regulating the IP pheromone is more "leaky" than the bacteriocin promoter (Table 1). By some unknown specific or nonspecific mechanism, the environmental response may influence the transcriptional activity in a way that increases the production of IP to the extent that the threshold concentration required to induce the regulatory cycle leading to bacteriocin production is reached. This does not have to be a dramatic increase, but it will allow some of the individual bacteria to start the autoinduction cycle and thereby induce bacteriocin production in all nearby members of the same species, as seen in Fig. 4.

It is of evolutionary interest to note the close similarity between the regulation of bacteriocin production, regulation of competence in *S. pneumoniae,* and regulation of virulence response in *S. aureus.* It is particularly interesting to note how the IP pheromones among both class II bacteriocins and competence factors of streptococci biosynthetically and physiochemically resemble the class II bacteriocins (see chapter 2 of this book).

Finally, it should be emphasized that these regulatory mechanisms discussed in this chapter have the potential to be used in industrial LAB strains. They could be applied to control the expression of specific genes at specific time points during fermentation processes in, for example, the food and feed industry (Eijsink et al., 1995; de Ruyter et al., 1997). This gives an extra dimension to further studies of the mechanism of regulation of bacteriocin production in LAB, since these studies may not only yield fundamental biological insight but may also give access to biotechnological application of the discoveries made.

REFERENCES

Anderssen, E. L., D. B. Diep, I. F. Nes, V. Eijsink, and J. Nissen-Meyer. 1998. Antagonistic activity of *Lactobacillus plantarum* C11: two new two-peptide bacteriocins, plantaricin EF and JK, and the induction factor plantaricin A. *Appl. Environ. Microbiol.* **64:**2269–2272.

Axelsson, L., and A. Holck. 1995. The genes involved in production of and immunity to sakacin A, a bacteriocin from *Lactobacillus sake* Lb706. *J. Bacteriol.* **177:**2125–2137.

Aymerich, T., H. Holo, L. S. Håvarstein, M. Hugas, M. Garriga, and I. F. Nes. 1996. Biochemical and genetic characterization of enterocin A from *Enterococcus faecium*, a new antilisterial bacteriocin in the pediocin family of bacteriocins. *Appl. Environ. Microbiol.* **62:**1676–1682.

Brurberg, M. B., I. F. Nes, and V. G. H. Eijsink. 1997. Pheromone-induced production of antimicrobial peptides in *Lactobacillus*. *Mol. Microbiol.* **26:**347–360.

Casaus, P., T. Nilsen, L. M. Cintas, I. F. Nes, P. E. Hernández, and H. Holo. 1997. Enterocin B, a new bacteriocin from *Enterococcus faecium* T136 which can act synergistically with enterocin A. *Microbiology* **143:**2287–2294.

Cintas, L. M., P. Casaus, L. S. Håvarstein, P. E. Hernández, and I. F. Nes. 1997. Biochemical and genetic characterization of enterocin P, a novel *sec*-dependent bacteriocin from *Enterococcus faecium* P13 with a broad antimicrobial spectrum. *Appl. Environ. Microbiol.* **63:**4321–4330.

Cserzo, M., E. Wallin, I. Simon, G. von Heijne, and A. Elofsson. 1997. Prediction of transmembrane alpha-helices in prokaryotic membrane pro-

teins: the Dense Alignment Surface method. *Protein Eng.* **10:**673–676.

de Ruyter, P. G., O. P. Kuipers, W. C. Meijer, and W. M. de Vos. 1997. Food-grade controlled lysis of *Lactococcus lactis* for accelerated cheese ripening. *Nat. Biotechnol.* **15:**976–979.

Diep, D. B. 1996. *Regulation of Bacteriocin Production in Lactobacillus plantarum C11.* Ph.D. thesis. Agricultural University of Norway, Ås, Norway.

Diep, D. B., L. S. Håvarstein, J. Nissen-Meyer, and I. F. Nes. 1994. The gene encoding plantaricin A, a bacteriocin from *Lactobacillus plantarum* C11, is located on the same transcription unit as an *agr*-like regulatory system. *Appl. Environ. Microbiol.* **60:**160–166.

Diep, D. B., L. S. Håvarstein, and I. F. Nes. 1995. A bacteriocin-like peptide induces bacteriocin synthesis in *Lactobacillus plantarum* C11. *Mol. Microbiol.* **18:**631–639.

Diep, D. B., L. S. Håvarstein, and I. F. Nes. 1996. Characterization of the locus responsible for the bacteriocin production in *Lactobacillus plantarum* C11. *J. Bacteriol.* **178:**4472–4483.

Diep, D. B., C. Grefsli, L. Axelsson, and I. F. Nes. The synthesis of the bacteriocin, sakacin A is regulated by a pheromone-peptide through a regulatory two-component system. Manuscript in preparation.

Dykes, G. A. 1995. Bacteriocins: ecological and evolutionary significance. *Trends Ecol. Evol.* **10:**186–189.

Ebina, Y., Y. Takahara, F. Kishi, and A. Nakazawa. 1983. Lex A protein is a repressor in repression of colicin E1 gene. *J. Biol. Chem.* **258:**13258–13261.

Eijsink, V. G. H., M. B. Brurberg, and I. F. Nes. 1995. System for gene expression in lactic acid bacteria. International patent application, PCT/NO96/00266.

Eijsink, V. G. H., M. B. Brurberg, P. H. Middelhoven, and I. F. Nes. 1996. Induction of bacteriocin production in *Lactobacillus sake* by a secreted peptide. *J. Bacteriol.* **178:**2232–2237.

Eijsink, V. G. H., M. Skeie, P. J. Middelhoven, M. B. Brurberg, and I. F. Nes. 1998. Comparative studies of class IIa bacteriocins of lactic acid bacteria. *Appl. Environ. Microbiol.* **64:**3275–3281.

Fath, M. J., L. H. Zhang, J. Rush, and R. Kolter. 1994. Purification and characterization of colicin V from *Escherichia coli* culture supernatants. *Biochemistry* **33:**6911–6917.

Hauge, H. H., J. Nissen-Meyer, I. F. Nes, and V. G. H. Eijsink. 1998. Amphiphilic α-helices are important structural motifs in the α and β peptides that constitute the bacteriocin lactococcin G and helicity is enhanced upon α-β interaction. *Eur. J. Biochem.* **251:**565–572.

Hauge, H. H., D. Mantzilas, G. N. Moll, W. N. Konings, A. J. M. Driessen, V. G. H. Eijsink, and J. Nissen-Meyer. Plantaricin A is an amphiphilic α-helical pheromone which exerts antimicrobial and pheromone activities through different mechanisms. *Biochemistry*, vol. 37, in press.

Håvarstein, L. S., H. Holo, and I. F. Nes. 1994. The leader peptide of colicin V shares consensus sequences with leader peptides that are common among peptide bacteriocins produced by Grampositive bacteria. *Microbiology* **140:**2383–2389.

Håvarstein, L. S., G. Coomaraswamy, and D. A. Morrison. 1995a. An unmodified heptadecapeptide pheromone induces competence for genetic transformation in *Streptococcus pneumoniae*. *Proc. Natl. Acad. Sci. USA* **92:**11140–11144.

Håvarstein, L. S., D. B. Diep, and I. F. Nes. 1995b. A family of bacteriocin ABC transporters carry out proteolytic processing of their substrates concomitant with export. *Mol. Microbiol.* **16:**229–240.

Håvarstein, L. S., P. Gaustad, P., I. F. Nes, and D. A. Morrison. 1996. Identification of the streptococcal competence-pheromone receptor. *Mol. Microbiol.* **21:**863–869.

Hühne, K., A. Holck, L. Axelsson, and L. Kroeckel. 1996. Analysis of the sakacin P gene cluster from *Lactobacillus sake* Lb674 and its expression in sakacin-negative *Lb. sake* strains. *Microbiology* **142:**1437–1448.

Jack, R. W., J. R. Tagg, and B. Ray. 1995. Bacteriocins of Gram-positive bacteria. *Microbiol. Rev.* **59:**171–200.

Janzon, L., S. Löfdahl, and S. Arvidson. 1989. Identification and nucleotide sequence of the delta-lysin gene, *hld*, adjacent to the accessory gene regulator (*agr*) of *Staphylococcus aureus*. *Mol. Gen. Genet.* **219:**480–485.

Ji, G., R. C. Beavis, and R. P. Novick. 1995. Cell density control of staphylococcal virulence mediated by an octapeptide pheromone. *Proc. Natl. Acad. Sci. USA* **92:**12055–12059.

Ji, G., R. Beavis, and R. P. Novick. 1997. Bacterial interference caused by autoinducing peptide variants. *Science* **276:**2027–2030.

Klaenhammer, T. R. 1993. Genetics of bacteriocins produced by lactic acid bacteria. *FEMS Microbiol. Rev.* **12:**39–86.

Konisky, J. 1982. Colicins and other bacteriocins with established modes of action. *Annu. Rev. Microbiol.* **36:**125–144.

Leer, R. L., J. M. B. M van der Vossen, M. van Giezen, J. M. van Noort, and P. H. Pouwels. 1995. Genetic analysis of acidocin B, a novel bacteriocin produced by *Lactobacillus acidophilus*. *Microbiology* **141:**1629–1635.

Magnuson, R., J. Solomon, and A. D. Grossman. 1994. Biochemical and genetic characterization of a competence pheromone from *B. subtilis*. *Cell* **77:**207–216.

Nes, I. F., L.-S. Håvarstein, and H. Holo. 1995. Genetics of non-lantibiotics bacteriocins. *Dev. Biol Stand.* **85:**645–651.

Nes, I. F., D. B. Diep, L. S. Håvarstein, M. B. Brurberg, V. G. H. Eijsink, and H. Holo. 1996. Biosynthesis of bacteriocins in lactic acid bacteria. *Antonie van Leeuwenhoek* **70:**113–128.

Nilsen, T., I. F. Nes, and H. Holo. 1998. An exported inducer peptide regulates bacteriocin production in *Enterococcus faecium* CTC492. *J. Bacteriol.* **180:**1848–1854.

Nissen-Meyer, J., A. Granly-Larsen, K. Sletten, M. Daeschel, and I. F. Nes. 1993. Purification and characterization of plantaricin A, a *Lactobacillus plantarum* bacteriocin whose activity depends on the action of two peptides. *J. Gen. Microbiol.* **139:**1973–1978.

O'Brian, G. J., and H. K. Mahanty. 1996. Complete nucleotide sequence of microcin 24 genetic region and analysis of a new ABC transporter. Accession number ECU47048.

Parkinson, J. S. 1993. Signal transduction schemes in bacteria. *Cell* **73:**857–871.

Pattus, A., D. Massotte, H. U. Wilmsen, J. Lakey, D. Tsernoglou, A. Tucker, and M. W. Parker. 1990. Colicins: prokaryotic killer-pores. *Experientia* **46:**180–192.

Pugsley, A. P. 1984. The ins and outs of colicins. Part I: Production, and translocation across membranes. *Microbiol. Sci.* **7:**168–175.

Quadri, L. E. N., M. Sailer, K. L. Roy, J. C. Vederas, and M. E. Stiles. 1994. Chemical and genetic characterization of bacteriocins produced by *Carnobacterium piscicola* LV17B. *J. Biol. Chem.* **269:**12204–12211.

Quadri, L. E. N., L. Z. Yan., M. E. Stiles, and J. C. Vederas. 1996. Effect of amino acid substitutions on the activity of carnobacteriocin B2: overproduction of the antimicrobial peptide, its engineered variants and its precursor in *Escherichia coli*. *J. Biol. Chem.* **272:**3384–3388.

Quadri, L. E. N., M. Kleerebezem, O. P. Kuipers, W. M. de Vos, K. L. Roy, J. C. Vederas, and M. E. Stiles. 1997. Characterization of a locus from *Carnobacterium piscicola* LV17B involved in bacteriocin production and immunity: evidence for global inducer-mediated transcriptional regulation. *J. Bacteriol.* **179:**6163–6171.

Risøen, P.-A., L. S. Håvarstein, D. B. Diep, and I. F. Nes. 1998. Identification of the DNA-binding sites of two response regulators involved in regulation of bacteriocin synthesis in *Lactobacillus plantarum* C11. *Mol. Gen. Genet.* **259**:224–232.

Sahl, H.-G., R. W. Jack, and G. Bierbaum. 1995. Biosynthesis and biological activities of lantibiotics with unique post-translational modifications. *Eur. J. Biochem.* **230**:827–853.

Saucier, L., A. Poon, and M. E. Stiles. 1995. Induction of bacteriocin in *Carnobacterium piscicola* LV17. *J. Appl. Bacteriol.* **78**:684–690.

Saucier, L., A. S. Paradkar, L. S. Frost, S. E. Jensen, and M. E. Stiles. 1997. Transcriptional analysis and regulation of carnobacteriocin production in *Carnobacterium piscicola* LV17. *Gene* **188**: 271–277.

Shirabe, K., Y. Ebina, T. Miki, T. Nakazawa, and A. Nakazawa. 1985. Positive regulation of the colicin E1 gene by cyclic AMP and cyclic AMP receptor protein. *Nucleic Acids Res.* **13**:4687–4698.

Skaugen, M., and I. F. Nes. 1994. Transposition in *Lactobacillus sake* and its abolition of lacocin S production by insertion IS*1163*, a new member of the IS*3* family. *Appl. Environ. Microbiol.* **60**:2818–2825.

Stephens, S. K., B. Floriano, D. P. Cathcart, S. A. Bayley, V. F. Witt, R. Jiménez-Díaz, P. J. Warner, and J. L. Ruiz-Barba. 1998. Molecular analysis of the locus responsible for the production of plantaricin S, a two-peptide bacteriocin produced by *Lactobacillus plantarum* LPCO10. *Appl. Environ. Microbiol.* **64**:1871–1877.

Stock, J. B., A. J. Ninfa, and A. M. Stock. 1989. Protein phosphorylation and regulation of adaptive responses in bacteria. *Microbiol. Rev.* **53**:450–490.

Stock, J. B., M. G. Surette, M. Levit, and P. Park. 1995. Two-component signal transduction systems: structure-function relationship and mechanisms of catalysis, p. 25–51. *In* J. A. Hoch and T. J. Silhavy (ed.), *Two-Component Signal Transduction.* ASM Press, Washington, D.C.

Tichaczek, P. S., J. Nissen-Meyer, I. F. Nes, R. F. Vogel, and W. P. Hammes. 1992. Characterization of the bacteriocins curvacin A from *Lactobacillus curvatus* LTH1174 and sakacin P from *L. sake* LTH673. *Syst. Appl. Microbiol.* **15**:460–468.

Van der Vossen, J. M., D. Van der Lelie, and G. Venema. 1987. Isolation and characterization of *Streptococcus cremoris* Wg2-specific promoters. *Appl. Environ. Microbiol.* **53**:2452–2457.

Venema, K., J. Kok, J. D. Marugg, M. Y. Toonen, A. M. Ledeboer, G. Venema, and M. L. Chikindas. 1995. Functional analysis of the pediocin operon of *Pediococcus acidilactici* PAC1.0: PedB is the immunity protein and PedD is the precursor processing enzyme. *Mol. Microbiol.* **17**: 515–522.

Wanner, B. L. 1992. Is cross regulation by phosphorylation of two-component response regulator proteins important in bacteria? *J. Bacteriol.* **174**: 2053–2058.

Worobo, R. W., M. J. Van Belkum, M. Sailer, K. L. Roy, J. C. Vederas, and M. E. Stiles. 1995. A signal peptide secretion-dependent bacteriocin from *Carnobacterium divergens*. *J. Bacteriol.* **177**:3143–3149.

γ-BUTYROLACTONES THAT CONTROL SECONDARY METABOLISM AND CELL DIFFERENTIATION IN *STREPTOMYCES*

Sueharu Horinouchi

13

The genus *Streptomyces*, which has probably evolved separately from *Bacillus subtilis* for more than 700 million years, comprises gram-positive soil bacteria with a complex life cycle (Chater, 1989a). Early in the life cycle of *Streptomyces griseus* on solid media, a spore germinates to grow as a branching, multinucleoid substrate mycelium mainly by cell-wall extension at hyphal tips (Fig. 1). As older parts of the substrate mycelium produce aerial mycelium in response to nutrient limitation, most cells in the substrate mycelium die. Septa are then produced at regular intervals along the hyphae to form uninucleoid compartments, each of which develops into a spore, thus forming long chains of spores. In addition to the complex morphological differentiation, *Streptomyces* spp. have the ability to produce a variety of secondary metabolites, including antibiotics and biologically active substances. Secondary metabolite formation is sometimes termed "physiological" differentiation because it occurs during the idiophase after the main period of rapid vegetative growth and assimilative metabolism. It is now apparent that these morphological and physiological differentiation processes are controlled by common regulatory genes and compounds. Some of these were reviewed by Chater (1989b, 1993).

The representative of regulatory compounds controlling both processes is an autoregulatory factor called A-factor (2-iso-capryloyl-3R-hydroxymethyl-γ-butyrolactone; see Fig. 1 for the structure), which was discovered by Khokhlov (1967; Khokhlov et al., 1988) as a compound that induced streptomycin biosynthesis and aerial mycelium formation in a mutant strain of *S. griseus*. Recombinant DNA techniques have enabled us to extend the pioneer work of Khokhlov and coworkers and to reveal molecular mechanisms by which A-factor controls streptomycin production and aerial mycelium formation. This chapter deals with how A-factor, in cooperation with its specific receptor protein, switches on the expression of genes required for secondary metabolism and cell differentiation. Preliminary research on A-factor biosynthesis is also included. It is apparent that similar autoregulatory systems via γ-butyrolactones and their specific receptors control secondary metabolism and cell differentiation by the same mechanism in almost all *Streptomyces* species.

Sueharu Horinouchi, Department of Biotechnology, Graduate School of Agriculture and Life Sciences, The University of Tokyo, Bunkyo-ku, Tokyo 113-8657, Japan.

Cell-Cell Signaling in Bacteria, Edited by Gary M. Dunny and Stephen C. Winans
©1999 American Society for Microbiology, Washington, D.C.

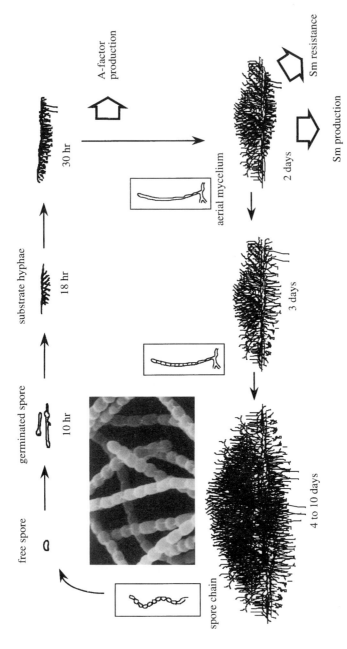

FIGURE 1 Life cycle of *S. griseus*. Spores germinate to form substrate (vegetative) mycelium, which grows by cell-wall extension at hyphal tips and branches, forming a coherent mat. A-factor produced in a growth-dependent manner reaches a certain critical concentration and triggers aerial mycelium formation and streptomycin (Sm) biosynthesis and resistance. A-factor freely diffuses in individual hyphae and also in neighboring mycelia. Aerial hyphae are eventually septated, which results in the generation of many unigenomic compartments, each destined to become a spore. Spore chains (photograph) consist of many tens of spores.

GENETICS AND BIOSYNTHESIS OF A-FACTOR

γ-Butyrolactones as Autoregulators

Hara and Beppu (1982a,b) confirmed the regulatory role of A-factor by using stereochemically synthesized A-factor and found that the 3*R* form of A-factor at a concentration of 10^{-9} M acts as a switch for streptomycin production, for Sm^r via streptomycin-6-phosphotransferase, and for aerial mycelium formation in *S. griseus*. Subsequent studies showed that A-factor also controls production of a diffusible yellow pigment, another secondary metabolite (Horinouchi and Beppu, 1992). Its remarkable pleiotropic effects at an extremely low concentration and its requirement of a specific receptor protein (see below) comparable to eukaryotic hormone receptors convince us of the identity of A-factor as a microbial hormone, rather than a pheromone (Khokhlov, 1982; Horinouchi and Beppu, 1990, 1992, 1994; Barabas et al., 1994). Since *Streptomyces* spp. grow by cell-wall extension at hyphal tips, it is conceivable that A-factor produced by a cell in the substrate mycelium at a certain stage of growth diffuses rapidly through an individual nonseptated hyphae and helps the cells in the hyphae to develop into aerial hyphae almost simultaneously. This system is analogous to eukaryotic hormones, which are produced in one organ, are transported to other organs containing their specific receptors, and exert their effects on various cellular functions there. The signaling system between physically separate individual cells in the same mycelium is in contrast to the quorum-sensing systems in the gram-negative single-cell bacteria growing in liquid culture. However, it is also possible and may be more important that A-factor serves as a signaling molecule for communication between neighboring mycelia, because *Streptomyces* spp. grow in the form of filamentous mycelia that are close enough to communicate to each other. Rapid sporulation of a whole population induced by the diffusible endogenous factor would be advantageous to the survival in the ecosystem, compared with piecemeal sporulation of individual hyphae induced by environmental stimuli such as nutrient limitation. In fact, A-factor is biosynthesized from the precursors derived from the two major biosynthetic pathways, a glycerol derivative from carbon metabolism and a β-keto acid from fatty acid metabolism, as described below.

The autoregulatory role of A-factor has been observed almost exclusively in the streptomycin-producing strain of *S. griseus*. However, the presence of A-factor or its homologs in a wide variety of *Streptomyces*, *Actinomyces*, and *Nocardia* spp. was revealed through the screening for compounds having A-factor activity on an A-factor-deficient mutant of *S. griseus*, on the assumption that regulatory compounds are not necessarily specific for a given strain (reviewed by Horinouchi and Beppu [1990, 1992]). The regulatory role of these compounds in the respective producers is not known. On the other hand, the following A-factor homologs exert regulatory roles in secondary metabolism and/or morphogenesis (Table 1): a series of virginiae butanolides, A to E, controlling virginiamycin production in *Streptomyces virginiae* (Yamada et al., 1987; Kondo et al., 1989); IM-2 controlling production of a blue pigment in a *Streptomyces* sp. (Sato et al., 1989; Hashimoto et al., 1992); a regulatory factor controlling both anthracycline production and cytodifferentiation in *Streptomyces bikiniensis* and *Streptomyces cyaneofuscatus* (Gräfe et al., 1983); and an inducer controlling anthracycline production in *Streptomyces viridochromogenes* (Gräfe et al., 1982). In addition to these autoregulators, butalactin was isolated as an antibiotic, but with no known autoregulatory role (Chatterjee et al., 1991). It may be synthesized primarily through a biosynthetic pathway similar to that of A-factor, because of the great structural similarity. Table 1 also includes an autoinducer [*N*-(3-oxohexanoyl-L-homoserine lactone] of bioluminescence in *Vibrio fischeri* as a reference for structural comparison and as a representative of homoserine lactone-type

TABLE 1 Chemical signaling molecules having a γ-butyrolactone ring

Factors	Producer	Biological activity	Reference
	S. griseus	Streptomycin Sporulation	Khokhlov (1982); Mori (1983)
	S. virginiae (VB-A)	Virginiamycin	Yamada et al. (1987)
	S. virginiae (VB-B)	Virginiamycin	Yamada et al. (1987)
	S. bikiniensis, S. cyaneofuscatus	Anthracycline	Gräfe et al. (1983)
	S. viridochromogenes	Anthracycline	Gräfe et al. (1982)
	Streptomyces sp. FRI-5	Blue pigment	Hashimoto et al. (1992)
	V. fischeri	Bioluminescence	Eberhard et al. (1981)
	P. syringae pv. *tomato*	Elicitor	Smith et al. (1993)

regulators that function as autoregulators for various biological aspects in a variety of single-cell bacteria. The homoserine lactones fall outside the scope of this chapter and are reviewed by Fuqua et al. (1994) and Dunny and Leonard (1997). An additional example is a signal molecule, syringolide, produced by *Pseudomonas syringae* pv. *tomato*, which elicits plant defense responses (Smith et al., 1993). This *Pseudomonas* strain expressing a gene for avirulence (*avrD*) produces two syringolides that cause the hypersensitive response specifically in soybean plants harboring *Rpg4* (Kobayashi et al., 1990), thus called "gene-for-gene" complementarity. The presence of autoregulators with a γ-butyrolactone controlling various examples of gene expression in the distantly related bacteria implies the generality of this type of compound as a chemical cellular signaling molecule not only in the filamentous bacterium *Streptomyces* but in many single-cell bacteria for the regulation of many biological processes.

AfsA as a Key Enzyme for A-Factor Biosynthesis

A-factor-deficient mutants of *S. griseus* that are phenotypically deficient in streptomycin production and sporulation are readily obtained spontaneously at a frequency as high as 1%. The frequency is further enhanced by UV irradiation and so-called plasmid-curing treatments, such as acridine orange treatments and protoplast regeneration (Hara and Beppu, 1982a). The instability of A-factor production can now be explained in terms of the location of *afsA* encoding a key enzyme for A-factor biosynthesis (see below); *afsA* is located near one end of the linear chromosome, the region of which is readily deleted (Lezhava et al., 1997). This is puzzling from the viewpoint of evolution, since *afsA* is essentially required for biosynthesis of A-factor, and A-factor is required for sporulation, which is advantageous to survival under various environmental conditions. The location of *afsA* near the end of the chromosome may be unusual among *Streptomyces* species, because *afsA* or its ho-

mologs in *Streptomyces coelicolor* A3(2) (Hara et al., 1983) and a *Streptomyces* sp. (Waki et al., 1997) appear to be stably maintained.

afsA encoding a protein of 301 amino acids was cloned as a gene that conferred A-factor production on an A-factor-deficient mutant strain, *S. griseus* HH1 (Horinouchi et al., 1984, 1989). Introduction of *afsA* on a high-copy-number plasmid into strain HH1, as well as all the *Streptomyces* strains tested, caused A-factor production in a very large amount owing to a gene dosage effect. In addition, *afsA* conferred A-factor production on *Streptomyces* strains that did not produce A-factor. We therefore speculated that AfsA was a key enzyme catalyzing formation of A-factor from precursors commonly present in *Streptomyces* spp. The precursors were presumably a glycerol derivative and a β-keto acid. Consistent with this, we recently found that expression of *afsA* under the control of the T7 promoter in *Escherichia coli* caused production of a compound with distinct A-factor activity on *S. griseus* (Ando et al., 1997a). The compound eluted from a high-pressure liquid chromatography column in the same fraction as authentic A-factor and restored the defect in streptomycin production and sporulation of strain HH1. Taking into consideration the strict ligand specificity of the A-factor receptor, as described below, the compound is A-factor or a homolog structurally very similar to A-factor. Production of the compound was inhibited in a dose-dependent manner by cerulenin, an inhibitor of fatty acid biosynthesis. Together with the biosynthetic study of virginiae butanolide (Sakuda et al., 1992, 1993), all of these data suggest that *afsA* catalyzes either condensation of a β-keto acid (C10 carbon unit), derived from the fatty acid biosynthetic pathway and a glycerol derivative (C3 carbon unit), or some modification of a compound which resulted from condensation of the putative precursors (Fig. 2). The presumptive β-keto acid may contain either acetyl-coenzyme A or acyl carrier protein at the end. Determination of the exact catalytic reaction by *afsA* and of the structures of

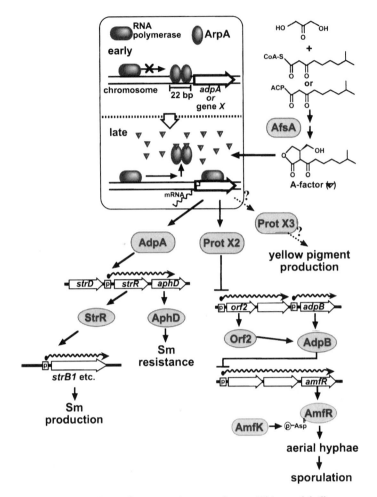

FIGURE 2 The A-factor regulatory pathway. This model illustrates how the positive A-factor signal determines the timing of streptomycin biosynthesis and aerial mycelium formation. A-factor is biosynthesized in a growth-dependent manner by the action of AfsA from a glycerol derivative and a β-keto acid derived from the fatty acid biosynthetic pathway. For triggering streptomycin production, A-factor binds the dimeric form of ArpA, which has bound the promoter region of *adpA* encoding a transcriptional activator for *strR*. Binding prevents the transcription of *strR* in early growth phase and dissociates it from the DNA, resulting in transcription and translation of *adpA*. AdpA then activates transcription of *strR* encoding a pathway-specific transcriptional activator for the streptomycin biosynthetic genes. StrR induced in this way activates all the streptomycin biosynthetic genes, resulting in biosynthesis of streptomycin from glucose. *aphD* encoding streptomycin-6-phosphotransferase, the major Sm^r determinant, is cotranscribed with *strR*. For triggering aerial mycelium formation, a still unidentified gene *X2* is under control of ArpA, just as *adpA* is. For induction of yellow pigment production by A-factor, gene *X3* under the control of ArpA may be present. Gene product X must be a repressor for *adpB* encoding a repressor for *amfR*. AmfR encoding a protein homologous to regulators of bacterial two-component regulatory systems plays a decisive role in aerial mycelium formation. A putative sensor kinase AmfK is still unidentified.

precursors by using chemically synthesized compounds is now in progress in our laboratory. The A-factor homologs with a hydroxyl group at the 6-position, such as virginiae butanolides and IM-2, are presumably synthesized by reduction of the keto group at this position of the A-factor-type compounds.

The biosynthesis of γ-butyrolactones in *Streptomyces* spp. is analogous to that of homoserine lactones involved in quorum sensing in a variety of single-cell bacteria (Fuqua et al., 1994; Dunny and Leonard, 1997), in that both use a β-keto acid derived from fatty acid biosynthesis. The homoserine lactones are produced from S-adenosylmethionine and a β-keto acid derived from the fatty acid biosynthetic pathway (Hanzelka and Greenberg, 1996; Moré et al., 1996).

A-FACTOR RECEPTOR PROTEIN (ArpA)

ArpA as a Repressor-Type Regulator

The similarity in the functional features of A-factor and eukaryotic hormones prompted us to search for a specific receptor protein by using the optically active $3R$ form of tritium-labeled A-factor. An A-factor-specific receptor protein, ArpA, was detected in the cytoplasmic fraction of *S. griseus* (Miyake et al., 1989). Scatchard analysis showed that ArpA bound A-factor in the molar ratio of 1:1 with a dissociation constant, K_d, of 0.7 nM. The extremely small K_d value is consistent with the extremely low effective concentration of A-factor (10^{-9} M in vivo) and is comparable to those of eukaryotic hormone receptors. The absence of A-factor-binding activity in the membrane supports the idea that A-factor diffuses freely through individual hyphae and into the neighboring mycelium without any surface receptor. ArpA was purified to homogeneity, and partial amino acid sequences were determined for the purpose of preparing oligonucleotides to clone the *arpA* gene by PCR (Onaka et al., 1995). Nucleotide sequencing of the cloned gene revealed that ArpA is a 276-amino-acid protein with a calculated molecular mass of 29.1 kDa. The *arpA* product expressed in *E. coli* under the control of the T7 promoter had distinct A-factor-specific binding activity. In addition, a homodimer of ArpA showed A-factor binding activity, although ArpA was readily aggregated, forming inactive precipitate. The NH_2-terminal portion of ArpA (amino acids 1 to 70) contained an α-helix-turn-α-helix DNA-binding motif that showed great similarity to those of many DNA-binding proteins, which suggested that ArpA would exert the regulatory role by directly binding to DNA.

Evidence to prove the functional role of ArpA detected in this way was obtained by genetic studies. *S. griseus* 2247, which is a mutant obtained as a stable and high-level streptomycin producer, taught us that ArpA might repress the expression of streptomycin production and sporulation; mutant 2247 required no A-factor to express both phenotypes and had a defect in ArpA (Miyake et al., 1990). The screening for mutants producing streptomycin and forming spores in the absence of A-factor among mutagenized A-factor-deficient *S. griseus* HH1 led to the isolation of four such mutants (KM5, KM7, KM12, and HO1). As expected, these mutants were deficient in both A-factor-binding activity and A-factor production. Furthermore, introduction of *arpA* even on a low-copy-number plasmid into those mutants completely abolished the ability to produce streptomycin and to form spores (Onaka et al., 1995, 1997). These genetic studies thus demonstrated that ArpA serves as a repressor-type regulator for secondary metabolism and morphogenesis. In terms of the timing of ArpA and A-factor production, ArpA is produced from a very early stage of growth, as determined by Western blotting with the anti-ArpA antibody, and A-factor produced in a growth-dependent manner is accumulated gradually until the end of exponential growth. In support of the repressor-type regulation of ArpA, the mutants deficient in ArpA begin to produce streptomycin and form spores 1 day earlier than the wild-type strain (Miyake et al.,

1990). In addition, exogenous supplementation of A-factor to both the wild-type strain and A-factor-deficient mutants at the time of inoculation causes streptomycin production earlier by 1 day than the wild-type strain but without any enhancement of the yield (Beppu, 1992).

Okamoto et al. (1995) cloned the virginiae butanolide receptor gene (barA) encoding a 232-amino-acid protein from S. virginiae. The IM-2 receptor gene (farA) was also cloned from a Streptomyces sp. by Waki et al. (1997). These two receptors and ArpA show end-to-end similarity, in agreement with the great similarity in chemical structure of their ligands. Notwithstanding the similarity among these receptors, their ligand specificities differ greatly; virginiae butanolides and IM-2 do not compete for the binding of A-factor to ArpA (Miyake et al., 1989; Sugiyama et al., in press), and vice versa (Okamoto et al., 1995). This is consistent with the observation that virginiae butanolides and IM-2 exert no in vivo effect on an A-factor-deficient S. griseus strain. It is likely that the COOH-terminal portions of these receptors recognize and bind their ligands, since the NH$_2$-terminal portions probably serve as DNA-binding domains. Our preliminary data have shown that mutant ArpA proteins having amino acid replacement in the α-helix-turn-α helix motif at the NH$_2$-terminal portion of ArpA lack DNA-binding activity (Sugiyama et al., in press). Concerning the ligand specificity of ArpA, the 6-ketone group of A-factor appears to play a decisive role. Reduction of this ketone leads to complete disappearance of the in vivo regulatory activity, although changes in the acyl chain length also cause a considerable decrease in activity (Khokhlov, 1982). All the other γ-butyrolactones so far isolated as regulators contain a hydroxyl group at this position (Table 1). A Streptomyces strain must have evolved using its own set of receptor and γ-butyrolactone as an autoregulatory system. Such a system would be useful for discriminating among different chemical signals and thereby for communicating to each other in the same group and preventing miscommunication among various actinomycetes in the ecosystem (Beppu, 1992).

A-Factor-Mediated Dissociation of ArpA from DNA

Because of the presence of an α-helix-turn-α-helix DNA-binding motif in ArpA, we determined whether ArpA bound to a specific DNA sequence. The method used in our binding-site selection assay was to recover specific DNA from a pool of random-sequence oligonucleotides by rounds of a binding/immunoprecipitation/PCR amplification procedure with histidine-tagged ArpA (ArpA-H) and anti-ArpA antibody (Onaka and Horinouchi, 1997). ArpA-H with six histidines at its COOH-terminal end was produced in E. coli and purified to near homogeneity. The pool of oligonucleotides contained a randomized region of 44 bp. By means of further binding-site selection experiments on the basis of the initially recovered oligonucleotide, a 22-bp palindromic binding site with the sequence 5'-GG(T/C)CGGT(A/T)(T/C)G(T/G)-3' as one-half of the palindrome was deduced as a consensus sequence recognized and bound by ArpA in the absence of A-factor. The nucleotides at positions 1, 2, 4, 5, 6, 7, and 10 were absolutely or almost absolutely conserved, and those at positions 3, 8, 9, and 11 preferred two nucleotides. In addition, ArpA required a minimum of 22 bp for binding. The dyad symmetry of the binding site is in agreement with the idea that ArpA in the form of a homodimer (Onaka et al., 1995) binds the DNA sequence; one subunit recognizes and binds to one-half of the palindrome and the other subunit binds to the other half. Transcriptional regulators with a homodimer form, such as cyclic AMP receptor protein, Cro of phage λ, and cI, bind to DNA with a twofold rotational symmetry, because, as determined by X-ray crystallography, one monomer of these proteins interacts with one-half of the DNA sites, and the other monomer interacts in a perfectly or nearly perfectly twofold symmetrical fashion with the other half of the DNA sites.

ArpA did not bind to the binding site in the presence of A-factor at a concentration between 32 and 160 nM. Furthermore, exogenous addition of A-factor to the DNA-ArpA complex caused immediate release of ArpA from the DNA. By analogy with the above transcriptional regulators, we assume that A-factor binding to the COOH-terminal domain of ArpA caused a subtle conformational change of the distal NH$_2$-terminal DNA-binding domain, resulting in dissociation of ArpA from DNA. In fact, ArpA seems to consist of two independently functional domains, one for A-factor binding and one for DNA binding, because a single amino acid replacement at proline-115 of ArpA results in abolishment of DNA-binding activity but not A-factor-binding activity (Onaka et al. 1997). A-factor is produced in a growth-dependent manner by the wild-type *S. griseus* strain and reaches 100 nM (25 ng/ml) in culture broth (Ando et al., 1997b; Hara and Beppu, 1982a). The intracellular concentration of A-factor is supposedly the same as the extracellular concentration, since A-factor diffuses freely through the membrane into the medium. This means that the target sites become free from ArpA gradually as the intracellular concentration of A-factor increases in a growth-dependent manner. This A-factor-mediated dissociation of ArpA from DNA accounts for the switching function of A-factor. We can thus imagine that ArpA prevents the expression of a gene, X, by binding to (presumably) the promoter during the early growth phase and that A-factor releases ArpA from the DNA, thus triggering the onset of ordered signal relay for secondary metabolism and aerial mycelium formation (Fig. 2). For streptomycin production, an A-factor-dependent protein (AdpA) that transmits the A-factor signal to the streptomycin biosynthetic genes has been identified as gene X (see below). Gene X for triggering morphogenesis is still unknown.

However, A-factor at a low nanomolar concentration is sufficient to cause streptomycin production and sporulation in the A-factor-deficient mutant strain HH1, when added at inoculation. Timing is also important for the switching role of A-factor, since streptomycin biosynthesis (Neumann et al., 1996) and sporulation (Barabas et al., 1994; Ando et al., 1997b) are determined at an early stage of the life cycle, usually not later than 10 to 12 h after inoculation. After this A-factor-sensitive stage, or the "decision phase" (Neumann et al., 1996), A-factor exogenously supplemented to A-factor-deficient mutants does not influence streptomycin production or sporulation. It is likely that gradual accumulation of A-factor during the A-factor-sensitive stage initiates an ordered sequence of metabolic events. In fact, addition of A-factor in a large amount at inoculation and during this phase to the A-factor-deficient mutants and the wild-type strain causes disturbance of growth and aerial mycelium formation (Ando et al., 1997b). This disturbance occurs in strains having the intact *arpA* but not in *arpA* mutants. Perhaps a large amount of A-factor during the A-factor-sensitive stage abruptly relieves ArpA repression and sets off an anomalous sequence of metabolic events. Although further studies of the A-factor-sensitive stage are necessary, it is evident that the A-factor-mediated dissociation of ArpA from the putative target gene, X, initiates the sequence of metabolic events that leads to streptomycin production and aerial mycelium and spore formation.

Receptor Proteins in *S. coelicolor* A3(2)

Similar to the wide distribution of A-factor homologs in *Streptomyces* spp., DNA sequences homologous to *arpA* are also widely distributed (Onaka et al., 1995). In fact, the genes encoding the virginiae butanolide-binding protein (Okamoto et al., 1995) and the IM-2-binding protein (Waki et al., 1997) have been cloned. *S. coelicolor* A3(2), which has been the most extensively and intensively studied strain, produces a series of A-factor homologs (Anisova et al., 1984). Recently, Bibb and his colleagues have identified at least four low-molecular-weight diffusible compounds structurally similar to A-factor that are able to cause precocious production of actinorhodin, and possibly undecylprodigiosin, in this strain (Bibb, personal communication). Our approach to studying a possible regulatory mechanism similar to the A-factor system in this strain was to characterize a putative ArpA-like receptor. PCR with a set of primers designed for an NH_2-terminal portion of ArpA and the chromosomal DNA of *S. coelicolor* A3(2) yielded an amplified fragment of 150 bp, the same size as that amplified with the chromosomal DNA of *S. griseus*. Standard DNA manipulation with the amplified DNA led to cloning of two genes, named *cprA* and *cprB*, each encoding an ArpA-like protein (Horinouchi, 1997; Onaka et al., 1998). CprA and CprB share 91% identity in amino acid sequence, and both show about 35% identity to ArpA. Both genes are located in the same cosmid clone among 319 cosmids covering the whole chromosome (Redenbach et al., 1996), suggesting that an ancestor gene has duplicated to yield these two genes.

Gene disruption of *cprB* resulted in precocious production of actinorhodin, leading to overproduction. In addition, aerial mycelium formation and, accordingly, sporulation of the *cprB* disruptants began earlier by 1 day, although the abundance of sporulation after long cultivation was the same as that in the parental strain. The *cprB* gene thus acts as a repressor-type regulator on actinorhodin production and sporulation. Consistent with this, the introduction of *cprB* even on a low-copy-number plasmid resulted in reduced production of actinorhodin and a delay of sporulation. CprB thus behaves as a repressor-type regulator, just as ArpA does in *S. griseus*. We assume that one or more of the A-factor homologs serve as the ligand of CprB and control the timing of differentiation in the same manner as A-factor in *S. griseus*.

On the other hand, disruption of *cprA* resulted in severe reduction of actinorhodin and undecylprodigiosin production. In addition, the timing of sporulation was delayed by 1 day, although the abundance of sporulation was the same as that in the parental strain. The *cprA* gene thus appeared to act as an activator or an accelerator for secondary metabolism and morphogenesis. As expected, extra copies of *cprA* led to overproduction of these secondary metabolites and accelerated the timing of sporulation. A simple explanation of the activator-like behavior of CprA is that CprA itself or a ligand-bound form of CprA is a transcriptional activator able to bind to an operator site in either the same or a different gene, which is different from that bound by CprB. However, great similarity in amino acid sequence in the α-helix-turn-α-helix DNA-binding motifs of CprA and CprB suggests that both recognize and bind the same operator site. In fact, both CprA and CprB recognize and bind the same target site as ArpA (Sugiyama et al., in press). For elucidation of the molecular mechanism by which CprA controls secondary metabolism and morphogenesis in a phenotypically positive way, further studies, such as determination of the ligand as well as the target gene(s) and identification of binding sites in the target genes, are required.

REGULATORY PROTEINS THAT RELAY A-FACTOR SIGNAL TO DOWNSTREAM GENES

AdpA, Which Transmits A-Factor Signal to Streptomycin Biosynthetic Gene Cluster

As described above, the timing of streptomycin production is determined by the intracellular concentration of A-factor. Every gene cluster for biosynthesis of secondary metabolites includes a transcriptional activator gene, called a pathway-specific regulator, that is responsible for transcription of the other biosynthetic genes. It is conceivable that various regulatory signals to start the biosynthesis of secondary metabolites are received by the pathway-specific regulatory genes. The streptomycin biosynthetic gene cluster in *S. griseus* contains *strR* as a pathway-specific regulatory gene. The A-factor signal is transmitted, via an A-factor-dependent transcriptional activator (AdpA), to an upstream activating sequence located about 300 bp upstream of the transcriptional start point of *strR* (reviewed by Horinouchi and Beppu, 1994). AdpA is produced in response to A-factor (Vujaklija et al., 1993). AdpA consists of 405 amino acids and contains an α-helix-turn-α-helix DNA-binding motif in the central portion (Ohnishi and Horinouchi, 1997). The nucleotide sequencing of the region upstream from the initiation codon of *adpA* predicted the presence of a palindrome of 22 bp that matches the consensus sequence of the ArpA binding site. As expected, ArpA recognizes and binds this palindrome in the absence of A-factor. We can thus imagine how the A-factor signal is transmitted to the streptomycin biosynthetic gene cluster (Fig. 2). At an early stage of growth, when the concentration of A-factor is still low, the expression of *adpA* is repressed by ArpA. *adpA* is then inducibly transcribed upon binding of A-factor to ArpA, and the transcriptional activator AdpA is produced, which in turn binds and activates the expression of *strR*. The pathway-specific transcriptional activator, StrR, then activates the expression of all the biosynthetic genes by binding multiple sites in the gene cluster (Retzlaff and Distler, 1995). Thus, all the major regulatory steps for relaying the A-factor signal to the streptomycin biosynthetic genes are elucidated.

AdpB, Which Transmits A-Factor Signal to Sporulation Genes

A-factor-deficient mutants cannot form spores. It was therefore assumed that overexpression of A-factor-controlled genes in the A-factor regulatory cascade in the absence of A-factor might lead to sporulation. Such genes might encode regulatory proteins that mediate the transfer of the A-factor signal to the downstream genes responsible for aerial mycelium and spore formation. Cloning experiments based on this strategy have identified four such genes (reviewed by Horinouchi, 1996). Among these genes, expression of *amfR* encoding a response regulator of two-component regulatory systems is apparently controlled by A-factor. AmfR plays an important role in aerial mycelium formation, because disruption of this gene results in complete loss of morphogenesis (Ueda et al., 1998). Alignment of AmfR with other regulators in this family predicted that Asp-54 might be phosphorylated (Ueda et al., 1993). Consistent with this, an amino acid replacement from Asp-54 to Asn by means of site-directed mutagenesis resulted in the loss of the ability of *amfR* to induce sporulation in the A-factor-deficient mutant *S. griseus* HH1. By analogy with many other two-component regulatory systems, a phosphorylated form of AmfR conceivably acts as a transcriptional activator for some genes required for aerial mycelium formation. A putative sensor kinase AmfK able to transfer a phosphate to AmfR has not been identified. AmfR may play a key role similar to that of Spo0A in the sporulation of *B. subtilis* (Burbulys et al., 1991).

amfR is mainly transcribed by a promoter further upstream from the AmfR-coding region (Fig. 2). Between the promoter and the coding sequence, two open reading frames are present. Promoter-probing experiments showed that this promoter is repressed in the presence of A-factor. In agreement with the A-factor control of the promoter, a protein (AdpB) able to bind the promoter region was detected by gel retardation assay only in the absence of A-factor. Recently, AdpB was purified and *adpB* was cloned. AdpB consists of 331 amino acids and contains an α-helix-turn-α-helix DNA-binding motif in the central region (Ueda et al., 1997). In vitro transcriptional analysis with the purified AdpB, the promoter region, and a σ^B-RNA polymerase preparation showed that the transcription was inhibited by AdpB in a concentration-dependent manner, indicating that AdpB is a repressor (Ueda et al., 1998). Consistent with this, introduction of *adpB* even on a low-copy-number plasmid into the wild-type strain repressed aerial mycelium and spore formation almost completely.

adpB appears to form a gene cluster with two other genes (Fig. 2). Transcriptional analysis of the gene cluster is now in progress. *adpB* is transcribed by two promoters, one just in front of the coding sequence and the other in front of the cluster. Although the regulation of the expression of this gene cluster has not yet been determined, the two promoters do not contain any sequence homologous to the consensus sequence of the ArpA binding site. Consistent with this, no binding of ArpA to the promoter regions has been detected. A simple model is that the A-factor signal serves in a repressive way for the expression of *adpB*, thus resulting in switching on of *amfR* transcription by A-factor. Another possibility is that AdpB is modified into an inactive form by a certain enzyme that is produced in response to A-factor. Further study will reveal the intermediate regulatory steps connecting protein X and AmfR.

CONCLUDING REMARKS AND BIOLOGICAL IMPLICATIONS

The filamentous morphology of *Streptomyces* spp. with differentiated mycelia and spores is obviously similar to that of fungi. In addition, the γ-butyrolactone-type autoregulators in *Streptomyces* spp. are similar in functional features to sex pheromones in fungi and yeasts. The regulation of morphological and physiological differentiation in filamentous *Streptomyces* spp. is conceivably the consequence of the convergent evolution, in which the diffusible γ-butyrolactones have emerged to act as chemical signals between the cells at a distance within individual hyphae and between different hyphae. For discrimination of the signals, *Streptomyces* strains must have adapted their receptor proteins for the ligands during evolution. Even in single-cell bacteria, γ-butyrolactones (homoserine lactones), in cooperation with specific receptors, serve as chemical signals that play a role in cell density-dependent regulation during interactions between bacteria. Both the homoserine lactone- and γ-butyrolactone-type signals are synthesized from derivatives from fatty acid metabolism as one of the precursors, although the former and the latter are composed of derivatives from amino acid metabolism and carbon metabolism, respectively, as the other precursor. In both cases, the precursors are derivatives from major metabolic pathways, from which γ-butyrolactones are biosynthesized by only two or more steps. This makes it feasible for cells to sense nutritional conditions or cell-density.

The A-factor/ArpA system in *S. griseus* phenotypically acts as a switch for both secondary metabolism and cell differentiation in an all-or-none fashion. However, the CprA and CprB system in *S. coelicolor* A3(2) appears to control the timing of antibiotic production and aerial mycelium formation. The virginiae butanolide/BarA system in *S. virginiae* also controls the timing of virginiamycin production. In other words, γ-butyrolactones act as an all-or-none switch in some strains but as a tuner in conjunction with other signals in other strains. This implies that most *Streptomyces* species contain redundant signal transduction pathways leading to the onset of differentiation.

Many regulatory steps must be required for morphogenesis and secondary metabolism in *Streptomyces*. It is conceivable that diffusible low-molecular-weight compounds control a step for normal and healthy differentiation, especially for cell-cell communication in individual hyphae. Probable candidates have been isolated, although no definitive evidence for their autoregulatory roles has been obtained because of a lack of genetic data in these cases (reviewed by Horinouchi and Beppu, 1990, 1992). Exploration and analysis of low-molecular-weight compounds as autoregulators will lead to a better understanding of the morphological and physiological differentiation of filamentous *Streptomyces* spp.

ACKNOWLEDGMENTS

The work from this laboratory was supported by the Nissan Science Foundation, the Proposal-Based

Advanced Industrial Technology R&D Program of the New Energy and Industrial Technology Development Organization (NEDO) of Japan, and the Research for the Future Program of the Japan Society for Promotion of Science (JSPS).

REFERENCES

Ando, N., N. Matsumori, S. Sakuda, T. Beppu, and S. Horinouchi. 1997a. Involvement of AfsA in A-factor biosynthesis as a key enzyme. *J. Antibiot.* **50:**847–852.

Ando, N., K. Ueda, and S. Horinouchi. 1997b. A *Streptomyces griseus* gene (*sgaA*) suppresses the growth disturbance caused by high osmolality and a high concentration of A-factor during early growth. *Microbiology* **143:**2715–2723.

Anisova, L. N., I. N. Blinova, O. V. Efremenkova, Y. P. Ko'zmin, V. V. Onoprienko, G. M. Smirnova, and A. S. Khokhlov. 1984. Regulators of the development of *Streptomyces coelicolor* A3(2). *Izv. Akad. Nauk SSSR, Ser. Biol.* **1:**98–108.

Barabas, G., A. Penyige, and T. Hirano. 1994. Hormone-like factors influencing differentiation of *Streptomyces* cultures. *FEMS Microbiol. Rev.* **14:**75–82.

Beppu, T. 1992. Secondary metabolites as chemical signals for cell differentiation. *Gene* **115:**159–165.

Bibb, M. J. Personal communication.

Burbulys, D., K. A. Trach, and J. A. Hoch. 1991. Initiation of sporulation in *B. subtilis* is controlled by a multicomponent phosphorelay. *Cell* **64:**545–552.

Chater, K. F. 1989a. Sporulation in *Streptomyces*, p. 277–299. *In* I. Smith, R. A. Slepecky, and P. Setlow (ed.), *Regulation of Procaryotic Development: Structural and Functional Analysis of Bacterial Sporulation and Germination*. American Society for Microbiology, Washington, D.C.

Chater, K. F. 1989b. Multilevel regulation of *Streptomyces* differentiation. *Trends Genet.* **5:**372–377.

Chater, K. F. 1993. Genetics of differentiation in *Streptomyces*. *Annu. Rev. Microbiol.* **47:**685–713.

Chatterjee, S., E. K. S. Vijayakumar, C. M. M. Franco, U. P. Borde, J. Blumbach, and B. N. Ganguli. 1991. Butalactin: a new butanolide antibiotic from *Streptomyces corchorusii*. *Tetrahedron Lett.* **32:**141–144.

Dunny, G. M., and B. A. B. Leonard. 1997. Cell-cell communication in gram-positive bacteria. *Annu. Rev. Microbiol.* **51:**527–564.

Eberhard, A. A., L. Burlingame, C. Eberhard, G. L. Kenyon, K. H. Nealson, and N. J. Oppenheimer. 1981. Structural identification of autoinducer of *Photobacterium fischeri* luciferase. *Biochemistry* **20:**2444–2449.

Fuqua, W. C., S. C. Winans, and E. P. Greenberg. 1994. Quorum sensing in bacteria: the LuxR-LuxI family of cell density-responsive transcriptional regulators. *J. Bacteriol.* **176:**269–275.

Gräfe, U., W. Shade, I. Eritt, and W. F. Fleck. 1982. A new inducer of anthracycline biosynthesis from *Streptomyces viridochromogenes*. *J. Antibiot.* **35:**1722–1723.

Gräfe, U., G. Reinhardt, W. Schade, I. Eritt, W. F. Fleck, and L. Radics. 1983. Interspecific inducers of cytodifferentiation and anthracycline biosynthesis in *Streptomyces bikiniensis* and *S. cyaneofuscatus*. *Biotechnol. Lett.* **5:**591–596.

Hanzelka, B. L., and E. P. Greenberg. 1996. Quorum sensing in *Vibrio fischeri*: evidence that S-adenosylmethionine is the amino acid substrate for autoinducer synthesis. *J. Bacteriol.* **178:**5291–5294.

Hara, O., and T. Beppu. 1982a. Mutants blocked in streptomycin production in *Streptomyces griseus*—the role of A-factor. *J. Antibiot.* **35:**349–358.

Hara, O., and T. Beppu. 1982b. Induction of streptomycin inactivating enzyme by A-factor in *Streptomyces griseus*. *J. Antibiot.* **35:**1208–1215.

Hara, O., S. Horinouchi, T. Uozumi, and T. Beppu. 1983. Genetic analysis of A-factor synthesis in *Streptomyces coelicolor* A3(2) and *Streptomyces griseus*. *J. Gen. Microbiol.* **129:**2939–2944.

Hashimoto, K., T. Nihira, S. Sakuda, and Y. Yamada. 1992. IM-2, a butyrolactone autoregulator, induces production of several nucleoside antibiotics in *Streptomyces* sp. FRI-5. *J. Ferment. Bioeng.* **73:**449–455.

Horinouchi, S. 1996. *Streptomyces* genes involved in aerial mycelium formation. *FEMS Microbiol. Lett.* **141:**1–9.

Horinouchi, S. 1997. A hormonal (A-factor) control for secondary metabolism and cell differentiation in *Streptomyces*, abstr. 3S7. Abstracts of 10th International Symposium on Biology of Actinomycetes, May, 1997, Beijing, China.

Horinouchi, S., and T. Beppu. 1990. Autoregulatory factors of secondary metabolism and morphogenesis in actinomycetes. *Crit. Rev. Biotechnol.* **10:**191–204.

Horinouchi, S., and T. Beppu. 1992. Autoregulatory factors and communication in actinomycetes. *Annu. Rev. Microbiol.* **46:**377–398.

Horinouchi, S., and T. Beppu. 1994. A-factor as a microbial hormone that controls cellular differentiation and secondary metabolism in *Streptomyces griseus*. *Mol. Microbiol.* **12:**859–864.

Horinouchi, S., Y. Kumada, and T. Beppu. 1984. Unstable genetic determinant of A-factor biosynthesis in streptomycin-producing organisms: cloning and characterization. *J. Bacteriol.* **158:**481–487.

Horinouchi, S., H. Suzuki, M. Nishiyama, and T. Beppu. 1989. Nucleotide sequence and transcriptional analysis of the *Streptomyces griseus* gene (*afsA*) responsible for A-factor biosynthesis. *J. Bacteriol.* **171**:1206–1210.

Khokhlov, A. S. 1982. Low molecular weight microbial bioregulators of secondary metabolism, p. 97–109. *In* V. Krumphanzl, B. Sikyta, and Z. Vanek (ed.), *Overproduction of Microbial Products*. Academic Press, London.

Khokhlov, A. S. 1988. Results and perspectives of actinomycete autoregulators studies, p. 338–345. *In* Y. Okami, T. Beppu, and H. Ogawara (ed.), *Biology of Actinomycetes '88*. Japan Scientific Societies Press, Tokyo.

Khokhlov, A. S., I. I. Tovarova, L. N. Borisova, S. A. Pliner, L. A. Schevchenko, E. Y. Kornitskaya, N. S. Ivkina, and I. A. Rapoport. 1967. A-factor responsible for the biosynthesis of streptomycin by a mutant strain of *Actinomyces streptomycini*. *Dokl. Akad. Nauk SSSR* **177**:232–235.

Kobayashi, D. Y., S. J. Tamaki, and N. T. Keen. 1990. Molecular characterization of avirulence gene D from *Pseudomonas syringae* pv. *tomato*. *Mol. Plant-Microbe Interact.* **3**:94–102.

Kondo, K., Y. Higuchi, S. Sakuda, T. Nihira, and Y. Yamada. 1989. New virginiae butanolides from *Streptomyces virginiae*. *J. Antibiot.* **42**:1873–1876.

Lezhava, A., D. Kameoka, H. Sugino, K. Goshi, H. Shinkawa, O. Nimi, S. Horinouchi, T. Beppu, and H. Kinashi. 1997. Chromosomal deletions in *Streptomyces griseus* that remove the *afsA* locus. *Mol. Gen. Genet.* **253**:478–483.

Miyake, K., S. Horinouchi, M. Yoshida, N. Chiba, K. Mori, N. Nogawa, N. Morikawa, and T. Beppu. 1989. Detection and properties of A-factor binding protein from *Streptomyces griseus*. *J. Bacteriol.* **171**:4298–4302.

Miyake, K., T. Kuzuyama, S. Horinouchi, and T. Beppu. 1990. The A-factor-binding protein of *Streptomyces griseus* negatively controls streptomycin production and sporulation. *J. Bacteriol.* **172**:3003–3008.

Moré, M. I., L. D. Finger, J. L. Stryker, C. Fuqua, A. Eberhard, and S. C. Winans. 1996. Enzymatic synthesis of a quorum-sensing autoinducer through use of defined substrates. *Science* **272**:1655–1658.

Mori, K. 1983. Revision of the absolute configuration of A-factor: the inducer of streptomycin biosynthesis, basing on the reconfirmed (*R*)-configuration of (+)-paraconic acid. *Tetrahedron* **39**:3107–3109.

Neumann, T., W. Piepersberg, and J. Distler. 1996. Decision phase regulation of streptomycin production in *Streptomyces griseus*. *Microbiology* **142**:1953–1963.

Ohnishi, Y., and S. Horinouchi. 1997. Regulation of streptomycin biosynthesis by A-factor-dependent protein in *Streptomyces griseus*, p. 63–64. Abstracts of Japan-UK Joint Study on Molecular Genetics of *Streptomyces*, October, 1997, Norwich, United Kingdom.

Okamoto, S., K. Nakamura, and Y. Yamada. 1995. Virginiae butanolide binding protein from *Streptomyces virginiae*: evidence that Vbr is not the virginiae butanolide binding protein and reidentification of the true binding protein. *J. Biol. Chem.* **270**:12319–12326.

Onaka, H., and S. Horinouchi. 1997. DNA-binding activity of the A-factor receptor protein and its recognition DNA sequences. *Mol. Microbiol.* **24**:991–1000.

Onaka, H., N. Ando, T. Nihira, Y. Yamada, T. Beppu, and S. Horinouchi. 1995. Cloning and characterization of the A-factor receptor gene from *Streptomyces griseus*. *J. Bacteriol.* **177**:6083–6092.

Onaka, H., M. Sugiyama, and S. Horinouchi. 1997. A mutation at proline-115 in the A-factor receptor protein of *Streptomyces griseus* abolishes DNA-binding activity but not ligand-binding activity. *J. Bacteriol.* **179**:2748–2752.

Onaka, H., T. Nakagawa, and S. Horinouchi. 1998. Involvement of two A-factor receptor homologues in *Streptomyces coelicolor* A3(2) in the regulation of secondary metabolism and morphogenesis. *Mol. Microbiol.* **28**:743–753.

Redenbach, M., H. M. Kieser, D. Denapaite, A. Eichner, J. Cullum, H. Kinashi, and D. A. Hopwood. 1996. A set of ordered cosmids and a detailed genetic and physical map for the 8 Mb *Streptomyces coelicolor* A3(2) chromosome. *Mol. Microbiol.* **21**:77–96.

Retzlaff, L., and J. Distler. 1995. The regulator of streptomycin gene expression, StrR, of *Streptomyces griseus* is a DNA binding activator protein with multiple recognition sites. *Mol. Microbiol.* **18**:151–162.

Sakuda, S., A. Higashi, S. Tanaka, T. Nihira, and Y. Yamada. 1992. Biosynthesis of virginiae butanolide, a butyrolactone autoregulator from *Streptomyces*. *J. Am. Chem. Soc.* **114**:663–668.

Sakuda, S., S. Tanaka, K. Mizuno, O. Sukcharoen, T. Nihira, and Y. Yamada. 1993. Biosynthetic studies on virginiae butanolide A, a butyrolactone autoregulator from *Streptomyces*. Part 2. Preparation of possible biosynthetic intermediates and conversion experiments in a cell-free system. *J. Chem. Soc. Perkin Trans.* **1993**:2309–2315.

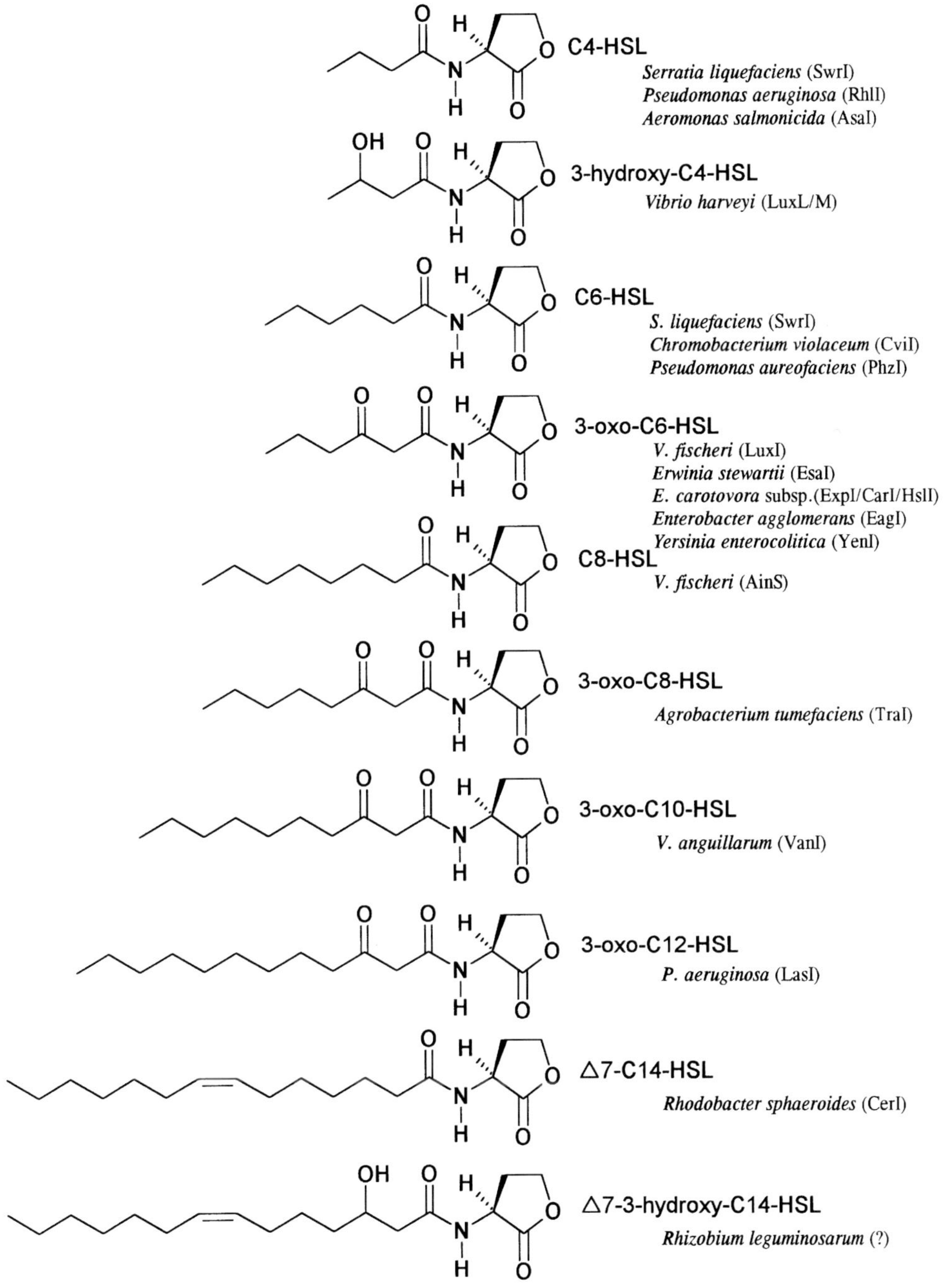

FIGURE 1 Comparison of acyl–HSL structures. Chemical structures of most of the characterized acyl–HSL molecules are arrayed by acyl chain length and derivation. Those LuxI-type proteins known to synthesize the depicted acyl–HSLs as their primary factor are included to the right of each structure. Only the primary acyl–HSL produced by a given LuxI-type protein is provided (Bainton et al., 1992; Beck von Bodman and Farrand, 1995; Cao and Meighen, 1989; Eberhard et al., 1981; Eberl et al., 1996; Gray et al., 1996; Kuo et al., 1994; Pearson et al., 1994, 1995; Puskas et al., 1997; Schripsema et al., 1996; Swift et al., 1993, 1997; Throup et al., 1995; Winson et al., 1995; Zhang et al., 1993).

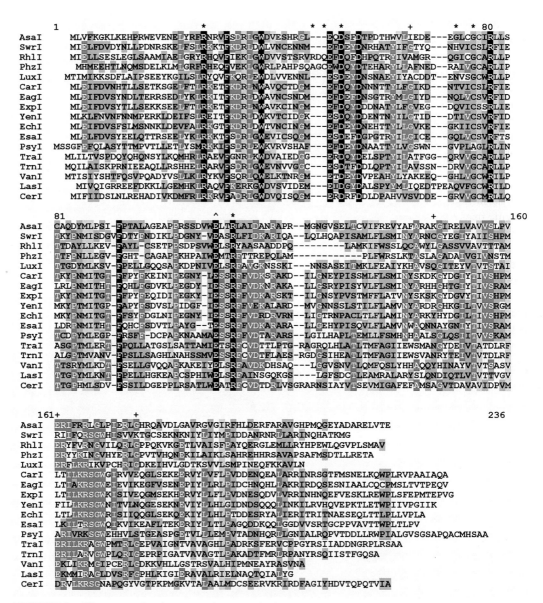

FIGURE 2 Alignment of acyl-HSL synthases. Multiple sequence alignment of 17 different (<65% identity) LuxI-type proteins arrayed in ascending order by the size of the primary acyl-HSL synthesized. Sequences were aligned using the Clustal alignment algorithm included in the LaserGene sequence analysis package. Black shading indicates invariant residues; yellow shading indicates functionally similar residues conserved in all sequences; green shading indicates residues present in at least seven of the sequences; and turquoise shading indicates functionally similar residues present in at least seven sequences. Positions where mutations have been identified that result in loss of enzyme activity in both LuxI and RhlI (★), in LuxI alone (+), and in RhlI alone (/\) are indicated (Hanzelka et al., 1997; Parsek et al., 1997).

some of the details of the reaction are now understood. More recently, the substrates for several LuxI-type proteins were identified, and these proteins were definitively proved to be acyl-HSL synthases (Hanzelka and Greenberg, 1996; Moré et al., 1996; Schaefer et al., 1996, Parsek and Greenberg, personal communication). For at least three LuxI-type proteins (LuxI, RhlI, TraI) the substrates are S-adenosylmethionine (AdoMet) and the appropriate fatty acid substrate conjugated to acyl carrier protein (ACP). It is expected that other LuxI-type proteins utilize the same or similar substrates, although variations on this scheme may emerge as more of these enzymes are studied.

Derivation of the Homoserine Moiety

Homoserine and related compounds are found in most bacteria as intermediates of the methionine-lysine-threonine biosynthetic pathway (Fig. 3). The simplest prediction was that the homoserine moiety in acyl-HSLs was incorporated directly from cellular pools of homoserine or homoserine lactone. However, experiments with cell cultures and with whole cell extracts from *V. fischeri* suggested that methionine was the most effective substrate for incorporation into 3-oxo-C6-HSL (Eberhard et al., 1991), while homoserine lactone and O-succinyl homoserine failed to be significantly incorporated. The incorporation of methionine suggested that methionine itself or AdoMet might be the substrate for homoserine lactone ring formation.

Owing to its important and varied roles in a number of cellular reactions, AdoMet is a difficult metabolic precursor to analyze genetically. It is the primary source of methyl groups for a variety of cellular reactions. AdoMet is also a substrate for a wide range of metabolic reactions outside of single carbon metabolism. AdoMet is synthesized from methionine and ATP, catalyzed by the MetK protein (AdoMet synthetase) (Fig. 3) (Hobson and Smith, 1973; Satishchandran et al., 1993). Perhaps not surprisingly, considering its diverse and essential roles, there are simply no

stable conditional mutants available in which consistent, significant reduction in AdoMet pool sizes can be achieved. Although there are a number of well-characterized mutants that are blocked upstream of methionine synthase (*metH*, *metE*), these are all methionine auxotrophs.

Hanzelka and Greenberg (1996) surmounted these problems by using well-defined *E. coli* methionine biosynthesis mutants and analyzing LuxI-dependent 3-oxo-C6-HSL synthesis in methionine auxotrophs after addition of methionine or homoserine to cultures maintained under starvation conditions. Mutations that block a step on the methionine branch of the lysine-threonine-methionine pathway (Fig. 3) resulted in strains unable to synthesize 3-oxo-C6-HSL unless provided with methionine, while provision of homoserine had no effect (a mutant blocked for homoserine synthesis synthesized 3-oxo-C6-HSL with exogenous methionine or homoserine). Addition of cycloleucine, an inhibitor of AdoMet synthetase, diminished this methionine-dependent 3-oxo-C6-HSL synthesis. These results implied that AdoMet was required for 3-oxo-C6-HSL synthesis. Furthermore, use of ^{15}N-methionine to supplement starving cultures led to efficient incorporation into 3-oxo-C6-HSL, while ^{15}N-homoserine did not. In clear agreement with the earlier work of Eberhard et al. (1991) using the *V. fischeri* crude extracts, the conclusion from this genetic analysis was that AdoMet serves as the precursor for the homoserine lactone moiety.

More recent work has employed the AdoMet hydrolase from bacteriophage T7 to deplete intracellular pools of AdoMet in *E. coli* (Val and Cronan, 1998). In this background, synthesis of 3-oxo-C8-HSL by the TraI protein was severely reduced in proportion to the observed reduction in AdoMet pools. This study again provides indirect evidence for the role of AdoMet as a substrate in acyl-HSL synthesis.

The most convincing evidence for the role of AdoMet in acyl-HSL synthesis has been ob-

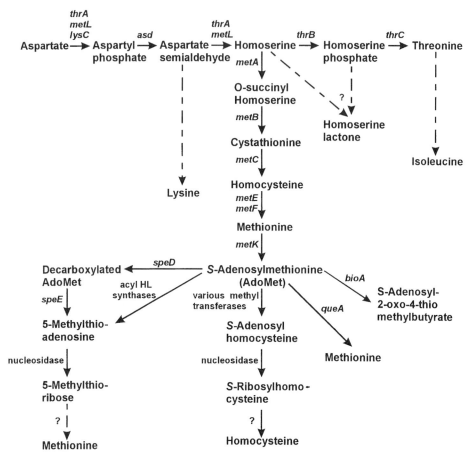

FIGURE 3 Metabolism of methionine and *S*-adenosylmethionine. Modified from Hanzelka and Greenberg (1996) and other references cited in the text.

tained from in vitro studies using purified 6-histidinyl-TraI (H_6-TraI) and maltose binding protein-LuxI (MalE-LuxI) fusion proteins (Moré et al., 1996; Schaefer et al., 1996). Using H_6-TraI and an *E. coli* S28 extract to supply fatty acid pools, Moré et al. (1996) demonstrated 3-oxo-C8-HSL synthesis with AdoMet, but no detectable synthesis with homoserine lactone. Furthermore, purified 3-oxo-C8-ACP (see below) and AdoMet were effective substrates for H_6-TraI. A MalE-LuxI fusion also catalyzed C6-HSL synthesis when provided with hexanoyl-ACP and AdoMet as substrates (Schaefer et al., 1996). With the evidence from these purified systems it seems likely that the lactonized homoserine moiety

in acyl-HSLs is derived from the methionine portion of AdoMet.

Many different enzymes, including acyl-HSL synthases, utilize AdoMet as a substrate (Fig. 3). Therefore, it may be possible to draw parallels between other known reactions of AdoMet and acyl-HSL synthesis. Although the most common cellular function for AdoMet is to supply methyl groups (Drummond et al., 1993; Matthews, 1996), it also serves as a substrate in biotin synthesis, polyamine synthesis, and several tRNA base modification reactions (Eisenberg and Stoner, 1971; Markham et al., 1983; Nishimura, 1983; Slany et al., 1993; Tabor and Tabor, 1983). For the methyl transferases, AdoMet binding motifs are composed of conserved, charged residues that mediate interactions with AdoMet (Kumar et al., 1994; Labahn et al., 1994). The acyl-HSL synthases have no discernible AdoMet binding motifs and no amino acid sequence homology with any of the enzymes that use AdoMet as a substrate. Therefore, it appears that acyl-HSL synthases have evolved a novel mechanism of interacting with the AdoMet molecule, or that the similarities between this active site and its nearest relative are so distant as to be undetectable.

Unacylated Homoserine Lactone: an Alternative Biosynthetic Path?

In *E. coli*, intracellular unacylated homoserine lactone has been implicated as a starvation signal involved in initiating *rpoS*-dependent gene expression (Huisman and Kolter, 1994). Homoserine lactone and homocysteine thiolactone are produced directly through the proofreading activity of the tRNA^{Met} amino acyl transfer RNA synthetase during tRNA charging (Greene, 1996; Jakubowski and Goldman, 1992; Patte, 1996; Kolter and Goodrich-Blair, personal communication), thus presumably coupling the levels of these intracellular signals to the nutritional status of the cell. Although it seemed plausible to hypothesize that the acyl-HSL synthases might utilize homoserine lactone derived in such a manner, the biochemical and genetic data for acyl-HSL synthase function clearly implicate AdoMet as the precursor.

Acyl Chain Derivation

The structure of the fatty acid portion of acyl-HSLs suggested early on that synthesis of these compounds might recruit fatty acids from lipid metabolism (Eberhard et al., 1981). Intracellular pools of fatty acids can be derived from either biosynthetic or degradative routes. Biosynthesis of fatty acids occurs by stepwise addition of two carbon units donated through malonyl-ACP units to nascent acyl chains covalently anchored to ACP via a thioester linkage to its 4′-phosphopantetheine prosthetic group (Fig. 4) (Cronan, 1996). Holo-ACP is one of the most abundant proteins in *E. coli* (Jackowski and Rock, 1983). During synthesis in *E. coli*, acyl-ACP chains may range from 4 to 18 carbons in length, with different oxidation states at the third position of the growing acyl chain (depending on the state of elongation), and may contain an unsaturated linkage at the 3,4-position (Fig. 4). In contrast, exogenously supplied or recycled fatty acids are activated by conjugation to coenzyme A (CoA) and are then subjected to stepwise β-oxidation (Clark and Cronan, 1996). Exogenously supplied fatty acids may serve as precursors for any number of metabolites, including but not restricted to fatty acids. There is also some evidence for transesterification from fatty acyl CoA conjugates directly to ACP, providing a link between fatty acid degradation and biosynthesis (Rock and Jackowski, 1985).

Eberhard et al. (1991) found that chemically synthesized 3-oxohexanoyl-CoA could act as a substrate when crude extracts from *V. fischeri* were used as a source of 3-oxo-C6-HSL synthase. The slow rate of reaction observed with this compound suggested that the whole extracts were converting the CoA conjugates to the appropriate substrates, the most likely of which was a fatty acyl-ACP conjugate. Further apparent support for this speculation was provided by the finding that the chiral carbon at the third position in preparations of in vivo-synthesized *Vibrio harveyi* autoinducer, 3-hydroxy-C4-HSL, was of the D conformation, consistent with a biosynthetic versus degradative origin for the acyl chain (Cao and Meighen, 1993). However, the relevance of this finding is now unclear since it is thought that the LuxL/M proteins, both unrelated to *V. fischeri* LuxI, act as the synthase for 3-hydroxy-C4-HSL (see below) (Bassler et al., 1993). More recent studies using *E. coli* conditional mutants deficient in fatty acid biosynthesis (*fab*) and fatty acid degradation (*fad*) have demonstrated that fatty acid biosynthetic capacity is required for 3-oxo-C8-HSL synthesis by the *A. tumefaciens* TraI protein,

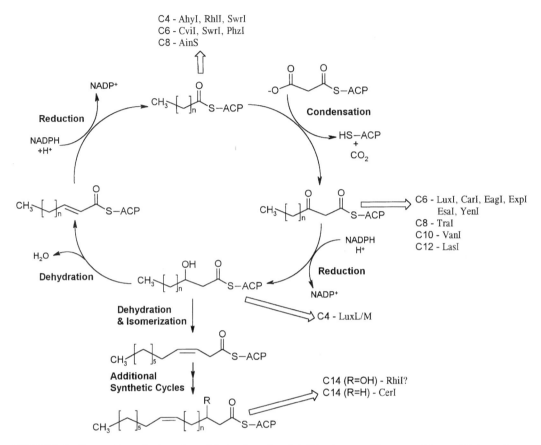

FIGURE 4 Fatty acid synthesis cycle and derivation of acyl-ACP substrates. The general fatty acid biosynthetic cycle is shown along with the shunt for synthesis of fatty acids with unsaturated bonds. Large open arrows indicate points where acyl-HSL synthases are likely to recruit their acyl-ACP substrates. Modified from Cronan (1996).

while fatty acid degradation is dispensable (Val and Cronan, 1998).

In vitro studies using purified MalE-LuxI and H$_6$-TraI fusion proteins have agreed with the prediction of Eberhard et al. (1991) and have implicated acyl-ACP intermediates from fatty acid biosynthesis as the substrates for LuxI-type acyl-HSL synthases. Whole, dialyzed E. coli (S28) extracts supplemented with malonyl-CoA and NADPH acted as effective acyl chain donors and together with AdoMet resulted in acyl-HSL synthesis (Moré et al., 1996). Provision of the acyl chain substrate from whole cell extracts was blocked with the

β-keto acyl synthase inhibitor cerulenin, indicating that active fatty acid biosynthesis is required in the reaction. Presumably, among the population of acylated ACPs in the cell extracts, the appropriate substrate for each cognate ACP is in sufficient supply only during active fatty acid synthesis. The compound 3-oxo-C8-ACP was synthesized by incubation of unacylated holo-ACP with synthetic 3-oxooctanoylthiocholine iodide (Moré et al., 1996). As expected, it was found to be an effective substrate for 3-oxo-C8-HSL synthesis by the H$_6$-TraI protein while 3-oxooctanoyl-CoA (synthesized similarly) was not. Finally, in contrast to the whole cell extracts, neither malonyl-CoA nor NADPH was required for 3-oxo-C8-HSL synthesis in vitro.

In similar studies, Schaefer et al. (1996) took advantage of an earlier observation that LuxI synthesizes the fully reduced C6-HSL as well as 3-oxo-C6-HSL in vivo (Kuo et al., 1994). Using the MalE-LuxI fusion protein, they found that the enzyme synthesized C6-HSL when provided with AdoMet and hexanoyl-ACP, but not when provided with hexanoyl-CoA. Furthermore, although MalE-LuxI utilized ACP charged with butanoyl and octanoyl fatty acids, it did so 30- to 50-fold less efficiently than with the substrate of the appropriate length (Schaefer et al., 1996).

Taken together, the two published studies of acyl-HSL synthesis in vitro clearly demonstrate a specificity for acylated ACPs and not acyl-CoA intermediates of fatty acid degradation. However, these results may not be generalizable to all members of the LuxI protein family. Recent preliminary studies with preparations of the RhlI protein of *P. aeruginosa* suggest that this protein may utilize either acyl-ACP or acyl-CoA (Jiang et al., 1998). Moreover, these studies also suggest that RhlI can utilize homoserine lactone or AdoMet as a precursor for the homoserine portion of the pheromone. It is unclear whether these findings, if correct, have relevance in vivo or are only due to enzymatic or nonenzymatic exchanges (Eberhard et al., 1991). Further work with the pure enzyme will determine whether there is variability in the substrate specificity of this and perhaps other acyl-HSL synthases.

ENZYMOLOGY OF ACYL-HSL SYNTHESIS

Identification of AdoMet and acyl-ACP as substrates for acyl-HSL synthesis suggested a plausible enzymatic mechanism (Eberhard et al., 1991). The amino group of AdoMet provides a strong nucleophile for attacking the carbonyl group of the fatty acid moiety of the acyl-ACP (Fig. 5). It remains unclear whether this nucleophilic attack would occur

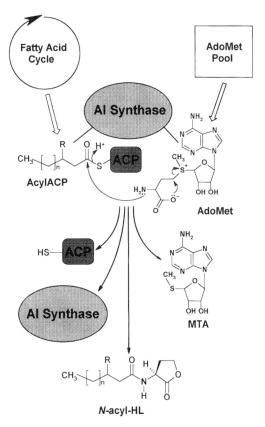

FIGURE 5 Catalytic mechanism. Putative enzymatic mechanism for acyl-HSL synthesis by LuxI-type proteins indicating precursors and probable products. MTA, 5-methylthioadenosine. Modified from Fuqua et al. (1996).

on the acyl-ACP conjugate itself or on a transient acyl-acyl HSL synthase reaction intermediate (see below). The reaction ligates the acyl chain to AdoMet via an amide bond. Lactonization of the acyl-AdoMet intermediate would generate the free acyl-HSL and 5'-methylthioadenosine. While this mechanism is chemically plausible and supported by current information, the details and order of catalysis remain active areas of investigation. For example, it is not yet clear whether lactonization is enzymatically driven or spontaneous upon amide bond formation. However, the MalE-LuxI protein is unable to catalyze acyl-HSL synthesis when provided with hexanoyl-AdoMet, and hexanoyl-AdoMet is apparently stable with respect to lactonization, suggesting that lactonization may be an integral component of the reaction mechanism (Hanzelka and Greenberg, 1996). Therefore, LuxI-type acyl-HSL synthases may direct lactonization as well as amide bond formation, a surprising range of activity for a relatively small protein. Clearly, more work is needed to fully understand the mechanism of catalysis.

Kinetics of Acyl-HSL Synthesis

Experiments with the purified MalE-LuxI and H_6-TraI proteins revealed an unusually slow rate of reaction (1.1 mol/min/mol protein for MalE-LuxI and C6-ACP; 1.03 mol/min/mol protein for H_6-TraI and 3-oxo-C8-ACP [Moré et al., 1996; Schaefer et al., 1996]). The C6-ACP substrate is clearly not identical to the normal substrate for LuxI (3-oxo-C6-ACP), and the kinetics with LuxI and its cognate substrate may be substantially different. The native LuxI and TraI proteins may, of course, have rates of reaction different from those of the fusion proteins. In fact, preliminary analysis of purified native LuxI suggests that the actual rates of reaction may be higher than those determined for the fusion derivative (Parsek and Greenberg, personal communication). However, it is not unreasonable that these enzymes, and by extension other LuxI homologs, catalyze acyl-HSL synthesis very slowly. The concentration of acyl-HSLs that results in half-maximal activation of target genes in a variety of bacteria is as low as 1 nM (Fuqua et al., 1996). Therefore, the catalytic rate of the acyl-HSL synthases in these systems must be low to maintain noninducing levels of acyl-HSL at low cell density. It remains to be determined whether this slow reaction rate is a general feature of acyl-HSL synthases.

The substrate affinities also provide insight into the function and enzymatic activity of acyl-HSL synthases (Moré et al., 1996; Schaefer et al., 1996). The K_m for acylated ACP in the presence of saturating AdoMet was 0.33 μM for H_6-TraI (using 3-oxooctanoyl-ACP) and 9.6 μM for MalE-LuxI (using hexanoyl-ACP). Likewise, the K_m values for AdoMet in the presence of saturating acyl-ACP were 48 μM and 130 μM for H_6-TraI and MalE-LuxI, respectively. The difference between the substrate affinities for the two proteins may be attributable to the use of the noncognate hexanoyl-ACP substrate for MalE-LuxI or may reflect a true difference. It is interesting to note that the acyl-HSL synthases have 10- to 100-fold less affinity for AdoMet than for acyl-ACP. This difference in K_m may reflect the intracellular pools for each substrate. In growing cells, AdoMet is relatively abundant and therefore the affinity of the enzymes can be low. However, the cognate acyl-ACP pool represents only a fraction of the total acyl-ACP and therefore a lower K_m for this substrate would allow synthesis of adequate amounts of acyl-HSL, while not significantly depleting any specific fatty acid pools within the cell.

Does Acyl-HSL Synthesis Proceed via a "Ping-Pong" Mechanism?

It is unclear whether acyl-HSL synthesis involves a stepwise binding of substrates, a random interaction, or a simultaneous interaction. Even before AdoMet and acyl-ACP had been identified as precursors, a stepwise model was proposed in which a reactive acyl-acyl-HSL synthase intermediate was formed by attack of a cysteine residue on the acyl-pantetheine thioester bond, followed by attack of the amino group of AdoMet (Sitnikov et al., 1995). This prediction was based on the enzymology of a number of acyl transferases involved in fatty acid metabolism and phospholipid biosynthesis, including the keto-acyl synthases as well as enzymes involved in phospholipid assembly. For these enzymes, the fatty acid chain is transiently linked to a cysteine or serine residue on the acyl transferase (D'Agnolo et al., 1973; Garwin et al., 1980; Prescott and Vagelos, 1972; Raetz et al., 1987).

Examination of protein sequence alignments with multiple LuxI homologs revealed a single conserved cysteine residue (C68 in LuxI) in all acyl-HSL synthases known at that time (Fuqua et al., 1996; Sitnikov et al., 1995). This cysteine residue was predicted to be the site of acylation through formation of a reactive thioester bond. However, site-directed mutagenesis of this cysteine in LuxI, RhlI, and TraI did not abolish the enzymatic activity of these proteins (Hanzelka et al., 1997; Parsek et al., 1997; Fuqua, unpublished data). In fact, a LuxI-type protein from *P. syringae*, PsyI, has no conserved cysteine residue at this position. Furthermore, for LuxI, RhlI, and TraI, all additional nonconserved cysteines are also dispensable (Hanzelka et al., 1997; Parsek et al., 1997; Fuqua, unpublished data). For the TraI protein, a conserved serine residue at position 106 is essential for activity, raising the possibility that it might be a site for acylation via an ester bond. Several members of the LuxI family of acyl-HSL synthases have threonine residues at this position rather than serines (Fig. 2). However, the corresponding residue in LuxI and RhlI is dispensable for activity (Hanzelka et al., 1997; Parsek et al., 1997). While these mutational studies exclude the possibility that the tested Cys and Ser act as sites for transient acylation, it remains possible that transient acylation at some additional nucleophilic residue (presumably one of the highly conserved positions) is an obligate step in catalysis. Attempts to identify acylated LuxI-type acyl-HSL synthases in vitro have not been successful. Pretreatment of H_6-TraI with thiol modification reagents does reduce but does not abolish its activity, and the degree of inhibition imparted is affected by the modification reagent used (Moré and Winans, personal communication). These observations suggest that, although Cys-adducts resulting from chemical modification are impaired for activity, the modified residues are not directly involved in catalysis. Additional in vitro studies with other acyl-HSL synthases are needed to thoroughly test this idea.

SUBSTRATE INTERACTIONS OF ACYL-HSL SYNTHASES

The active site of an acyl-HSL synthase must accommodate two very different substrates: AdoMet and an acyl chain of specific length and chemistry carried by ACP. The mechanism of the reaction ligating the amino acid portion of AdoMet with the acyl chain is likely to be conserved between different acyl-HSL synthases, and therefore the amino acid residues that direct this reaction are likely to be among those highly conserved between all acyl-HSL synthases.

Recognition and Specificity for Acyl-ACP Conjugates

The ACP protein is composed of four highly negatively charged α–helical segments that pack together into a soluble rodlike structure (Rock and Cronan, 1979). The 4′-phosphopantetheine prosthetic group is conjugated to a serine located between helices 2 and 3 (Kim and Prestegard, 1989). Nuclear magnetic resonance spectroscopic studies of acyl-ACP conjugates in solution suggest that acyl chains of six carbons and shorter remain buried within the cleft between helices 2 and 3 (Mayo and Prestegard, 1985). Yet this sequestration cannot be overly restrictive because the site of the thioester linkage between the acyl chain and the 4′-phosphopantetheine prosthetic group is accessible to many enzymes that use acylated ACP as a substrate (Cronan, 1996). Likewise, the third position is certainly accessible to enzymes of the fatty acid biosynthetic cycle (Fig. 4) as well as to acyl-HSL synthases. Recognition of specific acyl-ACP conjugates, therefore, probably involves both direct interactions with the acyl chain, whether buried or not, and might also include perception of structural changes in ACP caused by the buried portion of the acyl chain. Finally, there may be interactions with the fatty acid biosynthetic enzymes that direct the step prior to recruitment for acyl-HSL synthesis (Fig. 4). A suggestion by Jiang et al. (1998) that RhlI might utilize both butyryl-

ACP and butyryl-CoA is surprising in this light, considering the dramatic structural differences between these two types of fatty acid conjugates.

Acyl-HSL synthases exhibit specificity for the length of the acyl-ACP used as a fatty acid donor and for the chemistry of this acyl chain. While it can be assumed that those residues required for binding of AdoMet should be conserved between different acyl-HSL synthases, the interaction with the acylated ACP substrate must be specific for one of the many species of acylated ACP present in a bacterial cell. Therefore, it seems likely that at least some of the amino acid residues that contact the acyl-ACP might be shared between enzymes that synthesize the same acyl-HSL, but not conserved between enzymes that produce chemically different acyl-HSLs. Multiple sequence alignments of 17 unique acyl-HSL synthases (only proteins less than 65% identical were considered), grouped by the structure of the primary acyl-HSL they synthesize, fail to reveal any clear common signature sequences (Fig. 2) between those enzymes that synthesize similar acyl-HSLs. There are regions of the protein sequences that exhibit heightened similarity between subsets of acyl-HSL synthases. For example, those proteins that direct synthesis of 3-oxo-C6-HSL share increased similarity in restricted regions, particularly the carboxy-terminal portions of the proteins. However, *V. fischeri* LuxI, which also synthesizes 3-oxo-C6-HSL, is not among this group of proteins and is in fact most similar to VanI from *Vibrio anguillarum*, which synthesizes 3-oxo-C10-HSL (Milton et al., 1997). Therefore, the significance of these apparent clusters of greater sequence similarity is unclear. Those bacteria that synthesize acyl-HSLs with acyl chains of six carbons are members of the *Enterobacteriaceae* family, and thus the observed elevated similarities may be more reflective of overall phylogenetic relationships between related microorganisms than signature sequences for their acyl-HSL synthases.

A very crude expectation might be that the length of the LuxI-type protein would reflect the length of the acyl-HSL and that those proteins that direct the synthesis of long-chain acyl-HSLs possess additional sequences to accommodate what is surely a larger substrate molecule (see substrates above). However, this simple model is clearly not the case. All LuxI-type proteins identified thus far range from 194 to 226 amino acids in length, and the variation within this range bears no relationship to the factor synthesized (Fig. 1 and 2). In fact, by far the greatest amount of variation is in the carboxy-terminal ends of these proteins, and at least for those proteins that have been investigated, many of the terminal residues are dispensable for acyl-HSL synthesis (Singh and Fuqua, 1997; Swift et al., 1993). Furthermore, distances between conserved residues are extremely well conserved, and no obvious correlation occurs with gap sizes and acyl chain length: the largest internal gaps are 12 and 9 residues, found for the PhzI (C6-HSL) and RhlI (C4-HSL) sequences (Fig. 2).

Acyl-HSL Synthase Specificity: Cognate Products and Side Reactions

Several studies have analyzed the range of products generated by the in vivo activity of acyl-HSL synthases (Eberl et al., 1996; Jones et al., 1993; Kuo et al., 1994; Winson et al., 1995; Zhang et al., 1993). These studies generally have defined the primary product of the reaction as that acyl-HSL that best activates the corresponding LuxR-type protein. However, in many cases, additional products are generated at detectable levels (Jones et al., 1993; Shaw et al., 1997). The biological significance of these additional products remains an open question. The noncognate acyl-HSLs are often themselves inducers, albeit significantly less effective ones than the primary reaction product. As such, these side products may function as competitive inhibitors of the LuxR-type transcriptional regulators (see Eberhard et al., 1986). A wide range of acyl-HSL analogs are effective inhibitors of the *A. tumefaciens* TraR protein when it is expressed at wild-type levels (Zhu et al., 1998). It is also formally possible that minor products potentiate overall gene induction or are recognized by additional LuxR-type proteins, although there are no data to this effect.

There is at least one example of a biological role for competition between structurally similar acyl-HSLs. In *V. fischeri* LuxI directs the synthesis of 3-oxo-C6-HSL and small amounts of C6-HSL, while a separate acyl-HSL synthase called AinS (not homologous to LuxI; see below) directs the synthesis of C8-HSL. Strains that carry null mutations in *ainS* express the luminescence genes at lower cell densities than do wild-type strains, and this effect is abolished by addition of synthetic C8-HSL (Kuo et al., 1996). This suggests that in *V. fischeri* C8-HSL modulates activation of luminescence by competing with 3-oxo-C6-HSL for LuxR.

ACYL-HSL SYNTHASE STRUCTURE/FUNCTION STUDIES

There is currently very little information regarding the structures of LuxI-type proteins and how they relate to catalysis. The proteins have a typical ratio of hydrophobic to hydrophilic regions, with an even distribution of charged amino acid residues. Purified fusion derivatives of several acyl-HSL synthases demonstrate variable degrees of solubility. The *A. tumefaciens* TraI protein is estimated to be monomeric as judged by gel filtration chromatography (Moré et al., 1996), suggesting that other acyl-HSL synthases may also function as monomers.

Mutational analyses have been performed for three acyl-HSL synthases and have begun to identify residues essential for activity. For the LuxI protein itself, random mutagenesis and identification of loss-of-function alleles reveals two primary clusters of mutations (Hanzelka et al., 1997). One of these regions ranges between residues 24 and 104, while the other is in the carboxy-terminal half between residues 133 and 164. Many of those mutations that lead to complete loss of function correspond to acidic or basic residues conserved throughout the LuxI family of proteins (Fig. 2). The absolutely conserved residues are

localized in the first of these two regions, and therefore it was speculated that this might be the region that interacts with AdoMet and serves to catalyze amide bond formation. Similarly, residues in the second region of the protein, which exhibits less overall sequence conservation, were predicted to be involved in acyl chain selection. Therefore, LuxI might be tentatively divided into two domains, the N-terminal region that binds AdoMet and the C-terminal region that provides specificity for the appropriate acyl-ACP conjugate (Hanzelka et al., 1997). However, mutational studies using the RhlI protein from *P. aeruginosa* cast uncertainty on this model for acyl-HSL synthesis. In the RhlI study, random mutagenesis again identified requisite acidic and basic amino acids required for catalysis (Parsek et al., 1997). No loss-of-function mutants resulted from mutations in the carboxy-terminal region corresponding to the same region in LuxI (Fig. 2). While the RhlI data should generate skepticism regarding the simple two-domain model, it remains possible that both studies are valid reflections of acyl-HSL structure/function. RhlI and LuxI generate C4-HSL and 3-oxo-C6-HSL, respectively, and the mechanism of catalysis may vary between acyl-HSL synthases that synthesize structurally divergent acyl-HSLs.

These mutational analyses have identified conserved charged residues as essential for protein function. The requirement for these acidic and basic residues suggests that acyl-HSL synthesis may be mediated through acid-base chemistry within the active site. Conversely, those nonpolar residues conserved throughout the LuxI family are not essential. Site-directed mutations in the nonpolar conserved residues in RhlI caused significant reduction in activity, but not complete loss of function (Parsek et al., 1997). Taken together, these studies suggest that amide bond formation between AdoMet and acyl-ACP substrates is directed by the action of charged residues in the active site. Possible catalytic interactions include formation of a reactive anhydride between an aspartate or glutamate and the acyl group of the acyl-ACP or perhaps acid-base interactions between the charged residues and the amino, carboxyl, or sulfonium groups on AdoMet and the carbonyl group of the acyl-ACP. Conserved nonpolar residues are not absolutely required but may be involved in active site architecture or in positioning the substrates within the active site.

While this work has confirmed the importance of acidic and basic amino acid residues conserved throughout the LuxI protein family, little insight into the structure of the protein and its association with substrates has resulted. Several efforts are under way to generate structural information from crystals of acyl-HSL synthases as well as co-crystals of acyl-HSL synthases with AdoMet or acyl-ACP.

REGULATION OF ACYL-HSL SYNTHESIS

The basal level of acyl-HSL synthesized by bacteria at low cell density plays an essential role in acyl-HSL quorum sensing. Therefore, any regulatory factors or environmental conditions that impinge either negatively or positively on basal level acyl-HSL synthesis will have a significant effect on the overall regulation of target genes.

Regulation of Enzyme Activity

One simple prediction is that acyl-HSL synthesis should be inhibited by the reaction products. At elevated concentrations in vitro, both 5-methylthioadenosine and uncharged ACP inhibit the activity of H_6-TraI and the EsaI protein (Moré and Winans, personal communication). Surprisingly, at lower concentrations, uncharged holo-ACP stimulates the synthesis of 3-oxo-C8-HSL by H_6-TraI up to 50-fold when the cognate substrate 3-oxooctanoyl-ACP is limiting. The generality and physiological significance of the ACP-dependent stimulation remains unexplained but may reflect a heightened affinity for fatty acid pools during periods of starvation when charged acyl-ACPs are limiting.

Control of Acyl-HSL Synthase Gene Expression

Transcriptional control of LuxI-type protein gene expression is well documented for many *luxI*-type genes. Commonly, although not always, these genes are activated as targets of the LuxR-type transcriptional regulator, resulting in a positive feedback loop where transcriptional activation of the gene results in greater synthesis of the inducer. While most studies have demonstrated this positive feedback loop at the level of mRNA using gene fusions, Eberhard et al. (1991) reported that 3-oxo-C6-HSL synthesis is increased when the *luxI* gene is induced. The activation of the acyl-HSL synthase genes by the very inducer they synthesize might function as a desensitization switch; cells that are activated at a specific cell density remain activated at a significantly lower density. A consistent point of confusion is the idea that this positive feedback loop is itself the process of autoinduction. In fact, autoinduction does not require that the acyl-HSL synthase gene(s) be under the positive control of the LuxR-type protein; it only requires that the expression of this gene(s) be at a low basal level. There are now several examples where transcription of the *luxI*-type gene is not regulated by the acyl-HSL (Beck von Bodman and Farrand, 1995; Throup et al., 1995).

Regulating expression of the acyl-HSL synthase allows the cell to control the production of the acyl-HSL in response to environmental conditions. In addition, the sensitivity of the positive feedback switch dictates how rapidly overall gene induction occurs. For example, the *lasI* gene of *P. aeruginosa* is activated at significantly lower 3-oxo-C12-HSL concentrations than are other target genes, suggesting a hierarchy of activation (Seed et al., 1995). Early activation of acyl-HSL synthesis during induction may ensure rapid virulence factor production in the host animal. In several other systems where the *luxI*-type gene is positively regulated by its cognate acyl-HSL, the gene is carried within one of the target operons, thereby directly tying activation of acyl-HSL synthesis to activation of other target genes (Fuqua et al., 1996; Swift et al., 1996).

There are also several examples of additional transcriptional regulatory systems that control acyl-HSL synthase gene expression. In *P. aeruginosa*, expression of the *rhlI* gene (either directly or via induction of *rhlR*) is under the control of the LasR-LasI quorum sensor, thereby obligately tying C4-HSL synthesis to the upstream 3-oxo-C12-HSL signal (Latifi et al., 1996; Pesci et al., 1997). In addition, there is evidence that in *P. aeruginosa* PAO the GacA two-component-type response regulator is required for *rhlI* expression (Reimmann et al., 1997). Similar GacA-dependent expression is reported for the *P. aureofaciens phzI* gene (Chancey and Pierson, personal communication). The environmental signal(s) perceived directly or indirectly by GacA, presuming there are such signals, is still unknown, but at least in these two pseudomonads, they would impinge upon acyl-HSL synthesis.

A Second Class of Acyl-HSL Synthases

Recent studies analyzing regulation of *lux* genes in *V. fischeri* and *V. harveyi* have identified proteins that are likely acyl-HSL synthases but bear no sequence similarity to LuxI. In *V. harveyi*, synthesis of 3-hydroxy-C4-HSL requires the *luxM* gene (Bassler et al., 1993). Likewise, production of C8-HSL in *V. fischeri* is mediated by a protein called AinS (Gilson et al., 1995). The LuxM and AinS proteins share a region of strong conservation with each other but are not homologous to LuxI. Both of these presumptive acyl-HSL synthase genes are associated with genes for two-component-type sensor kinases (LuxN and AinR, respectively) that are also homologous to each other. The LuxN protein is required to maintain responsiveness to 3-hydroxy-C4-HSL in *V. harveyi* and is predicted to act through the LuxO response regulator (Bassler et al., 1993, 1994). Therefore, these acyl-HSL-dependent quorum sensors appear to

function by a dramatically different mechanism than LuxR-LuxI-type systems. Preliminary reports of AinS in vitro activity confirm that this protein is an acyl-HSL synthase (Hanzelka et al., personal communication). The relationship between LuxI-type and AinS-type acyl-HSL synthases remains unknown. There is no obvious sequence similarity between these proteins, and hence they may represent a remarkable case of convergent evolution. The membrane permeability of acyl-HSLs and their ready synthesis from primary metabolites could make them ideal cell-to-cell signals, such that they have evolved twice.

QUESTIONS AND FUTURE DIRECTIONS

There are clearly a great many questions that remain to be addressed regarding acyl-HSL synthesis in bacteria. While genetic and biochemical approaches have revealed the likely substrates for the reaction for several acyl-HSL synthases, the universality of these substrates for all members of the LuxI family remains in question. How each substrate is recognized is unclear, as are the mechanisms of formation of the amide linkage and of the homoserine lactone ring. Although support for the two-step or "ping-pong" model is dwindling, it has yet to be disproved.

Acyl-HSL synthases have a remarkable capacity for recognizing the correct acyl chain out of the fatty acid pools in the cell. How are the appropriate acyl-ACPs distinguished from inappropriate conjugates? What portions of the enzyme dictate acyl chain specificity? Thus far, mutational studies have identified only key residues conserved throughout the LuxI family. There must be contacts between the enzyme and the acyl substrates and ACP itself that confer fatty acid specificity. Additional mutant analyses with multiple acyl-HSL synthases, coupled with biochemical approaches, should begin to define those functions required for specificity, as well as further refine the general catalytic models. Generation of hybrid acyl-HSL synthases that combine catalytic specificities would also provide insights into the functional organization of the proteins. Structural characterization of acyl-HSL synthases, both in free form and in complex with substrates, will be essential to an understanding of how these proteins perform all of those steps required for biosynthesis.

Differences and similarities in acyl-HSL synthesis between LuxI-type and AinS/LuxM-type enzymes are sure to illuminate many details of each. Comparative analyses of these two classes of enzymes will allow us to determine the relatedness of the enzymes at the level of enzymatic structure and function.

With a detailed understanding of the architecture and action of acyl-HSL synthases, the prospects for directed pharmaceutical intervention, either inhibiting synthesis or controlling the time of target gene induction, will merit investigation. The acyl-HSL synthases appear to utilize relatively unique mechanisms for recruiting AdoMet and acyl-groups from normal cellular metabolism and therefore may be reasonable targets for drug development.

Overall regulation of acyl-HSL synthesis and the manner in which this is coordinated with normal cellular metabolism is a fertile area for future studies. While bacterial cell density clearly has overlap with nutrient starvation, the two conditions can be quite distinct, and the cell must have mechanisms to coordinate the two types of response but keep them independent. It is fascinating to witness the various ways in which different bacteria employ acyl-HSLs, and it surely will be equally informative to understand how each cell has evolved to regulate this process. There are certain to be multiple stages during acyl-HSL synthesis that are subject to complex regulatory control. This area has only begun to be investigated.

In short, our understanding of acyl-HSL synthesis in bacteria has advanced significantly over the last several years. This progress has been primarily due to renewed efforts to biochemically characterize the reaction as it became clear that acyl-HSL-dependent quorum sensing was not only an interesting form of regulation in bioluminescent marine bacteria,

but a widespread form of intercellular communication in diverse bacteria. Future investigation of acyl-HSL synthesis in diverse bacteria is sure to contribute to our rapidly evolving ideas of how bacteria coordinate their actions and the routes of communication by which this is achieved.

ACKNOWLEDGMENTS

We wish to thank our many colleagues for providing valuable input, with special thanks to Susanne Beck von Bodman, Arun Chatterjee, Heidi Goodrich-Blair, Pete Greenberg, Roberto Kolter, Sandy Pierson, Paul Williams, Steve Winans, and Margret Moré for sharing results prior to publication, and Jim Shinkle for commenting on the manuscript. Research funding is provided by the National Science Foundation Grant MCB-9723837 (C.F.).

REFERENCES

Bainton, N. J., B. W. Bycroft, S. R. Chhabra, P. Stead, L. Geldhill, P. J. Hill, C. E. D. Rees, M. K. Winson, G. P. C. Salmond, G. S. A. B. Stewart, and P. Williams. 1992. A general role for the *lux* autoinducer in bacterial cell signalling: control of antibiotic biosynthesis in *Erwinia*. *Gene* **116**:87–91.

Bassler, B. L., M. Wright, R. E. Showalter, and M. R. Silverman. 1993. Intercellular signalling in *Vibrio harveyi*: sequence and function of genes regulating expression of luminescence. *Mol. Microbiol.* **9**:773–786.

Bassler, B. L., M. Wright, and M. R. Silverman. 1994. Sequence and function of LuxO, a negative regulator of luminescence in *Vibrio harveyi*. *Mol. Microbiol.* **12**:403–412.

Beck von Bodman, S., and S. K. Farrand. 1995. Capsular polysaccharide biosynthesis and pathogenicity in *Erwinia stewartii* require induction by an N-acylhomoserine lactone autoinducer. *J. Bacteriol.* **177**:5000–5008.

Cao, J.-G., and E. A. Meighen. 1989. Purification and structural identification of an autoinducer for the luminescence system of *V. harveyi*. *J. Biol. Chem.* **264**:21670–21676.

Cao, J.-G., and E. A. Meighen. 1993. Biosynthesis and stereochemistry of the autoinducer controlling luminescence in *Vibrio harveyi*. *J. Bacteriol.* **175**:3856–3862.

Chancey, S. and L. L. Pierson III. Personal communication.

Clark, D. P., and J. E. Cronan, Jr. 1996. Two-carbon compounds and fatty acids as carbon sources, p. 343–357. *In* F. C. Niedhardt, R. Curtiss III, J. L. Ingraham, E. C. C. Lin, K. B. Low, Jr., B. Magasanik, W. S. Reznikoff, M. Riley, M. Schaechter, and H. E. Umbarger (ed.), *Escherichia coli and Salmonella: Cellular and Molecular Biology*, 2nd ed. ASM Press, Washington, D.C.

Cronan, J. E., Jr. 1996. Biosynthesis of membrane lipids, p. 612–636. *In* F. C. Niedhardt, R. Curtiss III, J. L. Ingraham, E. C. C. Lin, K. B. Low, Jr., B. Magasanik, W. S. Reznikoff, M. Riley, M. Schaechter, and H. E. Umbarger (ed.), *Escherichia coli and Salmonella: Cellular and Molecular Biology*, 2nd ed. ASM Press, Washington, D.C.

Cubo, M. T., A. Economou, G. Murphy, A. W. B. Johnston, and J. A. Downie. 1992. Molecular characterization and regulation of the rhizosphere-expressed genes *rhiABCR* that can influence nodulation by *Rhizobium leguminosarum* biovar *viciae*. *J. Bacteriol.* **174:**4026–4035.

D'Agnolo, G., I. S. Rosenfeld, J. Awaya, S. Omura, and P. R. Vagelos. 1973. Inhibition of fatty acid biosynthesis by the antibiotic cerulenin. Specific inactivation of the B-ketoacyl-acyl carrier protein synthetase. *Biochim. Biophys. Acta* **326:** 155–166.

Drummond, J. T., S. Huang, R. M. Blumenthal, and R. G. Matthews. 1993. Assignment of enzymatic function to specific protein regions of cobalamin-dependent methionine synthase from *Escherichia coli*. *Biochemistry* **32:**9290–9295.

Eberhard, A. 1972. Inhibition and activation of bacterial luciferase synthesis. *J. Bacteriol.* **109:**1101–1105.

Eberhard, A., A. L. Burlingame, C. Eberhard, G. L. Kenyon, K. H. Nealson, and N. J. Oppenheimer. 1981. Structural identification of autoinducer of *Photobacterium fischeri*. *Biochemistry* **20:** 2444–2449.

Eberhard, A., C. A. Widrig, P. McBath, and J. B. Schineller. 1986. Analogs of the autoinducer of bioluminescence in *Vibrio fischeri*. *Arch. Microbiol.* **146:**35–40.

Eberhard, A., T. Longin, C. A. Widrig, and S. J. Stranick. 1991. Synthesis of the *lux* gene autoinducer in *Vibrio fischeri* is positively autoregulated. *Arch. Microbiol.* **155:**294–297.

Eberl, L., M. K. Winson, C. Sternberg, G. S. A. B. Stewart, G. Christiansen, S. R. Chhabra, B. W. Bycroft, P. Williams, S. Molin, and M. Givskov. 1996. Involvement of *N*-acyl-L-homoserine lactone autoinducers in control of multicellular behavior of *Serratia liquefaciens*. *Mol. Microbiol.* **20:**127–136.

Eisenberg, M. A., and G. L. Stoner. 1971. Biosynthesis of 7,8-diaminopelargonic acid in cell free

extracts from 7-keto-8 aminopelargonic acid and S-adenosyl-L-methionine. *J. Bacteriol.* **108:**1135–1140.

Fuqua, C. Unpublished data.

Fuqua, W. C., and S. C. Winans. 1994. A LuxR-LuxI type regulatory system activates *Agrobacterium* Ti plasmid conjugal transfer in the presence of a plant tumor metabolite. *J. Bacteriol.* **176:**2796–2806.

Fuqua, C., S. C. Winans, and E. P. Greenberg. 1996. Census and consensus in bacterial ecosystems: the LuxR-LuxI family of quorum-sensing transcriptional regulators. *Annu. Rev. Microbiol.* **50:** 591–624.

Gambello, M. J., and B. H. Iglewski. 1991. Cloning and characterization of the *Pseudomonas aeruginosa lasR* gene, a transcriptional activator of elastase expression. *J. Bacteriol.* **173:**3000–3009.

Garwin, J. L., A. L. Klages, and J. E. Cronan, Jr. 1980. B-ketoacyl-acyl carrier protein synthase II of *Escherichia coli*. Evidence for function in the thermal regulation of fatty acid synthesis. *J. Biol. Chem.* **255:**3263–3265.

Gilson, L., A. Kuo, and P. V. Dunlap. 1995. AinS and a new family of autoinducer synthesis proteins. *J. Bacteriol.* **177:**6946–6951.

Gray, K. M., J. P. Pearson, J. A. Downie, B. E. A. Boboye, and E. P. Greenberg. 1996. Cell-to-cell signalling in the symbiotic nitrogen-fixing bacterium *Rhizobium leguminosarum*: autoinduction of a stationary phase and rhizosphere-expressed genes. *J. Bacteriol.* **178:**372–376.

Greene, R. C. 1996. Biosynthesis of methionine, p. 542–560. *In* F. C. Niedhardt, R. Curtiss III, J. L. Ingraham, E. C. C. Lin, K. B. Low, Jr., B. Magasanik, W. S. Reznikoff, M. Riley, M. Schaechter, and H. E. Umbarger (ed.), *Escherichia coli and Salmonella: Cellular and Molecular Biology*, 2nd ed. ASM Press, Washington, D.C.

Hanzelka, B. L., and E. P. Greenberg. 1996. Quorum sensing in *Vibrio fischeri*: evidence that S-adenosylmethionine is the amino acid substrate for autoinducer synthesis. *J. Bacteriol.* **178:**5291–5294.

Hanzelka, B. L., A. M. Stevens, M. R. Parsek, T. J. Crone, and E. P. Greenberg. 1997. Mutational analysis of the *Vibrio fischeri* LuxI polypeptide: critical regions of an autoinducer synthase. *J. Bacteriol.* **179:**4882–4887.

Hanzelka, B. L., P. V. Dunlap, and E. P. Greenberg. Personal communication.

Hobson, A. C., and D. A. Smith. 1973. S-adenosylmethionine synthetase in methionine regulating mutants of *Salmonella typhimurium*. *Mol. Gen. Genet.* **126:**7–18.

Huisman, G. W., and R. Kolter. 1994. Sensing starvation: a homoserine lactone-dependent signaling pathway in *Escherichia coli*. *Science* **265**:537–539.

Hwang, I., P.-L. Li, L. Zhang, K. R. Piper, D. M. Cook, M. E. Tate, and S. K. Farrand. 1994. TraI, a LuxI homologue, is responsible for production of conjugation factor, the Ti plasmid *N*-acylhomoserine lactone autoinducer. *Proc. Natl. Acad. Sci. USA* **91**:4639–4643.

Jackowski, S., and C. O. Rock. 1983. Ratio of active to inactive forms of acyl carrier protein in *Escherichia coli*. *J. Biol. Chem.* **258**:15186–15191.

Jakubowski, H., and E. Goldman. 1992. Editing errors in selection of amino acids for protein synthesis. *Microbiol. Rev.* **56**:412–429.

Jiang, Y., M. Camara, S. R. Chhabra, K. R. Hardie, B. W. Bycroft, A. Lazdunski, G. P. C. Salmond, G. S. A. B. Stewart, and P. Williams. 1998. In vitro biosynthesis of the *Pseudomonas aeruginosa* quorum-sensing signal molecule, *N*-butanoyl-L-homoserine lactone. *Mol. Microbiol.* **28**:193–204.

Jones, S., B. Yu, N. J. Bainton, M. Birdsall, B. W. Bycroft, S. R. Chhabra, A. J. R. Cox, P. Golby, P. J. Reeves, S. Stephens, M. K. Winson, G. P. C. Salmond, G. S. A. B. Stewart, and P. Williams. 1993. The *lux* autoinducer regulates the production of exoenzyme virulence determinants in *Erwinia carotovora* and *Pseudomonas aeruginosa*. *EMBO J.* **12**:2477–2482.

Kim, Y. M., and J. H. Prestegard. 1989. A dynamic model for the structure of acyl carrier protein in solution. *Biochemistry* **28**:8792–8797.

Kolter, R., and H. Goodrich-Blair. Personal communication.

Kumar, S., X. Cheng, S. Klimasauskas, S. Mi, J. Posfai, R. J. Roberts, and G. G. Wilson. 1994. The DNA (cytosine-5) methyltransferases. *Nucleic Acids Res.* **22**:1–10.

Kuo, A., N. V. Blough, and P. V. Dunlap. 1994. Multiple *N*-acyl-L-homoserine lactone autoinducers of luminescence genes in the marine symbiotic bacterium *Vibrio fischeri*. *J. Bacteriol.* **176**:7558–7565.

Kuo, A., S. M. Callahan, and P. V. Dunlap. 1996. Modulation of luminescence operon expression by *N*-octanoyl-L-homoserine lactone in *ainS* mutants of *Vibrio fischeri*. *J. Bacteriol.* **178**:971–976.

Labahn, J., J. Granzin, G. Schluckebier, D. P. Robinson, W. E. Jack, I. Schildkraut, and W. Saenger. 1994. Three-dimensional structure of the adenine-specific DNA methyltransferase M*Taq* I in complex with the cofactor *S*-adenosylmethionine. *Proc. Natl. Acad. Sci. USA* **91**:10957–10961.

Latifi, A., M. Foglino, K. Tanaka, P. Williams, and A. Lazdunski. 1996. A hierarchical quorum-sensing cascade in *Pseudomonas aeruginosa* links the transcriptional activators LasR and RhlR (VsmR) to expression of the stationary-phase sigma factor RpoS. *Mol. Microbiol.* **21**:1137–1146.

Markham, G. D., C. W. Tabor, and H. Tabor. 1983. *S*-adenosylmethionine decarboxylase (*Escherichia coli*). *Methods Enzymol.* **94**:228–230.

Matthews, R. W. 1996. One-carbon metabolism, p. 600–611. *In* F. C. Niedhardt, R. Curtiss III, J. L. Ingraham, E. C. C. Lin, K. B. Low, Jr., B. Magasanik, W. S. Reznikoff, M. Riley, M. Schaechter, and H. E. Umbarger (ed.), *Escherichia coli and Salmonella: Cellular and Molecular Biology*, 2nd ed. ASM Press, Washington, D.C.

Mayo, K. H., and J. H. Prestegard. 1985. Structural characterization of short-chain acylated acyl carrier proteins by NMR. *Biochemistry* **24**:7834–7838.

Milton, D. L., A. Hardman, M. Camara, S. R. Chhabra, B. W. Bycroft, G. S. A. B. Stewart, and P. Williams. 1997. Quorum sensing in *Vibrio anguillarum*: characterization of the *vanI/vanR* locus and identification of the autoinducer *N*-(3-oxodecanoyl)-L-homoserine lactone. *J. Bacteriol.* **179**:3004–3012.

Moré, M. I., and S. C. Winans. Personal communication.

Moré, M. I., L. D. Finger, J. L. Stryker, C. Fuqua, A. Eberhard, and S. C. Winans. 1996. Enzymatic synthesis of a quorum-sensing autoinducer through use of defined substrates. *Science* **272**:1655–1658.

Nishimura, S. 1983. Structure, biosynthesis, and function of queuosine in transfer RNA. *Prog. Nucleic Acid Res. Mol. Biol.* **28**:49–73.

Parsek, M. R., and E. P. Greenberg. Personal communication.

Parsek, M. R., A. L. Schaefer, and E. P. Greenberg. 1997. Analysis of random and site-directed mutations in *rhlI*, a *Pseudomonas aeruginosa* gene encoding an acylhomoserine lactone synthase. *Mol. Microbiol.* **26**:301–310.

Passador, L., J. M. Cook, M. J. Gambello, L. Rust, and B. H. Iglewski. 1993. Expression of *Pseudomonas aeruginosa* virulence genes requires cell-to-cell communication. *Science* **260**:1127–1130.

Patte, J.-C. 1996. Biosynthesis of threonine and lysine, p. 528–541. *In* F. C. Niedhardt, R. Curtiss III, J. L. Ingraham, E. C. C. Lin, K. B. Low, Jr., B. Magasanik, W. S. Reznikoff, M. Riley, M.

Schaechter, and H. E. Umbarger (ed.), *Escherichia coli and Salmonella: Cellular and Molecular Biology,* 2nd ed. ASM Press, Washington, D.C.

Pearson, J. P., K. M. Gray, L. Passador, K. D. Tucker, A. Eberhard, B. H. Iglewski, and E. P. Greenberg. 1994. Structure of the autoinducer required for expression of *Pseudomonas aeruginosa* virulence genes. *Proc. Natl. Acad. Sci. USA* **91:** 197–201.

Pearson, J. P., L. Passador, B. H. Iglewski, and E. P. Greenberg. 1995. A second *N*-acylhomoserine lactone signal produced by *Pseudomonas aeruginosa. Proc. Natl. Acad. Sci. USA* **92:** 1490–1494.

Pesci, E. C., J. P. Pearson, P. C. Seed, and B. H. Iglewski. 1997. Regulation of *las* and *rhl* quorum sensing in *Pseudomonas aeruginosa. J. Bacteriol.* **179:**3127–3132.

Piper, K. R., S. Beck von Bodman, and S. K. Farrand. 1993. Conjugation factor of *Agrobacterium tumefaciens* regulates Ti plasmid transfer by autoinduction. *Nature* **362:**448–450.

Pirhonen, M., D. Flego, R. Heikinheimo, and E. T. Palva. 1993. A small diffusible molecule is responsible for the global control of virulence and exoenzyme production in the plant pathogen *Erwinia carotovora. EMBO J.* **12:**2467–2476.

Prescott, D. J., and P. R. Vagelos. 1972. Acyl carrier protein. *Adv. Enzymol.* **36:**269–311.

Puskas, A., E. P. Greenberg, S. Kaplan, and A. L. Schaefer. 1997. A quorum-sensing system in the free-living photosynthetic bacterium *Rhodobacter sphaeroides. J. Bacteriol.* **179:**7530–7537.

Raetz, C. R. H., G. M. Carman, and W. Dowhan. 1987. Phospholipids chiral at phosphorous. Steric course of the reactions catalyzed by phosphatidylserine synthase from *Escherichia coli* and yeast. *Biochemistry* **26:**4022–4027.

Reimmann, C., M. Beyeler, A. Latifi, H. Winteler, M. Foglini, A. Lazdunski, and D. Haas. 1997. The global activator GacA of *Pseudomonas aeruginosa* PAO positively controls the production of the autoinducer *N*-butyryl-homoserine lactone and the formation of the virulence factors pyocyanin, cyanide, and lipase. *Mol. Microbiol.* **24:**309–319.

Rock, C. O., and J. E. Cronan, Jr. 1979. Reevaluation of the solution structure of the acyl carrier protein. *J. Biol. Chem.* **254:**9778–9785.

Rock, C. O., and S. Jackowski. 1985. Pathway for incorporation of exogenous fatty acids into phosphatidylethanolamine in *Escherichia coli. J. Biol. Chem.* **260:**12720–12724.

Satishchandran, C., J. C. Taylor, and G. D. Markham. 1993. Isozymes of S-adenosyl-

methionine synthetase are encoded by tandemly duplicated genes in *Escherichia coli. Mol. Microbiol.* **9:**835–846.

Schaefer, A. L., D. L. Val, B. L. Hanzelka, J. E. Cronan, and E. P. Greenberg. 1996. Generation of cell-to-cell signals in quorum sensing: acyl homoserine activity of purified *Vibrio fischeri* LuxI protein. *Proc. Natl. Acad. Sci. USA* **93:**9505–9509.

Schripsema, J., K. E. E. de Rudder, T. B. van Vliet, P. P. Lankhorst, E. de Vroom, J. W. Kijne, and A. A. N. van Brussel. 1996. Bacteriocin *small* of *Rhizobium leguminosarum* belongs to the class of *N*-acyl-L-homoserine lactone molecules, known as autoinducers and as quorum sensing co-transcription factors. *J. Bacteriol.* **178:** 366–371.

Seed, P. C., L. Passador, and B. H. Iglewski. 1995. Activation of the *Pseudomonas aeruginosa lasI* gene by LasR and the *Pseudomonas* autoinducer PAI—an autoinduction regulatory hierarchy. *J. Bacteriol.* **177:**654–659.

Shaw, P., G. Ping, S. L. Daly, C. Cha, J. E. J. Cronan, K. L. Rinehart, and S. K. Farrand. 1997. Detecting and characterizing *N*-acylhomoserine lactone signal molecules by thin-layer chromatography. *Proc. Natl. Acad. Sci USA* **94:** 6036–6041.

Singh, P. L., and W. C. Fuqua. 1997. Genetic dissection of the *Agrobacterium tumefaciens* autoinducer synthase TraI, abstr. H-170, p. 312. Abstr. Gen. Meet. Am. Soc. Microbiol. 1997.

Sitnikov, D., J. B. Schineller, and T. O. Baldwin. 1995. Transcriptional regulation of bioluminescence genes from *Vibrio fischeri. Mol. Microbiol.* **17:**801–812.

Slany, R. K., M. Bosl, P. F. Crain, and H. Kersten. 1993. A new function of S-adenosylmethionine: the ribosyl moiety of AdoMet is the precursor of the cyclopentenediol moiety of the tRNA wobble base queuine. *Biochemistry* **32:**7811–7817.

Swift, S., M. K. Winson, P. F. Chan, N. J. Bainton, M. Birdsall, P. J. Reeves, C. E. D. Rees, S. R. Chhabra, P. J. Hill, J. P. Throup, B. W. Bycroft, G. P. C. Salmond, P. Williams, and G. S. A. B. Stewart. 1993. A novel strategy for the isolation of *luxI* homologues: evidence for the widespread distribution of a LuxR:LuxI superfamily in enteric bacteria. *Mol. Microbiol.* **10:**511–520.

Swift, S., J. P. Throup, P. Williams, G. P. C. Salmond, and G. S. A. B. Stewart. 1996. Quorum sensing: a population density component in the determination of bacterial phenotype. *Trends Biochem. Sci.* **21:**214–219.

Swift, S., A. V. Karlyshev, L. Fish, E. L. Durant, M. K. Winson, S. R. Chhabra, P. Williams,

S. Macintyre, and G. S. A. B. Stewart. 1997. Quorum-sensing in *Aeromonas hydrophila* and *Aeromonas salmonicida*: identification of the LuxRI homologs AhyRI and AsaRI and their cognate *N*-acylhomoserine lactone signal molecules. *J. Bacteriol.* **179**:5271–5281.

Tabor, C. W., and H. Tabor. 1983. Putrescine aminopropyltransferase. *Methods Enzymol.* **94**:265–269.

Throup, J. P., M. Camara, G. S. Briggs, M. K. Winson, S. R. Chhabra, B. W. Bycroft, P. Williams, and G. S. A. B. Stewart. 1995. Characterisation of the *yenI* / *yenR* locus from *Yersenia enterocolitica* mediating the synthesis of two *N*-acylhomoserine lactone signal molecules. *Mol. Microbiol.* **17**:345–356.

Val, D. L., and J. E. J. Cronan. 1998. In vivo evidence that *S*-adenosylmethionine and fatty acid synthetic intermediates are the substrates for the LuxI family of autoinducer synthases. *J. Bacteriol.* **180**:2644–2651.

Wang, X., P. A. J. de Boer, and L. I. Rothfield. 1991. A factor that positively regulates cell division by activating transcription of the major cluster of essential cell division genes in *Escherichia coli*. *EMBO J.* **10**:3363–3372.

Winson, M. K., M. Camara, A. Latifi, M. Foglino, S. R. Chhabra, M. Daykin, M. Bally, V. Chapon, G. P. C. Salmond, B. W. Bycroft, A. Lazdunski, G. S. A. B. Stewart, and P. Williams. 1995. Multiple *N*-acyl-L-homoserine lactone signal molecules regulate production of virulence determinants and secondary metabolites in *Pseudomonas aeruginosa*. *Proc. Natl. Acad. Sci. USA* **92**:9427–9431.

Zhang, L., P. J. Murphy, A. Kerr, and M. E. Tate. 1993. Agrobacterium conjugation and gene regulation by *N*-acyl-homoserine lactones. *Nature* **362**:446–448.

Zhu, J., J. W. Beaber, M. I. Moré, C. Fuqua, A. Eberhard, and S. C. Winans. 1998. Analogs of the autoinducer 3-oxooctanoyl-homoserine lactone strongly inhibit activity of the TraR protein of *Agrobacterium tumefaciens*. *J. Bacteriol.* **180**:5398–5405.

within the LuxR superfamily. Since the discovery of LasR, this subfamily of LuxR homologs has grown to around 20 members (Fuqua et al., 1996). Included among these are TraR from *Agrobacterium tumefaciens*, which together with an acylhomoserine lactone signal regulates conjugal plasmid transfer (Piper et al., 1993; see chapter 8 of this volume), and ExpR from *Erwinia carotovora*, which together with an acylhomoserine lactone signal regulates exoenzyme production and antibiotic synthesis in this bacterium (Pirhonen et al., 1993; see chapter 7 of this volume). All LuxR homologs show a low overall amino acid conservation with LuxR (usually about 25% identity in pairwise comparisons). Although the overall amino acid similarity is low, there are two regions that show more identity (Fig. 3). One of these corresponds to the helix-turn-helix motif in the C-terminal domain, and the other corresponds to the region in the N-terminal domain (residues 79 to 127) (Shadel et al., 1990b; Slock et al., 1990) associated with autoinducer binding (Fuqua et al., 1996).

Nucleotide sequences similar to the *lux* box (Fig. 1) are present upstream of at least some promoters regulated by LasR and TraR (Gray et al., 1994) and are essential for transcriptional activation (Fuqua and Winans, 1996; Rust et al., 1996). In one study it was shown that the recognition sequences upstream of the *lasB* promoter (normally regulated by LasR) are sufficiently similar to the *lux* box that LuxR can activate this promoter to a certain degree in the presence of its cognate autoinducer (3-oxo-C6-HSL). In turn, LasR can activate transcription of the *luxI* promoter in the presence of its cognate autoinducer, 3-oxododecanoyl homoserine lactone (3-oxo-C12-HSL). Neither of the autoinducers functioned with the R-protein of the other system (Gray et al., 1994). Interestingly, the LuxR homolog EsaR, from *Erwinia stewartii*, is thought to regulate target genes by repression instead of activation (Beck von Bodman and Farrand, 1995; see chapter 7 of this volume). Based on the location of the DNA-binding sites of a number of other bacterial repressors (Collado-Vides et al., 1991), its site

of action would be predicted to be either overlapping or downstream of the −10 consensus site in promoters that it regulates.

MECHANISM OF TRANSCRIPTIONAL ACTIVATION

Efforts to genetically and biochemically define the mechanism by which LuxR activates transcription are in the early stages. The minimal upstream region necessary for activation of the *luxI* promoter has been defined to include sequences from the *lux* box through the start of transcription (Devine et al., 1989; Stevens and Greenberg, 1997). The *lux* box position at the *luxI* promoter is similar to the location of the CRP box in CRP class II promoters. Therefore, it is tempting to hypothesize that the interactions occurring between LuxR and RNAP might be similar to those observed between CRP and RNAP at class II-type promoters (Ishihama, 1993) (Fig. 4). It has been hypothesized that at class II promoters, CRP interacts with the RNAP α subunit C-terminal domain (CTD), the α subunit N-terminal domain (NTD), and the σ subunit (Busby and Ebright, 1997). However, it is not possible to predict interactions of RNAP subunits with transcriptional activators on the basis of either the position of the target DNA-binding site or the amino acid sequence of the polypeptide (Gussin et al., 1992; Danot et al., 1996). We have studied the involvement of the α CTD with LuxR in vitro and in vivo, and our results show that the α CTD is required for transcriptional activation by LuxR (Stevens and Greenberg, unpublished results), as is the case with a number of other transcriptional activators (Ishihama, 1993). However, little else is known about the transcription initiation complex at the *luxI* promoter or at quorum sensing-regulated genes in any other bacterium.

FUTURE DIRECTIONS

It has become increasingly clear over the past several years that LuxR represents an important family of cell density-responsive transcriptional regulators that, as a group, control expression of a range of target genes involved

```
        1                                                                            80
AsaR    MKQDQLLEYLEHFTSVTGDDRLAELIGRFTLGMCYDYYRFALIIPMSMQRPKVVLFN-QCPDSWVQAYTANHML
RhlR    MRNDGGFLLWWDGLRSEMQPIHDSQGVFAVLEKEVRRLCFDYYAYGVRHTIPFTRPKTEVHG-TYPKAWLERYQMQNYG
PhzR    MELGQQLGWDAYFYSIFARTMDMQEFTAVALRALRELRFDFFRYGMCSVTPFMRPRTYMYG-NYPEDWVQRYQAANYA
LuxR    MKNINADDTYRIINKIKACRAYDINQCLSDMTKMVHCEYYLTLAIIYPHSMVKSDISILDNYPKKWRQYYDDANLI
EsaR    MFSFFLENQTITDTLQTYIQRKLSPLCSPDYAYTVVSKKNPSNVLIIS---SYPDEWIRLYRANNFQ
ExpR    MSQLFYNNETISRIIKSQFDMALSHYCDIKYAMVLNKKKPTEILIIS---NHHDEWREIYQANNYQ
YenR    MIIDYFDNESINEDIKNYIQRRIKTYCDLCYSLVMNKKTPLHPTIIS---NYPLDWVKKYKKNSYH
TraR    MQHWLDKLTDLAAIEGDECILKTGLADIADHFCFTGAYLHIQHRHITAVT------NYHRQWQSTYFDKKFE
TrnR    MSVNGNLRSLIDMLEAAQDGHMIKIALRSFAHSCCYDRFALQKDGTQVRTFH------SYPGPWESIYLGSDYF
VanR    MYKILRLIQENQQITSHDDLENVLNGLNNLICHEFFLFGLSFQPTLKTSETLVTD-NYPNSWRQQYDESGFM
LasR    MALVDGFLELERSSGKLEWSAILQKMASDLCFSKILFGLLPKDSQDYENAFIVG-NYPAAWREHYDRAGYA
CerR    MDIIDLSTVATDDASFLDYIDQLCQKLCFDYASYATTSPMTGAVQGYA----NYPDSWKMHYMRRNLH
RhiR    VKEESSAVSNLVFDFLSESASAKSKDDVLLLFGKISQYFCFSYFAISGIPSPIERIDSYFVLG-NWSVGWFDRYRENNY
RaiR    MSPSHAEQFSFFLLSGPDLRIADIAGSGNDAGRSRPHLCDIAYGSPCDLAGATDSNPLLMLTYYPEWVKQYRDRDYF
SdiA    MQDKDFFSWRRTMLLRFQRMETAEEVYHEIELQAQQLEYDYYSLCVRHPVPFTRPKVAFYT-NYPEAWVSYYQAKNFL

        81                                                                           160
AsaR    ACDPIIQLARKQTLPIYWNRLDERARFLQEGSLDVMGLAAEFGL-RNGISFPLHGAA-GENGILSFITAERAS--SDLLL
RhlR    AVDPAILNGLRSSEMVVWS------DSLFDQSRMLWNEARDWGL-CVGATLPIRAPN-NLLSVLSVARDQQNI--SSFER
PhzR    VIDPTVKHSKVSSSPILAS------NELFRGCPDLWSEANDSNL-RHGLAQPSFNTQ-GRVGVLSLARKDNPI--SLQEF
LuxR    KYDPIVDYSNSNHSPINWN--IFENNAVNKKSPNVIKEAKTSGL-ITGFSFPIHTAN-NGFGMLSFAHSEKDNYIDSLFL
EsaR    LTDPVILTAFKRTSPFAWD--ENITLMSDLRFTKIFSLSKQYNI-VNGFTYVLHDHM-NNLALLSVIIKGNDQTALEQRL
ExpR    HIDPVVIAALNKITPFPWD--EDLLVSTQLKMSKIFNLSREHNI-TNGYTFVLHDHS-NNLVMLSIMIDESNVSNIDDVI
YenR    LIDPVILTAKDKVAPFAWD--DNSVINKKSTDSAVFKLAREYNI-VNGYTFVLHDNS-NNMATLNISNGSDDSISFDERI
TraR    ALDPVVKRARSRKHIFTWS-GEHERPTLSKDERAFYDHASDFGI-RSGITIPIKTAN-GFMSMFTMASDKPVIDLDREID
TrnR    NIDPVLAEAKRRRDVFFWT-ADAWPARGSSPLRRFRDEAISHGI-RCGVTIPVEGSY-GSAMMLTFASPERKV-DISGVL
VanR    HIDPIVKYSITNFLPIRWD----DAKRVNNDGRVIFEARCNGL-KAGFSIPIHGLR-GEFGMISFATSDTK---SYDLN
LasR    RVDPTVSHCTQSVLPIFWE-----PSIYQTRKQHEFFEASAAGL-VYGLTMPLHGAR-GELGALSLSVEAENRAEANRFI
CerR    RVDPTIHKSALSIAPVDWS----RFERDERFRAVFF-AAEDFGITPQGLTVPVRGPY-GDRGLLSVTRNCARPEWEKHKR
RhiR    HADPIVHLSKTCDHAFVWS-EALRDQKLDRQSRRVMDEAREFKL-IDGFSVPLHTAA-GFQSIVSFGA-----EKVELST
RaiR    SIDPVVRLGRRGFLPVEWS----ASGWDSGRAYGFFKEAMAFCVGRQGVTLPVRGPQ-GERSLFTVTSNHPDAY-WRQFR
SdiA    AIDPVLNPENFSQGHLMWN------DDLFSEAQPLWEAARAHGL-RRGVHSVFNAAQTGALGFLSFSRCSRRE--IPILS

        161                                                                          240
AsaR    ESSPILSWMSNYIFEAAIRIV-------RVSLREDDPQEALTDRETECLFWASEGKTSGEIACILGITERIVNYHLNQVT
RhlR    EEIRLRLRCMIELLTQKLTDL--------EHPMLMSNPVCLSHREREILQWTADGKSSGEIAIILSISESTVNFHHKNIQ
PhzR    EALKVVTKAFAAAVHEKISEL--------ESDVRVFNTDVEFSGRECDVLRWTADGKTSEIGVIMCVCTDIVNYHHRNIQ
LuxR    HACMNIPLIVPSLV-DNYRKI---------NIANNKSNNDLTKREKECLAWACEGKSSWDISKILGCSERIVTFHLTNAQ
EsaR    AAEQGTMQMLLIDFNEQMYRLAGTEGERAPALNQSADKTIFSSRENEVLYWASMGKTYAEIAAITGISVSIVKFHIKNVV
ExpR    ESNKDKLQMTLMTIHAETISL-YREMIRNKEDERSNDKDIFSQRENEILYWASMGKTYQEIALILDIKTGIVKFHIGNVV
YenR    EINKEKIQMLLIITHERMKLGLYQSNSDKNENRNTQIERDIFSPRENEILYWASMGKTYAEISILGIKRSIVKFHIGNVV
TraR    AVAAAAATIGQIHARISFLRTT-----------PTAEDAAWLDPKEATYLRWIAVGKTMEEIADVEGVKYNSVRVKLREAM
TrnR    DPKKAVQLLMMVHYQLKIIAA-----------KTVLNPKQMLSPRPMLCLVWASKGKTASVTANLTGINARIVQHYLDKAR
VanR    QQSIHTSQLIVPLLAHNIGNI-------TRYHKDAKPRAVLTAREVQCLAWAAEGKSAWEIATIINTSERIVKFHFSNAC
LasR    ESVLPTLWMLKDYALQSGAGL--------AFEHPVSKPVVLTSREKEVLQWCAIGKTSWEISVICNCSEANVNFHMGNIR
CerR    AVIGELQVAAVHLHDAVMRSD---------VISRALRQPRLSTREIEVQVAAAGKSQTDIGDILGISHRIVEVHLRSAR
RhiR    CDRSALYIMAAYAHSLLRAQI------GNDASRKIQALPMITTREREIIHWCAAGKTAIEIATILGRSHRIIQNVILNIQ
RaiR    MDSMRDLQFLAHHLHDRAMVL--------SGMRKVADLPRLSRRELQCLEMTANGLLAKQICARLSISVSAVQLYLASAR
SdiA    DELQLKMQLLVRESLMALMRL--------NDEIVMTPEMNFSKREKEILRWTAEGKTSAEIAMILSISENTVNFHQKNMQ

        241                      278
AsaR    RKTGSMNRYQAIAKGVSSGILLPNLEQVVVTNFPKLMQ
RhlR    KKFDAPNKTLAAYAAALGLI
PhzR    RKIGASNRVQASRYAVAMGYI
LuxR    MKLNTTNRCQSISKAILTGAIDCPYFKN
EsaR    VKLGVSNARQAIRLGVELDLIRPAASAAR
ExpR    KKLGVLNAKHAIRLGIELQLIRPVQS
YenR    RKLGVLNAKHAIRLGIELKLIKPI
TraR    KRFDVRSKAHLTALAIRRKLI
TrnR    AKLDAESVPQLVAIAKDRGLV
VanR    KKLGATNRYQAITKAILGGYINPYL
LasR    RKFGVTSRRVAAIMAVNLGLITL
CerR    EKLGTLSTVQAVGRAIGLGLVYPR
RhiR    RKLNVVNTPQMIAESFRLRIIR
RaiR    RKLTVATTSEQLLGPRRSN
SdiA    KKINAPNKTQVACYAAATGLI
```

in bacterial processes of scientific, medical, agricultural, and economic importance. Several of them, including LuxR, are involved in control of genes required for adaptation to conditions specific to animal or plant hosts (Fuqua et al., 1996). Most of our understanding of the structure-function of LuxR is derived from genetic studies. Difficulties in studying the biochemistry of LuxR have proved to be an obstacle in developing a more complete picture of this protein. A fragment of LuxR containing the C-terminal transcriptional activator domain has been purified and studied in vitro (Stevens et al., 1994; Stevens and Greenberg, 1997). This fragment, LuxRΔN, and RNAP require each other to interact with the target regulatory DNA sequences in vitro. This is unusual and suggests either that there are unique qualities to the interactions between LuxR and RNAP, or that LuxRΔN, because it consists of only one of the two LuxR domains, exhibits some unusual and artifactual features. Our understanding of the interaction of LuxR with RNAP and the regulatory DNA is in its infancy.

Understanding the interaction of LuxR with the autoinducer signal is even less refined. Because we have not yet succeeded in studying this interaction in vitro, all of our knowledge is derived from studies of the wild-type and mutant LuxR proteins in *V. fischeri* or *E. coli*. From these studies we have been able to infer that the autoinducer binds to LuxR, and the binding appears to be in a specific region of the LuxR N-terminal domain. From studies of analog binding to cells containing LuxR, we also have some information on the way in which the signal binds the protein, but our view is very crude. The devel-

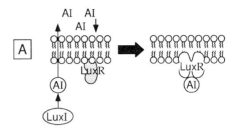

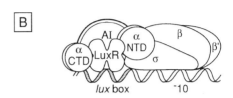

FIGURE 4 Model of the mechanism of transcriptional activation involving LuxR. (A) The quorum-sensing response. The LuxI polypeptide is responsible for autoinducer (3-oxo-C6-HSL) production. The cytoplasmic membrane is permeable to this autoinducer, which then accumulates at high cell densities. The autoinducer binds to the membrane-associated N-terminal domain of LuxR, which then multimerizes and becomes functional as an activator of transcription. (B) One possible arrangement of the transcription complex at the *luxI* promoter. The activator binding site is centered at −42.5. α CTD, α NTD, β, β′, and σ denote the RNA polymerase (RNAP) C-terminal and N-terminal domains of the two α subunits and the RNAP β, β′, and σ70 subunits, respectively.

opment of methods to study the N-terminal autoinducer-binding domain in vitro would be a major step forward. Efforts to manipulate quorum-sensing systems by the development of autoinducer antagonists that alter the functions of the transcription factors are currently under way. These investigations may lead to the discovery of means to either enhance a desired symbiotic function or eliminate an un-

FIGURE 3 Alignment of LuxR homologs. The multiple sequence alignment is of 15 different LuxR homologs (<60% identity; only one example of nearly identical homologs is shown) with the polypeptides arrayed in ascending order by the size of the primary acylhomoserine lactone sensed (the signals sensed by RaiR and SdiA have not been determined). The sequences were aligned by using the Clustal alignment algorithm included in the LaserGene sequence analysis package. Black shading indicates invariant residues; yellow shading indicates functionally similar residues conserved in all sequences; green shading indicates residues present in at least eight of the sequences; and turquoise shading indicates functionally similar residues in at least eight sequences.

wanted pathogenic one. A detailed knowledge of the precise molecular and biochemical function of LuxR and its homologs in their interactions with not only autoinducers but also DNA targets and RNAP is essential for this work to progress further.

Another area in which our knowledge is limited is the extent to which activation of transcription by LuxR or its homologs alters the cellular physiology of the bacterium. Efforts are under way to define the complete cellular response to quorum sensing in *V. fischeri*. It seems clear that genes other than those of the luminescence operon are controlled by LuxR (e.g., see Callahan and Dunlap, 1997). The total number of regulated genes and their functions remain to be determined. Identification of other LuxR target genes will provide useful tools for better defining the interactions of LuxR with regulatory DNA. There may also be unidentified factors in *V. fischeri* that either directly or indirectly alter LuxR activity. In *A. tumefaciens*, the LuxR homolog TraR is inhibited by two other polypeptides, TraM (Hwang et al., 1995; Fuqua et al., 1995) and TrlR (Winans, unpublished results; see chapter 8 of this volume). This type of antagonism has not been described in other quorum-sensing systems, but it is too early to say whether the phenomenon is unique to *A. tumefaciens* or a more common component of quorum-sensing systems.

ACKNOWLEDGMENTS

We thank our colleagues who have shared unpublished results with us. We particularly thank C. Fuqua for preparing the sequence alignment shown in Fig. 3. We are grateful for the support of our research by the following agencies: the Biology Department and College of Arts and Sciences at Virginia Polytechnic Institute and State University (A.M.S.), the Office of Naval Research, grant N00014-5-0190 (E.P.G.), and the Cystic Fibrosis Foundation (E.P.G.).

REFERENCES

Adar, Y. Y., and S. Ulitzer. 1993. GroESL proteins facilitate binding of externally added inducer by LuxR protein-containing *E. coli* cells. *J. Biolumin. Chemilumin.* **8:**261–266.

Adar, Y. Y., M. Simaan, and S. Ulitzer. 1992. Formation of the LuxR protein in the *Vibrio fischeri lux* system is controlled by HtpR through the GroESL proteins. *J. Bacteriol.* **174:**7138–7143.

Batchelor, S. E., M. Cooper, S. R. Chhabra, L. A. Glover, G. S. A. B. Stewart, P. Williams, and J. I. Prosser. 1997. Cell density-regulated recovery of starved biofilm populations of ammonia-oxidizing bacteria. *Appl. Environ. Microbiol.* **60:**2281–2286.

Beck von Bodman, S., and S. K. Farrand. 1995. Capsular polysaccharide biosynthesis and pathogenicity in *Erwinia stewartii* require induction by an *N*-acylhomoserine lactone autoinducer. *J. Bacteriol.* **177:**5000–5008.

Boucher, P. E., and S. Stibitz. 1995. Synergistic binding of RNA polymerase and BvgA phosphate to the pertussis toxin promoter of *Bordetella pertussis*. *J. Bacteriol.* **177:**6486–6491.

Busby, S., and R. H. Ebright. 1997. Transcriptional activation at Class II CAP-dependent promoters. *Mol. Microbiol.* **23:**853–859.

Callahan, S. M., and P. V. Dunlap. 1997. A 14-kD non-Lux protein whose production is positively regulated by autoinducer in *Vibrio fischeri*, abstr. H-127, p. 305. Abstr. Gen. Meet. Am. Soc. Microbiol. 1997.

Choi, S. H., and E. P. Greenberg. 1991. The C-terminal region of the *Vibrio fischeri* LuxR protein contains an inducer-independent *lux* gene activating domain. *Proc. Natl. Acad. Sci. USA* **88:**11115–11119.

Choi, S. H., and E. P. Greenberg. 1992a. Genetic dissection of DNA binding and luminescence gene activation by the *Vibrio fischeri* LuxR protein. *J. Bacteriol.* **174:**4064–4069.

Choi, S. H., and E. P. Greenberg. 1992b. Genetic evidence for multimerization of LuxR, the transcriptional activator of *Vibrio fischeri* luminescence. *Mol. Mar. Biol. Biotechnol.* **1:**408–413.

Collado-Vides, J., B. Mabasanik, and J. D. Gralla. 1991. Control site location and transcriptional regulation in *Escherichia coli*. *Microbiol. Rev.* **55:**371–394.

Danot, O., D. Vidal-Ingigliardi, and O. Raibaud. 1996. Two amino acid residues from the DNA-binding domain of MalT play a crucial role in transcriptional activation. *J. Mol. Biol.* **262:**1–11.

Davies, D. G., M. R. Parsek, J. P. Pearson, B. H. Iglewski, J. W. Costerton, and E. P. Greenberg. 1998. The involvement of cell-to-cell signals in the development of a bacterial biofilm. *Science* **280:**295–298.

Devine, J. H. 1993. Ph.D. thesis. Texas A & M University, College Station.

Devine, J. H., C. Countryman, and T. O. Baldwin. 1988. Nucleotide sequence of the *luxR* and *luxI* genes and structure of the primary regulatory region of the *lux* regulon of *Vibrio fischeri* ATCC 7744. *Biochemistry* **27**:837–842.

Devine, J. H., G. S. Shadel, and T. O. Baldwin. 1989. Identification of the operator of the *lux* regulon from the *Vibrio fischeri* strain ATCC 7744. *Proc. Natl. Acad. Sci. USA* **86**:5688–5692.

Dolan, K. M., and E. P. Greenberg. 1992. Evidence that GroEL, not σ^{32}, is involved in transcriptional regulation of the *Vibrio fischeri* luminescence genes in *Escherichia coli*. *J. Bacteriol.* **179**:5132–5135.

Dunlap, P. V., and E. P. Greenberg. 1985. Control of *Vibrio fischeri* luminescence gene expression in *Escherichia coli* by cyclic AMP and cyclic AMP receptor protein. *J. Bacteriol.* **164**:45–50.

Dunlap, P. V., and E. P. Greenberg. 1988. Control of *Vibrio fischeri lux* gene transcription by a cyclic AMP receptor protein-LuxR protein regulatory circuit. *J. Bacteriol.* **179**:4040–4060.

Egland, K. A., and E. P. Greenberg. Unpublished results.

Engebrecht, J., and M. Silverman. 1984. Identification of genes and gene products necessary for bacterial bioluminescence. *Proc. Natl. Acad. Sci. USA* **81**:4154–4158.

Engebrecht, J., and M. Silverman. 1987. Nucleotide sequence of the regulatory locus controlling expression of bacterial genes for bioluminescence. *Nucleic Acids Res.* **15**:10455–10467.

Engebrecht, J., K. H. Nealson, and M. Silverman. 1983. Bacterial bioluminescence: isolation and genetic analysis of the functions from *Vibrio fischeri*. *Cell* **32**:773–781.

Fuqua, C., M. Burbea, and S. C. Winans. 1995. Activity of the *Agrobacterium* Ti plasmid conjugal transfer regulator TraR is inhibited by the product of the *traM* gene. *J. Bacteriol.* **177**:1367–1373.

Fuqua, W. C., and S. C. Winans. 1996. Conserved *cis*-acting promoter elements are required for density-dependent transcription of *Agrobacterium tumefaciens* conjugal transfer genes. *J. Bacteriol.* **178**:435–440.

Fuqua, W. C., S. C. Winans, and E. P. Greenberg. 1994. Quorum sensing in bacteria: the LuxR-LuxI family of cell density-response transcriptional regulators. *J. Bacteriol.* **176**:269–275.

Fuqua, W. C., S. C. Winans, and E. P. Greenberg. 1996. Census and consensus in bacterial ecosystems: the LuxR-LuxI family of quorum-sensing transcriptional regulators. *Annu. Rev. Microbiol.* **50**:727–751.

Gambello, M. J., and B. H. Iglewski. 1991. Cloning and characterization of the *Pseudomonas aeruginosa lasR* gene, a transcriptional activator of elastase expression. *J. Bacteriol.* **173**:3000–3009.

Gilson, L., A. Kuo, and P. V. Dunlap. 1995. AinS and a new family of autoinducer synthesis proteins. *J. Bacteriol.* **177**:6946–6951.

Gray, K. M., L. Passador, B. H. Iglewski, and E. P. Greenberg. 1994. Interchangeability and specificity of components from the quorum-sensing regulatory systems of *Vibrio fischeri* and *Pseudomonas aeruginosa*. *J. Bacteriol.* **176**:3076–3080.

Greenberg, E. P. 1997. Quorum sensing in gram-negative bacteria. *ASM News* **63**:371–377.

Gussin, G. N., C. Olson, K. Igarashi, and A. Ishihama. 1992. Activation defects caused by mutations in *Escherichia coli rpoA* are promoter specific. *J. Bacteriol.* **174**:5156–5160.

Hanzelka, B. L., and E. P. Greenberg. 1995. Evidence that the N-terminal region of the *Vibrio fischeri* LuxR protein constitutes an autoinducer-binding domain. *J. Bacteriol.* **177**:815–817.

Henikoff, S., J. C. Wallace, and J. P. Brown. 1990. Finding protein similarities with nucleotide sequence databases. *Methods Enzymol.* **183**:111–132.

Hwang, I., D. M. Cook, and S. K. Farrand. 1995. A new regulatory element modulates homoserine lactone-mediated autoinduction of Ti plasmid conjugal transfer. *J. Bacteriol.* **177**:449–458.

Ishihama, A. 1993. Protein-protein communication within the transcription apparatus. *J. Bacteriol.* **175**:2483–2489.

Kahn, D., and G. Ditta. 1991. Modular structure of FixJ: homology of the transcriptional activator domain with the -35 binding domain of sigma factors. *Mol. Microbiol.* **5**:987–997.

Kaiser, D., and R. Losick. 1997. How and why bacteria talk to each other. *Sci. Am.* **276**:68–73.

Kaplan, H. B., and E. P. Greenberg. 1985. Diffusion of autoinducer is involved in regulation of the *Vibrio fischeri* luminescence system. *J. Bacteriol.* **163**:1210–1214.

Kaplan, H. B., and E. P. Greenberg. 1987. Overproduction and purification of the *luxR* gene product: transcriptional activator of the *Vibrio fischeri* luminescence system. *Proc. Natl. Acad. Sci USA* **84**:6639–6643.

Kolibachuk, D., and E. P. Greenberg. 1993. The *Vibrio fischeri* luminescence gene activator LuxR is a membrane-associated protein. *J. Bacteriol.* **175**:7307–7312.

Kuo, A., S. M. Callahan, and P. V. Dunlap. 1996. Modulation of luminescence operon expressions by *N*-octanoyl-L-homoserine in *ainS* mutants of *Vibrio fischeri*. *J. Bacteriol.* **178:**971–976.

McLean, R. J. C., M. Whiteley, D. J. Stickler, and W. C. Fuqua. 1997. Evidence of autoinducer activity in naturally occurring biofilms. *FEMS Microbiol Lett.* **154:**259–263.

Muller-Breitkreutz, K., and U. K. Winkler. 1993. Anaerobic expression of the *Vibrio fischeri lux* regulon in *E. coli* is FNR-dependent, p. 142–146. *In* A. Szalay, L. Kricka, and P. Stanley (ed.), *Bioluminescence and Chemiluminescence Status Report.* John Wiley & Sons, Chichester, U.K.

Nealson, K. H., T. Platt, and J. W. Hastings. 1970. Cellular control of the synthesis and activity of the bacterial luminescent system. *J. Bacteriol.* **104:**313–322.

Piper, K. R., S. B. von Bodman, and S. K. Farrand. 1993. Conjugation factor of *Agrobacterium tumefaciens* regulates Ti plasmid transfer by autoinduction. *Nature* **362:**448–450.

Pirhonen, M., D. Flego, R. Heikinheimo, and E. T. Palva. 1993. A small diffusible signal molecule is responsible for the global control of *Erwinia carotovora*. *EMBO J.* **12:**2467–2476.

Poellinger, K. A., J. P. Lee, J. V. Parales, Jr., and E. P. Greenberg. 1995. Intragenic suppression of a *luxR* mutation: characterization of an autoinducer-independent LuxR. *FEMS Microbiol Lett.* **129:**97–102.

Puskus, A., E. P. Greenberg, S. Kaplan, and A. L. Schaefer. 1997. A quorum sensing system in the free-living photosynthetic bacterium *Rhodobacter sphaeroides*. *J. Bacteriol.* **179:**7530–7537.

Ren, Y. L., S. Garges, S. Adhya, and J. S. Krakow. 1988. Cooperative DNA binding of heterologous proteins: evidence for contact between the cyclic AMP receptor protein and RNA polymerase. *Proc. Natl. Acad. Sci. USA* **85:**4138–4142.

Rust, L., E. C. Pesci, and B. H. Iglewski. 1996. Analysis of the *Pseudomonas aeruginosa* elastase (*lasB*) regulatory region. *J. Bacteriol.* **178:**1134–1140.

Schaefer, A. L., D. L. Val, B. L. Hanzelka, J. E. Cronan, Jr., and E. P. Greenberg. 1996a. Generation of cell-to-cell signals in quorum sensing: acylhomoserine lactone synthase activity of a purified *Vibrio fischeri* LuxI protein. *Proc. Natl. Acad. Sci. USA* **93:**9505–9509.

Schaefer, A. L., B. L. Hanzelka, A. Eberhard, and E. P. Greenberg. 1996b. Quorum sensing in *Vibrio fischeri*: probing autoinducer-LuxR interactions with autoinducer analogs. *J. Bacteriol.* **178:**2897–2901.

Shadel, G. S., and T. O. Baldwin. 1991. The *Vibrio fischeri* LuxR protein is capable of bidirectional stimulation of transcription and both positive and negative regulation of the *luxR* gene. *J. Bacteriol.* **173:**568–574.

Shadel, G. S., and T. O. Baldwin. 1992. Positive autoregulation of the *Vibrio fischeri luxR* gene. *J. Biol. Chem.* **267:**7696–7702.

Shadel, G. S., J. H. Devine, and T. O. Baldwin. 1990a. Control of the *lux* regulon of *Vibrio fischeri*. *J. Biolumin. Chemilumin.* **5:**99–106.

Shadel, G. S., R. Young, and T. O. Baldwin. 1990b. Use of regulated cell lysis in a lethal genetic selection in *Escherichia coli*: identification of the autoinducer-binding region of the LuxR protein from *Vibrio fischeri* ATCC 7744. *J. Bacteriol.* **172:**3980–3987.

Sitnikov, D. M., J. B. Schineller, and T. O. Baldwin. 1995. Transcriptional regulation of bioluminescence genes from *Vibrio fischeri*. *Mol. Microbiol.* **17:**801–812.

Sitnikov, D. M., G. S. Shadel, and T. O. Baldwin. 1996. Autoinducer-independent mutants of the LuxR transcriptional activator exhibit differential effects on the two *lux* promoters of *Vibrio fischeri*. *Mol. Gen. Genet.* **252:**622–625.

Slock, J., D. VanRiet, D. Kolibachuk, and E. P. Greenberg. 1990. Critical regions of the *Vibrio fischeri* LuxR protein defined by mutational analysis. *J. Bacteriol.* **172:**3974–3979.

Stevens, A. M., and E. P. Greenberg. 1997. Quorum sensing in *Vibrio fischeri*: essential elements for activation of the luminescence genes. *J. Bacteriol.* **179:**557–562.

Stevens, A. M., and E. P. Greenberg. Unpublished results.

Stevens, A. M., K. M. Dolan, and E. P. Greenberg. 1994. Synergistic binding of the *Vibrio fischeri* LuxR transcriptional activator domain and RNA polymerase to the *lux* promoter region. *Proc. Natl. Acad. Sci. USA* **91:**12619–12623.

Swartzman, E., S. Kapoor, A. F. Graham, and E. A. Meighen. 1990. A new *Vibrio fischeri lux* gene precedes a bidirectional termination site for the *lux* operon. *J. Bacteriol.* **172:**6797–6802.

Swift, S., J. P. Throup, P. Williams, G. P. C. Salmond, and G. S. A. B. Stewart. 1996. Quorum sensing: a population-density component in the determination of bacterial phenotype. *Trends Biochem. Sci.* **21:**214–219.

Ulitzer, S., and P. V. Dunlap. 1995. Regulatory circuitry controlling luminescence autoinduction in *Vibrio fischeri*. *Photochem. Photobiol.* **62:**625–632.

Winans, S. Unpublished results.

You, Z., J. Fukushima, T. Ishiwata, B. Chang, M. Kurata, S. Kawamoto, P. Williams, and K. Okuda. 1996. Purification and characterization of LasR as a DNA-binding protein. *FEMS Microbiol Lett.* **142:**301–307.

SELF-SIGNALING BY Phr
PEPTIDES MODULATES
BACILLUS SUBTILIS DEVELOPMENT

Marta Perego

16

SIGNAL TRANSDUCTION IN BACTERIA

Bacterial signal transduction through the so-called two-component system is a widespread mechanism that couples a wide variety of stimuli to a diverse array of adaptive responses. Indeed, a large number of bacterial responses involve the interaction of two regulatory proteins, a sensor or receiver, and a response regulator or transmitter. These paired regulators define two families of homologous proteins highly conserved among bacteria (for reviews, see Hoch and Silhavy, 1995). Similar communication systems have been identified in lower eukaryotes and plants as well, indicating that this could be a widespread strategy for building signaling circuits (Chang et al., 1993; Ota and Varshavsky, 1993; Posas et al., 1996; Alex et al., 1996).

Typically, signal recognition is accomplished by a sensor histidine kinase, which autophosphorylates at a conserved histidine in response to the signal. Histidine protein kinases represent a class of enzymes distinct from the serine/threonine and tyrosine kinases generally used by eukaryotic signal transduction systems (Marshall, 1995). The phosphorylated histidine kinase is mated to a specific response regulator that catalyzes the transfer of the phosphoryl group to an aspartyl of the response regulator, normally resulting in activation of its transcription or enzymatic properties. Generally, the phosphorylated response regulator is a transcription factor for genes encoding proteins that allow a response to the original signal that activated the system. There are many variations on this theme that may add one or more kinases or more response regulators to customize the signal transduction pathway for a specific purpose (Parkinson and Kofoid, 1992; Stock et al., 1989).

TWO-COMPONENT REGULATORY SYSTEMS IN *BACILLUS SUBTILIS* DEVELOPMENT

In bacteria, depletion of essential nutrients or drastic variation in environmental conditions is interpreted as a form of stress that results in cessation of exponential growth. In the time between exponential growth and stationary phase, the so-called transition state, sporulating bacteria in general, and *Bacillus subtilis* in particular, can adopt several responses, includ-

Marta Perego, Department of Molecular and Experimental Medicine, Division of Cellular Biology, The Scripps Research Institute, 10550 N. Torrey Pines Road, NX-1, La Jolla, CA 92037.

Cell-Cell Signaling in Bacteria, Edited by Gary M. Dunny and Stephen C. Winans
©1999 American Society for Microbiology, Washington, D.C.

ing induction of chemotaxis and motility, competence for genetic transformation, synthesis of macromolecule-degrading enzymes, and antibiotic production. These responses can help the cell to overcome the stress condition and eventually decide whether to resume vegetative growth or commit to sporulation. Each of these postexponential-phase responses is controlled by at least one two-component signal transduction system from the 35 identified by the sequencing of the *B. subtilis* chromosome (Kunst et al., 1997). These systems are likely to function as a network that senses physiological and environmental conditions and integrates all the information in order to determine which adaptive response is most appropriate.

SPORULATION INITIATION

Commitment to sporulation is the most extreme decision that the cell can make in order to allow long-term survival of the species under the most adverse conditions. Spore formation involves an energy-intensive pathway and the activation of numerous specific genes in a temporal sequence in order to produce a complex differentiated structure. Therefore, it is not surprising that the decision to abandon cell growth and division in order to initiate the sporulation process is subject to complex control mechanisms. The decision requires integration of a large number of synergistic and opposing regulatory activities responding to different input signals from the environment, the metabolism, and the cell cycle. The nature of these signals is still unclear, but the mechanisms of signal recognition, integration, and transduction that activate the sporulation-specific program of gene expression are becoming more and more clear. The bulk of information now available on the mechanism of sporulation initiation came from studies of sporulation mutants blocked at the very early stage of sporulation, stage 0 (Piggot and Coote, 1976). Stage 0 mutants are unable to form any of the characteristic morphological structures of sporulation and are defective in the production of enzymes and antibiotics

typically synthesized at the onset of sporulation. The *spo0* mutants are somehow locked in exponential growth and do not recognize and/or respond to signals for sporulation initiation. Such mutants are able to maintain growth until nutrients are exhausted, whereupon cell lysis occurs. Through mapping studies, mutations were placed in several loci, and through cloning and sequencing work, the composition of these loci and the nature of their gene products were determined. *spo0* mutations mapped in eight genetically distinct loci, *spo0A*, *spo0B*, *spo0E*, *spo0F*, *spo0H*, *spo0J*, *spo0K*, and *spo0L*, that ultimately were found to form the central processing unit of the signal transduction system for sporulation initiation (Hoch, 1993).

This signal transduction system for sporulation, the phosphorelay, differs from other two-component systems by the mechanism of phosphate flow and types of accessory proteins that control phosphate flow (Burbulys et al., 1991) (Fig. 1). There are two major sensor kinases, KinA and KinB, that input phosphate into the system, and two response regulators, Spo0F and Spo0A. Also peculiar to this system is the presence of the Spo0B phosphotransferase, which reversibly transfers the phosphoryl group between the two response regulators. Both KinA and KinB are involved in interpreting environmental signals and transducing this information into the formation of an autophosphorylated kinase (Hoch, 1993). However, KinA and KinB play distinct roles in the initiation of sporulation and are likely to be regulated by different input signals (Dartois et al., 1996). KinB, which is expressed during exponential phase, is responsible for the low level of sporulation observed during vegetative growth, while KinA is the major kinase for sporulation initiation and exerts its activity mainly at the onset of stationary phase. Therefore, sequential activation of kinases in response to multiple signals is one of the key concepts for understanding the significance of the phosphorelay signal transduction system and the mechanisms that trigger *B. subtilis* development (Trach and Hoch, 1993).

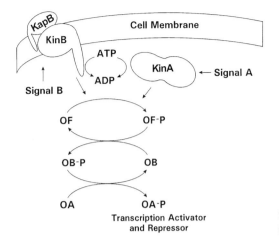

FIGURE 1 The phosphorelay signal transduction system (Burbulys et al., 1991).

In the phosphorelay, the phosphoryl group of the kinases is transferred to the Spo0F response regulator, generating Spo0F~P. The Spo0F protein consists only of the conserved amino-terminal domain of response regulators similar to CheY of the chemotaxis system (Stock et al., 1989). However, unlike CheY and despite their close resemblance in size and structure (Madhusudan et al., 1996), Spo0F appears not to have an active regulatory role by binding to a target protein. Instead, Spo0F serves as a secondary messenger for this system, being a substrate for both KinA and KinB, therefore allowing different signals to activate the same pathway. Spo0F~P is also the substrate for the Spo0B phosphotransferase. The phosphoryl group from the aspartyl phosphate of Spo0F~P is transferred to a histidine residue of Spo0B and from there again to an aspartyl residue of the second response regulator of the system, the Spo0A protein. The final goal of the phosphorelay is to produce Spo0A~P, the activated form of the transcription factor responsible for transcriptional regulation of genes involved in the initial stages of sporulation. Spo0A~P may serve as a repressor or an activator of transcription, depending on the particular promoter affected, and it acts by binding to the target promoters at a 7-bp 0A-box consensus sequence (TGA/TCGAA) (Strauch et al., 1990). One

of the major targets for Spo0A~P repressor activity is the *abrB* gene. AbrB is a transition-state regulator that prevents gene expression during exponential growth by binding to promoters of genes that are usually activated during stationary phase or the early part of sporulation (Strauch et al., 1989). AbrB somehow functions to indicate the level of Spo0A~P in the cell. As the Spo0A~P accumulates, the level of AbrB decreases, thereby allowing transcription of genes whose products are required in late exponential phase and early stationary phase. Among the promoters whose transcription is activated by Spo0A~P are those for the *spoIIA* and *spoIIG* operons, which encode σ^F and σ^E, respectively (Satola et al., 1992). These sigma factors are part of the mechanism for compartmentalized gene expression in the forespore and mother cell (Stragier and Losick, 1996).

Accumulation of Spo0A~P represents the central element determining whether a cell will divide or sporulate. Since the cellular level of Spo0A~P reflects the activity of the phosphorelay, this plays the role of a central processing unit for the integration of a myriad of signals that results in the decision to continue the vegetative growth or to initiate the differentiation process. Thus, it is not surprising that phosphate flow through the phosphorelay is subject to a variety of regulatory checkpoints.

The complexity of the system provides multiple levels of control on the production of Spo0A~P. These controls can be at the level of both transcriptional regulation of the phosphorelay components and regulation of their enzymatic activities.

REGULATION OF PHOSPHATE FLOW IN THE PHOSPHORELAY

The activity of the phosphorelay and the level of Spo0A~P in the cell are controlled by three known mechanisms: (i) transcriptional regulation of the proteins comprising the phosphorelay, (ii) regulation of phosphate input in the phosphorelay by control of kinase activity, and (iii) negative regulation of phosphate level by specific phosphatases.

Transcriptional Control

Regulation of the cellular concentration of the components of the phosphorelay occurs by complex interaction of transcriptional controls. The product of the phosphorelay, Spo0A~P, acts through a positive feedback loop to stimulate Spo0F expression and its own synthesis by a promoter switching mechanism. Such a regulatory circuit presents the advantage of a rapid response from the cell to stress conditions conducive to sporulation. In addition, Spo0A~P acts by an autoregulation mechanism: when the Spo0A~P concentration reaches a level sufficient to accomplish its function in the cell, it represses any further expression of both Spo0A and Spo0F (Strauch et al., 1992, 1993). Other transcriptional controls such as catabolite repression also play important roles in this regulation. The details of such controls have been reviewed (Hoch, 1993).

Control of Phosphate Input

The first level of control of the flux of phosphate through the phosphorelay begins with the histidine kinases, KinA and KinB. As previously mentioned, KinB is responsible for the initial activation of the phosphorelay and production of Spo0A~P sufficient for inhibition of *abrB* transcription (Trach and Hoch, 1993).

KinB is an integral membrane protein with the kinase domain in the cytoplasm of the cell. Effector molecules modulating KinB activity are still not well defined. However, its requirement for the KapB lipoprotein, which is anchored to the outer surface of the cytoplasmic membrane, suggests a model in which KapB functions as a ligand-binding protein and interacts with the transmembrane domain of KinB to transport or otherwise process a signaling compound, thereby affecting the activity of the kinase domain (Dartois et al., 1997). Furthermore, several mutations in genes that appear to be involved in KinB activation have been isolated and are indicative of a possible complex mechanism of kinase activity control (Dartois et al., 1996).

The mechanism of activation of KinA is still a mystery. The search for upstream genes activating KinA has been unsuccessful so far. This may indicate that there are no genes upstream or that, if they exist, they are lethal if mutated, or that more than one pathway exists for KinA activation.

On the other hand, an inhibitor of the autophosphorylation activity of KinA has been identified. This inhibitor, KipI, is specific for the catalytic domain of the kinase and does not affect the amino-terminal domain presumably involved in signal perception. KipI is encoded in an operon of genes of unknown function but is regulated by the availability of fixed nitrogen. Thus, the inhibitory activity of KipI represents a novel regulatory mechanism for two-component signal transduction systems and another way by which the phosphorelay is affected by a negative regulator under control of metabolic and environmental conditions (Wang et al., 1997).

Negative Regulation by Phosphatases

It is widely recognized that protein phosphorylation/dephosphorylation is an extremely important aspect of the mechanism that regulates a variety of prokaryotic and eukaryotic cellular functions, including cell proliferation, differentiation, and activation. Disturbance of these mechanisms in higher organisms is known to be a cause of uncontrolled growth of tumors (Charbonneau and Tonks, 1992). It has now become evident that, in bacteria as well as in eukaryotes, protein phosphatases play a key role in signal transduction systems. We have shown that in *B. subtilis* development the major enzymes controlling phosphate flow in the phosphorelay and the phosphorylation level of Spo0A are the phosphatases of Spo0A~P and Spo0F~P (Perego et al., 1994). These are aspartyl-phosphate phosphatases since they promote the removal of the phosphoryl group from the phosphorylated aspartic residue of response regulators.

The first regulatory molecule of this type identified as being involved in draining phosphate from the phosphorelay was the product of the *spo0E* gene. The *spo0E* gene was genetically identified as a negative regulator of sporulation. When overexpressed, it inhibited sporulation, and when deleted, the cells were phenotypically hypersporulating and unstable (Perego and Hoch, 1991). Subsequently, the Spo0E protein was found to be a specific phosphatase for the Spo0A~P transcription factor (Ohlsen et al., 1994).

A new dimension for protein phosphatases in bacterial signal transduction came from the discovery in *B. subtilis* of a family of sequence-related phosphatases (Rap), some members of which are phosphatases in the phosphorelay leading to sporulation (Perego et al., 1994). Although the activity of the phosphatases of the Rap family resembles the activity of Spo0E in dephosphorylating aspartyl-phosphate residues, there is no amino acid sequence homology between Spo0E and the Rap phosphatase proteins or with the CheZ phosphatase acting on the CheY~P response regulator of the chemotaxis system (Blat and Eisenbach, 1996).

THE Rap PHOSPHATASES AS PARTNERS IN THE PHOSPHORELAY

Identification of RapA and RapB

The first response regulator aspartyl-phosphate phosphatase (Rap) identified was RapA.

RapA is the product of a gene originally defined by the *spo0L* stage 0 sporulation mutation and by the glucose starvation-inducible gene *gsiA* (Perego et al., 1994; Mueller et al., 1992). As for the previously characterized Spo0E protein, genetic evidence suggested a negative regulatory role for RapA in the control of sporulation initiation. Inactivation of the *rapA* gene resulted in a sporulation-proficient strain that nevertheless was unstable and segregated sporulation-deficient mutants. This suggested that inactivation of *rapA* resulted in increased pressure for the cell to sporulate (confirmed by a higher efficiency of sporulation compared to the wild-type strain). On the other hand, hyperexpression of the *rapA* gene resulted in a sporulation-deficient phenotype. These results were consistent with the idea that the product of the *rapA* gene was a negative regulator of sporulation. Simultaneously with the characterization of *rapA*, a gene was identified within the *B. subtilis* chromosome sequencing project whose gene product showed 50% identical residues with RapA. Inactivation of this gene, called *rapB*, did not affect sporulation, but when the gene was present on a multicopy plasmid, cells were severely impaired in their ability to sporulate and appeared to be blocked at stage 0. Therefore, the product of *rapB* appeared to be functionally similar to RapA as a negative regulator of sporulation (Perego et al., 1994).

Purification of the RapA and RapB proteins from an *Escherichia coli* expression system allowed us to show in vitro that RapA and RapB are specific phosphatases of the Spo0F~P response regulator intermediate of the phosphorelay. They are inactive as phosphatases on the Spo0A~P response regulator or on any other component of the phosphorelay. In confirmation of these results, a mutation suppressing the sporulation phenotypes of both *rapA* and *rapB* overexpression was found to be a missense mutation (Y13S) in Spo0F. The Y13S mutation abolished the phosphatase activity of both RapA and RapB on Spo0F~P in vitro as well as in vivo (Perego et al., 1994). Because the Spo0B phosphotransferase activity is readily reversible, phosphatase activity of RapA and RapB on Spo0F~P not only prevents accumulation of Spo0A~P but also has the effect of draining phosphate already accumulated on Spo0A, thereby inhibiting the initiation of sporulation. Thus, induction of Rap phosphatase activity can control the cells' commitment to sporulation by a reversible modulation of the level of Spo0A~P. In this view, Rap phosphatases become essential components of the phosphorelay signal transduction system (Fig. 2). Like their eukaryotic counterparts tyrosine and serine/threonine phosphatases (Brautigan, 1997), the negative output of Rap phosphatases counteracts the positive input of the kinases; therefore, the activating signals are extinguished and the system is reset to its baseline activity.

How Are the Phosphatases Regulated?

Since the Rap phosphatases are downregulators of the phosphorelay by counteracting the input from the kinases, the balance between these two activities must be precisely controlled by the cell in order to induce the correct response at the appropriate time. With the phosphatases being critical partners in the overall regulation of the phosphorelay signal transduction system, their own regulation becomes of central importance.

Transcriptional regulation plays a major role in determining whether a phosphatase will be present to exert its action. The transcription factor ComA and its cognate histidine kinase ComP are required for induction of transcription of the *rapA* gene at the end of exponential growth (Mueller et al., 1992; Dubnau et al., 1994). Therefore, ComA prevents sporulation by inducing the RapA phosphatase that deactivates the phosphorelay.

In contrast, *rapB* gene transcription occurs mainly during early exponential growth, and by the end of exponential growth its transcription is off. Expression of *rapB* is induced by high levels of carbon sources or growth media that do not sustain the sporulation process, it is independent of ComA/ComP, and it is not under postexponential control. Therefore, *rapB* expression appears to be regulated by physiological conditions that favor vegetative growth and prevent sporulation (Perego, unpublished results).

Thus, RapA and RapB are differentially induced by environmental or physiological conditions that are alternative to sporulation. This indicates that Rap phosphatases represent a mechanism evolved to recognize additional signals to be integrated by a complex developmental signal transduction system such as the phosphorelay.

A second control on Rap phosphatases was found at the level of protein activity. Inactivation of a small gene coding for a 44-aminoacid protein translationally coupled with *rapA* caused a sporulation-deficient phenotype. Simultaneous inactivation of *rapA* restored sporulation proficiency. This result was interpreted to indicate that the small gene played a role in regulation of RapA activity. This gene was

FIGURE 2 Rap phosphatases as partners in the phosphorelay. Positive inputs from the kinases are counteracted by the negative signals interpreted by the phosphatases, and all are integrated by the phosphorelay to modulate the output of the system, i.e., the phosphorylated form of the Spo0A response regulator. (From Hoch and Silhavy, 1995, with permission.)

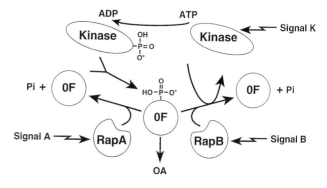

named *phrA* (for phosphate regulator) (Perego and Hoch, 1996). PhrA resembles exported proteins by the presence of a hydrophobic N-terminal signal peptide sequence followed by a hydrophilic C-terminal region with a potential signal peptidase cleavage site separating the two domains. This arrangement suggests that the C-terminal hydrophilic domain of 19 residues is exported from the cell. In fact, the sporulation-defective phenotype of a mutant lacking the *phrA* gene can be complemented by exogenously provided synthetic peptides corresponding to the last five or more of the C-terminal residues of the exported domain (Perego and Hoch, 1996). A peptide consisting of the last four residues is not active. The oligopeptide transport system (Opp) (Perego et al., 1991; Rudner et al., 1991) is required for the complementation activity. This suggests that the original 19-amino-acid exported peptide might be processed to something smaller in order to accommodate the transport capabilities of the Opp system (Tame et al., 1994). Furthermore, the sporulation deficiency of *opp* mutants was shown to be suppressed by deletion of the *rapA* and *rapB* genes or by the Spo0F (Y13S) protein insensitive to the Rap phosphatase activity. These results indicated that the sporulation deficiency of *opp* mutants was due to their inability to re-import the peptides required for controlling *rap* gene expression or Rap phosphatase activity (Perego and Hoch, 1996).

The *rapB* gene is also followed by a small gene that could code for a small protein. However, this putative *phrB* gene does not cause any sporulation phenotype when inactivated. Moreover, studies reveal that *phrB* does not appear to be transcribed and therefore does not seem to participate in the regulation of RapB activity (Perego, unpublished results).

The available data suggested that RapA induction produced an exported peptide whose subsequent import was required for regulation of the expression or activity of RapA. This model raised several questions: At what level (expression or activity) does the peptide act?

How specific is the peptide? Does it regulate both RapA and RapB? Why is the peptide exported?

INTERACTION OF Rap PHOSPHATASES WITH Phr PEPTIDES

How Does the Peptide Inhibit?

The mechanism of modulation of the RapA phosphatase activity by the PhrA peptide has been uncovered by means of in vitro experiments using synthetic peptides (Perego, 1997). A peptide comprising the last five amino acids at the C-terminal end of the PhrA protein (ARNQT) directly inhibited RapA-catalyzed dephosphorylation of the Spo0F~P response regulator. The PhrA hexapeptide was also tested and some activity was observed, but inhibition of RapA activity was fivefold less efficient than that observed with the pentapeptide (Fig. 3A). The PhrA pentapeptide is specifically active on RapA, since RapB enzymatic activity on Spo0F~P is totally insensitive to this peptide. RapA is not inhibited by another peptide, the PhrC pentapeptide (ERGMT), naturally produced by *B. subtilis* cells. However, the PhrC pentapeptide was found to slightly inhibit the phosphatase activity of RapB (Fig. 3B). Thus, the PhrA pentapeptide is the active molecule specifically and directly inhibiting the RapA phosphatase activity.

Sequence–Dependent Peptide Specificity

The observations that the PhrA pentapeptide is active on RapA but inactive on RapB, and that the PhrC pentapeptide weakly inhibits RapB but not RapA, raised the question of what mechanism rules the specificity of target recognition. Since pentapeptides are unlikely to have a stable quaternary structure, a structure-based mechanism seemed to be the less probable. Therefore, it seemed more likely that target specificity is amino acid sequence-dependent. PhrA (ARNQT) and PhrC (ERGMT) have in common an arginine residue at position 2 and a threonine at position

FIGURE 3 Inhibition of Rap phosphatases by Phr peptides. (A) SpoOF~P (lane 1) is dephosphorylated by RapA (lane 2), and the reaction is inhibited by addition of increasing concentrations of the PhrA pentapeptide (lanes 3 to 9). The PhrA hexapeptide is fivefold less active than the pentapeptide (lanes 10 to 13). (B) Sp0F~P (lane 1) is dephosphorylated by RapB (lane 2), and the reaction is moderately inhibited by increasing concentrations of the PhrC pentapeptide (lanes 3 to 9). The PhrC hexapeptide is not active (lanes 10 to 13) (for details, see Perego, 1997).

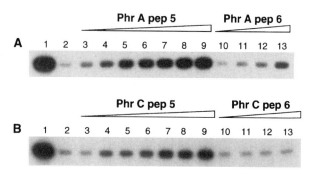

5, and these same residues at the same positions are present in some of the other Phr proteins (see below). Hence, the amino acid residues at the remaining three positions were candidates for being the determinants of peptide specificity. To test this possibility, six synthetic peptides were designed to contain single amino acid substitutions at the three variable positions by replacing the residue on one peptide with the corresponding one on the other peptide. In vitro experiments using the six modified peptides showed that a single amino acid substitution results in inactive peptides in all cases but one. In the latter case, the peptide (ERNMT) was found to be more active in inhibiting RapB than the original poorly active PhrC (ERGMT), but it was still inactive toward RapA. In another case, a modified

PhrA peptide (ARNMT) partially lost activity for RapA but acquired some activity toward RapB (Perego, 1997) (Fig. 4).

Therefore, Phr specificity for target recognition depends upon the amino acid sequence of the peptides, and single amino acid changes of the variable residues remarkably affect peptide activity and specificity toward the target phosphatase.

PEPTIDE PROCESSING AND THE EXPORT-IMPORT CONTROL CIRCUIT

The mechanism regulating the RapA phosphatase activity by the product of the *phrA* gene is depicted in Fig. 5. The *rapA* operon is induced by the ComA regulator at the end of exponential phase in order to prevent spor-

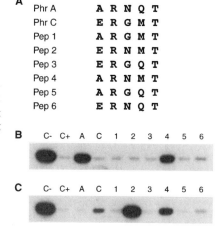

FIGURE 4 Sequence-dependent specificity for Phr peptide target recognition. (A) Amino acid sequences of the PhrA and PhrC pentapeptides and their derivatives obtained by singularly replacing residues at positions 1, 3, and 4 of PhrA with the corresponding one of PhrC and vice versa. Inhibition of RapA (B) or RapB (C) phosphatase activity on SpoOF~P by the mutant peptides is shown. The control level of SpoOF~P is shown in the lanes labeled C−, whereas the dephosphorylation of SpoOF~P by RapA or RapB is shown in the lanes labeled C+. PhrA (A), PhrC (C), and peptides 1 to 6 were added at equal concentrations (for details, see Perego, 1997).

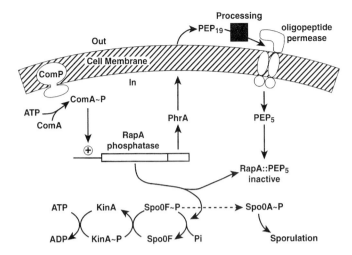

FIGURE 5 The export-import control circuit for PhrA pentapeptide activation modulates *B. subtilis* differentiation through regulation of the RapA phosphatase activity. RapA is produced after induction by ComA~P and prevents sporulation during competence development by dephosphorylating Spo0F~P. PhrA is first exported, processed, and then reimported by the oligopeptide transport system as a pentapeptide that directly inhibits RapA. This allows the accumulation of Spo0A~P and the initiation of the differentiation process.

ulation while competence develops. Dephosphorylation of Spo0F~P by RapA impedes accumulation of Spo0A~P, thereby inhibiting sporulation initiation. Induction of the *rapA* operon results in production of the PhrA peptide in its preinhibitor form of 44 amino acids. To reach its final form of peptide inhibitor, a sequence of proteolytic events is required. The preinhibitor peptide is exported to the outside of the cell membrane by the secretion machinery, and the signal peptide is removed, presumably by type I signal peptidases. This results in a putative proinhibitor peptide consisting of the C-terminal 19 amino acids of the preinhibitor. The proinhibitor must be subject to a further proteolytic event in order to produce the 5-amino-acid active inhibitor. This cleavage is likely to occur on the outside surface of the membrane by a yet unknown enzyme. The uptake of the active peptide inhibitor depends on the oligopeptide permease, Opp, the structure of which is designed for peptides up to five amino acids long and is very unlikely to bind and transport a peptide the size of the proinhibitor (Tame et al., 1994).

Internalization of the pentapeptide inhibitor results in direct inhibition of the phosphatase activity and allows accumulation of Spo0A~P and the initiation of sporulation. Therefore, regulation of the *B. subtilis* differ-

entiation process occurs via modulation of the RapA protein phosphatase of the phosphorelay signal transduction system by a peptide export-import circuit (Perego, 1997). We have postulated this circuit to be a control mechanism: the time it takes to export the proinhibitor, process it to an inhibitor, and reimport it through the Opp system determines how long the phosphatase can act in preventing sporulation while other ComA-dependent functions develop.

RATIONALE FOR THE EXPORT-IMPORT CONTROL CIRCUIT

The reason behind the export processing and re-import of the peptide before it becomes active as an inhibitor is a very intriguing question. We previously proposed a quorum-sensing role for the inhibitor peptide, although at the moment not much of the evidence supports this interpretation (Perego and Hoch, 1996; Perego et al., 1996). The term quorum sensor is used to indicate signaling molecules that regulate specific gene expression in a cell density-dependent fashion (Swift et al., 1996). The molecules are constitutively produced and diffuse in the medium, and the cells are freely permeable to such quorum sensors as homoserine-lactone derivatives. Thus, high cell density is required to achieve a critical concentration of the molecule necessary to trigger the cell response. On the other hand, pheromones are chemicals produced by an organism that can alter behavior or gene expression of other organisms of the same species. This definition might be applicable to the cCF10 molecule inducing plasmid conjugation in *Enterococcus faecalis* and to the ComX and CSP peptides inducing competence in *B. subtilis* and *Streptococcus pneumoniae*. Their constitutive transcription results in accumulation in the medium, thereby justifying their role as cell density signals (Dunny et al., 1995; Magnuson et al., 1994; Pestova et al., 1996).

For the PhrA peptide, however, the scenario might be different. PhrA is produced and exported upon induction only in a limited time during cell growth and when the Opp system is maximally present in the cell. This makes it less likely that PhrA might be free to diffuse in the medium and be re-imported by a cell other than the one that produced it. This is supported by the observation that the free peptide in the medium is detected only in *opp* mutants, where it accumulates owing to the lack of a re-import mechanism (Perego and Hoch, 1996). Furthermore, the protease susceptibility of peptides makes them poor candidates for being freely diffusible molecules with a quorum sensor or pheromone role in organisms such as *Bacillus* spp. that produce large quantities of extracellular proteases (Pero and Sloma, 1993). It is still possible, however, that the proinhibitor plays a role in cell-cell communication. Export of such a peptide to the outside surface of the cell membrane (not necessarily followed by secretion into the medium) may result in communication between interacting cells and over short distances. This possibility is suggested by the observation that a *phrA* mutant can be complemented by culture of the wild-type strain only when the two types of cells are mixed in the same medium, indicating that cell-cell contact may be required for transfer of information (Perego and Hoch, 1996). Thus, if Phr peptides have a cell-cell communication function, they may do so in the microenvironment of the colony.

We now believe that the PhrA peptide more likely acts as a timing device whose processing from the preinhibitor form to the final active pentapeptide inhibitor represents a control circuit in which the processing enzymes are regulatory checkpoints that determine the time and rate of formation of the final product (Perego, 1997). The processing steps may, in turn, be controlled by a regulatory hierarchy so that intra- or extracytoplasmic events may regulate the length of time that the Rap phosphatase is active. Therefore, the peptide may be seen as a regulator acting as a communication pathway from the inside to the outside of the cell and vice versa.

We also believe that the PhrA peptide is exported only to the outside of the cellular membrane, not necessarily secreted into the medium, and that its activity is mainly a self-regulatory mechanism for timing intracellular developmental events. The regulatory hierarchy controlling the peptide processing may involve temporal control of the enzymatic activities responsible for production of the peptide active form. For example, removal of the signal peptide is presumably carried out by type I signal peptidases. In *B. subtilis*, type I signal peptidases are redundantly encoded by five genes (*sipS, T, U, V,* and *W*) (Tjalsma et al., 1997) with similar substrate specificities but different preferences for the protein precursors. Genetic studies have revealed that the capacity for protein secretion in *B. subtilis* may be modulated through temporally controlled expression of the signal peptidase genes and that efficient secretion is dependent on the particular signal peptidase being expressed. Interestingly, the *sipS* and *sipT* genes are regulated by the DegS-DegU signal transduction system that controls degradative enzyme production (Bolhuis et al., 1996; Tjalsma et al., 1997; Msadek et al., 1995). These observations suggest that the processing of the PhrA preinhibitor to proinhibitor may be dependent on the presence of signal peptidases whose control can ultimately affect the time interval in which the phosphatase is active. It is noteworthy to remember at this point that the

DegS-DegU regulatory system is associated with later events in the competence pathway while the *rapA* and *phrA* genes are induced by the ComA regulator of competence in the earliest step of the process. Therefore, it is possible that later events in the pathway, regulated by DegS and DegU, may be required to activate the proteolytic events needed to generate the phosphatase inhibitor. In this manner, timing regulation of inhibitor production provides an interval during which the phosphatase is free to inhibit sporulation and allow competence and/or other transition-state events to develop.

It is in this view that the peptide's journey through the export-import circuit represents a timing device for the coordination of post-exponential events such as competence and sporulation. Transcription of the *rapA* operon has the immediate effect of blocking sporulation, and this block is not released until the PhrA peptide is re-imported. Therefore, any delay in importation of the pentapeptide will prolong the block of sporulation initiation by increasing the time interval in which the phosphatase is active. At this time, there is no evidence (or need) for a pheromone or quorum sensor kind of activity associated with the peptide. Providing exogenous peptide and monitoring the resulting changes in gene expression is not evidence for cell-cell communication but may only result in misinterpretation of alterations caused by timing disturbances. Furthermore, the PhrA peptide is doubtless not required for sporulation as long as the RapA phosphatase is inactive. In fact, the double mutant *rapA phrA* sporulates perfectly well and even better and more precociously than the wild-type strain, while competence development is reduced in length and efficiency. On the other hand, in a mutant lacking only the *phrA* gene, while sporulation is inhibited, competence is prolonged in time and is more efficient (Perego, unpublished results). A further argument against the PhrA peptide acting as pheromone or quorum sensor is provided by the following observation: direct expression of the pentapeptide inside the cell, in a condition that maintains temporal control but bypasses the export-import circuit, also results in a strain that sporulates precociously and very efficiently as in the case of the *rapA-phrA* double mutant (Perego, unpublished results). This is consistent with the idea that PhrA and the export-import circuit are mainly a self-regulatory timing device required for coordination of intracellular developmental events.

The reader may also refer to chapter 3 in this volume, which summarizes recent work on the PhrC peptide and presents an interpretation of the role of Phr peptides in *B. subtilis* cell signaling.

THE FAMILIES OF Rap PHOSPHATASES AND Phr PEPTIDES

With *rapA* and *rapB*, a total of 11 genes (*rapA* to *rapK*) coding for proteins of the Rap family have been identified on the *B. subtilis* chromosome (Kunst et al., 1997). Three additional genes have been identified on plasmids isolated from *Bacillus* strains of industrial interest. The presence of these plasmids in an otherwise wild-type strain results in impaired sporulation (Meijer et al., 1995a,b,c). Thus, it is reasonable to believe that these genes have been selected and maintained on plasmids to prevent sporulation in order to allow synthesis of advantageous postexponential products such as proteases or antibiotics for extended periods of time. Homologs of the *B. subtilis* Rap phosphatases have been identified only in *Clostridium acetobutilicum*, in which the partial chromosome sequence has revealed the presence of three genes coding for proteins with sequence similarities ($\approx$20% identical residues) to Rap phosphatases (Perego, unpublished results).

The products of the *B. subtilis rap* chromosomal genes are very similar. RapB, RapC, RapE, RapF, RapH, RapI, and RapJ share between 40 and 50% identical residues with RapA, whereas RapD, RapG, and RapK are more distantly related, with 23, 26, and 27% identical residues with RapA, respectively (Fig. 6). Not all of these phosphatases have a

phrA-like gene translationally coupled downstream of the phosphatase gene. As previously mentioned, a sequence coding for a putative PhrB protein is detectable downstream of *rapB*, but this gene is not transcribed (Perego, unpublished results). RapD and RapJ do not have an associated PhrA-like peptide, whereas the *phrC*, *phrE*, *phrF*, and *phrG* genes are not only translationally coupled to their corresponding *rap* gene but are also independently transcribed from a sigma H-dependent promoter. The *rapH*, *rapI*, and *rapK* genes may also be followed by a Phr peptide coding gene.

In contrast to the high level of homology displayed by the Rap phosphatases, the sequence of the proteins coded by *phr* genes revealed little homology (Fig. 7). Phr proteins, however, are more similar in the distribution of residues: the amino-terminal domain is very hydrophobic while the C-terminal end is hydrophilic. The two domains are generally separated by a putative signal peptidase cleavage site. It is intriguing that in some Phr pentapeptides there is a reoccurrence of an arginine residue at position 2 from the amino-terminal end and a threonine residue at position 5, suggesting a putative XRXXT consensus motif for Phr peptides.

Not all of these phosphatases have activity on the SpOF~P response regulator. Although little characterization of *rapH*, *rapI*, *rapJ*, and *rapK* has been carried out at this time, we have in vivo and in vitro evidence that RapE targets Spo0F~P while RapC, RapD, RapF, and RapG do not (Perego, unpublished results). RapC is postulated to act on the ComA~P response regulator, based on genetic evidence. In fact, deletion of *rapC* or *phrC* affects transcription of some ComA-dependent genes in an opposite manner (Solomon et al., 1996; Perego, unpublished

results). However, our in vitro experiments failed to show any phosphatase activity of purified RapC on purified ComA~P. This suggests that the effect of RapC and PhrC on ComA-dependent gene expression may be indirect, implying the existence of a still unknown intermediate regulator of competence.

CONCLUSIONS

The discovery of *B. subtilis* Rap phosphatases adds a new dimension to the regulation of two-component signal transduction systems. These systems may have been adapted to control complex processes, such as the initiation of development in bacteria, which are subject to a variety of intra- and extracellular regulatory controls. The recruitment of phosphatases regulated by unique mechanisms allows the recognition of additional signals, other than the ones controlling the kinases, to influence the signal transduction pathways. In the phosphorelay signal transduction system for sporulation initiation, it is now clear that the phosphorylation level of the response regulator Spo0A, which is the output of the system, is the result of the balance of two opposing activities: the phosphorylation reaction catalyzed by the kinases and the dephosphorylation reaction carried out by the phosphatases. In this view, Rap phosphatases become integral components of two-component systems and must be considered in the overall context of signal transduction regulation.

The discovery that the Phr exported peptides are involved in regulation of Rap phosphatase activity widens the complexity of these regulatory systems. In recent years, it has become more evident that extracellular factors represent a widespread mechanism among bacteria to regulate a variety of physiological processes. Extracellular factors may be viewed

FIGURE 6 Amino acid sequence alignment of the 11 Rap phosphatases coded by *B. subtilis* chromosomal genes (Kunst et al., 1997). 100% identical residues are indicated by the dark shading, whereas the light shading represents at least 60% identity.

```
PhrA:  MKSK---WMSGLLLVAVGFSFTQVM---VHAGETANTEGKTFHIA--ARNQT  :44
                               *

PhrC:  MKLKSK-LFVICLAAAAIFTAAGVS---ANAEALDFHVT-------ERGMT  :40
                              *

PhrE:  MKSK---LFISLSAVLIGLAFFGSMY--NGEMKEASRNVTLAPTHEFLV  :44
                               *

PhrF:  MKLKSK-LLLSCLALSTVFVATTIA---NAPTHQIEVAQRGMI  :39
                              *

PhrG:  MKRF---LIGAGVAAVILSGWFIAD---HQTHSQEMKVAEKMIG   :38
                              *

PhrI:  MKISR--ILLAAVILSSVFSITYL----QSDHNTEIKVAADRGVGA  :40
                             *

PhrK:  MKK----LVLCVSILAVILSGVALTQL-STDSPSNIQVAERPVGGD  :41
                             *
```

FIGURE 7 Amino acid sequence of seven Phr proteins identified on the *B. subtilis* chromosome and associated with corresponding Rap phosphatases. The residues in the hydrophobic domain are shaded, whereas pentapeptides that were shown to be physiologically active are in bold. Asterisks indicate the putative cleavage site by type I signal peptidases as determined by the SignalP program (Nielsen et al., 1997). The available nucleotide sequence of the *rapH* operon does not allow a defined identification of a *phrH* gene; therefore, the sequence of PhrH has been omitted.

as inter- or intracellular communication signals that play key roles in modulating bacterial physiology. In this regard, the family of Phr exported peptides is of particular interest. The discovery of the export-import control circuit modulating the activation of the PhrA pentapeptide inhibitor, in fact, indicates that Phr peptides might be a class of peptides mainly involved in self-regulatory mechanisms for timing intracellular events and not necessarily implicated in intercellular communication with quorum sensor or pheromone-like functions.

ACKNOWLEDGMENTS

Work in this laboratory was supported in part by grants GM19416 and GM55594 from the National Institute of General Medical Sciences, National Institutes of Health, USPHS. I thank Tom Bray for computer assistance.

REFERENCES

Alex, L. A., K. A. Borkovich, and M. I. Simon. 1996. Hyphal development in *Neurospora crassa*: involvement of a two-component histidine kinase. *Proc. Natl. Acad. Sci. USA* **93**:3416–3421.

Blat, Y., and M. Eisenbach. 1996. Oligomerization of the phosphatase CheZ upon interaction with the phosphorylated form of CheY. *J. Biol. Chem.* **271**:1226–1231.

Bolhuis, A., A. Sorokin, V. Azevedo, S. D. Ehrlich, P. G. Braun, A. de Jong, G. Venema, S. Bron, and J. M. van Dijl. 1996. *Bacillus subtilis* can modulate its capacity and specificity for protein secretion through temporally controlled expression of the *sipS* gene for signal peptidase I. *Mol. Microbiol.* **22**:605–618.

Brautigan, D. L. 1997. Signaling by kinase cascade, p. 113–124. *In* J. Corbin and S. Francis (ed.), *Phosphatases as Partners in Signaling Networks*. Lippincott-Raven, Philadelphia.

Burbulys, D., K. A. Trach, and J. A. Hoch. 1991. The initiation of sporulation in *Bacillus subtilis* is controlled by a multicomponent phosphorelay. *Cell* **64**:545–552.

Chang, C., S. F. Kwok, A. B. Bleecker, and E. M. Meyerowitz. 1993. Arabidopsis ethylene-response gene *ETR1*: similarity of product to two-component regulators. *Science* **262**:539–544.

Charbonneau, H., and N. K. Tonks. 1992. 1002 protein phosphatases? *Annu. Rev. Cell Biol.* **8**:463–493.

Dartois, V., T. Djavakhishvili, and J. A. Hoch. 1996. Identification of a membrane protein involved in activation of the KinB pathway to sporulation in *Bacillus subtilis*. *J. Bacteriol.* **178**:1178–1186.

Dartois, V., T. Djavakhishvili, and J. A. Hoch. 1997. KapB is a lipoprotein required for KinB signal transduction and activation of the phosphorelay to sporulation in *Bacillus subtilis*. *Mol. Microbiol.* **26:** 1097–1108.

Dubnau, D., J. Hahn, M. Roggiani, F. Piazza, and Y. Weinrauch. 1994. Two-component regulators and genetic competence in *Bacillus subtilis*. *Res. Microbiol.* **145:**403–411.

Dunny, G. M., B. A. B. Leonard, and P. J. Hedberg. 1995. Pheromone-inducible conjugation in *Enterococcus faecalis*: interbacterial and host-parasite chemical communication. *J. Bacteriol.* **177:**871–876.

Hoch, J. A. 1993. Regulation of the phosphorelay and the initiation of sporulation in *Bacillus subtilis*. *Annu. Rev. Microbiol.* **47:**441–465.

Hoch, J. A., and T. J. Silhavy (ed.). 1995. *Two-Component Signal Transduction*. American Society for Microbiology, Washington, D.C.

Kunst, F., et al. 1997. The complete genome sequence of the Gram positive model organism *Bacillus subtilis* (strain 168). *Nature* **390:**249.

Madhusudan, J. W. Zapf, J. M. Whiteley, J. A. Hoch, N. H. Xuong, and K. I. Varughese. 1996. Crystal structure of a phosphatase–resistant mutant of sporulation response regulator Spo0F from *Bacillus subtilis*. *Structure* **4:**679–690.

Magnuson, R., J. Solomon, and A. D. Grossman. 1994. Biochemical and genetic characterization of a competence pheromone from *B. subtilis*. *Cell* **77:**207–216.

Marshall, C. J. 1995. Specificity of receptor tyrosine kinase signaling: transient versus sustained extracellular signal-regulated kinase activation. *Cell* **80:** 179–185.

Meijer, W. J. J., A. J. de Boer, S. van Tongeren, G. Venema, and S. Bron. 1995a. Characterization of the replication region of the *Bacillus subtilis* plasmid pLS20: a novel type of replicon. *Nucleic Acids Res.* **23:**3214–3223.

Meijer, W. J. J., A. de Jong, G. Bea, A. Wisman, H. Tjalsma, and G. Venema. 1995b. The endogenous *Bacillus subtilis* (*natto*) plasmids pTA1015 and pTA1040 contain signal peptidase-encoding genes: identification of a new structural module on cryptic plasmids. *Mol. Microbiol.* **17:**621–631.

Meijer, W. J. J., G. Venema, and S. Bron. 1995c. Characterization of single strand origins of cryptic rolling-circle plasmids from *Bacillus subtilis*. *Nucleic Acids Res.* **23:**612–619.

Msadek, T., F. Kunst, and G. Rapoport. 1995. A signal transduction network in *Bacillus subtilis* includes the DegS/DegU and ComP/ComA two-component systems, p. 447–471. *In* J. A. Hoch and T. J. Silhavy (ed.), *Two-Component Signal*

Transduction. American Society for Microbiology, Washington, D.C.

Mueller, J. P., G. Bukusoglu, and A. L. Sonenshein. 1992. Transcriptional regulation of *Bacillus subtilis* glucose starvation-inducible genes: control of *gsiA* by the ComP-ComA signal transduction system. *J. Bacteriol.* **174:**4361–4373.

Nielsen, H., J. Engebrecht, S. Brunak, and G. von Heijne. 1997. Identification of prokaryotic and eukaryotic signal peptides and prediction of their cleavage sites. *Protein Eng.* **10:**1–6.

Ohlsen, K. L., J. K. Grimsley, and J. A. Hoch. 1994. Deactivation of the sporulation transcription factor Spo0A by the Spo0E protein phosphatase. *Proc. Natl. Acad. Sci. USA* **91:**1756–1760.

Ota, I. M., and A. Varshavsky. 1993. A yeast protein similar to bacterial two-component regulators. *Science* **262:**566–569.

Parkinson, J. S., and E. C. Kofoid. 1992. Communication modules in bacterial signaling proteins. *Annu. Rev. Genet.* **26:**71–112.

Perego, M. 1997. A peptide export-import control circuit modulating bacterial development regulates protein phosphatases of the phosphorelay. *Proc. Natl. Acad. Sci. USA* **94:**8612–8617.

Perego, M. Unpublished results.

Perego, M., and J. A. Hoch. 1991. Negative regulation of *Bacillus subtilis* sporulation by the *spo0E* gene product. *J. Bacteriol.* **173:**2514–2520.

Perego, M., and J. A. Hoch. 1996. Cell-cell communication regulates the effects of protein aspartate phosphatases on the phosphorelay controlling development in *Bacillus subtilis*. *Proc. Natl. Acad. Sci. USA* **93:**1549–1553.

Perego, M., C. F. Higgins, S. R. Pearce, M. P. Gallagher, and J. A. Hoch. 1991. The oligopeptide transport system of *Bacillus subtilis* plays a role in the initiation of sporulation. *Mol. Microbiol.* **5:**173–185.

Perego, M., C. G. Hanstein, K. M. Welsh, T. Djavakhishvili, P. Glaser, and J. A. Hoch. 1994. Multiple protein aspartate phosphatases provide a mechanism for the integration of diverse signals in the control of development in *Bacillus subtilis*. *Cell* **79:**1047–1055.

Perego, M., P. Glaser, and J. A. Hoch. 1996. Aspartyl-phosphate phosphatases deactivate the response regulator components of the sporulation signal transduction system in *Bacillus subtilis*. *Mol. Microbiol.* **19:**1151–1157.

Pero, J., and A. Sloma. 1993. Proteases, p. 939–952. *In* A. L. Sonenshein, J. A. Hoch, and R. Losick (ed.), *Bacillus subtilis and Other Gram-Positive Bacteria*. American Society for Microbiology, Washington, D.C.

Pestova, E. V., L. S. Håvarstein, and D. A. Morrison. 1996. Regulation of competence for genetic transformation in *Streptococcus pneumoniae* by an auto-induced peptide pheromone and a two-component regulatory system. *Mol. Microbiol.* **21:** 853–862.

Piggot, P. J., and J. G. Coote. 1976. Genetic aspects of bacterial endospore formation. *Bacteriol. Rev.* **40:**908–962.

Posas, F., S. M. Wurgler-Murphy, T. Maeda, E. A. Witten, T. C. Thai, and J. Saito. 1996. Yeast HOG1 MAP kinase cascade is regulated by a multistep phosphorelay mechanism in the SLN1-YPD1-SSK1 "two-component" osmosensor. *Cell* **86:**865–875.

Rudner, D. Z., J. R. Ladeaux, K. Breton, and A. D. Grossman. 1991. The *spo0K* locus of *Bacillus subtilis* is homologous to the oligopeptide permease locus and is required for sporulation and competence. *J. Bacteriol.* **173:**1388–1398.

Satola, S. W., J. M. Baldus, and C. P. Moran. 1992. Binding of Spo0A stimulates *spoIIG* promoter activity in *Bacillus subtilis*. *J. Bacteriol.* **174:** 1448–1453.

Solomon, J. M., B. A. Lazazzera, and A. D. Grossman. 1996. Purification and characterization of an extracellular peptide factor that affects two different developmental pathways in *Bacillus subtilis*. *Genes Dev.* **10:**2014–2024.

Stock, A. M., J. M. Mottonen, J. B. Stock, and C. E. Schutt. 1989. Three-dimensional structure of CheY, the response regulator of bacterial chemotaxis. *Nature* **337:**745–749.

Stock, J. B., A. J. Ninfa, and A. M. Stock. 1989. Protein phosphorylation and regulation of adaptive response in bacteria. *Microbiol. Rev.* **53:**450–490.

Stragier, P., and R. Losick. 1996. Molecular genetics of sporulation in *Bacillus subtilis*. *Annu. Rev. Genet.* **30:**297–341.

Strauch, M. A., G. B. Spiegelman, M. Perego, W. C. Johnson, D. Burbulys, and J. A. Hoch. 1989. The transition state transcription regulator *abrB* of *Bacillus subtilis* is a DNA binding protein. *EMBO J.* **8:**1615–1621.

Strauch, M. A., V. Webb, G. Spiegelman, and J. A. Hoch. 1990. The Spo0A protein of *Bacillus subtilis* is a repressor of the *abrB* gene. *Proc. Natl. Acad. Sci. USA* **87:**1801–1805.

Strauch, M. A., K. A. Trach, J. Day, and J. A. Hoch. 1992. Spo0A activates and represses its own synthesis by binding at its dual promoters. *Biochimie* **74:**619–626.

Strauch, M. A., J.-J. Wu, R. H. Jonas, and J. A. Hoch. 1993. A positive feedback loop controls transcription of the *spo0F* gene, a component of the sporulation phosphorelay in *Bacillus subtilis*. *Mol. Microbiol.* **7:**967–974.

Swift, S., J. P. Throup, P. Williams, G. P. C. Salmond, and G. S. A. B. Stewart. 1996. Quorum sensing: a population-density component in the determination of bacterial phenotype. *Trends Biochem. Sci.* **21:**214–219.

Tame, J. R. H., G. N. Murshudov, E. J. Dodson, T. K. Neil, G. G. Dodson, C. F. Higgins, and A. J. Wilkinson. 1994. The structural basis of sequence-independent peptide binding by OppA protein. *Science* **264:**1578–1581.

Tjalsma, H., M. A. Noback, S. Bron, G. Venema, K. Yamane, and J. M. van Dijl. 1997. *Bacillus subtilis* contains four closely related type I signal peptidases with overlapping substrate specificities. *J. Biol. Chem.* **272:**25983–25992.

Trach, K. A., and J. A. Hoch. 1993. Multisensory activation of the phosphorelay initiating sporulation in *Bacillus subtilis*: identification and sequence of the protein kinase of the alternate pathway. *Mol. Microbiol.* **8:**69–79.

Wang, L., R. Grau, M. Perego, and J. A. Hoch. 1997. A novel histidine kinase inhibitor regulating development in *Bacillus subtilis*. *Genes Dev.* **11:** 2569–2579.

A MULTICHANNEL TWO-COMPONENT SIGNALING RELAY CONTROLS QUORUM SENSING IN *VIBRIO HARVEYI*

Bonnie L. Bassler

17

BIOLUMINESCENCE AND QUORUM SENSING

Bioluminescent organisms (organisms capable of emitting visible light) are widespread in nature and include bacteria, fungi, fish, shrimp, squid, and insects (Meighen, 1991). These organisms inhabit terrestrial, marine, and freshwater environments, and they all have in common the requirement for oxygen to produce light. However, the individual light-producing enzymes (luciferases) and substrates (luciferins) differ dramatically between bioluminescent species, indicating that bioluminescence may have evolved several times (Hastings and Morin, 1991). Among the bioluminescent marine bacteria, both free-living and symbiotic species exist. The function of bioluminescence for the free-living bacteria is not known. Symbiotic bioluminescent bacteria are assumed to acquire a nutrient-rich habitat in the specialized light organs of host animals in exchange for the light that they produce. The eukaryotic hosts in these partnerships use the light produced by their bioluminescent symbionts for an array of processes, including attracting prey, avoiding or frightening predators, and attracting mates (Hastings and Morin, 1991).

The study of bacterial quorum sensing originated with analyses of the density-dependent regulation of luminescence (*lux*) in the marine symbiotic bacterium *Vibrio fischeri* and its free-living relative *Vibrio harveyi* (Nealson et al., 1970; Nealson and Hastings, 1979; chapter 18 of this volume). The major breakthrough in the understanding of quorum sensing occurred when Engebrecht et al. (1983) and Engebrecht and Silverman (1984) showed that in *V. fischeri*, the luciferase operon, *luxICDABE*, is regulated by a positive feedback loop controlled by the two regulatory proteins LuxI and LuxR. LuxI is responsible for production of an acyl homoserine lactone autoinducer, and LuxR is responsible for autoinducer detection and transcriptional activation of the *luxICDABE* operon. This simple regulatory circuit has become the foundation for all subsequent quorum-sensing studies, and several chapters of this volume are devoted to the continuing investigation of the quorum-sensing mechanism used by *V. fischeri* (see chapters 14, 15, and 18).

We now understand that a wide variety of gram-negative bacteria use acyl homoserine lactone autoinducers to control different density-dependent functions (see chapter 19 of

Bonnie L. Bassler, Department of Molecular Biology, Princeton University, Princeton, NJ 08544-1014.

Cell-Cell Signaling in Bacteria, Edited by Gary M. Dunny and Stephen C. Winans
©1999 American Society for Microbiology, Washington, D.C.

this volume). In all known cases except that of *V. harveyi*, acyl homoserine lactone-mediated quorum sensing is controlled by proteins homologous to the LuxI and LuxR proteins first identified in *V. fischeri*. This chapter will describe how the free-living bacterium *V. harveyi* accomplishes the same task as *V. fischeri*, that is, the density-dependent expression of bioluminescence mediated by an acyl homoserine lactone autoinducer, using completely different regulatory components.

ENVIRONMENTAL CONTROL OF LUMINESCENCE IN *V. HARVEYI*

The ecological niches of *V. harveyi* and *V. fischeri* differ dramatically. *V. harveyi* is found free-living in the ocean at depths less than 200 m, in shallow sediments, and on the surfaces and in the intestines of various marine animals (Nealson and Hastings, 1979). In each of these locations, *V. harveyi* resides in microbial communities in association with other species of bacteria. In contrast, although found free-living in the ocean, *V. fischeri* primarily inhabits the specialized light organs of host animals, where it lives in a monospecific association. Because *V. harveyi* exists in and interacts with a changing environment, and because it lives in the presence of other species of bacteria, the regulatory circuitry controlling the expression of luminescence in this bacterium was hypothesized to be more complicated, and to have the capacity to detect more extracellular inputs than that of the symbiont *V. fischeri* (Bassler et al., 1993).

It is known that a number of environmental factors influence the expression of luminescence in *V. harveyi* and other luminous bacteria. Specifically, *V. harveyi* increases light production at high cell density, under iron, oxygen, or carbohydrate limitation, and in the presence of toxins or DNA-damaging agents (Nealson and Hastings, 1979; Makemson, 1986). Possibly, in the ocean, several of these conditions are met simultaneously; as the *V. harveyi* cells grow and reach a critical cell density, they deplete the available iron, oxygen, and carbohydrate. Furthermore, under these conditions the bacteria themselves could produce toxic by-products, or alternatively, other, competitor bacteria could also release toxic by-products or antimicrobial compounds. The response of *V. harveyi* to this situation is an exponential increase in light production. Why? What is the benefit *V. harveyi* derives from producing light under these circumstances? The answer remains unknown, but at present we are beginning to understand how the cues controlling cell density-dependent regulation of luminescence are detected and processed and how iron limitation affects *lux* expression in *V. harveyi*. A fuller understanding of these regulatory processes may enable us to determine the ecological advantage of producing light for *V. harveyi*.

Like *V. fischeri*, *V. harveyi* was shown to produce an acyl homoserine lactone autoinducer that is secreted into the extracellular culture fluid to control the density-dependent production of light (Nealson et al., 1970). Cao and Meighen (1989) isolated and characterized the *V. harveyi* autoinducer and identified it as N-(3-hydroxybutanoyl)-L-homoserine lactone (referred to as 3-hydroxy-C4-HSL). The *V. harveyi* autoinducer was chemically synthesized, and Cao and Meighen demonstrated that light production in *V. harveyi* could increase as much as 1,000-fold in response to the exogenous addition of autoinducer. Attempts to clone the *V. harveyi* density-dependent *lux* circuit into *Escherichia coli* were unsuccessful. Although the first *luxA* and *luxB* genes ever cloned from luminous bacteria were those of *V. harveyi* (Belas et al., 1982), no chromosomal fragment of *V. harveyi* DNA has been identified that enables recombinant *E. coli* to express density-dependent luminescence. Fortunately, unlike *V. fischeri*, which until recently remained almost completely intractable to genetic manipulation, many techniques for the genetic analysis of undomesticated bacteria have been successfully adapted for use in *V. harveyi*. These techniques facilitated an in vivo approach to the identification of the components responsible for light production and its control in *V. harveyi*.

THE *V. HARVEYI luxCDABEGH* STRUCTURAL OPERON AND THE LuxR TRANSCRIPTIONAL ACTIVATOR

To locate the genes required for light production and its regulation in *V. harveyi*, Martin et al. (1989) used random insertion mutagenesis of the *V. harveyi* chromosome with transposon mini-Mu*lac* to identify loci in which mutations caused a dark or Lux⁻ phenotype. The resulting insertions that produced Lux⁻ strains of *V. harveyi* were mapped to two unlinked regions of the chromosome. The first region was shown by phenotypic analysis of mutants and hybridization to *V. fischeri lux* DNA probes (Martin et al., 1989), and by sequence analysis (Miyamoto et al., 1988a,b) to contain the *luxCDABE* operon similar to that of *V. fischeri*; i.e., the genes were arranged in the same order and encoded functions identical to those previously described for the symbiont *V. fischeri*. The second locus in which insertions resulted in a dark phenotype was about 0.6 kb in size and encoded a single regulatory function. No *luxCDABE* mRNA was transcribed in *V. harveyi* strains containing transposon insertions in this second locus, indicating that it encoded a positive transcription factor necessary for the expression of the *luxCDABE* operon. The gene was named *luxR* because initially it appeared to carry out the same function as *luxR* from *V. fischeri*. However, after cloning and sequencing of *luxR* from *V. harveyi*, Showalter et al. (1990) demonstrated that it had sequence similarity to the helix-turn-helix class of DNA binding proteins, but it shared no DNA sequence homology to *luxR* from *V. fischeri*. Additionally, LuxR from *V. harveyi* did not appear to be involved in detection of the autoinducer. Cotransformation of *E. coli* with *V. harveyi luxR* and *V. harveyi luxCDABE* increased the level of light produced by the recombinant 10,000-fold over the level of light produced by the recombinant carrying only the *luxCDABE* operon; however, this effect was not density dependent. Furthermore, the recombinant carrying both the *luxR* and the *luxCDABE* loci did not

respond to the addition of cell-free culture fluids from *V. harveyi*, indicating that the *V. harveyi* LuxR protein was a transcriptional activator of the *luxCDABE* operon but that it was not the autoinducer sensor (Showalter et al., 1990). Lastly, the *V. harveyi* LuxR protein did not regulate autoinducer biosynthesis, because *V. harveyi luxR* null strains were not defective in autoinducer production. In an effort to locate the *V. harveyi* genes that encoded the LuxI-like and the LuxR-like functions of *V. fischeri*, Miyamoto et al. (1988b) sequenced the DNA region upstream of the *V. harveyi luxCDABE* operon, and Swartzman et al. (1990) mapped the region downstream of *luxCDABE*. Neither analysis revealed a linked regulatory gene, but two new genes located downstream of *luxE* were identified. These genes were named *luxG* and *luxH* and were shown to be involved in flavin biosynthesis. With this finding, the *lux* structural operon was shown to consist of the genes *luxCDABEGH*. Mutations in either *luxG* or *luxH* did not confer a Lux phenotype.

Although the regulatory components that controlled density sensing in *V. harveyi* remained unknown, the function and regulation of the transcriptional activator LuxR came under active investigation. In the initial study identifying LuxR, Martin et al. (1989) used *lacZ* fusions and Northern blot analyses to show that transcription of *luxCDABEGH* was controlled by cell density, as expected, but that the transcription of *luxR* was not. Their results showed that the β-galactosidase activity in *V. harveyi* strains with *luxR-lacZ* transcriptional fusions was not dependent on cell density and did not increase upon the addition of cell-free culture fluids containing the *V. harveyi* autoinducer. However, since the *luxR*::mini-Mu*lac* insertions prevented the expression of wild-type LuxR protein, the authors could not rule out the possibility that LuxR was necessary for its own induction by autoinducer. Miyamoto et al. (1996) used a dark *V. harveyi* mutant that was defective in autoinducer production but was still capable of responding to exogenously supplied autoinducer to test whether the addition of autoinducer in the presence of LuxR altered the transcription of *luxR*. Primer extension assays were used to show that the amount of *luxR* transcript increased twofold when autoinducer was added. Deletion of the -35 promoter and upstream region eliminated this increase, indicating that the autoinducer acted directly at the *luxR* promoter. Swartzman et al. (1992) and Swartzman and Meighen (1993) analyzed the DNA binding of the purified LuxR protein and showed conclusively that it exhibited its positive transcriptional control of the *lux* operon by binding at two sites in the *luxCDABEGH* promoter. Additionally, Chatterjee et al. (1996) demonstrated that LuxR not only bound upstream of *luxCDABEGH* to activate transcription, but it also bound at its own promoter, where it acted as a repressor to prevent transcription. Two similar affinity binding sites were identified for LuxR at the *luxR* promoter, and binding of the protein caused a three- to fourfold reduction in *luxR* transcription. The results of Chatterjee et al. (1996) indicated that this repressive effect was mediated by the LuxR protein preventing the binding of RNA polymerase at the *luxR* promoter.

QUORUM-SENSING REGULATORY COMPONENTS IN *V. HARVEYI*: A MULTICHANNEL DENSITY-SENSING CONTROL CIRCUIT

Following the identification and cloning of the *V. harveyi* transcriptional activator gene *luxR*, it was clear that other regulatory components of the *lux* sensing circuit remained to be identified. At a minimum, *V. harveyi* had to possess an autoinducer synthesis gene and a gene encoding a sensor for detection of the autoinducer. However, mutagenesis procedures followed by screens for Lux$^-$ strains were unsuccessful in identifying additional genes involved in density-dependent Lux regulation. These results indicated that the *V. harveyi lux* regulatory genes were not obligatory for the production of light. Bassler et al. (1993, 1994a) showed that multiple independ-

ent density-sensing systems exist in *V. harveyi*, and that two different autoinducers and two different sensors were present. Null mutations in either autoinducer-detection system left a second system operational, resulting in strains with Lux$^+$ phenotypes.

In aged cultures of *V. harveyi* cells, spontaneous "variants" arise that are dim and produce only very low levels of light (Keynan and Hastings, 1961). Bassler et al. (1993) used such *V. harveyi* variants to identify and clone *lux* regulatory genes. Since the variants remain capable of producing some light, it was presumed that they had functional *luxCDABE* and *luxR* loci, because each of those genes is necessary for light production. However, since the dim variants did not increase light production at high cell densities, it was hypothesized that these strains contained defects in regulatory genes that rendered them unable to either make or respond to cell density cues. A collection of *V. harveyi* spontaneous dim variants was assembled and used for cloning *lux* regulatory genes. A wild-type library of *V. harveyi* genomic DNA carried on a broad-host-range cosmid was conjugated into this collection of dim variants, and the resulting exogenote colonies were screened for restoration of a wild-type Lux$^+$ phenotype. A family of cosmids harboring an identical subset of genomic DNA fragments was shown to restore Lux$^+$ function in one complementation class of dim variants. Furthermore, following recovery of the cosmids and transformation into recombinant *E. coli*, cell-free culture fluids from the recombinant *E. coli* were shown to contain a substance that stimulated wild-type *V. harveyi* cells to produce light at low cell density, suggesting that a *V. harveyi* autoinducer biosynthetic gene(s) resided on the cloned DNA (Bassler et al., 1993).

One representative cosmid that restored Lux$^+$ function in the dim variants and that appeared to direct production of autoinducer in *E. coli* was chosen for further analysis. The cloned *V. harveyi* DNA on the cosmid was mutagenized by transposon Tn*5* insertion, and the region of the chromosomal DNA responsible for complementing the dim variant was mapped. Each of the Tn5 insertions that inactivated the *lux* function was introduced into the corresponding location on the *V. harveyi* chromosome for in vivo analysis of the Lux phenotype. All of the *V. harveyi* Tn5 insertion mutants expressed wild-type levels of luminescence, indicating that, as predicted, the putative regulatory function(s) was not required for light production. A comprehensive analysis of the mutant phenotypes showed that two classes of mutants had been obtained, corresponding to Tn5 insertions positioned in two adjacent regions of the *V. harveyi* chromosome. *V. harveyi* strains carrying the first class of Tn5 insertions (called LuxM mutants) were partially defective in autoinducer production but were wild type in their ability to respond to exogenously supplied autoinducer. *V. harveyi* stains carrying the second class of Tn5 insertions (called LuxN mutants) produced wild-type levels of autoinducer, but these mutants were partially defective in their response to autoinducer.

The phenotypic analysis of the mutants suggested that at least two distinct autoinducers were produced by *V. harveyi*, and each was detected by a specific cognate sensor (Bassler et al., 1993). The autoinducers were termed HAI-1 and HAI-2 (for *V. harveyi* autoinducer). HAI-1 was shown to be the known autoinducer 3-hydroxy-C4-HSL, while HAI-2 is of unknown structure. The cognate sensors for HAI-1 and HAI-2 were termed sensor 1 and sensor 2. As will be shown below, the LuxM class of mutants do not produce HAI-1 but do produce HAI-2. LuxM-type mutants respond to both autoinducers, so the LuxM phenotype is HAI-1$^-$, HAI-2$^+$, Sensor 1$^+$, Sensor 2$^+$. The LuxN class of mutants produce both autoinducers but only respond to HAI-2, so that phenotype is HAI-1$^+$, HAI-2$^+$, Sensor 1$^-$, Sensor 2$^+$.

A quantitative analysis of the LuxM and LuxN mutant Lux phenotypes, along with that of the wild-type parent, is shown in Fig. 1. The leftmost panel depicts the density-dependent Lux response phenotype for wild-type *V. harveyi*. The no-addition control curve shows that luminescence per *V. harveyi* cell decreased rapidly following dilution of dense cells into fresh medium. During the time course of the experiment, the cells grew and produced autoinducer, which accumulated in the environment. When a critical concentration of autoinducer had been reached, the wild-type *V. harveyi* cells responded to its presence by increasing light production roughly 1,000-fold. Addition of cell-free supernatants from dense wild-type cells (containing HAI-1 and HAI-2) resulted in a high level of light production throughout the experiment. Addition of synthetic *V. harveyi* autoinducer, 3-hydroxy-C4-HSL (i.e., HAI-1), resulted in only a partial stimulation of light production in the wild-type recipient strain. The same was true for the addition of supernatant prepared from dense cultures of the LuxM mutants (i.e., HAI-2), demonstrating that the LuxM mutant was defective for the production of an autoinducer. In the middle panel of Fig. 1, the luminescence response phenotype of a LuxN mutant is shown. Similar to the wild-type strain, a LuxN mutant underwent density-dependent luminescence and responded to autoinducer present in cell-free culture fluids obtained from wild-type cells. However, unlike the wild-type strain, a LuxN mutant did not respond to the synthetic *V. harveyi* autoinducer 3-hydroxy-C4-HSL (HAI-1). The LuxN mutant did respond to the remaining autoinducer produced by the LuxM mutants (HAI-2). The sensor 1 locus was sequenced, and the gene responsible for the Sensor 1$^-$ phenotype was named *luxN* and the sensor 1 protein was called LuxN. DNA sequence analysis revealed that two genes were responsible for the LuxM phenotype and the synthesis of HAI-1. These genes were named *luxL* and *luxM*. Transposon insertion in either *luxL* or *luxM* resulted in no production of HAI-1. The analysis of the *V. harveyi* mutants predicted the existence of a second independent density-sensing system composed of distinct but analogous functions, i.e., a sensor and an extracellular autoinducer pair (sensor 2,

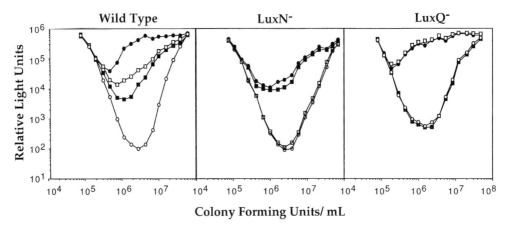

FIGURE 1 Response phenotypes of *V. harveyi lux* regulatory mutants. The luminescence responses of *V. harveyi* strains to the addition of autoinducers is shown. The left panel shows the wild-type strain (Tn*5* insertion outside the *lux* locus), the middle panel shows a *luxN*::Tn*5* mutant, and the right panel shows a *luxQ*::Tn*5* mutant. Relative light units or the light emission per cell is plotted as a function of the culture density. Dense, bright overnight cultures of the three strains were diluted 1:5,000 into fresh medium, and light output was measured as the cells grew. At the first time point, cell-free culture fluids (10%), synthetic HAI-1 (3-hydroxy-C4-HSL) (1 μg/ml), or nothing was added. ○, control (no addition); ●, wild-type cell-free culture fluid (HAI-1 + HAI-2); □, synthetic autoinducer (HAI-1); ■, LuxM cell-free culture fluid (HAI-2). Relative light units are defined as cpm × 10³/CFU/ml). The data for this figure are from Bassler et al. (1994a).

HAI-2). A model consistent with these results is presented in Fig. 2.

To identify and clone components of this putative second signaling system, double system 1–system 2 mutants were needed (Bassler et al. 1994a). Such mutants would be defective in both density-sensing systems, so their phenotypes were predicted to be very dim or dark. A *luxL*::Tn*5* mutant (Sensor 1⁺, Sensor 2⁺, HAI-1⁻, HAI-2⁺) was subjected to nitrosoguanidine (NTG) mutagenesis, and the dim and dark mutants obtained were analyzed. It was reasoned that the NTG-induced mutation, conferring the dim or dark phenotype, could reside in either *luxCDABE* or *luxR* or alternatively in the hypothetical density-sensing system 2, i.e., in the genes encoding either HAI-2 production or sensor 2 function. Mutations in system 2 were differentiated from those in *luxCDABE* or *luxR* by taking advantage of the functional system 1 sensor LuxN. An autoinducer cross-feeding assay was used to test whether the dim and dark mutants

could produce light in response to exogenously supplied HAI-1. Autoinducer did not stimulate mutants containing NTG-induced defects in *luxCDABE* or *luxR*; however, some dim and dark mutants did respond to the addition of HAI-1, and these mutants were predicted to have suffered mutations in some function involved in density-sensing system 2 (Bassler et al., 1994a). Further analysis showed that all of the system 2 mutants obtained in this experiment appeared to have defects in sensor 2 function. No mutant was obtained with a defect in the production of HAI-2.

The library of cosmids containing the wild-type *V. harveyi* genome was conjugated into a putative system 2 Sensor 2⁻ mutant in order to clone the gene(s) encoding the sensor 2 function. The family of cosmids that restored wild-type Lux function in the mutant were analyzed by restriction analysis and Tn*5* mutagenesis. One locus composed of two genes, which were called *luxP* and *luxQ*, encoded sensor 2. The Tn*5* insertions that disrupted

Signalling System 1   **Signalling System 2**

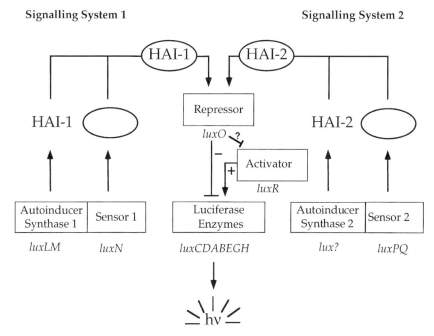

FIGURE 2   Model for regulation of quorum sensing in *V. harveyi*. Two independent density-sensing systems control the expression of the luminescence operon *luxCDABEGH* in *V. harveyi*. Signaling system 1 is composed of an autoinducer HAI-1 and its cognate receptor sensor 1. HAI-1 is 3-hydroxy-C4-HSL, and the *luxL* and *luxM* genes encode the autoinducer synthase enzymes responsible for its production. Sensor 1 is called LuxN and is encoded by the *luxN* gene. Signaling system 2 is also composed of a cognate sensor-autoinducer pair. The genes for synthesis of the second autoinducer HAI-2 have not been identified, and the structure of this autoinducer remains unknown. Sensor 2 is encoded by two genes, *luxP* and *luxQ*, and the proteins LuxP and LuxQ are proposed to act together to detect the presence of HAI-2. Sensory information from sensor 1 and sensor 2 is relayed to an integrator protein LuxO encoded by the *luxO* gene. LuxO acts negatively to control luminescence, and interaction with the sensors LuxN and LuxQ is proposed to inactivate its repressor activity. A positive transcription factor encoded by *luxR* is also necessary for expression of luminescence.

luxP and *luxQ* function were introduced onto the chromosome of *V. harveyi*, and the phenotypes of the resulting *V. harveyi* Sensor 2⁻ mutants were examined to determine their response to HAI-1 and HAI-2. A representative LuxQ mutant is presented in the right-hand panel of Fig. 1 to show the Sensor 2⁻ phenotype. As expected, and similar to a LuxN mutant, the LuxQ mutant was bright and underwent density sensing. Again, similar to a LuxN mutant, supernatants from dense wild-type *V. harveyi* cells containing HAI-1 and HAI-2 stimulated the LuxQ mutant to produce light at low cell density. In contrast

to a LuxN mutant, which did not respond to HAI-1, addition of HAI-1 to the LuxQ mutant stimulated it to produce light. And again, in contrast to a LuxN mutant, cell-free culture fluids from a LuxM mutant, which contained only HAI-2, had no effect on the LuxQ mutant.

These results show that the LuxN (Sensor 1⁻) mutant and the LuxQ (Sensor 2⁻) mutant have complementary phenotypes. A LuxN mutant responds to HAI-2 but not to HAI-1, and a LuxQ mutant responds to HAI-1 but not to HAI-2. The wild-type response appears to be the summation of the responses of the two mutants, that is, wild-type *V. harveyi* is partially stimulated by either HAI-1 or HAI-2 and fully stimulated by addition of both autoinducers. The model in Fig. 2 shows the interaction of each cognate sensor-autoinducer pair and the genes encoding the known *lux* regulatory functions. As shown in the figure, the gene(s) for biosynthesis of HAI-2 remain to be identified and the structure of this autoinducer is unknown. There are apparently no other density-sensing systems in *V. harveyi*, because a double *luxN*, *luxQ* null mutant produces light constitutively (i.e., maximal light production occurs even at very low cell densities), indicating that all density sensing has been abolished (Freeman and Bassler, in press).

TWO-COMPONENT SIGNAL RELAY CONTROLS QUORUM SENSING IN *V. HARVEYI*

The DNA encoding the *luxL*, *luxM*, *luxN*, *luxP*, and *luxQ* genes was sequenced and used for database analysis to identify similarities with other known regulatory genes and proteins. At that time, no DNA or protein sequence similarity was found for the HAI-1 production genes *luxL* and *luxM*. This result indicated that the autoinducer biosynthetic functions encoded by these two genes was novel, and that LuxL and LuxM might carry out different enzymatic steps than that performed by the LuxI homologs (Bassler et al., 1993). Sequence analysis of the two autoinducer sensor genes *luxN* and *luxQ* showed that they encoded proteins that had a high degree of similarity to one another and also to members of the family of bacterial two-component signal transducers (Bassler et al., 1993, 1994a). This family of proteins exists in a wide variety of both gram-negative and gram-positive bacteria, with 28 known pairs in *E. coli*.

Two-component signal transducers detect environmental stimuli and relay this information internally to elicit changes in gene expression in response to fluctuating conditions. In general, two-component systems are composed of a membrane-bound sensor kinase and a cognate response regulator pair. Interaction of the periplasmic region of a sensor kinase with a ligand causes autophosphorylation of a conserved histidine on the protein. This phosphoryl group is subsequently transferred to a conserved aspartate residue on the second component, the response regulator. This second phosphorylation event results in a change in activity of the response regulator, which is relayed downstream to induce a change in expression of a gene or genetic regulon. Usually, response regulators are DNA binding proteins, and they directly affect gene expression by activation or repression of the target locus. Many variations of this general two-component mechanism exist; for example, often more than two components are involved, more than one stimulus may be detected, or dephosphorylation of the sensor kinase and the response regulator may be the event that triggers the change in gene expression. For a review of two-component signaling, see Parkinson and Kofoid (1992).

In the *lux* signaling systems, the sensors LuxN and LuxQ are called hybrid-kinase proteins because each of these proteins contains a membrane-spanning region, a conserved histidine sensor kinase domain, and a response regulator domain. Other examples of hybrid-kinase proteins containing both the sensor kinase and response regulator domains in a single protein are also known (Leroux et al., 1987; McCleary and Zusman, 1990; Iuchi et al., 1990). The amino terminus of the system 1 sensor, LuxN, contains eight putative transmembrane spanning segments forming four periplasmic loops. One or more of these loops could interact with HAI-1. The periplasmic domain is followed by the histidine sensor kinase domain, containing the conserved His at residue 441 as well as each of the other known conserved regions of sequence homology in sensor kinases (Bassler et al., 1993; Bassler and Silverman, 1995). These regions are called the H, N, G1, F, and G2 homology blocks (Stock et al., 1989; Parkinson and Kofoid, 1992). The C-terminal third of the LuxN protein encodes the response regulator domain, which includes the Asp 12, 13, Asp 57, and Lys 109 regions of homology characteristic of response regulator proteins (numbering refers to amino acid positions in the response regulator CheY).

The system 2 sensor, LuxQ, is predicted to contain two N-terminal transmembrane regions forming one periplasmic loop. Similar to LuxN, the membrane-spanning domain of LuxQ is followed by the histidine sensor kinase domain, and the His at residue 492 in the H homology block is the proposed site of autophosphorylation. LuxQ also contains the conserved N, G1, F, and G2 homology blocks. The response regulator domain of LuxQ is at the carboxyl terminus and contains the Asp 12, 13, Asp 57, and Lys 109 homology regions (Bassler et al., 1994a; Bassler and Silverman, 1995). LuxP is also involved in sensor 2 function, and sequence analysis of *luxP* shows that it encodes a protein with high similarity to the ribose binding protein of *E. coli* and *Salmonella typhimurium* (Groarke et al. 1983). Lux signaling is proposed to occur by the interaction of the autoinducers HAI-1 and HAI-2 in the periplasm with their cognate sensors LuxN and LuxPQ, respectively. These interactions are hypothesized to initiate phosphorelay reactions, which transduce a signal to the *luxCDABEGH* operon resulting in light emission.

In system 2 signaling, the LuxP protein is proposed to act as the primary receptor for HAI-2, then together the LuxP-HAI-2 complex interacts with the periplasmic domain of LuxQ to initiate the phosphorylation cascade. In the *Agrobacterium tumefaciens* two-component system that responds to the signal galactose, a similar mechanism has been proposed. A protein called ChvE, which has homology to the ribose binding protein of *E. coli* and *S. typhimurium,* acts as the receptor for the signal galactose. The galactose-ChvE complex stimulates VirA, a two-component hybrid-kinase protein, to relay a signal to a response regulator called VirG (Heath et al., 1995).

The use of multiple autoinducers to control quorum sensing has now been shown in several other systems (see chapter 10 of this volume). Furthermore, two-component control of quorum sensing might not be restricted to *V. harveyi*. Recently, Gilson et al. (1995) reported the cloning of a locus called *ainS* from *V. fischeri*. The locus controls the production of a second *V. fischeri* autoinducer, C8-HSL, and the gene *ainS* shows homology in the 3′ region to the autoinducer synthase gene *luxM* of *V. harveyi*. A two-component sensor kinase protein (called *ainR*) was identified downstream of *ainS* in *V. fischeri*. Gilson et al. (1995) proposed that, given the similarity in the genetic organization of *luxM* and *luxN* in *V. harveyi* and *ainS* and *ainR* in *V. fischeri*, the AinR protein could be the sensor that responds to the autoinducer produced by *ainS*. These results suggest that, in addition to the LuxI family of autoinducer synthases, there could exist a second family of proteins involved in autoinducer biosynthesis homologous to LuxM and AinS. And, autoinducers produced by this class of synthases could be recognized by two-component sensors. Finally, there is apparently no gene similar to *luxL* of *V. harveyi* at the *ainSR* locus of *V. fischeri*, so the role of LuxL in autoinducer production remains unclear.

TWO SIGNALS, ONE OUTPUT: SIGNAL INTEGRATION BY THE LuxO PROTEIN

The similarity of the LuxN and LuxQ sensors to other two-component histidine sensor kinases indicated that signal recognition probably occurred in the periplasm and that the mechanism of signal relay was a phosphorylation/dephosphorylation cascade. However, the sequence analysis did not predict a DNA binding motif in either LuxN or LuxQ, so it was not obvious how the initial signaling events were transduced to effect a change in expression of the *luxCDABEGH* operon. Possibly another component of the regulatory circuit existed that fulfilled the function of coupling autoinducer detection directly to changes in the output, light emission.

To clone and characterize this putative *lux* regulatory component, a strategy similar to that described for *luxLMN* was used (Bassler et al., 1994b). A library of recombinant cosmids containing *V. harveyi* chromosomal DNA was introduced into one complementation group of spontaneous dim variants for identification of recombinant cosmids that restored wild-type *lux* function. Analysis of the DNA that encoded the complementing locus showed that one gene, called *luxO*, was responsible for the phenotype. To determine what the phenotype would be of a mutant that could detect the autoinducer signal but could not transduce the signal to the *lux* operon, Tn5 insertions were generated throughout the cloned *luxO* gene, and these insertions were introduced into the *luxO* locus in the *V. harveyi* chromosome. The *luxO* null mutants produced light constitutively (Bassler et al., 1994b). As shown in Fig. 3, the phenotype of a LuxO mutant was maximally bright, because addition of either HAI-1 or HAI-2 or both HAI-1 and HAI-2 simultaneously did not increase light production above the no-addition control. This result was interpreted to mean that LuxO acted as a repressor to negatively regulate *lux* expression, and also that the

LuxO protein integrated and relayed both the system 1 and system 2 signals. It was argued that if LuxO relayed signal only from system 1 or only from system 2, a *luxO* null mutant would have had a partial *lux* phenotype and would have remained responsive to one of the two autoinducers. LuxO is shown in the model in Fig. 2 downstream of both LuxN and LuxQ, where it integrates and transduces their respective cell density signals.

The *luxO* locus was sequenced and shown to encode a two-component protein of the response regulator class (Bassler et al., 1994b; Bassler and Silverman, 1995). The Asp 12, 13, Asp 57, and Lys 109 residues are at the corresponding positions of Glu 3, Asp 4, Asp 47, and Lys 97 in LuxO. The Asp at position 47 is predicted to be the site of phosphorylation in LuxO. Although LuxO has similarity to the C termini of LuxN and LuxQ, unlike these proteins it also contains a predicted helix-turn-helix DNA binding motif near its C terminus, which could directly interact with *lux* DNA to influence gene expression. Sequence analysis of the *luxO* gene in the original dim *V. harveyi* mutant revealed that this locus encodes a LuxO protein that is "locked" in a form that mimics the phosphorylated state of

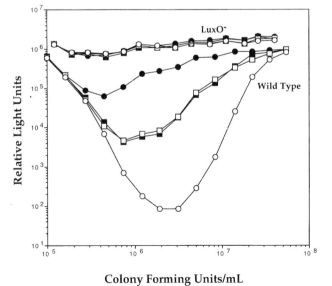

FIGURE 3 The response phenotype of a *luxO* null mutant. The luminescence phenotypes of wild-type *V. harveyi* and a *V. harveyi* strain containing a *luxO*::Tn5 insertion mutation are shown. Cells were grown and diluted, and luminescence was measured as described in the legend for Fig. 1. ○, control (no addition); ●, wild-type cell-free culture fluid (HAI-1 + HAI-2); □, synthetic autoinducer (HAI-1); ■, LuxM cell-free culture fluid (HAI-2). These data are from Bassler et al. (1994b).

the protein (see below), so it encodes a constitutive LuxO repressor (Lilley and Bassler, unpublished results).

A schematic depiction of the Lux signaling circuit is shown in Fig. 4. The autoinducers HAI-1 (3-hydroxy-C4-HSL) and HAI-2 (unknown structure) are proposed to interact with their cognate sensors LuxN and LuxPQ in the periplasm. These interactions alter the autophosphorylation activity of the sensors LuxN and LuxQ, and they in turn interact with the integrator protein LuxO, to change its level of phosphorylation at Asp 47. The change in phosphorylation state of LuxO is hypothesized to inactivate its repressor activity by impairing its ability to bind DNA. This event could enable LuxR to bind at the *luxCDABEGH* promoter and activate transcription, resulting in light emission in response to cell density. LuxO could exert its negative effect by binding at the *luxCDABEGH* operon, or alternatively, LuxO could act indirectly by preventing the transcription of *luxR*. The results of Martin et al. (1989) and Miyamoto et al. (1996) show that expression of *luxR* changes at most only two-fold in response to cell density. If LuxO acted at the *luxR* promoter, one would predict that

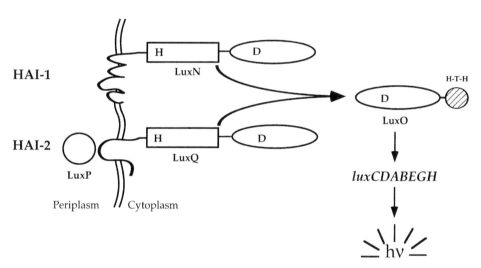

FIGURE 4 Model for two-component control of quorum sensing in *V. harveyi*. The proteins that control quorum sensing in *V. harveyi* are members of the bacterial family of two-component adaptive response regulators. LuxN and LuxQ contain both a sensor kinase domain and a response regulator domain. LuxO is a member of the response regulator class of two-component proteins, and it also contains a DNA binding domain. LuxP, which acts in conjunction with LuxQ, is similar to the ribose binding protein of *E. coli* and *S. typhimurium*. Quorum sensing in *V. harveyi* is hypothesized to occur as follows. At low cell density, and in the absence of the autoinducing signals, the sensors LuxN and LuxQ are proposed to be active kinases. In this state, the sensors autophosphorylate at conserved histidine residues and transfer the phosphoryl group to the shared response regulator protein, LuxO. LuxO is phosphorylated at a conserved aspartate residue. In the phosphorylated state, LuxO actively represses *luxCDABEGH* expression, and no light is emitted. At high cell density, when the autoinducers are present, interaction of HAI-1 and HAI-2 with their cognate sensors LuxN and LuxPQ, respectively, is proposed to inhibit the kinase activity of the sensors. Binding of autoinducer then results in dephosphorylation of LuxN and LuxQ and a cessation of the transfer of phosphate to LuxO. Dephosphorylation of LuxO inactivates its repressor activity, which results in expression of *luxCDABEGH* and light emission in response to cell density. Preliminary evidence indicates that the sensors LuxN and LuxQ act as protein phosphatases to accelerate the dephosphorylation of LuxO. H, histidine; D, aspartate; H-T-H, helix-turn-helix DNA binding domain.

luxR expression would show a strong dependence on cell density. These results indicate, therefore, that LuxO acts directly on the *luxCDABEGH* promoter.

The effect of phosphorylation in the Lux circuit is currently being investigated. Two possibilities are under consideration. The presence of the autoinducer signals could stimulate the autophosphorylation activity of the sensors LuxN and LuxQ, which subsequently act as protein kinases to phosphorylate LuxO and to inactivate its repressor activity. Alternatively, the default state of the system could be that LuxN and LuxQ have high constitutive autophosphorylation activity. In this case, the presence of the autoinducers would cause a decrease in the autophosphorylation activity of LuxN and LuxQ, and the sensors could in turn act as protein phosphatases to dephosphorylate LuxO. If this situation existed, then dephosphorylation of LuxO would inactivate its DNA binding capability and lead to derepression of the *lux* operon. The phenotypic analysis of locked LuxO mutants mimicking the phosphorylated or dephosphorylated form of the protein indicates that the latter possibility is correct. We find that mutations conferring locked LuxO mutants that mimic the phosphorylated state of the protein act as constitutive repressors, while mutant LuxO proteins locked in the unphosphorylated state are inactive and therefore unable to repress *lux* expression (Freeman and Bassler, in press).

QUORUM SENSING IN
V. HARVEYI AND
INTERSPECIES COMMUNICATION

It is clear that *V. harveyi* has two independent density-sensing systems that carry out redundant signaling functions in regulating *lux*. And, each of the systems is composed of analogous functions, i.e., an extracellular autoinducer and its cognate sensor. The signals merge at the LuxO protein to activate expression of the same output, light production. Why does *V. harveyi* need two systems to accomplish this one task? One explanation is that in the marine environment there exist situations when only system 1 or system 2 is operational. However, this condition has never been mimicked in the laboratory. In every growth medium tested, both of the density-sensing systems have been expressed. Alternatively, although both systems are used for receiving and transmitting cell density cues, it is possible that system 1 or system 2 is used to couple other environmental cues (iron, oxygen, carbohydrate) to the *lux* operon. The regulatory components required for *V. harveyi* to monitor the levels of iron, oxygen, and carbohydrate and the mechanism *V. harveyi* uses to relay this information to the *lux* operon have not yet been characterized, so whether these environmental cues are transduced exclusively through system 1 or system 2 or some other system remains to be determined.

One possible use for the two *V. harveyi* signaling systems is to differentiate between various species of bacteria in the environment (Bassler et al., 1997). To test this hypothesis, cell-free culture fluids were prepared from a large number of both marine and terrestrial bacteria and examined for the presence of substances that induced *V. harveyi* to produce light. Three reporter strains with different phenotypes were analyzed for their responses: the wild type (Sensor 1^+, Sensor 2^+), a Sensor 1^+, Sensor 2^- strain, and a Sensor 1^-, Sensor 2^+ strain. This series of tester strains was used to determine whether any signaling substances were produced by the various bacteria, and if so, whether *V. harveyi* detected the substance(s) through signaling system 1, signaling system 2, or both signaling systems. Under the conditions used in the experiment, most bacteria, including *E. coli*, *S. typhimurium*, *Pseudomonas aeruginosa*, and *Bacillus subtilis*, did not produce substances that induced the expression of luminescence in *V. harveyi*. However, *Vibrio cholerae*, *Vibrio parahaemolyticus*, *Vibrio anguillarum*, *Vibrio alginolyticus*, *Vibrio natriegens*, *Photobacterium phosphoreum*, and *Yersinia enterocolitica* did produce substances that mimicked the autoinducer activity of *V. harveyi*. Furthermore, all of these species produced substances that were HAI-2-like and stimulated the Sensor 1$^-$, Sensor 2$^+$ reporter strain. The only species identified that produced an HAI-1-like activity was *V. parahaemolyticus*, which apparently produces two substances, one that mimics the action of HAI-1 and one that mimics the action of HAI-2.

The culture fluids of many of the bacterial strains that produced an HAI-2-like activity did not stimulate the wild-type reporter strain that was Sensor 1$^+$, Sensor 2$^+$ but only stimulated the system 2 reporter strain, indicating that when system 1 is functional, system 2 has reduced sensitivity. Additionally, since many species produced substances that activated system 2, but only *V. parahaemolyticus* made a substance that could activate system 1, it appears that not only does system 1 have higher sensitivity than system 2, but it also appears to be relatively species specific. Perhaps *V. harveyi* uses the higher-sensitivity, species-specific system 1 to monitor the environment for the presence of other *V. harveyi*, and it uses the lower-sensitivity, species nonspecific system 2 to monitor the environment for the presence of other species of bacteria. Coordination of the inputs from both of these signal-response systems could allow *V. harveyi* to appropriately control density-dependent functions, like bioluminescence, in response to the cumulative, multispecies cell density.

CONCLUSIONS: A COMPLEX, MULTICHANNEL SIGNALING CIRCUITRY IS APPROPRIATE FOR *V. HARVEYI*

Regulation of the expression of bioluminescence in *V. harveyi* is complex and consists of multiple interconnected pathways of signal transduction. *V. harveyi* and *V. fischeri* use a different collection of regulatory components and a different biochemical signaling mechanism to carry out an identical function. Perhaps each organism's regulatory circuit is ideal for the niche in which it resides. *V. fischeri* produces density-dependent light while living in pure culture in a constant environment in the light organ of its eukaryotic host, so a simple positive feedback circuit could be suffi-

cient for this mode of existence. *V. harveyi*, in contrast, exists free-living in the ocean, in greatly fluctuating conditions, and in the presence of other species of bacteria. A more complicated signaling circuitry might be necessary to survive in these conditions. Additionally, in *V. harveyi*, a system that is responsive to the presence of other species of bacteria might also be critical for appropriately controlling density-regulated functions.

Studies are under way that should demonstrate that the use of a phosphorylation signaling cascade permits both amplification and dampening of the *V. harveyi* density-dependent signal. Use of a two-component mechanism could enable *V. harveyi* to fine-tune the sensory cascade in order to adjust output with changing environmental inputs. This sensory relay, consisting of multiple channels, may also enable *V. harveyi* to integrate additional cues into the same circuit. Perhaps two-component sensor kinases other than LuxN and LuxQ transduce signals through the integrator protein LuxO. Alternatively, LuxO could receive cues directly from the environment, or it could interact with regulators that are not two-component proteins. For example, LuxO could interact directly with LuxR or some other as yet unidentified protein. There is preliminary evidence indicating that LuxO is involved in channeling the cue from limitation for iron to the expression of *luxCDABEGH*. As mentioned, missense mutations in LuxO that result in locked mutants mimicking the phosphorylated state of LuxO result in the constitutive repression of expression of luminescence. Surprisingly, these mutants also constitutively secrete siderophore, indicating that a cross-wiring event between the regulation of iron acquisition and the regulation of light production occurs at LuxO (Freeman et al., unpublished results). Other such signal integration events may occur at LuxO or at the other Lux regulators. For example, two cues could pass through the LuxP, LuxQ sensor; i.e., the HAI-2 cue could pass through LuxP then LuxQ, and some other environmental cue could bypass LuxP and act directly through LuxQ.

Many other possibilities exist for building additional complexity into the *V. harveyi* Lux signaling circuitry, as has been shown in other two-component circuits (Ninfa et al., 1988; Wanner and Wilmes-Riesenberg, 1992). At present, however, understanding of both the Lux signaling circuit and the Lux signaling mechanism is limited. Other regulatory proteins in the density-sensing apparatus remain to be identified (for example, the gene or genes for production of AI-2), and the proteins responsible for reception and transduction of the other environmental cues remain entirely unknown. Lux expression in *V. harveyi* appears, therefore, to be an excellent model for understanding how bacteria perceive their environment, how they adapt to fluctuating conditions, and how they respond to multiple, temporally coincident cues. We still do not know why *V. harveyi* makes light; however, the circuitry controlling its expression promises to be as interesting as the phenomenon of bioluminescence itself.

REFERENCES

Bassler, B. L., and M. R. Silverman. 1995. Intercellular communication in marine *Vibrio* species: density-dependent regulation of bioluminescence, p. 431–445. *In* J. A. Hoch and T. J. Silhavy (ed.), *Two-Component Signal Transduction.* American Society for Microbiology, Washington, D.C.

Bassler, B. L., M. Wright, R. E. Showalter, and M. R. Silverman. 1993. Intercellular signalling in *Vibrio harveyi*: sequence and function of genes regulating expression of luminescence. *Mol. Microbiol.* **9**:773–786.

Bassler, B. L., M. Wright, and M. R. Silverman. 1994a. Multiple signalling systems controlling expression of luminescence in *Vibrio harveyi*: sequence and function of genes encoding a second sensory pathway. *Mol. Microbiol.* **13**:273–286.

Bassler, B., M. Wright, and M. Silverman. 1994b. Sequence and function of *luxO*, a negative regulator of luminescence in *Vibrio harveyi*. *Mol. Microbiol.* **12**:403–412.

Bassler, B. L., E. P. Greenberg, and A. M. Stevens. 1997. Cross-species induction of luminescence in the quorum-sensing bacterium *Vibrio harveyi*. *J. Bacteriol.* **179**:4043–4045.

Belas, R., A. Mileham, D. Cohn, M. Hilmen, M. Simon, and M. Silverman. 1982. Bacterial bioluminescence: isolation and expression of the luciferase genes from *Vibrio harveyi*. *Science* **218**:791–793.

Cao, J., and E. A. Meighen. 1989. Purification and structural identification of an autoinducer for the luminescence system of *Vibrio harveyi*. *J. Biol. Chem.* **264**:21670–21676.

Chatterjee, J., C. M. Miyamoto, and E. A. Meighen. 1996. Autoregulation of *luxR*: the *Vibrio harveyi lux*-operon activator functions as a repressor. *Mol. Microbiol.* **20**:415–425.

Engebrecht, J., and M. Silverman. 1984. Identification of genes and gene products necessary for bacterial bioluminescence. *Proc. Natl. Acad. Sci. USA* **81**:4154–4158.

Engebrecht, J., K. Nealson, and M. Silverman. 1983. Bacterial bioluminescence: isolation and genetic analysis of functions from *Vibrio fischeri*. *Cell* **32**:773–781.

Freeman, J. A., and B. L. Bassler. A genetic analysis of the function of LuxO, a two-component response regulator involved in quorum sensing in *Vibrio harveyi*. *Mol. Microbiol.*, in press.

Freeman, J. A., B. N. Lilley, and B. L. Bassler. Unpublished results.

Gilson, L., A. Kuo, and P. V. Dunlap. 1995. AinS and a new family of autoinducer synthesis proteins. *J. Bacteriol.* **177**:6946–6951.

Groarke, J. M., W. C. Mahoney, J. N. Hope, C. E. Furlong, F. T. Robb, H. Zalkin, and M. A. Hermodson. 1983. The amino acid sequence of the D-ribose binding protein from *Escherichia coli* K12. *J. Biol. Chem.* **258**:12952–12956.

Hastings, J. W., and J. G. Morin. 1991. Bioluminescence. *In* C. L. Prosser (ed.), *Neural and Integrative Animal Physiology.* Wiley-Interscience Press, New York.

Heath, J. D., T. C. Charles, and E. W. Nester. 1995. Ti plasmid and chromosomally encoded two-component systems important in plant cell transformation by *Agrobacterium* species, p. 367–386. *In* J. A. Hoch and T. J. Silhavy (ed.), *Two-Component Signal Transduction.* American Society for Microbiology, Washington, D.C.

Iuchi, S., Z. Matsuda, T. Fujiwara, and E. C. C. Lin. 1990. The *arcB* gene of *Escherichia coli* encodes a sensor-regulator protein for anaerobic repression of the *arc* modulon. *Mol. Microbiol.* **4**:715–727.

Keynan, A., and J. W. Hastings. 1961. The isolation and characterization of dark mutants of luminescent bacteria. *Biol. Bull.* **121**:375.

Leroux, B., M. F. Yanofsky, S. C. Winans, J. E. Ward, S. F. Ziegler, and E. W. Nester. 1987. Characterization of the *virA* locus of *Agrobacterium* *tumefaciens*: a transcriptional regulator and host range determinant. *EMBO J.* **6:**849–856.

Lilley, B. N., and B. L. Bassler. Unpublished results.

Makemson, J. C. 1986. Luciferase-dependent oxygen consumption by bioluminescent vibrios. *J. Bacteriol.* **165:**461–466.

Martin, M., R. Showalter, and M. Silverman. 1989. Identification of a locus controlling expression of luminescence genes in *Vibrio harveyi*. *J. Bacteriol.* **171:**2406–2414.

McCleary, W. R., and D. R. Zusman. 1990. FrzE of *Myxococcus xanthus* is homologous to both CheA and CheY of *Salmonella typhimurium*. *Proc. Natl. Acad. Sci. USA* **87:**5898–5902.

Meighen, E. A. 1991. Molecular biology of bacterial bioluminescence. *Microbiol. Rev.* **55:**123–142.

Miyamoto, C. M., M. Boylan, A. F. Graham, and E. A. Meighen. 1988a. Organization of the *lux* structural genes of *Vibrio harveyi*. Expression under the T7 bacteriophage promoter, mRNA analysis and nucleotide sequence of the *luxD* gene. *J. Biol. Chem.* **263:**13393–13399.

Miyamoto, C. M., A. F. Graham, and E. A. Meighen. 1988b. Nucleotide sequence of the *luxC* gene and the upstream DNA from the bioluminescent system of *Vibrio harveyi*. *Nucleic Acids Res.* **16:**1551–1562.

Miyamoto, C. M., J. Chatterjee, E. Swartzman, R. Szittner, and E. A. Meighen. 1996. The role of the *lux* autoinducer in regulating luminescence in *Vibrio harveyi*; control of *luxR* expression. *Mol. Microbiol.* **19:**767–775.

Nealson, K. H., and J. W. Hastings. 1979. Bacterial bioluminescence: its control and ecological significance. *Microbiol. Rev.* **43:**496–518.

Nealson, K. H., T. Platt, and J. W. Hastings. 1970. Cellular control of the synthesis and activity of the bacterial luminescent system. *J. Bacteriol.* **104:**313–322.

Ninfa, A. J., E. G. Ninfa, A. N. Lupas, A. Stock, B. Magasanik, and J. Stock. 1988. Crosstalk between bacterial chemotaxis signal transduction proteins and regulators of transcription of the Ntr regulon: evidence that nitrogen assimilation and chemotaxis are controlled by a common phosphotransfer mechanism. *Proc. Natl. Acad. Sci. USA* **85:**5492–5496.

Parkinson, J. S., and E. C. Kofoid. 1992. Communication modules in bacterial signaling proteins. *Annu. Rev. Genet.* **26:**71–112.

Showalter, R. E., M. O. Martin, and M. R. Silverman. 1990. Cloning and nucleotide sequence of *luxR*, a regulatory gene controlling luminescence in *Vibrio harveyi*. *J. Bacteriol.* **172:**2946–2954.

Stock, J. B., A. J. Ninfa, and A. M. Stock. 1989. Protein phosphorylation and regulation of adaptive responses in bacteria. *Microbiol. Rev.* **53:**450–490.

Swartzman, E., and E. A. Meighen. 1993. Purification and characterization of a poly(dA-dT) *lux*-specific DNA-binding protein from *Vibrio harveyi* and identification as LuxR. *J. Biol. Chem.* **268:**16706–16716.

Swartzman, E., C. Miyamoto, A. Graham, and E. A. Meighen. 1990. Delineation of the transcriptional boundaries of the *lux* operon of *Vibrio harveyi* demonstrates the presence of two new *lux* genes. *J. Biol. Chem.* **265:**3513–3517.

Swartzman, E., M. Silverman, and E. A. Meighen. 1992. The *luxR* gene product of *Vibrio harveyi* is a transcriptional activator of the *lux* promoter. *J. Bacteriol.* **174:**7490–7493.

Wanner, B. L., and M. R. Wilmes-Riesenberg. 1992. Involvement of phosphotransacetylase, acetate kinase, and acetyl phosphate synthesis in control of the phosphate regulon in *Escherichia coli*. *J. Bacteriol.* **174:**2124–2130.

PAST AND FUTURE

EARLY OBSERVATIONS DEFINING QUORUM-DEPENDENT GENE EXPRESSION

Kenneth H. Nealson

18

Autoinduction, or quorum sensing, as it is now called, was initially visualized as a mechanism whereby luminous marine bacteria communicate with each other, allowing them to turn off the bioluminescent system in the dilute and energy-limited conditions of the open ocean. The mechanism was proposed entirely on the basis of physiological studies in batch cultures, supported by the isolation and characterization of the first autoinducer and strongly supported by chemostat experiments. The central features of this original hypothesis have been largely confirmed by more recent molecular studies.

Early in the studies of bacterial bioluminescence, it was recognized by several workers that the production of light could be uncoupled from the growth of the cells (Farghaly, 1950; Harvey, 1952), but the nature of this uncoupling was not well defined. Salt concentration, temperature, pH, and types and amounts of nutrients all affected light emission, but few generalizations could be made other than that the level of luminescence was generally much higher in complex as compared to minimal defined media. During this early period, none of the enzymes involved in light emission (luciferase and accessory enzymes) had been purified, and the taxonomy of the luminous bacteria was not yet established.

The first autoinduction experiments were done with a strain of luminous bacteria called *Photobacterium fischeri* MAV, now recognized as *Vibrio harveyi* B-392 (Nealson, 1969; Nealson and Hastings, 1991; Nealson et al., 1970). This strain showed many of the features reported by earlier workers and was examined for the development of luminescence as a function of time and growth in a variety of growth media (Nealson, 1969; Nealson and Markovitz, 1970; Nealson et al., 1970). After a series of experiments involving mixing of used (conditioned) media, autoinduction began to be viewed as a process that involved an extracellular inducer activity that was produced by the cells, accumulated in the growth medium, and acted as the specific inducer for the luminous system (Eberhard, 1972; Nealson, 1969, 1977; Nealson et al., 1970). The ideas were supported by studies of growth and luminescence in minimal media, and by the presence of several groups of mutants whose phenotypes were consistent with the autoinduction model. Although the mechanism was met

Kenneth H. Nealson, Jet Propulsion Laboratory, California Institute of Technology, 4800 Oak Grove Drive, Pasadena, CA 91109.

Cell-Cell Signaling in Bacteria, Edited by Gary M. Dunny and Stephen C. Winans
©1999 American Society for Microbiology, Washington, D.C.

with some skepticism (to markedly understate the actual response), no other explanations could account for the data as well, and after several years of work, no disproof of the hypothesis was obtained. In the late 1970s a natural, weakly luminescent variant of *Vibrio fischeri* was isolated which served as an adequate bioassay for autoinducer (Nealson, 1977). Using this assay, Eberhard et al. (1981a,b) purified the first homoserine lactone autoinducer, synthesized the compound, and showed that it was indeed a specific inducer of *lux*. Rosson and Nealson (1981) used a carbon-limited chemostat to control cell density and demonstrated that, when cell density was kept to 10^7 cells per ml or less, luminescence was greatly diminished and this luminescence could be reestablished by the addition of the purified autoinducer. This led to a more quantitatively based, although still descriptive, physiological model of autoinduction, one that has in the broadest sense stood the test of time. In 1984, when the *lux* genes of *V. fischeri* were successfully cloned and expressed in *Escherichia coli* (Engebrecht et al., 1983), the basic features of the "simple" system that describes autoinduction in this species were revealed at the molecular level. This left little doubt that the model originally proposed was a crude but biologically accurate depiction of quorum-dependent bioluminescence. In this chapter I briefly review the early history of quorum sensing, discussing the observations and experiments that, in the absence of genetic data, were used to formulate and defend the model.

THE EARLY DAYS

In the late 1950s and early 1960s, several workers noted that the expression of bioluminescence in various marine bacterial species was not an easily defined process. A variety of factors were purported to control the luminescence, including salt concentration, media nutrients, temperature, etc. (Coffey, 1967; Farghaly, 1950; Harvey, 1952) and it was noted that growth and luminescence were not always in concert. At that time the concepts

of inducible genes, operons, and control mechanisms were new even to those working on *E. coli*, and little thought was given to similar mechanisms in undefined marine species in which the enzyme for light emission, luciferase, had not yet been purified. An example of such an early experiment is shown in Fig. 1, in which data of Farghaly (1950) are replotted to show the subtle difference between growth and luminescence during the early stages of growth. In this work the investigators focused on the dramatic drop in luminescence after stationary phase was reached. In Fig. 1B, the increase in luminescence as a function of growth shows a similar pattern, although there are not many data points to choose from at these early stages of growth. In these experiments, the investigators were more concerned with the rapid decrease in luminescence late in the growth phase, while the factors responsible for the inducible nature of the increase were not discussed.

THE GERM OF THE MODEL

In the mid-1960s, work from several laboratories began to suggest that some control mechanisms were responsible for what was reported as an early lag and subsequent synthetic phase of bioluminescence in the luminous bacteria, as shown in Fig. 2. Kempner and Hanson (1968), on the basis of physiological experiments using conditioned medium, reached the conclusion that rich medium contained an inhibitor of luminescence, and that if this inhibitor was removed by growing cells in the medium, then the lag in luminescence was also removed. The inhibitor was not identified, but the activity was retained even after ashing of the media (removal of organic carbon by heating for several hours at 300°C), suggesting that it might be a metal of some kind (Kempner and Hanson, 1968). Subsequent experiments (Haygood and Nealson, 1984; Makemson and Hastings, 1982) would eventually suggest that the inhibitory activity described by Kempner and Hanson was probably due to iron acting as a transcriptional down-regulator, but acting distinctly from the

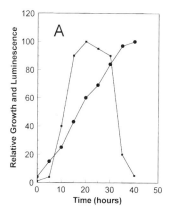

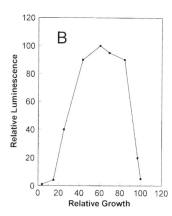

FIGURE 1 Growth and luminescence of bacteria as a function of time (A) and luminescence per milliliter as a function of growth (B). These data, replotted from Farghaly (1950), demonstrate the basic pattern of bioluminescence in which a lag in synthesis relative to growth is seen, followed by an increase in luminous activity at a rate greater than growth. The values are relative values of luminescence (and growth) normalized to the maximum values during the experiment.

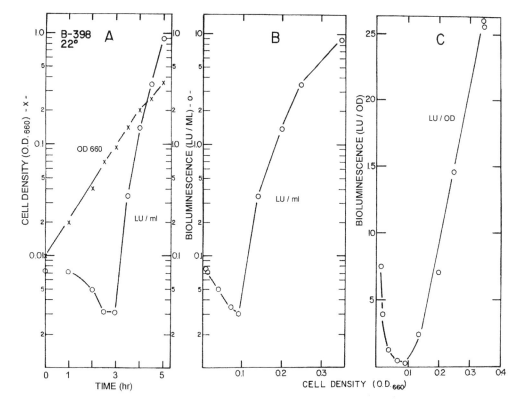

FIGURE 2 Growth and luminescence of *V. fischeri* B-398. These plots show the same data presented in three different formats. (A) Growth and luminescence per milliliter plotted as a function of time (shown in Fig. 1A). (B) Luminescence per milliliter of culture as a function of cell number (similar to data of Fig. 1B). (C) Specific activity of bioluminescence (LU/OD) as a function of cell number. Data from Nealson (1977) are expressed as LU (light units), a relative measure of luminescence, and OD_{660}, a measure of cell density.

autoinducer. In contrast, Coffey (1967) proposed that the cells responded positively to an inducer, and that the inducer was arginine. In complex media containing arginine, the cells exhibited very bright luminescence, and the addition of arginine to minimal medium resulted in cells with much higher specific activity of luminescence. The data of both authors were reproducible, although the interpretations of the data were in conflict with each other and with several of the subsequent experiments done by other laboratories. It was out of this latter work that the autoinduction model grew.

In the 1960s the purification of the enzyme bacterial luciferase (Hastings et al. 1965) and its characterization as a heterodimer (Friedland and Hastings, 1967) was published, setting the stage for the study of the regulation of the system. The luciferase work showed that early in the growth phase there was virtually no bacterial luciferase in the dim cells, and that only a few hours later luciferase accounted for several percent of the cellular protein in the brilliantly luminous bacteria (Hastings et al. 1965); clearly, some inducible system must be operating. Initial experiments were in complex media, which yielded results similar to those shown in Fig. 2 for *V. fischeri*. Those results clearly showed that growth and luminescence were uncoupled in the early stages of growth, with luciferase synthesis lagging for several hours and in vivo light emission actually decreasing during this lag. After several hours, however, the lag ended and was followed by a period of synthesis at a rate much faster than growth, resulting in cells that were highly enriched for luciferase and extremely bright. That this was an induction was established by immunological methods using anti-luciferase antibodies to measure the synthesis of cross-reacting material and by the use of inhibitors of protein and mRNA synthesis (Nealson, 1969; Nealson and Markovitz, 1970; Nealson et al., 1970). Thus, it appeared that the cells induced luciferase synthesis at a fairly low cell density well before stationary phase with no added inducer. This was the beginning of the use of the term autoinduction in bacterial bioluminescence.

Given the vagaries of complex media, and the possibility that this induction might be due simply to the removal of an inhibitor, the experiments were moved to a minimal medium, where the observations of Kempner and Hanson (1968) and Coffey (1967) were confirmed. In minimal medium, the lag in the synthesis of the luminous system was greatly shortened. However, it was not eliminated, indicating that while some inhibitory activity was present in the complex medium, an induction still occurred in the defined minimal medium. Second, the greatly reduced luminescence seen in minimal medium could be restored by adding back various components of the complex medium. Arginine alone could nearly restore the cells to maximum luminescence levels, with little effect on growth rate. However, three observations suggested that arginine was not the inducer of the luminous system: (i) the concentrations (~10 mM) of arginine required were very high for an inducer, (ii) low levels of autoinduction occurred even in the absence of exogenous arginine addition (Fig. 3), and (iii) arginine stimulation did not occur until autoinduction had occurred, so that it appeared to be an enhancer, rather than a specific inducer (Fig. 3). Since the low level of autoinduction occurs at a cell density of 0.01 (OD_{660}), it is easily missed, and if arginine is added any time after autoinduction occurs, the effect is rapid and inducer-like in the sense that it is blocked by inhibitors of protein and mRNA synthesis (Nealson, 1969; Nealson and Markovitz, 1970; Nealson et al., 1970). The minimal medium with arginine proved to be a good medium in which to study autoinduction, as growth was slow, luminescence was high, and no inhibitors characteristic of the complex medium were present. It should be mentioned that the arginine stimulation, which is specific for a few strains of luminous bacteria, has never been fully explained at the molecular level.

In the original formulation of the autoinduction model, it was strongly suggested that

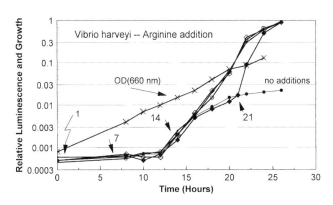

FIGURE 3 Growth (OD$_{660}$) and luminescence of *V. harveyi* in minimal medium, with and without arginine (1 mg/ml) added. The control and arginine additions at 1, 7, and 14 h show times of induction, with light beginning to increase in all cases, at an OD$_{660}$ of approximately 0.01. When arginine was added, induction was enhanced, and light intensity kept rising for many hours longer, reaching levels of approximately 100 times higher. When arginine was added at 21 h, a rapid response was seen, with light reaching the other arginine-enhanced levels only a few hours later. Data are replotted from Nealson et al. (1970).

autoinduction was due to a positive effect, but no mechanism was proposed (Nealson et al., 1970). However, the search for an inducer or inhibitor was launched, and a series of experiments of the type outlined in Fig. 4 were done in several species of luminous bacteria. These experiments pointed to the possible existence of a positive inducer activity being present in the used (conditioned) medium from cells of *V. fischeri* (Eberhard, 1972; Nealson, 1969, 1977). In these experiments, the wild-type cells were grown to various densities, the cells were removed by centrifugation, and the used medium was reinoculated with fresh cells. Alternatively, various amounts of fresh medium were added to the used medium. In comparative experiments, the induced cells were added to fresh medium with various amounts of conditioned medium added to them. The rationale of these experiments was that if an inhibitor was present, then its removal should allow induction, and

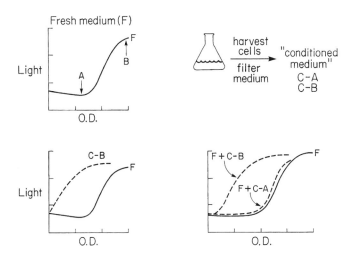

FIGURE 4 Conditioned medium experiments. This diagram shows the types of experiments that were done to try to distinguish between the removal of an inhibitor and the addition of an active substance (inducer). Cells were grown in fresh medium (F) to various levels (A or B) of luminescence and were harvested to yield conditioned medium (C-A and C-B). If C-A was added to fresh medium, very little if any effect was noted, while if C-B was added to fresh medium, it moved the time of induction significantly earlier. If different amounts of conditioned medium were added, the time of induction was earlier in rough proportion to the amount of C-B added. These experiments suggested that an inducer was present, although they tended to be very difficult to quantitatively reproduce and were abandoned as a quantitative bioassay.

after this point there would be little difference in the activity of the conditioned media of various ages. Alternatively, if a positive activity were present, it might accumulate in the used medium and thus show different responses with time and with dilution.

A major problem with the experiments described in Fig. 4 was that, while activity could be easily demonstrated, it was virtually impossible to develop a dependable and reproducible bioassay. This problem precluded using such an approach as a good quantitative bioassay for purification of the inducer. Thus, while such experiments strongly supported the model of autoinduction rather than that of an inhibitor, they were not entirely satisfactory because of a constantly changing background of luminescence in the control cells, something that was not possible to understand until many years later. One important piece of information gleaned from these early experiments, however, was that the inducer activity of *V. fischeri* was more stable than that of *V. harveyi* (Eberhard, 1972), and later experiments focused on the *V. fischeri* system.

ISOLATION OF THE AUTOINDUCER: THE APPEARANCE OF HOMOSERINE LACTONE

The isolation of autoinducer was accomplished after the availability of a good bioassay, which was developed by virtue of having access to natural variants of luminous bacteria with variable levels of autoinduction activity (Nealson, 1977). These strains were discovered during a survey of various strains of *V. fischeri* that exhibited very different levels of bioluminescence (Table 1) (Nealson, 1977). A strain called MJ-1, isolated as a symbiont from the luminous fish *Monocentris japonicus* (Ruby and Nealson, 1976), was nearly 10-fold brighter than the type strain previously studied, while a strain called B-61, obtained from John Reichelt, was only 1% as bright as the type strain. Media mixing experiments of the type shown in Fig. 4 suggested that differences in autoinducer levels were responsible for these differences (Nealson and Hastings,

1979), suggesting that these two strains might be used as a source and bioassay for autoinducer, respectively. In fact, this approach was quite successful. Strain B-61 exhibited a very low level of luminescence, and cells diluted into fresh minimal medium showed no increase in luminescence for up to 1 h. Ethyl acetate extracts of the conditioned medium from MJ-1, when added back to the B-61 cells, showed that a reasonably good and reproducible quantitative response could be obtained (Fig. 5). Thus, a good source of autoinducer and an acceptable bioassay for the activity were available.

Using these tools, Anatol Eberhard, on sabbatical leave at Scripps Institution of Oceanography, accomplished the purification of the autoinducer (AI). The purification involved extraction of the activity with ethyl acetate, several drying and redissolution steps, silica gel chromatography, and two high-pressure liquid chromatography purifications, resulting in 2.7 mg of a pure compound from an original 6-liter culture of MJ-1 (Eberhard, 1972; Eberhard et al., 1981a). This material was identified as 3-oxo-*N*-(tetrahydro-2-oxo-3-furanyl)hexanamide, the first of the homoserine lactone (HSL) autoinducers (Fig. 6). Other, less systematic names for this AI are *N*-(*β*-ketocaproyl)homoserine lactone, *N*-(3-oxohexanoyl)homoserine lactone, and *N*-(3-oxohexanoyl)-3-aminodihydro-2(3H)-furanone. Synthetic HSL was prepared and shown to have activity almost identical to that of the naturally purified component. Using B-61 as the bioassay, it was determined that addition of HSL at concentrations of 3×10^{-10} M could stimulate bioluminescence, qualifying this compound as an inducer of very high activity. It was further shown that AI stimulated luciferase synthesis activity within minutes of addition, and that this stimulation could be blocked by inhibitors of protein and mRNA synthesis (Fig. 7). It was also determined that the material induced other strains of *V. fischeri* but not other strains of luminous bacteria. Subsequently, another autoinducer was identified in *V. harveyi* (Cao and Meighen, 1989)

TABLE 1 Bioluminescence and autoinducer activity in natural isolates of *V. fischeri*

Strain	Source of isolate	Maximum in vivo luminescence	Maximum in vitro luminescence	Maximum autoinducer activity (%)
MJ-1	Symbiont of *Monocentris japonicus*	100	100	100
1009	Shrimp parasite	80	67	90
B-398	ATCC 7744	17	14	5–14
MB-65	Seawater isolate	2	5	6
B-61	John Reichelt	0.01	1.4	0.12

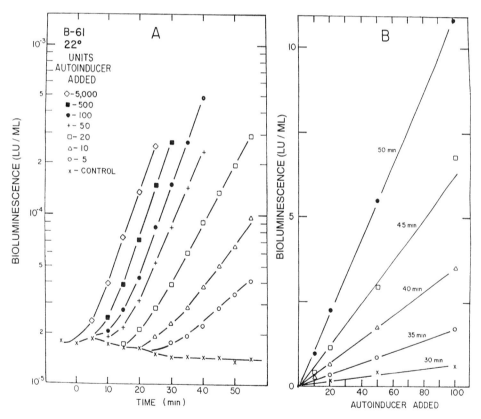

FIGURE 5 Assay for autoinducer. Cells of the dim variant B-61 (see Table 1) were suspended in a minimal growth medium in which light was stable for periods of 1 to 2 h. Under these conditions, if various amounts of used medium were added, a quantitative bioassay was achieved, with autoinducer response being linear to the amount added over a factor of approximately 10-fold (from 10 to 100 relative units). Using appropriate dilutions and back calculations, it was possible to use this assay for quantitative determination of autoinducer. Data are from Nealson (1977).

FIGURE 6 Autoinduction model and details. (A) The basic model of autoinduction that dominated thinking for a decade. Here it is imagined that the cells produce autoinducer (AI), which is equilibrated with the exterior growth medium, and that as long as it stays below a certain critical concentration, the cells will be dark. If the cells accumulate to a critical volume, or AI accumulates for other reasons, then the luminous system will be turned on. (B) The structures of the first two HSL autoinducers that were identified (Cao and Meighen, 1989; Eberhard, 1972; Eberhard et al., 1981a). (C) The structure of the *V. fischeri lux* operon, derived from Engebrecht et al. (1983), with the linked genes for autoinducer production (*luxI*) and autoinducer responding element (*luxR*). As discussed in the text, this is the only *lux* operon known to date in which the regulatory genes (*luxR* and *luxI*) are closely linked to the structural genes for bioluminescence (*luxA-E*).

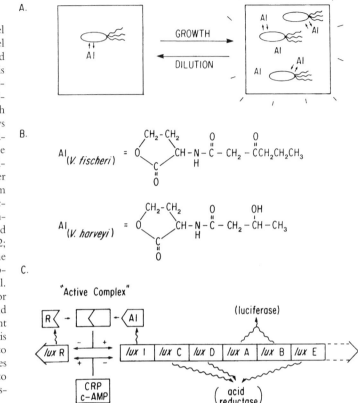

(Fig. 6), with a different structure and specificity.

At this point, several key components of the model had been verified. An inducer had been identified that could be extracted from the external growth medium. This inducer acted at the level of transcription, stimulating the synthesis of luciferase and the enhancement of luminescence, and could act at very low concentrations. But what was the use of such a mechanism? Given that luminous bacteria have two "life-styles," symbiotic and planktonic, it was proposed that the ability to "spatially sense the environment" (Nealson and Hastings, 1979, 1991) would be of advantage to the planktonic bacteria, allowing them to avoid synthesis of the metabolically expensive luminous system when it served no purpose. This might be especially true in the nutrient-limiting oceanic waters. This led to the concept of the model shown in Fig. 6A, in which it was proposed that autoinduction conferred an advantage to planktonic luminous bacteria by turning off synthesis and expression of bioluminescence.

FURTHER EVIDENCE IN SUPPORT OF THE MODEL: CHEMOSTATS, CELL DENSITY, AND LIGHT EMISSION

As mentioned above, while the model of autoinduction seemed to be consistent with all of the available data, its acceptance was less than universal by the microbiological community. Several times our laboratory attempted to publish a model of the type shown in Fig. 6A. Each time, the peer reviewers suggested that the diagram be removed. The

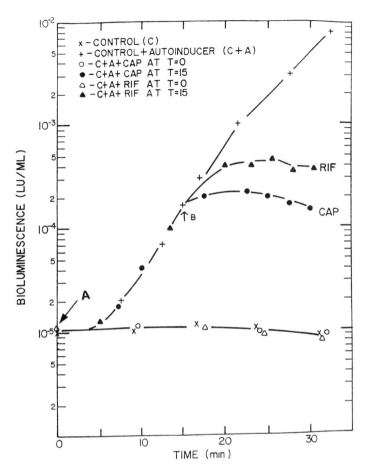

FIGURE 7 Induction of bioluminescence by autoinducer. In this experiment, autoinducer (ethyl acetate extracts of conditioned medium) is added to cells of B-61 at zero time, resulting in an increase in luminescence within minutes. This increase is completely blocked by inhibitors of protein and mRNA synthesis (CAP, chloramphenicol; RIF, rifampin). Addition of inhibitors at 15 min results in a rapid cessation of increase in luminescence, as shown. Data are from Nealson (1977).

comment "bacteria just don't do this," which appeared in several of these reviews, can be attributed to a reluctance to accept the notion of intercellular communication by bacteria. The concept that bacteria might put something into the growth medium that could act in some form of intercellular communication was not ready for acceptance in the early 1980s.

A set of experiments that nearly got the model published involved the use of carbon-limited chemostats to control cell density (Rosson and Nealson, 1981). A key prediction of the model was that if inducer acted to communicate between cells, then there should be a critical cell density (specific for each strain, and dependent upon the rate of AI production) below which light would not be produced. Furthermore, based on the activity of the autoinducer and the amount produced per cell, it was speculated that this number would actually be relatively high (on the order of 10^6 to 10^7 cells per ml), predicting that virtually all planktonic luminous bacteria in the ocean (where they are on the order of 1 to 100 cells per ml) should be dark. A direct test of this hypothesis was accomplished using a carbon-limited chemostat system in which cell density could be controlled by the concentration of glycerol, as demonstrated in Fig. 8. Using this approach, cells could be maintained in a healthy growing state at a known cell concentration and dilution rate for extended periods. A variety of experiments were done in these studies (25), but two critical results are shown in Fig. 9. (i) Below a cell density of 10^6 to 10^7

FIGURE 8 Chemostat design for the study of autoinduction. In these experiments, growth was maintained under constant conditions in the 100-ml growth chamber. The 2-liter media reservoir contained glycerol at a known concentration (3 g/liter) and was used to supply the growth chamber for 10 dilutions (1 liter total). At this time, the dilution reservoir was used to refill the media reservoir to 2 liters, thus halving the concentration of carbon in the medium. After several dilutions, carbon in the medium becomes limiting, and the cell density is determined by the glycerol concentration in the media reservoir. Under these conditions, continuously growing cells (2.4-h doubling time) can be maintained at specified cell densities for unlimited time periods, barring contamination. The graph shows data from Rosson and Nealson (1981), in which the glycerol concentrations of the media reservoir are shown as closed squares, and the decrease in cell density and luminescence per milliliter are shown as closed and open circles, respectively. The closed stars show that over the range of cell densities studied in this experiment, no decrease in specific activity of luminescence occurred. Cell density is expressed as OD_{660} (times 0.01), so that the value shown as 10 is actually an OD = 0.1. Luminescence is expressed as relative units (LU) per ml or per OD.

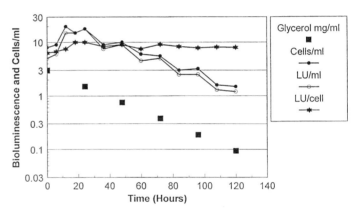

the specific activity of bioluminescence (luminescence per cell) dropped by more than four orders of magnitude to undetectable levels, and (ii) addition of pure AI to these cells resulted in the reestablishment of bioluminescence, even at low cell density (Fig. 9B).

Additional strains and species of luminous bacteria were examined for autoinduction, revealing the existence of several very bright, apparently constitutive strains, as well as many different species showing density-dependent

luminescence. Only those of the *V. fischeri* group were sensitive to the pure autoinducer.

ACCEPTANCE OF THE MODEL: THE *lux* OPERON AND ITS CONTROL ELEMENTS

Widespread acceptance of the model of autoinduction came with the molecular confirmation of the genetic elements involved. This was accomplished through the thesis work of Joanne Engebrecht (Engebrecht et al. 1983),

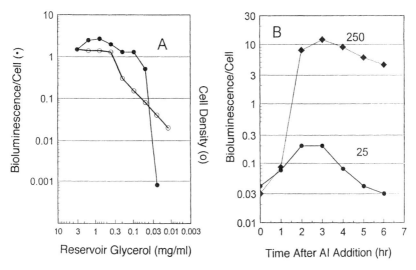

FIGURE 9 Effect of cell density and autoinducer addition on bioluminescence of *V. fischeri* cells. (A) These data, replotted from Rosson and Nealson (1981), are from chemostat growth experiments, in which either cells/ml (OD) or luminescence (LU) per OD are plotted as a function of glycerol concentration of the reservoir. Using this approach, the low glycerol concentrations are used to maintain the cell density at a low value; cells maintained at densities below an OD of 0.01 (about 10^7 ml^{-1}) are essentially nonluminescent. These cells remain dark at all densities below this level unless autoinducer (AI) is added. (B) When AI is added (either 25 or 250 μl), a resultant but transient increase in luminescence is seen. These cultures eventually become dark again as AI is lost via dilution.

who isolated the *lux* genes from *V. fischeri* MJ-1 and found, quite fortuitously, that the control elements were linked to these genes. Thus, autoinduction occurred even in the strains of *E. coli* into which the genes had been cloned (Engebrecht et al. 1983). Considering that this gene arrangement is known (so far) only in *V. fischeri*, the discovery of the regulatory system was greatly aided by the choice of bacterium used. At any rate, these studies led to the identification of the *luxI* and *luxR* genes, their location with regard to the *lux* structural genes, and a refinement of a model for autoinduction that was published in 1983 (Fig. 6C).

By mutating the isolated DNA, it was possible to verify the various gene functions (e.g., autoinducer production by *luxI*, autoinducer response by *luxR*, and structural genes involved with light emission [*luxAB*] and alde-

hyde production [*luxC, D,* and *E*]). This work was all aided greatly by the availability of pure autoinducer to test mutants and verify phenotypes.

CAUTIONS, PRECAUTIONS, TRAPS, AND SPECULATIONS

In the years since the publication of the first model, a virtual explosion of information has occurred, and the mechanism of autoinduction is now readily accepted as a way in which microbes deal with their environment and communicate with each other (Fuqua et al., 1994; Greenberg, 1997; Losick and Kaiser, 1997). It has even led to a change of the name of the mechanism to quorum sensing, a name that I found quite amusing considering the relative chaos that exists in faculty meetings with a sufficient quorum present! Perhaps one might prefer a name that implies that the decision is intelligent or useful rather than simply legal.

I close with a couple of observations and speculations that are made by one now not in this area of research, one who has seen virtually the entire history of this mechanism as it has matured.

First, there is now the tendency to identify quorum sensing as a mechanism on the basis of the presence of the *luxI* and *luxR* genes. Whether or not this is advisable is an interesting question. Studies of the regulation of luminescence have revealed that a variety of mechanisms operate and that there are a lot of physiological "traps" that could lead to misinterpretation. For example, studies by several workers (Haygood and Nealson, 1984; Makemson and Hastings, 1982), have shown that iron is a strong repressor of the *lux* system and that addition of iron chelators (like siderophores or synthetic chelators) can lead to early induction of luminescence. Thus, the interpretation of autoinducer activity produced by other cells (Greenberg et al., 1979) could well be the production of iron-binding compounds, especially if the experiments are done in a complex medium. In the absence of pure autoinducer compounds, it is difficult if not impossible to conclude that the proposed mechanism is the one that is actually controlling the system. A variety of other physiological regulators (nutrient type and amount, oxygen, pH, etc.) (Nealson and Hastings, 1979, 1991) can have major effects as well, and should be considered.

Second, the presence of the *luxI* and *luxR* genes does not necessarily mean that they are functional, although it may seem reasonable that they could be operating in a "quorum-sensing" mode. In almost none of the recent studies of quorum sensing has there been any attempt to link the behavioral physiology of the organisms to the proposed genetic control mechanisms by the appropriate activity experiments. It would thus seem that a mechanism that was impossible to accept as existing just a few years ago on the basis of extensive physiological and biochemical evidence is now ac-

cepted without scrutiny of either physiology or biochemistry, simply on the basis of the presence of regulatory genes. This trend may be scientifically as unwise as was the original unwillingness to accept new ideas or models.

Finally, in the early days of autoinduction, it was impossible to predict or comprehend the impact of the work and/or the model. As far as was known, the mechanism was developed by a marine bacterium to allow survival in a nutrient-limited ocean when away from its symbiotic host(s). In those days intercellular bacterial communication was not easy to accept, and a mechanism to allow it to occur seemed less than important. As is so elegantly demonstrated in this treatise, quorum sensing is now commonplace in many different symbiotic and pathogenic systems. With its acceptance, the idea that bacteria might gain considerable advantage by communicating with one another has become accepted as well—a decided shift in our perceptions in just a few years. Such a scenario makes a strong case for basic research. The autoinduction mechanism was discovered because a few curious investigators wanted to find out how an "unimportant" marine bacterium regulated its light emission in response to culture density. No one would have dreamed that it could have uncovered a general mechanism of regulation, let alone a conceptual shift in the way we view bacterial physiology and ecology.

REFERENCES

Cao, J., and E. A. Meighen. 1989. Purification and structural identification of an autoinducer for the luminescence system of *Vibrio harveyi. J. Biol. Chem.* **264:**21670–21676.

Coffey, J. J. 1967. Inducible synthesis of bacterial luciferase: specificity and kinetics of induction. *J. Bacteriol.* **94:**1638–1647.

Doudoroff, M. 1942. Studies on the luminous bacteria: nutritional requirements of some species. *J. Bacteriol.* **44:**451–459.

Eberhard, A. 1972. Inhibition and activation of bacterial luciferase synthesis. *J. Bacteriol.* **109:**1101–1105.

Eberhard, A., C. Eberhard, A. L. Burlingame, G. L. Kenyon, N. J. Oppenheimer, and K. H.

Nealson. 1981a. Purification, identification and synthesis of *Photobacterium fischeri* autoinducer, p. 113–120. *In* M. DeLuca (ed.), *Bioluminescence and Chemiluminescence,* Academic Press, New York.

Eberhard, A., A. L. Burlingame, C. Eberhard, G. L. Kenyon, K. H. Nealson, and J. J. Oppenheimer. 1981b. Structural identification of autoinducer of *Photobacterium fischeri* luciferase. *Biochemistry* **20:**2444–2449.

Engebrecht, J., K. H. Nealson, and M. Silverman. 1983. Bacterial bioluminescence: isolation and genetic analysis of functions from *Vibrio fischeri*. *Cell* **32:**773–781.

Farghaly, A. H. 1950. Factors influencing the growth and light production of luminous bacteria. *J. Cell. Comp. Physiol.* **36:**165–184.

Friedland, J., and J. W. Hastings. 1967. Nonidentical subunits of bacterial luciferase. Their isolation and recombination to form active enzyme. *Proc. Natl. Acad. Sci. USA* **58:**2336–2342.

Fuqua, W. C., S. C. Winans, and E. P. Greenberg. 1994. Quorum sensing in bacteria: the LuxR-LuxI family of cell density-responsive transcriptional regulators. *J. Bacteriol.* **176:**2796–2802.

Greenberg, E. P. 1997. Quorum sensing in gram-negative bacteria. *ASM News* **63:**371–377.

Greenberg, E. P., J. W. Hastings, and S. Ulitzur. 1979. Induction of luciferase synthesis in *Beneckea harveyi* by other marine bacteria. *Arch. Microbiol.* **120:**87–91.

Harvey, E. N. 1952. *Bioluminescence.* Academic Press, New York.

Hastings, J. W., W. H. Riley, and J. Massa. 1965. The purification, properties and chemiluminescent quantum yield of bacterial luciferase. *J. Biol. Chem.* **240:**1473–1481.

Haygood, M. G., and K. H. Nealson. 1984. Mechanisms of iron regulation of luminescence in *Vibrio fischeri*. *J. Bacteriol.* **162:**209–216.

Kempner, E. S., and F. E. Hanson. 1968. Aspects of light production by *Photobacterium fischeri*. *J. Bacteriol.* **95:**975–979.

Losick, R., and D. Kaiser. 1997. Why and how bacteria communicate. *Sci. Am.* **276:**68–73.

Makemson, J. C., and J. W. Hastings. 1982. Iron represses bioluminescence and affects catabolite repression of luminescence in *Vibrio harveyi*. *Curr. Microbiol.* **7:**181–186.

Nealson, K. H. 1969. *Mutational and Biochemical Studies of Bacterial Bioluminescence*. Ph.D. thesis, University of Chicago.

Nealson, K. H. 1977. Autoinduction of bacterial luciferase: occurrence, mechanism and significance. *Arch. Microbiol.* **112:**73–79.

Nealson, K. H., and J. W. Hastings. 1979. Bacterial bioluminescence: its control and ecological significance. *Microbiol. Rev.* **43:**496–518.

Nealson, K. H., and J. W. Hastings. 1991. The luminous bacteria, p. 625–639. *In* A. Balows, H. G. Trueper, M. Dworkin, W. Harder, and K.-H. Schleifer (ed.), *The Prokaryotes,* 2nd ed., vol. 1. Springer-Verlag, New York.

Nealson, K. H., and A. Markovitz. 1970. Mutant analysis and enzyme subunit complementation in bacterial bioluminescence in *Photobacterium fischeri*. *J. Bacteriol.* **104:**300–312.

Nealson, K. H., T. Platt, and J. W. Hastings. 1970. Cellular control of the synthesis and activity of the bacterial luminescent system. *J. Bacteriol.* **104:**313–322.

Rosson, R. A., and K. H. Nealson. 1981. Autoinduction of bacterial bioluminescence in a carbon limited chemostat. *Arch. Microbiol.* **129:**299–304.

Ruby, E. G., and K. H. Nealson. 1976. Symbiotic association of *Photobacterium fischeri* with the marine luminous fish *Monocentris japonica*: a model of symbiosis based on bacterial studies. *Biol. Bull.* **151:**574–586.

N-ACYLHOMOSERINE LACTONES AND QUORUM SENSING IN PROTEOBACTERIA

Simon Swift, Paul Williams, and Gordon S. A. B. Stewart

19

The purpose of this chapter is to review the remarkable diversity of applications to which the proteobacteria are now known to employ the *N*-acyl homoserine lactone (AHL) language for communication. It is a measure of the rate of advances in this area that establishing the existence of the AHL language has in many cases preceded the capacity of molecular microbiology to define a cell density-dependent phenotype.

Cell-cell communication enables an individual bacterium to gather information from its environment, allowing an evaluation of its own population size. Transduction of this information to effectors of gene expression leads to the elaboration of an appropriate phenotype. The production and subsequent release of a small signal molecule provides the bacterium with an environmental parameter that can be measured, and hence it is this signal that provides the crucial element for which the concentration is reflective of population size. The dynamics of signal generation, accumulation, recognition, and response are, therefore, central to the process whereby the bacterium can sense self and modulate phenotype accordingly.

The regulatory activity of small molecule signals in bacteria was well documented before 1990 (Stephens, 1986; Shapiro, 1988), including a role for *N*-3-oxohexanoylhomoserine lactone (3-oxo-C6-HSL) in the regulation of bioluminescence in *Vibrio* (*Photobacterium*) *fischeri* (Nealson, 1977; Eberhard et al., 1981). At that time, however, 3-oxo-C6-HSL was thought to be unique to this marine bacterium and its close phylogenetic neighbors (Greenberg et al., 1979). The evidence published in 1992 that *Erwinia carotovora* also produced this signal molecule and furthermore used it in regulating the production of a *β*-lactam antibiotic, 1-carbapen-2-em-3-carboxylic acid (carbapenem) (Bainton et al., 1992b), prompted a reevaluation of the distribution of this signal in eubacteria (Bainton et al., 1992a). Subsequently, it was demonstrated that cell-cell communication using AHL signals is a widespread phenomenon in gram-negative bacteria (Tables 1 and 2) (Bainton et al., 1992a; Williams et al., 1992; Fuqua et al., 1994, 1996; Swift et al., 1994; 1996; Salmond et al., 1995).

Simon Swift and Paul Williams, Institute of Infections and Immunity, Queen's Medical Centre, University of Nottingham, Nottingham, NG7 2UH, United Kingdom, and School of Pharmaceutical Sciences, University of Nottingham, Nottingham, NG7 2RD, United Kingdom. *Gordon S. A. B. Stewart*, School of Pharmaceutical Sciences, University of Nottingham, Nottingham, NG7 2RD, United Kingdom.

Cell-Cell Signaling in Bacteria, Edited by Gary M. Dunny and Stephen C. Winans
©1999 American Society for Microbiology, Washington, D.C.

Such has been the ensuing interest and pace of discovery in this new field termed quorum sensing (Fuqua et al., 1994; Greenberg, 1997) that what was previously regarded as a control mechanism of esoteric interest in a squid or fish symbiont has become a candidate for the development of novel anti-infective agents applicable to the control of pathogens in the postantibiotic age (Swift et al., 1994; Williams, 1994).

DETECTION OF AHLs

The use of biosensors for AHLs has been instrumental in the discovery of AHL-mediated quorum sensing in a number of gram-negative bacteria. Accumulation of AHLs results in the activation of transcription and the expression of a given characteristic. Several laboratories have described bacterial strains designed to detect AHLs. Such strains do not express an AHL synthase, but do express an AHL-responsive transcriptional activator (LuxR homolog) and contain an AHL-activated promoter fused to a reporter gene(s) such as *lacZ* or *luxCDABE*. Since such a strain does not produce any AHL, expression of the reporter requires an exogenous source of AHL. The simplest sensors are AHL synthase mutants of strains regulating a measurable characteristic by quorum sensing. For example, *V. fischeri luxI* null mutant strains bioluminesce in the presence of exogenous 3-oxo-C6-HSL and closely related compounds (Nealson, 1977; Greenberg et al., 1979), *Pseudomonas aureofaciens phzI* null mutant strains produce an orange halo of phenazine antibiotic in the presence of agonistic AHLs (Pierson and Pierson, 1996), and *Chromobacterium violaceum cviI* null mutant strains produce the purple pigment violacein in the presence of *N*-hexanoyl homoserine lactone (C6-HSL) and C6-HSL agonists (Fig. 1) (McClean et al., 1997).

The realization that structurally diverse AHLs exist in nature, having variation in the acyl side chain and having different activation profiles for given LuxR homologs, has prompted the construction of in vivo reporters, which can, in combination, detect a wide

spectrum of AHL structures. Plasmid-based biosensors carrying *luxRI'*::*luxAB*(*Vibrio harveyi*) (Swift et al., 1993), *luxRI'*::*luxCDABE*(*Photorhabdus luminescens*) (Eberl et al., 1996; Swift et al., 1997; Winson et al., 1998a), *rhlRI'* (*vsmR'*)::*luxCDABE* (Winson et al., 1995, 1998a), *ahyRI'*::*luxCDABE* (Swift et al., 1997), *lasRI'*::*luxCDABE* (Winson et al., 1998a), *traG'*::*lacZ*/*traR* (Piper et al., 1993), *rhlA'*::*lacZ* (Pesci et al., 1997), and *lasB'*::*lacZ* (Gray et al., 1994) fusions have been described. *Escherichia coli* biosensors containing a single chromosomal copy of *lasB'*::*lacZ* and *lasI'*::*lacZ* fusions (Seed et al., 1995) have been constructed using lysogens of recombinant λ bacteriophages. The strength of activation for the given AHL is dependent upon the LuxR homolog. Figures 1 and 2 highlight some of the recombinant biosensors and their utilization.

From T-streaks or assays of conditioned media, AHL biosensors greatly facilitate the characterization of quorum-sensing signal molecule(s) produced by a given organism. Other compounds produced by the target organism may give false-negative results in these assays because of bactericidal or bacteriostatic effects on the biosensor. The extraction of AHLs from spent culture medium using organic solvents (Eberhard et al., 1981; Bainton et al., 1992b; Throup et al., 1995a) can overcome this problem and also allow for the concentration of any AHL present. Where possible, the transformation of target organisms with biosensor plasmids can also circumvent the problem of antimicrobial activity (Swift et al., 1993, 1997; Eberl et al., 1996). Furthermore, the assay of reporter gene expression throughout growth in these transformed strains enables any cell density-dependent production to be determined. Broad-host-range vectors such as pSB403 (Winson et al., in press [A]; Fig. 2A), which is based upon pRK415 (Keen et al., 1988), are best suited to this type of study (Fig. 2C). The use of *luxCDABE* from *P. luminescens* as the reporter allows for the non-destructive measurement of promoter activity in a single

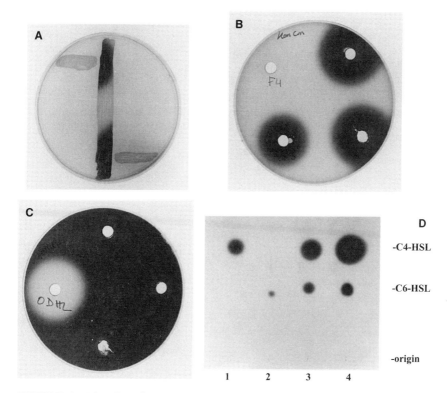

FIGURE 1 The *C. violaceum cviI* mutant, CV026, as an *N*-acylhomoserine lactone biosensor. Biosensor *C. violaceum* CV026 (McClean et al., 1997) produces the purple pigment violacein in the presence of exogenous AHLs. (A) T-streaks of *C. violaceum* CV026 (vertical) against *Aeromonas hydrophila* (horizontal, top) and *A. salmonicida* (horizontal, bottom). (B) Forward assay of high-pressure liquid chromatography (HPLC) fractions. Wells were cut into an L-broth agar plate seeded with CV026 and 5 μl of concentrated HPLC fractions from a dichloromethane supernatant extract of *Escherichia coli* JM109 (pAHH1) (Swift et al., 1997) added to each well. Pigmentation indicates those fractions containing activating AHL. (C) Reverse assay of HPLC fractions. Wells were cut into an L-broth agar plate seeded with CV026 and supplemented with 5 μg ml^{-1} 3-oxo-C6-HSL. Five microliters of concentrated HPLC fractions from a dichloromethane supernatant extract of *E. coli* JM109 (pAHH1) (Swift et al., 1997) was added to each well. Longer-chain (C8 and greater) AHLs antagonize the production of pigment, as can be seen with the 3-oxo-C10-HSL (marked ODHL) control; the fractions tested here are deemed AHL-negative. (D) The tentative identification of AHLs produced by *Aeromonas hydrophila ahyI* using thin-layer chromatography (TLC) overlays. Spots of pigmentation allow visualization of activating AHLs after TLC separation; here spots from both *A. hydrophila* (lane 4) and *E. coli* JM109 (pAHH1) (lane 3) supernatant extracts are shown corresponding to C4-HSL (lane 1) and C6-HSL (lane 2) controls. The origin is marked. From Swift et al. (1997).

sample throughout growth. This is now particularly useful given the availability of automated systems for the measurement of bioluminescence and optical density (Winson et al., 1998b).

AHL biosensors have also been used effectively to screen for recombinant clones of AHL synthase genes in *E. coli*. Genomic libraries prepared from organisms activating the biosensor can be introduced into an *E. coli*

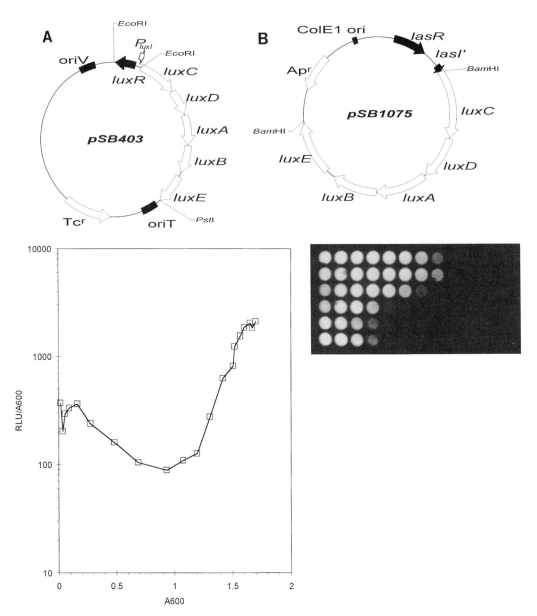

FIGURE 2 Promoter::*luxCDABE* fusions as *N*-acylhomoserine lactone biosensors. The plasmid pSB403 (IncP) (A) contains a *luxRI'* (*V. fischeri*)::*luxCDABE*(*Photobacterium fischeri*) fusion and has a broad host range (Winson et al., 1998a). pSB1075 (B) carries a *lasRI'* (*Pseudomonas aeruginosa*)::*luxCDABE*(*P. luminescens*) fusion and is responsive to long-chain AHLs (Winson et al., 1998a). (C) The use of pSB403. Cell density–dependent light emission by *Aeromonas hydrophila* (pSB403) is presented as relative light units (RLU) per unit A_{600} relative to increasing cell density (A_{600}) and indicates a cell density–dependent production of AHL. From Swift et al. (1997). (D) The relative sensitivity of *E. coli* JM109 (pSB1075) to *N*-(3-oxoacyl)-L-homoserine lactones. Decreasing acyl chain length (14, 12, 10, 8, 6, 4) is shown from top to bottom; 10-fold serial dilutions of AHLs starting at 4×10^{-4} M are shown from left to right. One hundred microliters of a 1:10 dilution in L-broth of an overnight culture of the biosensor was added to 100 μl of the AHL dilution, and bioluminescence was detected after 4 h using a Berthold LB980 (E. G. & G. Berthold U.K. Ltd., Milton Keynes, England) photon video camera.

strain containing an AHL reporter plasmid, and the resulting transformants can be screened for reporter activation (Swift et al., 1993). If the plate is incubated for a sufficient time interval and the transformant density is high, a ring of cross-fed colonies will be seen around the AHL synthase clone. Alternatively, patched libraries can be screened with biosensor overlays. The use of this strategy has identified *luxI* and genetically linked *luxR* homologs from *Aeromonas hydrophila* (Swift et al., 1997), *Aeromonas salmonicida* (Swift et al., 1997), *Agrobacterium tumefaciens* (Fuqua and Winans, 1994; Hwang et al., 1994), *C. violaceum* (Throup et al., 1995b; McClean et al., 1997), *Enterobacter agglomerans* (Swift et al., 1993), *Obesumbacterium proteus* (Prest, 1996), *Pseudomonas aeruginosa* (Latifi et al., 1995), *Serratia liquefaciens* (Eberl et al., 1996), *Vibrio anguillarum* (Milton et al., 1996), *Yersinia enterocolitica* (Throup et al., 1995a), *Yersinia pseudotuberculosis* (Atkinson et al. 1997), *Yersinia pestis* (Isherwood et al., 1997), and *Yersinia ruckeri* (Atkinson et al., submitted). A summary of the known AHL quorum-sensing organisms is given in Table 1, and LuxRI homolog names, GenBank accession numbers, cognate signal molecules, and any phenotypic characters controlled are given in Table 2. As yet, analysis of protein primary structure alignments has failed to identify motifs that could be assigned to the biosynthesis of a given AHL in LuxI homologs, or the response to a given AHL in LuxR homologs.

Given the profound sequence divergence of these homologs at both the nucleic acid and protein levels, any putative evolutionary relationship between homologs is not at all apparent and is further complicated by the presence of an entirely independent route to AHL synthesis identified in at least two marine bacteria. The *luxMN* (Bassler et al., 1993) and *ainS* (Gilson et al., 1995) loci provide an as yet uncharacterized AHL synthase capacity in *V. harveyi* and *V. fischeri*, respectively.

IDENTIFICATION OF AHLs

The signal AHL can be purified by fractionating concentrated supernatant extracts using high-pressure liquid chromatography (HPLC) (Bainton et al., 1992a; Camara et al., 1998) or thin-layer chromatography (TLC) (Shaw et al., 1997; Swift et al., 1997). Separation of molecules extracted from spent culture supernatant with organic solvents (e.g., dichloromethane, ethyl acetate) is made on the basis of differences in mass and polarity (Williams and Fleming, 1987; Kemp, 1991; Camara et al., 1998).

HPLC is an effective method for the fractionation and preparation of AHLs for structural analysis. Typically, C_8 reverse-phase columns are employed with sample elution using either gradient or isocratic mobile phases (e.g., acetonitrile-water). Biosensors can be used to identify active HPLC fractions (Fig. 1B and C), which can then be subjected to mass spectrometry (MS) and nuclear magnetic resonance spectroscopy. The structure of the predicted molecule can then be confirmed by chemical synthesis (Bainton et al., 1992b; Chhabra et al., 1993).

TLC overlay procedures (Shaw et al., 1997; Swift et al., 1997; McClean et al., 1997) (Fig. 1D) can be a valuable prelude to the detailed chemical characterization described above. Fractionation and detection in 24 h is possible, with spots of pigmentation or bioluminescence corresponding to the position of the AHL. These spots can be imaged and compared with known standards on the basis of R_f and overlaying biosensor activation. For preparative TLC, samples can be recovered from the silica matrix with acetone, enabling further chemical analysis.

High-resolution MS techniques involve the ionization (either by a beam of high-energy electrons in electron-impact MS or by fast Xe or Ar atoms in fast-atom-bombardment MS) and fragmentation of the sample. In some cases, despite extensive subfractionation, bioactive fractions are contaminated and the spectra obtained are difficult to interpret. By using tandem MS, peaks of the expected mass-to-charge ratio (m/z) can be selected such that the product ion spectrum can be unequivocally assigned by comparison with a synthetic

TABLE 1 A summary of organisms screened for *N*-acylhomoserine lactone production

Subdivision	AHL-positive species[a]			Species currently reported AHL-negative[b]		
	Species	Screening method[c]	Reference[d]	Species	Screening method[c]	Reference[d]
α	*Agrobacterium* spp.		9	*Brevundimonas diminuta*	C	3
	Rhizobium spp.		4, 9			
	Rhodobacter sphaeroides		9			
β	*Burkholderia cepacia*	C	3, 5	*Neisseria meningitidis*	A	3
	Chromobacterium violaceum		9			
	Nitrosomonas europaea		9			
	Ralstonia solanacearum		9			
γ	*Aeromonas* spp.		9	*Actinobacillus pleuropneumoniae*	A	3
	Citrobacter spp.	A	1	*Escherichia coli*	C	3, 9
	Enterobacter spp.		9	*Enterobacter cloacae*	C	3
	Erwinia spp.		9	*Haemophilus influenzae*	C	3
	Hafnia spp.	A	8	*Klebsiella pneumoniae*	C	3
	Obesumbacterium spp.		7	*Proteus vulgaris*	C	3
	Pseudomonas aeruginosa		9	*Providencia stuartii*	B	3
	Pseudomonas aureofaciens		9	*Salmonella* spp.	C	3
	Pseudomonas fluorescens		9	*Serratia marcescens*	A	1, 9
	Pseudomonas syringae		9	*Pseudomonas alcaligenes*	C	3
	Pseudomonas putida	A	3	*Pseudomonas mendocina*	C	3
	Rahnella aquatilis	A	8	*Pseudomonas stutzeri*	C	3
	Serratia spp.	A	1, 9	*Pseudomonas fragi*	C	3
	Proteus spp.	A	1	*Pseudomonas putida*	C	3
	Vibrio fischeri		9	*Pseudomonas chlororaphis*	C	3
	Vibrio anguillarum		9	*Vibrio cholerae*	C	3
	Vibrio harveyi		9	*Vibrio alginolyticus*	B	3
	Vibrio fluvialis	C	3	*Vibrio hollisiae*	B	3
	Vibrio logei		6	*Vibrio natrigens*	B	3
	Vibrio metschnikovii	C	3	*Vibrio parahaemolyticus*	B	3
	Xenorhabdus nematophilus		9	*Vibrio mimucus*	B	3
	Yersinia spp.		9	*Vibrio vulnificus*	B	3
				Vibrio proteolyticus	B	3
				Vibrio diazotrophicus	B	3
				Xanthomonas campestris	C	2
δ and ε				*Campylobacter coli*	A	3
				Campylobacter jejuni	A	3
				Helicobacter pylori	A	3

[a] Genera or species identified as AHL producers using screening procedures are listed.

[b] Individual species in which screening procedures have, to date, failed to identify signals that complement AHL biosensors.

[c] For organisms not listed in Table 2, A denotes those organisms screened using supernatant or T-streak assays with biosensors detecting short-chain AHLs, e.g., *E. coli* (pSB401) and *C. violaceum* CV026 (forward assay). B denotes those organisms screened using supernatant or T-streak assays with biosensors detecting both short-chain AHLs and long-chain AHLs, e.g., *E. coli* (pSB1075) and *C. violaceum* CV026 (reverse assay). C denotes those organisms screened using supernatant or T-streak assays and assays of dichloromethane-extracted culture supernatant with biosensors detecting short-chain and long-chain AHLs. The reader should note examples where the screening of different strains of a given species has suggested a differential production of AHLs, e.g., *Serratia marcescens* and *Pseudomonas putida*.

[d] References: 1, Bainton et al., 1992b; 2, Barber et al., 1997; 3, Camara et al., unpublished results; 4, Gray, 1997; 5. Greenberg, 1997; 6, Greenberg et al., 1979; 7, Prest, 1996; 8, Swift et al., 1993; 9, Table 2.

TABLE 2 A summary of N-acylhomoserine lactone quorum sensing in proteobacteria[a]

Bacterium	Phenotype	Quorum sensing apparatus[b]			GenBank accession number	Reference[c,d]
		LuxI homolog	LuxR homolog	Major AHL signal		
Aeromonas hydrophila	Extracellular protease, biofilm formation	AhyI	AhyR	C4-HSL	X89469	41, 42
Aeromonas salmonicida	Extracellular protease	AsaI	AsaR	C4-HSL	U65741	41
Agrobacterium tumefaciens[e]	Conjugation	TraI	TraR	3-oxo-C8-HSL	L17024, U43674, L22207	16, 21, 34, 48, 49
Chromobacterium violaceum	Antibiotics, violacein, exoenzymes, cyanide	CviI	CviR	C6-HSL	X74300	26, 43, 45
Enterobacter agglomerans	?	EagI	EagR	3-oxo-C6-HSL		7, 40
Erwinia carotovora subsp. carotovora[f]	Carbapenem antibiotic, exoenzymes	ExpI (CarI)	ExpR and CarR	3-oxo-C6-HSL	U17224, X72891, X74299, X80475	3, 7, 9, 23, 35, 40, 50
Erwinia chrysanthemi	?	ExpI (EchI)	ExpR (EchR)	?	X96440, U45854	
Erwinia stewartii	Exopolysaccharide	EsaI	EsaR	3-oxo-C6-HSL	L32183, L32184	5
Escherichia coli	Cell division	?	SdiA	?	X03691	39
Nitrosomonas europaea	Emergence from lag phase	?	?	3-oxo-C6-HSL		4
Obesumbacterium proteus	?	OprI	OprR	3-oxo-C6-HSL		36
Pseudomonas aeruginosa	Multiple exoenzymes, Xcp, RhlR, biofilm formation	LasI	LasR	3-oxo-C12-HSL	M59425, L04681	8, 10, 17, 24, 30, 31, 51
	Multiple exoenzymes, cyanide, RpoS, lectins, pyocyanin, rhamnolipid	RhlI (VsmI)	RhlR (VsmR)	C4-HSL	L08962, U11811, U15644	6, 24, 25, 28, 29, 46, 51
Pseudomonas aureofaciens	Phenazine antibiotic	PhzI	PhzR	C6-HSL	L32729, L33724	32, 33, 47
Pseudomonas fluorescens	Phenazine antibiotic	PhzI	PhzR	?	L48616	
Pseudomonas syringae pv. tabaci	?	PsyI	PsyR	?	U39802	
Ralstonia solanacearum	?	SolI	SolR	C8-HSL	AF021840	15
Rhizobium etli	Restriction of nodule number	RaiI	RaiR	?	U92712, U92713	38
Rhizobium leguminosarum	Nodulation, bacteriocin small	?	RhiR	3-hydroxy-7-cis-C14-HSL	M98835	19, 20, 52

Continued on following page

TABLE 2 (Continued)

Bacterium	Phenotype	Quorum sensing apparatus[b]				Reference[c,d]
		LuxI homolog	LuxR homolog	Major AHL signal	GenBank accession number	
Rhodobacter sphaeroides	Community escape	CerI	CerR	7-*cis*-C14-HSL	AF016298	37
Serratia liquefaciens	Swarming, protease	SwrI	SwrR	C4-HSL	U22823	13, 18
Vibrio anguillarum	?	VanI	VanR	3-oxo-C10-HSL	U69677	27
Vibrio (Photobacterium) fischeri	Bioluminescence	LuxI	LuxR	3-oxo-C6-HSL	M19039, M96844, M25752	12, 14, 53
Xenorhabdus nematophilus	Virulence, bacterial lipase	?	?	3-hydroxy-C4-HSL or an agonist		11
Yersinia enterocolitica	?	YenI	YenR	C6-HSL	X76082	44
Yersinia pestis	?	YpeI	YpeR	?		22
Yersinia pseudotuberculosis	?	YesI	YesR	3-oxo-C6-HSL		1, 2
Yersinia ruckeri	?	YtbI	YtbR	C8-HSL	AF079136	
		YukI	YukR	?	AF079135	2

[a] The reader should also note the LuxLMN system of *V. harveyi* (Bassler et al., 1993) and the AinSR system of *V. fischeri* (Gilson et al., 1995) in which gene regulation via AHLs is provided by proteins with no significant sequence homology to LuxRI.

[b] Alternative nomenclature for effectively the same LuxRI homologs is given in parentheses. Signal molecule abbreviations: C4-HSL, N-butanoylhomoserine lactone; 3-oxo-C8-HSL, N-(3-oxo)-octanoylhomoserine lactone; C6-HSL, N-hexanoylhomoserine lactone; 3-oxo-C6-HSL, N-(3-oxo)-hexanoylhomoserine lactone; 3-oxo-C12-HSL, N-(3-oxo)-dodecanoylhomoserine lactone; 3-hydroxy-7-*cis*-C14-HSL, N-(3-hydroxy)-7,8-*cis*-tetradecenoylhomoserine lactone; 7-*cis*-C14-HSL, N-7,8-*cis*-tetradecenoylhomoserine lactone; ODHL, N-(3-oxo)-decanoylhomoserine lactone; 3-hydroxy-C4-HSL, N-(3-hydroxy)-butanoylhomoserine lactone.

[c] Where no reference is given, data were obtained from the GenBank deposition.

[d] References: 1, Atkinson et al., 1997; 2, Atkinson et al., submitted; 3, Bainton et al., 1992b; 4, Batchelor et al., 1997. Note here that the AHL signal was demonstrated to be 3-oxo-C6-HSL or a closely related compound; 5, Beck von Bodman and Farrand, 1995; 6, Brint and Ohman, 1995; 7, Chan, 1995; 8, Chapon-Hervé et al., 1997; 9, Chatterjee et al., 1995; 10, Davies et al., 1998; 11, Dunphy et al., 1997; 12, Eberhard et al., 1981; 13, Eberl et al., 1996; 14, Engebrecht and Silverman, 1987; 15, Flavier et al., 1997; 16, Fuqua and Winans, 1994; 17, Gambello and Iglewski, 1991; 18, Givskov et al., 1997; 20, Gray et al., 1996; 21, Hwang et al., 1994; 22, Isherwood et al., 1994; 23, Jones et al., 1993; 24, Latifi et al., 1995; 25, Latifi et al., 1995; 26, McClean et al., 1997; 28, Oschner and Reiser, 1995; 29, Oschner et al., 1994; 30, Passador et al., 1993; 31, Pearson et al., 1995; 32, Pierson and Pierson, 1996; 33, Pierson et al., 1994; 34, Piper et al., 1993; 35, Pirhonen et al., 1993; 36, Prest, 1996; 37, Puskas et al., 1997; 38, Rosemeyer et al., 1998; 39, Sharma et al., 1986; 40, Swift et al., 1993; 41, Swift et al., 1997; 42, Swift et al., unpublished results; 43, Taylor, 1997; 44, Throup et al., 1995a; 45, Winson et al., 1994; 46, Winson et al., 1995; 47, Wood et al., 1997; 48, Zhang et al., 1993; 49, chapter 8 of this volume; 50, chapter 7; 51, chapter 10; 52, chapter 7; 53, chapters 14 and 15.

[e] TraRI homologs have been identified on octopine-type and nopaline-type Ti plasmids, as summarized in Swift et al. (1996) and chapter 8.

[f] LuxRI homologs have been described in three strains of *E. carotovora* subsp. *carotovora*: ExpRI in strain SCC3193, CarI/RexR and CarR in strain SCRI193, and GS101 and HsII have been identified in 71. Essentially, ExpI = CarI = HsII; ExpR = RexR; CarR = a separate LuxR homolog dedicated to carbapenem regulation. Additionally, ExpRI homologs (EcbRI) have been identified in *E. carotovora* subsp. *betavascularum* (GenBank AF001050).

Source: This table is expanded from Swift et al. (1996) with permission.

standard. Figure 3 shows the use of MS techniques to identify *N*-butanoylhomoserine lactone (C4-HSL) as the major signal molecule produced by *A. hydrophila* (Swift et al., 1997).

On their own, MS techniques are unable to identify completely unknown compounds. Proton and ^{13}C nuclear magnetic resonance (Williams and Fleming, 1987; Kemp, 1991; Camara et al., 1998) provides key additional information on the structure of an unknown molecule. The quality of the spectrum obtained, and thus its interpretation, will be dependent upon the quantity and purity of the compound available. Bainton et al. (1992b) exemplify the approach to characterize fully the chemistry and the chirality of an AHL from a bacterial culture supernatant.

A second group of bioactive small molecules able to activate *luxRI′::luxCDABE* biosensors has been identified. Cyclic dipeptides (diketopiperazines, DKPs) have a detection limit of about 0.2 mM (cf. 1 nM for 3-oxo-C6-HSL) and have been detected in spent media from *P. aeruginosa*, *Pseudomonas alcaligenes*, *Pseudomonas fluorescens*, *Citrobacter freundii*, *E. agglomerans,* and *Proteus mirabilis* (Chhabra et al., submitted). In addition, in competition studies, some DKPs were found to antagonize the 3-oxo-C6-HSL/LuxR-mediated induction of bioluminescence, suggesting that these cyclic dipeptides are probably competing for the same binding site on the LuxR protein target. Interestingly, DKPs of the same or similar structure to those isolated from bacterial culture supernatants have been attributed to have host-specific phytotoxicity when produced by the fungi *Alternaria alternata* (Stierle et al., 1988; Bobylev et al., 1996) and to have pharmacological effects in the central nervous system of higher mammals (Prasad, 1995). Although the physiological role of the bacterial DKPs has yet to be established, they may represent a novel class of diffusible signal molecules.

Elucidation of the phenomenon we now know as quorum sensing has come from studying, first, the biology of bioluminescence (Nealson, 1977; Hastings and Nealson, 1977; Eberhard et al., 1981) in *V. fischeri*, and subsequently the production of carbapenem (Bainton et al., 1992b) and exoenzymes (Jones et al., 1993; Pirhonen et al., 1993) by *E. carotovora*, conjugation (Zhang et al., 1993; Piper et al., 1993; Fuqua and Winans, 1994) in *A. tumefaciens,* and production of elastase (Jones et al., 1993; Passador et al., 1993) by *P. aeruginosa*. Thus, the study of the molecular genetics, biochemistry, and chemistry of AHLs has been characterized by the ability to compare and contrast a mechanistically conserved regulatory system across diverse bacterial genera.

COORDINATION OF VIRULENCE

The regulation of virulence determinants by pathogenic bacteria throughout the infection and transmission cycle is an important consideration for the study of quorum sensing. A major objective of an invading bacterium is the evasion of host defenses. Therefore, the premature expression of bacterial toxins might be a poor strategy for pathogens, because it could alert the host and elicit defensive responses. Where the bacterium is able to evade host defenses and find a suitable niche, it can then proliferate to a level where the coordinated production of virulence determinants may overwhelm the host. In this respect, quorum sensing can be used for a concerted activation of a regulon coding for the components of an aggressive phenotype only when sufficient bacteria exist to make the phenotype effective. Furthermore, a positive feedback loop, when the AHL synthase is part of the induced regulon, allows for rapid signal amplification (Engebrecht et al. 1983; Hwang et al. 1994; Chan et al., 1995; Seed et al., 1995; Winson et al., 1995; Swift et al., 1997) and thus an accelerated induction of the aggressive phenotype to full potential. In addition to intercellular signaling to ensure the presence of effective numbers, it is now apparent that a range of other regulatory mechanisms interface with quorum sensing to ensure that a suitable niche is occupied before the individual bacterium activates its virulence determinants.

One of the clearest examples from which to appreciate the biological advantages of employing such a quorum-sensing mechanism is found in the phytopathogen *E. carotovora* (see also chapter 7 of this volume). Here, the apparent objective for successful pathogenesis is to release nutrients for bacterial growth from the target plant via the production of tissue-degradative enzymes (pectate lyases, polygalacturonase, cellulase, and a protease). A single cell of *E. carotovora* could never produce sufficient enzymes to have a biological effect on the plant. Indeed, the plant uses the production of small amounts of macromolecular degradation products as a signal to elicit the

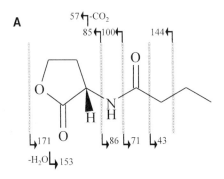

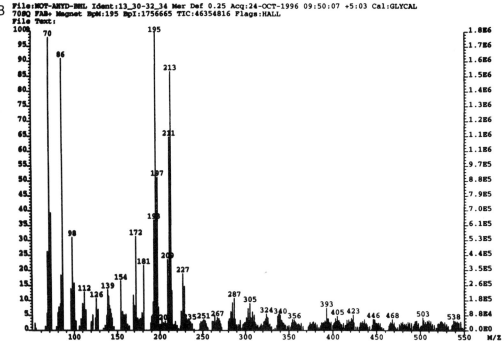

FIGURE 3 Identification of AHLs using mass spectrometry (MS). (A) The predicted daughter ion fragmentation species for C4-HSL, with the fragments giving prominent peaks in fast-atom-bombardment MS, which may gain H$^+$ or 2H$^+$. (B) The MS spectrum for C4-HSL. The peak at 172 is indicative of C4-HSL, and the tandem MS spectrum of the peak at 172 (C) identifies the molecule as C4-HSL with comparison to a synthetic standard (D). Panels B through D are from Swift et al. (1997).

C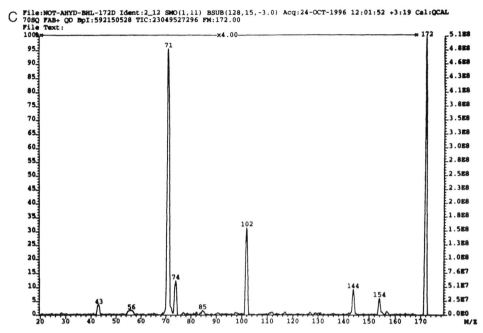

D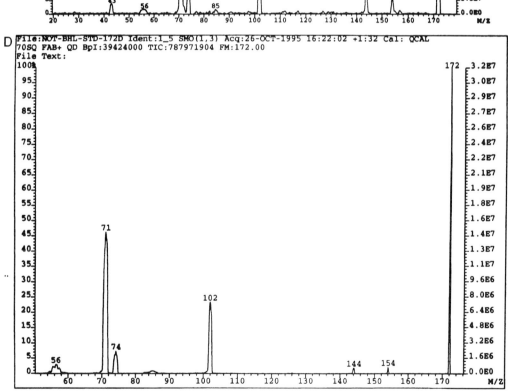

FIGURE 3   (*Continued*)

phytodefense mechanisms that antagonize the activity of the bacterial pathogen (Palva et al., 1993). The only way to successfully attack plant tissue is to coordinate the production of the degradative enzymes such that it is only when the combined production from a biologically relevant number of bacteria is available that the plant defenses are overcome and tissue dissolution is achieved. The bacterium therefore requires a mechanism with which to determine the biologically critical number of cells (a quorum). The cell-cell communication system provided via the AHL–mediated quorum-sensing mechanism seems perfect for the purpose, and indeed this is precisely the mechanism used by this phytopathogen (Jones et al., 1993; Pirhonen et al., 1993). The biological subtlety of this control system has a further level of sophistication. The sudden release from the plant of a nutrient soup provides a potential opportunity for the growth of competing microflora. We speculate that *E. carotovora* takes care of this potential disadvantage by coordinating the release of a broad-spectrum antibiotic (carbapenem) along with the degradative enzymes. This appears to be a clear case of cellular coordination to provide the nutrient source and protect the ecological niche from competitors, although as yet firm experimental evidence is lacking. The package is apparently evolutionarily successful, as the formula appears to be repeated by *P. aeruginosa* and *C. violaceum*, which both coordinate antibiotic and cyanide production (cyanide is a potent biocide; Flaishman et al., 1996; Schippers et al., 1990) with exoenzymes via quorum sensing (Winson et al., 1994; Latifi et al., 1995; Taylor, 1997).

Carbapenem synthesis is regulated via the dedicated LuxR homolog CarR and the cognate signal molecule 3-oxo-C6-HSL (Bainton et al., 1992b; McGowan et al., 1995). In vitro it is possible to advance the expression of carbapenem production simply by the addition of exogenous 3-oxo-C6-HSL (Williams et al., 1992; Chan et al., 1995), a phenomenon also observed with bioluminescence in *V. fischeri* (Nealson, 1977). However, exogenous AHL does not advance the production of pectate lyase, polygalacturonase, cellulase, or protease, indicating that these systems, which have their own LuxR homolog ExpR (RexR) (Jones et al., 1993; Pirhonen et al., 1993), must be subject to an additional and higher level of regulation than quorum sensing alone. Clearly, a high population density is not the sole requirement for the expression of a virulent phenotype by *E. carotovora*; a number of factors need to be satisfied, for example, the presence of factors from a suitable host (Liu et al., 1993; Flego et al., 1997) before exoenzymes are produced.

It now appears that the inability to advance quorum sensing-regulated phenotypes by the simple addition of the cognate AHL is more the rule than the exception when studying the induction of virulence determinants. For example, in the human pathogen *P. aeruginosa*, the expression of many virulence factors is controlled through a hierarchy of quorum-sensing systems (LasRI, which uses *N*-3-oxododecanoylhomoserine lactone [3-oxo-C12-HSL], and RhlRI, which uses C4-HSL) in coordination with other regulatory mechanisms (see chapter 10 of this volume; Latifi et al., 1996; Pearson et al., 1997). As with *E. carotovora*, it is not possible to simply advance the expression of elastase (Jones et al., 1993) or other virulence factors such as lectins PA-I and PA-II (Falconer et al., 1997) by adding exogenous AHLs.

In *P. aeruginosa*, quorum sensing exerts a further level of control over virulence through the control of exoproduct secretion. The expression of the components of the general secretory pathway (XcpPQ and XcpR-Z), responsible for the export of the majority of exoenzymes from the *P. aeruginosa* cell (Tommassen et al., 1992), is subject to regulation by quorum sensing as part of the regulons controlled by LasRI/3-oxo-C12-HSL and RhlRI/C4-HSL (Chapon-Hervé et al., 1997). This feature has not been observed in *Aeromonas* spp. (Swift et al., 1997).

Although the addition of AHLs cannot advance the expression of virulence determinants in *E. carotovora* or *P. aeruginosa*, in the absence of the quorum sensing circuit(s) there appears to be no significant expression either in vitro or in vivo at any population density. Figure 4 shows the effect of a *carI* (*expI*) mutation on the ability of *E. carotovora* to macerate potato tubers. This absolute linkage between quorum sensing and virulence gene expression is seen in other proteobacteria.

S. *liquefaciens* MG1 was identified as a producer of a wide spectrum of extracellular enzymes, including two proteases, at least one chitinase, a lipase, a phospholipase, and a nuclease (Givskov et al., 1997). All of these exoenzymes show a growth phase-dependent expression with an induction occurring at the transition between exponential and stationary phase. Insertional inactivation of the chromosomal *luxI* homolog *swrI* significantly reduced extracellular protease activity (Eberl et al., 1996) with at least the 51- and 57-kDa proteases being under quorum-sensing control (Givskov et al., 1997).

An equivalent situation is observed in *A. hydrophila*, an important pathogen of fish (Fryer and Bartholomew, 1996), which also infects humans (Thornley et al., 1997) and produces a spectrum of extracellular enzymes similar to that of *S. liquefaciens*. Both serine and metalloprotease activities have been described for *A. hydrophila* and have been demonstrated to make an important contribution to the virulence of this bacterium (Leung and Stevenson, 1988; Thornley et al., 1997). In an *ahyI* chromosomal mutant, both activities are virtually abolished (Swift et al., unpublished). Interestingly, in both *Serratia* and *Aeromonas* spp., the other exoenzymes tested (Eberl et al., 1996; Givskov et al., 1997; Swift et al., unpublished results) are unaffected. Clearly, there is still much to learn about the integration of quorum sensing with other regulatory mechanisms.

Thus far, we have considered AHLs only in the context of transcriptional activation of bacterial genes. Although it is clear that AHLs regulate a number of virulence determinants in vitro, a direct demonstration of their presence in body tissues and activity in vivo during infection has yet to be made. However, mutants with defects in quorum sensing do exhibit reduced virulence (Jones et al., 1993; Pirhonen et al., 1993; Tang et al., 1996). In addition, it is conceivable that host defenses may be capable of sensing AHLs and so be alerted to the presence of an invader. For ex-

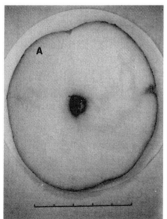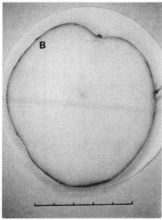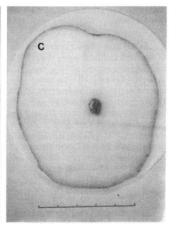

FIGURE 4   The effect of 3-oxo-C6-HSL on the virulence of a *carI* (*expI*) mutant of *E. carotovora*. Potato tubers were injected with 10 μl of stationary-phase broth cultures of *E. carotovora* subsp. *carotovora* SCRI193 (A), an *expI* mutant (B), and the same *expI* mutant coinjected with 10 μg ml^{-1} 3-oxo-C6-HSL (C) and incubated at 22°C for 7 days. The bar represents 5 cm. Reprinted from Jones et al. (1993) with permission.

ample, Telford et al. (1998) have demonstrated that 3-oxo-C12-HSL has potent immunomodulatory activity and is capable of influencing both macrophage and T-helper-cell responses by suppressing the lipopolysaccharide-stimulated release of interleukin-12, interferon-γ and tumor necrosis factor alpha. Since T-cell responses are an important component of the host defense against *P. aeruginosa* in experimental animal models of infection (Morissette et al., 1996), 3-oxo-C12-HSL may function as a virulence determinant per se by modulating host inflammatory responses to a degree that promotes the survival and growth of the pathogen. Such observations may also provide an explanation for the difficulties encountered in identifying quorum sensing-dependent phenotypes in bacteria such as *Yersinia* spp. (discussed below).

STATIONARY PHASE

Stationary-phase populations are substantially more resistant to environmental stresses, including free radicals, heat, cold, starvation, high osmolarity, and acid, than are exponentially growing cells. The mechanisms responsible for this increased resistance are often encoded by genes recognized by an alternative sigma factor (Loewen and Hengge-Aronis, 1994; DeMaio et al., 1996; Voelker et al., 1997). In *P. aeruginosa*, RpoS is under the quorum-sensing control through RhlRI and C4-HSL (Latifi et al., 1996) and fulfills a role for resistance to heat stress, acid stress, oxidative stress, ethanol stress, and osmotic stress (Jørgensen et al., unpublished results).

In *P. aeruginosa* we propose that those genes that are under RpoS regulation will be subject to a hierarchical quorum-sensing control. Disruption of *rhlRI* would, therefore, lead to the inactivation of all the genes within the RpoS regulon. We have already mentioned that many of the quorum sensing-regulated phenotypes in proteobacteria cannot be advanced by the simple addition of the cognate AHL. One explanation for this could be the secondary regulation of such systems by downstream response regulators such as RpoS. In *E. coli*,

RpoS is under both transcriptional and translational control, which restricts the availability of functional RpoS to the stationary phase irrespective of the activity of P_{rpoS} (Muffler et al., 1997). If this regulatory control were also in *P. aeruginosa*, the fact that the addition of C4-HSL could advance *rpoS* transcription would not necessarily result in RpoS availability. Genes hierarchically regulated via quorum sensing but directly regulated by RpoS could therefore be nonresponsive to the addition of C4-HSL. Furthermore, it is now becoming apparent that quorum sensing is part of a complex regulatory network, where additional environmental information is transduced through other pleiotropic regulators of gene expression, e.g., GacA/LemA (Reimmann et al., 1997), *alg* regulators (Latifi et al., 1997), and Vfr (Albus et al., 1997).

MULTICELLULARITY

In nature, bacteria are most commonly associated with surfaces, as opposed to the artificial liquid environment of the laboratory culture flask. The biofilm is a dynamic bacterial growth phenomenon, the maintenance of which was predicted to be dependent upon cell-cell communication (Williams and Stewart, 1994). It is exciting therefore to find that in situations of multicellular activity, bacterial communication via quorum sensing is apparently essential for controlling dynamic biofilm structures.

Nitrosomonas europaea is an ammonia-oxidizing, auxotrophic bacterium that is subject to temporal changes in nutrient supply in its natural environment. It is important for this organism to be able to remain viable during long periods of starvation and to be able to respond quickly to a new nutrient supply. The report that this bacterium exhibited a much-reduced lag in growth after prolonged nitrogen (ammonium) starvation as a biofilm when compared to cells growing in suspension (planktonic cells) suggested that cell-cell signaling may be involved. Studies with a *luxRI::luxAB* biosensor demonstrated that *N. europaea* produces 3-oxo-C6-HSL or a closely related compound (Batchelor et al., 1997). Furthermore, starved planktonic cells recovered substantially more quickly when supplied with ammonium and 3-oxo-C6-HSL, highlighting the important advantage of being at a high cell density, e.g., in a biofilm, for a rapid recovery from starvation to compete for intermittently available nutrients (Batchelor et al., 1997). This work also implicates quorum sensing in the control of primary metabolism fundamental to the growth and activity of bacteria in the natural environment.

Studies with *S. liquefaciens* have revealed that the multicellular activity of swarming motility, which allows surface translocation of specially differentiated bacteria, requires activation by quorum sensing (Eberl et al., 1996; Givskov et al., 1997). In liquid media and media containing low concentrations of agar (0.2 to 0.4%) cells exist as 1.5- to 3-μm motile rods carrying 3 to 15 flagellae per cell. The degree of flagellation is growth phase dependent and is controlled ultimately by the activity of the flagella master operon *flhDC*, which also regulates phospholipase expression and acts in response to nutritional and other environmental stresses. Cells transferred to rich media (e.g., L-broth) solidified with 0.4 to 1.2% agar undergo a significant morphological change. Cells at the periphery of the colony differentiate into long (5- to 50-μm), multinucleated, and hyperflagellated cells, which move rapidly (approximately 6 mm/hour; Givskov et al., 1996) outward to colonize the entire surface of the plate in a short period of time. Swarming motility is not seen on a minimal medium solidified with the same amount of agar unless a mixture of amino acids is included. Unlike swarming in *Proteus mirabilis*, in which a single amino acid (glutamine) is able to initiate swarming (Allison et al., 1993), no single amino acid alone is able to promote swarming motility (Eberl et al., 1996). *S. liquefaciens* cells with a mutation in the *luxI* homolog *swrI* are unable to swarm on a minimal agar supplemented with casamino acids (Eberl et al., 1996). The addition of spent supernatant or the cognate signal molecule C4-

HSL restores swarming motility. On a rich medium the *swrI* null mutant does swarm, but the initiation of swarming is severely compromised. The control of swarming motility in *S. liquefaciens* thus requires the integration of different regulatory mechanisms depending upon nutritional status, environmental stress, surface viscosity and cell density to elicit the correct morphological differentiation to allow swarming (Givskov et al., 1997).

Two roles can be identified for quorum sensing in the control of swarming motility. First, the requirement for an active *swrI* to give full activation of protease activity may be a feature in the production of amino acids for the activation of swarming motility. Second, the production of a wetting agent(s) by *Serratia* sp. appears to be dependent upon the presence of an active *swrI* gene (Givskov et al., 1997). Wetting agents reduce the surface tension of the medium and may correlate with a slime-like layer observed preceding the migration front in swarming motility (Eberl et al., 1996). Interestingly, the biosynthetic enzymes for the rhamnolipid surfactant produced by *P. aeruginosa* are part of the quorum-sensing regulon of this organism (Oschner et al., 1994; Oschner and Reiser, 1995).

QUORUM-SENSING CIRCUITS WITH NO KNOWN TARGET GENES

As indicated earlier, the use of biosensor complementation has facilitated the cloning of *luxI* homologs from a number of proteobacteria. In *E. agglomerans*, *V. anguillarum*, *Y. enterocolitica*, and *Y. pseudotuberculosis*, marker exchange has been used to make chromosomal mutations of *luxI* homologs, but screening for candidate phenotypes has so far failed to identify any regulated traits. In *Y. enterocolitica* two-dimensional polyacrylamide gel electrophoresis identified some differences between the *yenI* mutant and its parent strain (Throup et al., 1995a); however, none of the relevant proteins have yet been identified. Furthermore, as the *yenI* mutant strain activated bioluminescence when transformed with a *luxRI::luxCDABE* reporter in a density-dependent manner, there is probably a second *luxI* homolog in *Yersinia* spp. (Throup et al., 1995a). Investigation of *Y. pseudotuberculosis ypsI* (formerly *yepI*) negative strains has shown a similar phenomenon, and evidence for multiple or hierarchical quorum-sensing circuits in this organism has been provided by the cloning of a second locus, *ytbI* (formerly *yesI*) (Atkinson et al., 1997, submitted). Additional layers of complexity are also evident within *V. anguillarum*, in which the *vanI* negative, *vanR* negative, and *vanRI* negative mutants all produce AHLs capable of activating the *C. violaceum cviI* negative mutant, strongly suggesting the presence of multiple quorum-sensing regulons in this organism (Milton et al., 1997). A role for the product of VanI (*N*-3-oxodecanoyl-homoserine lactone; 3-oxo-C10-HSL) from *V. anguillarum* can be proposed as an antagonist of quorum-sensing signals and will be discussed in the next section.

In contrast to the *luxI* homolog *yenI*, which can be insertionally inactivated, it appears that the *luxR* homolog *yenR* is an essential gene in *Y. enterocolitica*. Marker exchange mutagenesis of chromosomal *yenR* proved possible only when a plasmid copy was supplied in *trans*. Subsequent curing of the *yenR* plasmid was then lethal, although the plasmid was easily cured from the parent *Y. enterocolitica* strain (Throup et al., unpublished results). The complexity and apparently essential nature of quorum sensing in *Yersinia* spp. suggests it plays a key role and is not a redundant feature inherited by vertical gene transfer. It may be that another *luxI* homolog must be inactivated before phenotypic control will be visible. Alternatively, production of the AHL itself may be at least part of the phenotype. As has been demonstrated with *P. aeruginosa*, 3-oxo-C12-HSL has an effect upon the immune system that impairs the host's response to bacterial infection. AHLs produced by *Yersinia* spp. may have similar effects, and it is worth noting that the type III secretion system (Cornelis and Wolf-Watz, 1997) of *Yersinia* spp. has evolved to deliver polypeptide virulence determinants directly into the eukaryotic cell. We speculate that this channel could also, for example, deliver AHLs or AHL synthases.

APPLICATIONS

The predictions made after discovering that 3-oxo-C6-HSL controls antibiotic and virulence gene expression in *E. carotovora* (Bainton et al., 1992a,b; Jones et al., 1993; Pirhonen et al., 1993; Swift et al., 1993) are beginning to be realized. The tremendous progress that has been made has generated a broad base of knowledge that is applicable to biotechnology (e.g., for the optimization of the antibiotic production) and to the food and other industries (e.g., for the rapid detection of contaminating organisms) and that has great therapeutic potential (e.g., for novel antibacterial targets and immunomodulatory agents).

QUORUM-SENSING BLOCKERS

A range of agonist and antagonist activities has been assigned to a variety of AHL analogs in empirical assays against LuxR, LasR, RhlR, and CarR activity (Eberhard et al., 1986; Chhabra et al., 1993; Winson et al., 1995; Passador et al., 1996; Schaefer et al., 1996; Winson et al., 1998a). These experiments showed that analogs closely related to the natural ligand for a given LuxR homolog were agonists and that as the structures diverged, AHLs with antagonistic activity could be identified. The discovery that longer-chain (C8 onward) AHLs inhibit the production of pigment by *C. violaceum* CV026 in the presence of an activating AHL (McClean et al., 1997) provides a useful assay of these molecules and has been used to show that the VanI protein of *V. anguillarum* synthesizes 3-oxo-C10-HSL (Milton et al., 1997).

This concept has been taken a stage further in studies with *Aeromonas* spp., in which it has been possible to antagonize the production of quorum sensing-controlled virulence factors by the addition of longer-chain AHLs (Swift et al., unpublished results). The C4-HSL-dependent production of exoprotease activity in a natural isolate of *A. hydrophila* and its *ahyI* negative mutant supplied with exogenous C4-HSL is significantly reduced in cultures treated with 3-oxo-C10-HSL, 3-oxo-C12-HSL, *N*-3-oxotetradecanoylhomoserine lactone (3-oxo-C14-HSL), *N*-3-decanoylhomoserine lactone (C10-HSL), or *N*-3-dodecanoyl-homoserine lactone (C12-HSL). A similar phenomenon has been observed for 3-oxo-C10-HSL with environmental isolates of *A. salmonicida* (Swift et al., 1997).

As *V. anguillarum*, a fish pathogen that inhabits the same ecological niche as *Aeromonas* spp., produces 3-oxo-C10-HSL, it is possible that this *Vibrio* sp. employs this AHL to outcompete *Aeromonas* sp. Indeed, this certainly seems to be the case for the macroalga *Delisea pulchra*, which produces furanones that specifically inhibit the quorum sensing-dependent swarming motility of *S. liquefaciens* (Givskov et al., 1996). No effects are seen on the other factors that could affect swarming, i.e., flagellar synthesis, cell elongation, or growth rate, and the effect is dependent upon the respective concentrations of C4-HSL and furanone in experiments employing an *swrI* negative strain. Although these experiments were performed in the artificial conditions of the laboratory, this seaweed is in competition with swarming bacteria (including other species of *Serratia*) in nature. The swarming motility and attachment of these bacteria to surfaces is strongly inhibited by furanones at concentrations in the range of 10 ng to 5 μg/cm^2 of surface. Interfering with quorum sensing by the production of furanones is, therefore, a defense against bacteria detrimental to the marine algae. Interestingly, the DKP molecules described earlier, which are capable of activating LuxR-based biosensors, are also able to act as competitive inhibitors of 3-oxo-C6-HSL (Chhabra et al., submitted) and may also conceivably function as quorum-sensing blockers.

CONCLUDING REMARKS

Quorum sensing controls a range of phenotypic traits in the proteobacteria in which AHLs function as the intercellular signals. The extent to which quorum sensing influences the modulation of the phenotype for a given organism depends upon the niche being oc-cupied, and the function of the AHL as a phenotype itself should not be overlooked. The signals used by other proteobacteria where (as yet) AHL or other signal molecules await discovery will be an important step forward, especially for important human pathogens like *E. coli*, *Salmonella* sp., *Campylobacter* sp., *Neisseria* sp., *Haemophilus* sp., and *Legionella* sp. The discovery of non–AHL-signaling molecules, for example, in *Xanthomonas* sp. (Barber et al., 1997; Chun et al., 1997), and the discovery of the DKPs (Chhabra et al., submitted) suggest that the identification of novel intercellular signals may not be too far away.

ACKNOWLEDGMENTS

We thank the Biotechnology and Biological Sciences Research Council, the Medical Research Council, the Wellcome Trust, and the European Union for funding and the assistance of M. Camara, S. R. Chhabra, L. Fish, M. T. G. Holden, and D. I. Pritchard in the preparation of this paper.

REFERENCES

Albus, A. M., E. C. Pesci, L. J. Runyen-Janecky, S. E. H. West, and B. H. Iglewski. 1997. Vfr controls quorum sensing in *Pseudomonas aeruginosa*. *J. Bacteriol.* **179:**3928–3935.

Allison, C., H. C. Lai, D. Gygi, and C. Hughes. 1993. Cell differentiation of *Proteus mirabilis* is initiated by glutamine, a specific chemoattractant for swarming cells. *Mol. Microbiol.* **8:**53–60.

Atkinson, S., J. P. Throup, P. Williams, and G. S. A. B. Stewart. 1997. Cloning and characterization of a quorum sensing locus in *Yersinia pseudotuberculosis*, abstr. B-398, p. 97. Abstr. 97th Gen. Meet. Am. Soc. Microbiol. 1997. American Society for Microbiology, Washington, D.C.

Atkinson, S., J. P. Throup, P. Williams, and G. S. A. B. Stewart. A hirarchical quorum sensing system in *Yersinia pseudotuberculosis* is involved in the regulation of motility and clumping. Submitted for publication.

Bainton, N. J., B. W. Bycroft, S. R. Chhabra, P. Stead, L. Gledhill, P. J. Hill, C. E. D. Rees, M. K. Winson, G. P. C. Salmond, G. S. A. B. Stewart, and P. Williams. 1992a. A general role for the *lux* autoinducer in bacterial cell signalling: control of antibiotic synthesis in Erwinia. *Gene* **116:**87–91.

Bainton, N. J., P. Stead, S. R. Chhabra, B. W. Bycroft, G. P. C. Salmond, G. S. A. B. Stewart, and P. Williams. 1992b. *N*-(3-Oxohexanoyl)-L-homoserine lactone regulated carbapenem antibiotic production in *Erwinia carotovora*. *Biochem. J.* **288**:997–1004.

Barber, C. E., J. L. Tang, J. X. Feng, M. Q. Pan, T. J. G. Wilson, H. Slater, J. M. Dow, P. Williams, and M. J. Daniels. 1997. A novel regulatory system required for pathogenicity *Xanthomonas campestris* is mediated by a small diffusible signal molecule. *Mol. Microbiol.* **24**:555–566.

Bassler, B. L., M. Wright, R. E. Showalter, and M. R. Silverman. 1993. Intercellular signalling in *Vibrio harveyi*: sequence and function of genes regulating expression of luminescence. *Mol. Microbiol.* **9**:773–786.

Batchelor, S. E., M. Cooper, S. R. Chhabra, L. A. Glover, G. S. A. B. Stewart, P. Williams, and J. I. Prosser. 1997. Cell-density regulated recovery of starved biofilm populations of ammonia oxidizing bacteria. *Appl. Environ. Microbiol.* **63**:2281–2286.

Beck von Bodman, S., and S. K. Farrand. 1995. Capsular polysaccharide biosynthesis and pathogenicity in *Erwinia stewartii* require induction by an *N*-acylhomoserine lactone autoinducer. *J. Bacteriol.* **177**:5000–5008.

Bobylev, M. M., L. I. Bobylev, and G. A. Strobel. 1996. Synthesis and bioactivity of analogs of maculosin, a host-specific phytotoxin produced by *Alternaria alternata* on spotted knapweed (*Centaurea maculosa*). *J. Agric. Food Chem.* **44**:3960–3964.

Brint, J. M., and D. E. Ohman. 1995. Synthesis of multiple exoproducts in *Pseudomonas aeruginosa* is under control of RhlR-RhlI, another set of regulators in strain PAO-1 with homology to the autoinducer-responsive LuxR-LuxI family. *J. Bacteriol.* **177**:7155–7163.

Camara, M., M. M. Daykin, and S. R. Chhabra. 1998. Detection, purification and synthesis of *N*-acylhomoserine lactone quorum sensing signal molecules. *Methods Microbiol.* **27**:319–330.

Camara, M., A. Hardman, M. T. G. Holden, S. Swift, M. K. Winson, P. J. Hill, R. J. Mole, J. P. Throup, B. W. Bycroft, P. Williams, and G. S. A. B. Stewart. Unpublished results.

Chan, P. F. 1995. Ph.D. thesis. University of Nottingham, Nottingham, United Kingdom.

Chan, P. F., N. J. Bainton, M. M. Daykin, M. K. Winson, S. R. Chhabra, G. S. A. B. Stewart, G. P. C. Salmond, B. W. Bycroft, and P. Williams. 1995. Small molecule mediated autoinduction of antibiotic biosynthesis in the plant pathogen *Erwinia carotovora*. *Biochem. Soc. Trans.* **23**:S127.

Chapon-Hervé, V., M. Akrim, A. Latifi, P. Williams, A. Lazdunski, and M. Bally. 1997. Regulation of *xcp* secretion pathway by multiple quorum sensing modulons in *Pseudomonas aeruginosa*. *Mol. Microbiol.* **24**:1169–1178.

Chatterjee, A., Y. Cui, Y. Liu, C. K. Dumenyo, and A. K. Chatterjee. 1995. Inactivation of *rsmA* leads to overproduction of extracellular pectinases, cellulases and proteases in *Erwinia carotovora* subsp. *carotovora* in the absence of the starvation cell density-signal, *N*-(3-oxohexanoyl)-L-homoserine lactone. *Appl. Environ. Microbiol.* **61**:1959–1967.

Chhabra, S. R., P. Stead, N. J. Bainton, G. P. C. Salmond, P. Williams, G. S. A. B. Stewart, and B. W. Bycroft. 1993. Autoregulation of carbapenem biosynthesis in *Erwinia carotovora* ATCC 39048 by analogues of *N*-(3-oxohexanoyl)-L-homoserine lactone. *J. Antibiot.* **46**:441–454.

Chhabra, S. R., M. T. G. Holden, P. Stead, N. J. Bainton, P. J. Hill, G. P. C. Salmond, G. S. A. B. Stewart, B. W. Bycroft, and P. Williams. Isolation and chemical characterization of cyclic dipeptides from *Pseudomonas aeruginosa* and other gram-negative bacteria: evidence for a family of diffusible prokaryotic signal molecules. Submitted for publication.

Chun, W., J. Cui, and A. Poplawsky. 1997. Purification, characterisation and biological role of a pheromone produced by *Xanthomonas campestris* pv. *campestris*. *Physiol. Mol. Plant Pathol.* **51**:1–14.

Cornelis, G. R., and H. Wolf-Watz. 1997. The *Yersinia* Yop virulon: a bacterial system for subverting eukaryotic cells. *Mol. Microbiol.* **23**:861–867.

Davies, D. G., M. R. Parsek, J. P. Pearson, B. H. Iglewski, J. W. Costerton, and E. P. Greenberg. 1998. The involvement of cell-to-cell signals in the development of a bacterial biofilm. *Science* **280**:295–298.

DeMaio, J., Y. Zhang, C. Ko, D. B. Young, and W. R. Bishai. 1996. A stationary-phase stress-response sigma factor from *Mycobacterium tuberculosis*. *Proc. Natl. Acad. Sci. USA* **93**:2790–2794.

Dunphy, G., C. Miyamoto, and E. Meighen. 1997. A homoserine lactone autoinducer regulates virulence of an insect-pathogenic bacterium, *Xenorhabdus nematophilus* (Enterobacteriaceae). *J. Bacteriol.* **179**:5288–5291.

Eberhard, A., A. L. Burlingame, G. L. Kenyon, K. H. Nealson, and N. J. Oppenheimer. 1981. Structural identification of autoinducer of *Photobacterium fischeri* luciferase. *Biochemistry* **20**:2444–2449.

Eberhard, A., C. A. Widrig, P. McBath, and J. B. Schineller. 1986. Analogs of the autoinducer of bioluminescence in *Vibrio fischeri*. *Arch. Microbiol.* **146**:35–40.

Eberl, L., M. K. Winson, C. Sternberg, G. S. A. B. Stewart, G. Christiansen, S. R. Chhabra, B. W. Bycroft, P. Williams, S. Molin, and M. Giskov. 1996. Involvement of N-acyl-L-homoserine lactone autoinducers in controlling the multicellular behaviour of *Serratia liquefaciens. Mol. Microbiol.* **20:**127–136.

Engebrecht, J., and M. Silverman. 1987. Nucleotide sequence of the regulatory locus controlling expression of bacterial genes for bioluminescence. *Nucleic Acids Res.* **15:**10455–10467.

Engebrecht, J., K. Nealson, and M. Silverman. 1983. Bacterial bioluminescence: isolation and genetic analysis of functions from *Vibrio fischeri. Cell* **32:**773–781.

Falconer, C., K. Winzer, N. C. Garber, N. Gilboa-Garber, M. Camara, A. Lazdunski, G. S. A. B. Stewart, and P. Williams. 1997. Lectin expression in *Pseudomonas aeruginosa* is regulated via quorum sensing, abstr. B-344, p. 88. Abstr. 97th Gen. Meet. Am. Soc. Microbiol. 1997. American Society for Microbiology, Washington, D.C.

Flaishman, M. A., Z. Eyal, A. Zilberstein, C. Voisard, and D. Haas. 1996. Suppression of *Septoria tritici* blotch and leaf rust of wheat by recombinant cyanide-producing strains of *Pseudomonas putida. Mol. Plant-Microbe Interact.* **9:**642–645.

Flavier, A. B., L. M. Ganova-Raeva, M. A. Schell, and T. P. Denny. 1997. Hierarchical autoinduction in *Ralstonia solanacearum*: control of acyl-homoserine lactone production by a novel autoregulatory system responsive to 3-hydroxypalmitic acid methyl ester. *J. Bacteriol.* **179:**7089–7097.

Flego, D., M. Pirhonen, H. Saarilahti, T. K. Palva, and E. T. Palva. 1997. Control of virulence gene expression by plant calcium in the phytopathogen *Erwinia carotovora. Mol. Microbiol.* **25:**831–838.

Fryer, J. L., and J. L. Bartholomew. 1996. Established and emerging infectious diseases of fish. *ASM News* **62:**592–594.

Fuqua, C., and S. C. Winans. 1994. A LuxR-LuxI type regulatory system activates *Agrobacterium* Ti plasmid conjugal transfer in the presence of a plant tumor metabolite. *J. Bacteriol.* **176:**2796–2806.

Fuqua, C., S. C. Winans, and E. P. Greenberg. 1996. Census and consensus in bacterial ecosystems: the LuxR-LuxI family of quorum-sensing regulators. *Annu. Rev. Microbiol.* **50:**727–751.

Fuqua, W. C., S. C. Winans, and E. P. Greenberg. 1994. Quorum sensing in bacteria—the LuxR-LuxI family of cell density-responsive

transcriptional regulators. *J. Bacteriol.* **176:**269–275.

Gambello, M. J., and B. H. Iglewski. 1991. Cloning and characterization of the *Pseudomonas aeruginosa lasR* gene, a transcriptional activator of elastase expression. *J Bacteriol.* **173:**3000–3009.

Gilson, L., A. Kuo, and P. V. Dunlap. 1995. AinS and a new family of autoinducer synthesis proteins. *J. Bacteriol.* **177:**6946–6951.

Givskov, M., R. de Nys, M. Manefield, L. Gram, R. Maximilien, L. Eberl, S. Molin, P. D. Steinberg, and S. Kjelleberg. 1996. Eukaryotic interference with homoserine lactone-mediated prokaryotic signaling. *J. Bacteriol.* **178:**6618–6622.

Givskov, M., L. Eberl, and S. Molin. 1997. Control of exoenzyme production, motility and cell differentiation in *Serratia liquefaciens. FEMS Microbiol. Lett.* **148:**115–122.

Gray, K. M. 1997. Intercellular communication and group behaviour in bacteria. *Trends Microbiol.* **5:**184–188.

Gray, K. M., L. Passador, B. H. Iglewski, and E. P. Greenberg. 1994. Interchangeability and specificity of components from the quorum-sensing regulatory systems of *Vibrio fischeri* and *Pseudomonas aeruginosa. J. Bacteriol.* **176:**3076–3080.

Gray, K. M., J. P. Pearson, J. A. Downie, B. E. A. Boboye, and E. P. Greenberg. 1996. Cell-to-cell signaling in the symbiotic nitrogen-fixing bacterium *Rhizobium leguminosarum*: autoinduction of a stationary phase and rhizosphere-expressed genes. *J. Bacteriol.* **178:**372–376.

Greenberg, E. P. 1997. Quorum sensing in gram-negative bacteria. *ASM News* **63:**371–377.

Greenberg, E. P., J. W. Hastings, and S. Ulitzur. 1979. Induction of luciferase synthesis in *Beneckea harveyi* by other marine bacteria. *Arch. Microbiol.* **120:**87–91.

Hastings, J. W., and K. H. Nealson. 1977. Bacterial bioluminescence. *Annu. Rev. Microbiol.* **31:**549–595.

Hwang, I., L. Pei-Li, L. Zhang, K. R. Piper, D. M. Cook, M. E. Tate, and S. K. Farrand. 1994. TraI, a LuxI homologue, is responsible for production of conjugation factor, the Ti plasmid N-acylhomoserine lactone autoinducer. *Proc. Natl. Acad. Sci. USA* **91:**4639–4643.

Isherwood, K., S. Atkinson, J. P. Throup, R. W. Titball, and P. C. F. Oyston. 1997. Quorum sensing in *Yersinia pestis*: identification and characterization of the *ypsI/ypsR* locus, abstr. H-165, p. 312. Abstr. 97th Gen. Meet. Am. Soc. Microbiol. 1997. American Society for Microbiology, Washington, D.C.

Jones, S., B. Yu, N. J. Bainton, M. Birdsall, B. W. Bycroft, S. R. Chhabra, A. J. R. Cox, P. Golby, P. J. Reeves, S. Stephens, M. K. Winson, G. P. C. Salmond, G. S. A. B. Stewart, and P. Williams. 1993. The lux autoinducer regulates the production of exoenzyme virulence determinants in *Erwinia carotovora* and *Pseudomonas aeruginosa*. *EMBO J.* **12**:2477–2482.

Jørgensen, F., M. Bally, V. Chapon-Hervé, A. Lazdunski, P. Williams, and G. S. A. B. Stewart. Unpublished results.

Keen, N. T., S. Tamaki, D. Kobayashi, and D. Trollinger. 1988. Improved broad-host range plasmids for DNA cloning in Gram-negative bacteria. *Gene* **70**:191–197.

Kemp, W. 1991. *Organic Spectroscopy*, 3rd ed. Macmillan, Houndmills, United Kingdom.

Latifi, A., M. K. Winson, M. Foglino, B. W. Bycroft, G. S. A. B. Stewart, A. Lazdunski, and P. Williams. 1995. Multiple homologues of LuxR and LuxI control expression of virulence determinants and secondary metabolites through quorum sensing in *Pseudomonas aeruginosa* PA01. *Mol. Microbiol.* **17**:333–343.

Latifi, A., M. Foglino, T. Tanaka, P. Williams, and A. Lazdunski. 1996. A hierarchical quorum sensing cascade in *Pseudomonas aeruginosa* links the transcriptional activators LasR and VsmR to expression of the stationary phase sigma factor RpoS. *Mol. Microbiol.* **21**:1137–1146.

Latifi, A., M. Foglino, and A. Lazdunski. 1997. The chain of command in *Pseudomonas* quorum sensing: response. *Trends Microbiol.* **5**:134–135.

Leung, K. Y., and R. M. W. Stevenson. 1988. Tn5-induced protease-deficient strains of *Aeromonas hydrophila* with reduced virulence for fish. *Infect. Immun.* **56**:2639–2644.

Liu, Y., H. Murata, A. Chatterjee, and A. K. Chatterjee. 1993. Characterization of a novel regulatory gene aepA that controls extracellular enzyme production in the phytopathogenic bacterium *Erwinia carotovora* subsp. *carotovora*. *Mol. Plant-Microbe Interact.* **6**:299–308.

Loewen, P. C., and R. Hengge-Aronis. 1994. The role of the sigma factor σ^s (KatF) in bacterial global regulation. *Annu. Rev. Microbiol.* **48**:53–80.

McClean, K. H., M. K. Winson, L. Fish, A. Taylor, S. R. Chhabra, M. Camara, M. Daykin, J. H. Lamb, S. Swift, B. W. Bycroft, G. S. A. B. Stewart, and P. Williams. 1997. Quorum sensing and *Chromobacterium violaceum*: exploitation and violacein production and inhibition for the detection of *N*-acylhomoserine lactones. *Microbiology* **143**:3703–3711.

McGowan, S, M. Sebaihia, S. Jones, B. Yu, N. Bainton, P. F. Chan, B. W. Bycroft, G. S. A. B. Stewart, P. Williams, and G. P. C. Salmond. 1995. Carbapenem antibiotic production in *Erwinia carotovora* is regulated by CarR, a homologue of the LuxR transcriptional activator. *Microbiology* **141**:541–550.

Milton, D. L., A. Hardman, M. Camara, S. R. Chhabra, B. W. Bycroft, G. S. A. B. Stewart, and P. Williams. 1997. Quorum sensing in *Vibrio anguillarum*: characterization of the vanI/R locus and identification of the autoinducer *N*-(3-oxododecanoyl)-L-homoserine lactone. *J. Bacteriol.* **179**:3004–3012.

Morissette C., C. Francoeur, C. Darmond-Zwaig, and F. Gervais. 1996. Lung phagocyte bactericidal function in strains of mice resistant and susceptible to *Pseudomonas aeruginosa*. *Infect. Immun.* **64**:4984–4992.

Muffler, A., M. Barth, C. Marschall, and R. Hengge-Aronis. 1997. Heat shock regulation of σ^s turnover: a role for DnaK and relationship between stress responses mediated by σ^s and σ^{32} in *Escherichia coli*. *J. Bacteriol.* **179**:445–452.

Nealson, K. H. 1977. Autoinduction of bacterial luciferase: occurrence, mechanism and significance. *Arch. Microbiol.* **112**:73–79.

Oschner, U. A., and J. Reiser. 1995. Autoinducer-mediated regulation of rhamnolipid biosurfactant synthesis in *Pseudomonas aeruginosa*. *Proc. Natl. Acad. Sci. USA* **92**:6424–6428.

Oschner, U. A., A. K. Koch, A. Fiechter, and J. Reiser. 1994. Isolation and characterization of a regulatory gene affecting rhamnolipid biosurfactant synthesis in *Pseudomonas aeruginosa*. *J. Bacteriol.* **176**:2044–2054.

Palva, T. K., K. O. Holstrom, P. Heino, and E. T. Palva. 1993. Induction of plant defense response by exoenzymes of *Erwinia carotovora* subsp. *carotovora*. *Mol. Plant-Microbe Interact.* **6**:190–196.

Passador, L., J. M. Cook, M. J. Gambello, L. Rust, and B. H. Iglewski. 1993. Expression of *Pseudomonas aeruginosa* virulence genes requires cell-to-cell communication. *Science* **260**:1127–1130.

Passador, L., K. D. Tucker, K. R. Guertin, M. P. Journet, A. S. Kende, and B. H. Iglewski. 1996. Functional analysis of the *Pseudomonas aeruginosa* autoinducer PAI. *J. Bacteriol.* **178**:5995–6000.

Pearson, J. P., L. Passador, B. H. Iglewski, and E. P. Greenberg. 1995. A second *N*-acylhomoserine lactone signal produced by *Pseudomonas aeruginosa*. *Proc. Natl. Acad. Sci. USA* **92**:1490–1494.

Pearson, J. P., E. C. Pesci, and B. H. Iglewski. 1997. Roles of *Pseudomonas aeruginosa las* and *rhl* quorum-sensing systems in control of elastase and rhamnolipid biosynthesis genes. *J. Bacteriol.* **179:** 5756–5767.

Pesci, E. C., J. P. Pearson, P. C. Seed, and B. I. Iglewski. 1997. Regulation of *las* and *rhl* quorum sensing in *Pseudomonas aeruginosa. J. Bacteriol.* **179:**3127–3132.

Pierson, L. S., III, and E. A. Pierson. 1996. Phenazine antibiotic production in *Pseudomonas aureofaciens*: role in rhizosphere ecology and pathogen suppression. *FEMS Microbiol. Lett.* **136:**101–108.

Pierson, L. S., III, V. D. Keppenne, and D. W. Wood. 1994. Phenazine antibiotic biosynthesis by *Pseudomonas aureofaciens* 30-84 is regulated by PhzR in response to cell density. *J. Bacteriol.* **176:** 3966–3974.

Piper, K., S. Beck von Bodman, and S. Farrand. 1993. Conjugation factor of *Agrobacterium tumefaciens* regulates Ti plasmid transfer by autoinduction. *Nature* **362:**448–450.

Pirhonen, M., D. Flego, R. Heikinheimo, and E. T. Palva. 1993. A small diffusible signal molecule is responsible for the global control of virulence and exoenzyme production by the plant pathogen *Erwinia carotovora. EMBO J.* **17:**2467–2476.

Prasad, C. 1995. Bioactive cyclic dipeptides. *Peptides* **16:**151–164.

Prest, A. 1996. Ph.D. thesis. University of Nottingham, Nottingham, United Kingdom.

Puskas, A., E. P. Greenberg, S. Kaplan, and A. L. Schaefer. 1997. A quorum-sensing system in the free-living photosynthetic bacterium *Rhodobacter sphaeroides. J. Bacteriol.* **179:**7530–7537.

Reimmann, C., M. Beyeler, A. Latifi, H. Winteler, M. Foglino, A. Lazdunski, and D. Haas. 1997. The global activator GacA of *Pseudomonas aeruginosa* PAO positively controls the production of the autoinducer *N*-butyryl-homoserine lactone and the formation of the virulence factors pyocyanin, cyanide, and lipase. *Mol. Microbiol.* **24:**309–319.

Rosemeyer, V., J. Michiels, C. Verreth, and J. Vanderleyden. 1998. *luxI-* and *luxR-*homologous genes of *Rhizobium etli* CNPAF512 contribute to synthesis of autoinducer molecules and nodulation of *Phaseolus vulgaris. J. Bacteriol.* **180:**815–821.

Salmond, G. P. C., B. W. Bycroft, G. S. A. B. Stewart, and P. Williams. 1995. The bacterial enigma: cracking the code of cell-cell communication. *Mol. Microbiol.* **16:**615–624.

Schaefer A. L., B. L. Hanzelka, A. Eberhard, and E. P. Greenberg. 1996. Quorum sensing in *Vibrio fischeri*: probing autoinducer-LuxR interactions with autoinducer analogs. *J. Bacteriol.* **178:** 2897–2901.

Schippers, B., A. W. Bakker, P. A. H. M. Bakker, and R. van Peer. 1990. Beneficial and deleterious effects of HCN-producing pseudomonads on rhizosphere interactions. *Plant Soil* **129:**75–83.

Seed, P. C., L. Passador, and B. H. Iglewshi. 1995. Activation of the *Pseudomonas aeruginosa lasI* gene by LasR and the *Pseudomonas* autoinducer PAI: an autoinduction regulatory hierarchy. *J. Bacteriol.* **177:**654–659.

Shapiro, J. A. 1988. Bacteria as multicellular organisms. *Sci. Am.* **256:**82–89.

Sharma, S., T. F. Stark, W. G. Beattie, and R. E. Moses. 1986. Multiple control elements for the *uvrC* gene unit of *Escherichia coli. Nucleic Acids Res.* **14:**2301–2318.

Shaw, P., G. Ping, S. L. Daly, C. Cha, J. E. Cronan, Jr., K. L. Rinehart, and S. K. Farrand. 1997. Detecting and characterizing acyl-homoserine lactone signal molecules by thin layer chromatography. *Proc. Natl. Acad. Sci. USA* **94:** 6036–6041.

Stephens, K. 1986. Pheromones among the procaryotes. *Crit. Rev. Microbiol.* **13:**309–334.

Stierle, A. C., J. H. Cardellina, and G. A. Strobel. 1988. Maculosin, a host-specific phytotoxin for spotted knapweed from *Alternaria alternata. Proc. Natl. Acad. Sci. USA* **85:**8008–8011.

Swift, S., M. K. Winson, P. F. Chan, N. J. Bainton, M. Birdsall, P. J. Reeves, C. E. D. Rees, S. R. Chhabra, P. J. Hill, J. P. Throup, B. W. Bycroft, G. P. C. Salmond, P. Williams, and G. S. A. B. Stewart. 1993. A novel strategy for the isolation of luxI homologues: evidence for the widespread distribution of a LuxR:LuxI superfamily in enteric bacteria. *Mol. Microbiol.* **10:**511–520.

Swift, S., N. J. Bainton, and M. K. Winson. 1994. Gram-negative bacterial communication by *N*-acyl homoserine lactones: a universal language? *Trends Microbiol.* **2:**193–198.

Swift, S., J. P. Throup, G. P. C. Salmond, P. Williams, and G. S. A. B. Stewart. 1996. Quorum sensing: a population-density component in the determination of bacterial phenotype. *Trends Biochem. Sci.* **21:**214–219.

Swift, S., A. V. Karlyshev, E. L. Durant, M. K. Winson, S. R. Chhabra, P. Williams, S. Mac-

intyre, and G. S. A. B. Stewart. 1997. Quorum sensing in *Aeromonas hydrophila* and *Aeromonas salmonicida*: identification of the LuxRI homologues AhyRI and AsaRI and their cognate signal molecules. *J. Bacteriol.* **179**:5271–5281.

Swift, S., L. Fish, M. J. Lynch, D. F. Kirke, P. Williams, and G. S. A. B. Stewart. Unpublished results.

Tang, H. B., E. DiMango, R. Bryan, M. Gambello, B. H. Iglewski, J. B. Goldberg, and A. Prince. 1996. Contribution of specific *Pseudomonas aeruginosa* virulence factors to pathogenesis of pneumonia in a neonatal mouse model of infection. *Infect. Immun.* **64**:37–43.

Taylor, A. 1997. Ph.D. thesis. University of Nottingham, Nottingham, United Kingdom.

Telford, G., D. Wheeler, P. Williams, P. T. Tomkins, P. Appleby, H. Sewell, G. S. A. B. Stewart, B. W. Bycroft, and D. I. Pritchard. 1998. The *Pseudomonas aeruginosa* quorum-sensing signal molecule, *N*-(3-oxododecanoyl)-L-homoserine lactone has immunomodulatory activity. *Infect. Immun.* **66**:36–42.

Thornley, J. P., J. G. Shaw, I. A. Gryllos, and A. Eley. 1997. Virulence properties of clinically significant *Aeromonas* species: evidence for pathogenicity. *Rev. Med. Microbiol.* **8**:61–72.

Throup, J. P., M. Camara, G. S. Briggs, M. K. Winson, S. R. Chhabra, B. W. Bycroft, P. Williams, and G. S. A. B. Stewart. 1995a. Characterisation of the *yenI/yenR* locus from *Yersinia enterocolitica* mediating the synthesis of two *N*-acyl-homoserine lactone signal molecules. *Mol. Microbiol.* **17**:345–356.

Throup, J., M. K. Winson, N. J. Bainton, B. W. Bycroft, P. Williams, and G. S. A. B. Stewart. 1995b. Signalling in bacteria beyond bioluminescence, p. 89–92. *In* A. Campbell, L. Kricka, and P. Stanley (ed.), *Bioluminescence: Fundamentals and Applied Aspects*. John Wiley & Sons, Chichester, United Kingdom.

Throup, J. P., P. Williams, and G. S. A. B. Stewart. Unpublished data.

Tommassen, J., A. Filloux, M. Bally, M. Murgier, and A. Lazdunski. 1992. Protein secretion in *Pseudomonas aeruginosa*. *FEMS Microbiol. Rev.* **103**:73–90.

Voelker, U., T. Q. Luo, N. Smirnova, and W. Haldenwang. 1997. Stress activation of *Bacillus subtilis* σ^B can occur in the absence of the σ^B negative regulator RsbX. *J. Bacteriol.* **179**:1980–1984.

Williams, D. H., and I. Fleming. 1987. Spectro-

scopic Methods in Organic Chemistry, 4th ed. McGraw-Hill, London.

Williams, P. 1994. Compromising bacterial communication skills. *J. Pharm. Pharmacol.* **46**:252–260.

Williams, P., and G. S. A. B. Stewart. 1994. Cell density dependent control of gene expression in bacteria—implications for biofilm development and control, p. 9–13. *In* W. W. Nichols, J. Wimpenny, D. Stickler, and H. Lappin-Scott (ed.), *Bacterial Biofilms and Their Control in Medicine and Industry*. Bioline, Cardiff, United Kingdom.

Williams, P., N. J. Bainton, S. Swift, M. K. Winson, S. R. Chhabra, G. S. A. B. Stewart, G. P. C. Salmond, and B. W. Bycroft. 1992. Small molecule mediated density dependent control of gene expression in prokaryotes: bioluminescence and the biosynthesis of carbapenem antibiotics. *FEMS Microbiol. Lett.* **100**:161–168.

Winson, M. K., N. J. Bainton, S. R. Chhabra, B. W. Bycroft, G. P. C. Salmond, P. Williams, and G. S. A. B. Stewart. 1994. Control of *N*-acylhomoserine lactones-regulated expression of multiple phenotypes in *Chromobacterium violaceum*, abstr. H-71, p. 212. Abstr. 94th Gen. Meet. Am. Soc. Microbiol. 1994. American Society for Microbiology, Washington, D.C.

Winson, M. K., M. Camara, A. Latifi, M. Foglino, S. R. Chhabra, M. Daykin, M. Bally, V. Chapon, G. P. C. Salmond, B. W. Bycroft, A. Lazdunski, G. S. A. B. Stewart, and P. Williams. 1995. Multiple *N*-acyl-L-homoserine lactone signal molecules regulate production of virulence determinants and secondary metabolites in *Pseudomonas aeruginosa*. *Proc. Natl. Acad. Sci. USA* **92**:9427–9431.

Winson, M. K., S. Swift, L. Fish, J. P. Throup, F. Jørgensen, S. R. Chhabra, B. W. Bycroft, P. Williams, and G. S. A. B. Stewart. 1998a. Construction and analysis of *luxCDABE*-based plasmid sensors for investigating *N*-acylhomoserine lactone-mediated quorum sensing. *FEMS Microbiol. Lett.* **163**:185–192.

Winson, M. K., S. Swift, P. J. Hill, C. M. Sims, G. Griesmayr, B. W. Bycroft, P. Williams, and G. S. A. B. Stewart. 1998b. Engineering the *luxCDABE* genes from *Photorhabdus luminescens* to provide a bioluminescent reporter for constitutive and promoter probe plasmids and mini-Tn5 constructs. *FEMS Microbiol. Lett.* **163**:193–202.

Wood, D. W., F. Gong, M. M. Daykin, P. Williams, and L. S. Pierson III. 1997. *N*-Acylhomoserine lactone-mediated regulation of phenazine gene expression by *Pseudomonas aureofaciens* 30-84 in the wheat rhizosphere. *J. Bacteriol.* **179:** 7663–7670.

Zhang, L., P. J. Murphy, A. Kerr, and M. E. Tate. 1993. *Agrobacterium* conjugation and gene regulation by *N*-acyl-L-homoserine lactones. *Nature* **362:**446–448.

EMERGING DENSITY-DEPENDENT CONTROL SYSTEMS IN GRAM-POSITIVE COCCI

Bettina A. B. Leonard and Andreas Podbielski

20

Traditionally, density signals were described in terms of determining bacterial quorum (reviewed by Gray [1997] and Kaiser [1996]). In these systems, the number of the same bacterial species occupying a niche are sensed by the production of small diffusible molecules by individual cells. These molecules accumulate and come back to the source when space is limited and cell numbers reach a critical mass. Three components are necessary to sense cell density: (i) a signal, (ii) a means of recognizing the signal, and (iii) accumulation of the signal (Fig. 1). In the classical model of quorum sensing, the signaling molecule is produced at a low-level constitutive rate by all cells present. If the cells are located in a limited space without high flowthrough, the signal will accumulate in proportion to cell number and will eventually reach a critical threshold. However, it should be kept in mind that, theoretically, signaling molecule density can also be increased by limiting the space around the cells rather than by increasing the cell num-

bers. Critical signal density also can be reached if the peptide signal is generated by protease action of a bacterial cell that is in intimate contact with its protease substrate. The local peptide concentration at the cleavage site could be magnitudes higher than that in the surroundings and result in a localized region where threshold concentrations are reached. Therefore, in this chapter, the emerging density-dependent signaling systems presented are defined as potential peptide regulatory systems in which the signaling peptide can accumulate and induce global changes in the bacterial cells.

To survive in the human host, pathogenic gram-positive cocci must express a number of virulence factors. For these bacteria, the high-density, low-flowthrough conditions that are a prerequisite for density-dependent signaling can be envisioned to occur in infection foci or intracellularly. In tissue infection foci, bacteria are quickly multiplying and the immune system is attempting to control bacterial spread by the formation of granulomas. This limits both space and the influx of fresh nutrients. Engulfment of bacteria in eukaryotic cell vacuoles could increase the relative cell density by limiting space and nutrients. The bacterial response to such conditions can be either to switch virulence factor expression to favor

Bettina A. B. Leonard, Department of Microbiology and Immunology, Temple University School of Medicine, 3400 N. Broad St., Philadelphia, PA 19140.   *Andreas Podbielski,* Department of Medical Microbiology and Hygiene, University of Ulm Clinic, D-89081 Ulm, Germany.

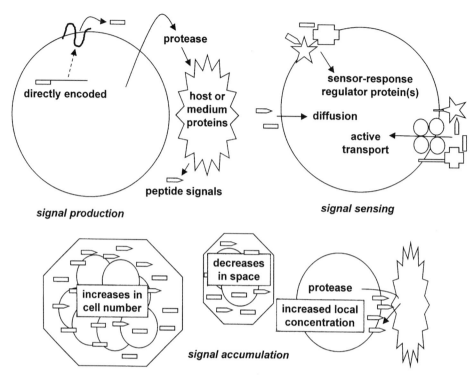

FIGURE 1 Three requirements for cell density-dependent signaling. Density-dependent signals can be either directly encoded by the cells or produced by the action of a cellular protease(s) on environmental proteins. Signal sensing occurs outside the cell by sensor-response two-component systems or by import of the signal molecule into the cell, followed by binding to intracellular effector molecules. Import of the signal into the cell can occur by diffusion or active transport mediated by ATP-dependent peptide permeases. Signal accumulation results either from increases in the cell number in space with limited flowthrough or, theoretically, by enclosing the cells in a smaller space or production of the signal by protease action at the site of contact between the substrate and cell, which would result in high localized signal concentrations.

spreading to new, less crowded niches (i.e., decrease adhesins and increase lytic factors) or to enter a quiescent nonreplicative state.

The purpose of this chapter is to highlight emerging density-dependent systems in gram-positive bacteria, especially *Streptococcus pyogenes* (group A streptococci [GAS]). Many of the data presented here have emerged in the last 3 years or represent new interpretations of older data. Hence, this chapter contains a fair amount of speculation. There is still much to be learned, and the working models presented here will have to be confirmed or rejected upon further characterization of the systems described below.

EMERGING DENSITY-DEPENDENT SYSTEMS IN GROUP A STREPTOCOCCI

GAS are obligate human pathogens that cause a spectrum of infections ranging from asymptomatic carriers to necrotizing fasciitis (Curtis, 1996; Stevens, 1996). The expression of virulence factors by GAS is controlled by a number of specific and often complex regulatory networks designed to integrate host environmental and bacterial signals. GAS are able to survive on tissue surfaces, in deep tissue sites, in the blood, and intracellularly. Each of these compartments poses different challenges in terms of host environment and bacterial load.

Many observations of virulence factor expression patterns suggest that different stages of in vitro batch culture may mimic different basic environments GAS could face in the host in terms of cell density and nutrient limitation. The emerging roles of cell-density signaling in GAS will be discussed in the context of these in vitro growth stages and their possible relevance to survival in the human host (Fig. 2).

Structural Components for Density Signaling

PEPTIDE PERMEASES

The transport of cell-density-signaling peptides into the bacterial cell through the oligopeptide permeases has been demonstrated in gram-positive bacteria. Examples of the processes affected include *Bacillus subtilis* sporulation (Lazazzera et al., 1997), *Enterococcus faecalis* conjugative plasmid transfer (Leonard et al., 1996), *Staphylococcus aureus* virulence factor regulation (Ji et al., 1995), *Streptococcus pneumoniae* competence development (Alloing et al., 1998), and *Streptococcus gordonii* adhesin expression (McNab and Jenkinson, 1998).

To date, the peptide permease systems identified in GAS are the dipeptide and oligopeptide permeases (Dpp and Opp, respectively) (Podbielski et al., 1996a; Podbielski and Leonard, 1998). Both transporters belong to the class of ABC transporters that are made up of a complex of five proteins, normally encoded in an operon (Higgins, 1995; Tam and Saier, 1993). Transport across the membrane is accomplished by a complex of four proteins: two transmembrane proteins (DppB/OppB and DppC/OppC) allowing for translocation of the substrate across the cytoplasmic membrane and two ATPases (DppD/OppD and DppE/OppF) that supply the energy for uptake. The peptide-binding proteins are lipoproteins in gram-positive bacteria and periplasmic proteins in gram-negative bacteria that interact with the transport complex. The source of peptide-binding proteins can be the DppA/OppA protein encoded with the *dpp/opp* operons or other OppA-like peptide-binding lipoproteins encoded in other locations in the chromosome or on plasmids (Alloing et al., 1994; Jenkinson et al., 1996; Leonard et al., 1996; Podbielski and Leonard, 1998).

It has been suggested that an excess ratio of binding proteins to transport proteins facilitates efficient peptide uptake since the peptide-binding step is predicted to be the rate-limiting step (Abouhamad and Manson, 1994; Podbielski et al., 1996a). The individual mechanisms for accomplishing this disproportionate gene expression vary between transport systems and bacterial species. For the *opp* and *dpp* operons of GAS, this is achieved by introducing an intraoperon transcription attenuator that results in a >20-fold excess of *dppA* and an approximately fivefold excess of *oppA*. In *Escherichia coli*, the relative abundance of *dppA* is proposed to be accomplished by an mRNA-stabilizing RIP element located within the *dppA-dppB* intergenic region (Kashiwagi et al., 1990; Abouhamad and Manson, 1994). Additional substrate-binding proteins encoded by separate monocistronically transcribed genes provide a source for additional specific or nonspecific substrate-binding proteins (Alloing et al., 1994; Jenkinson et al., 1996; Leonard et al., 1996). Finally, in lactococci, *oppA* is transcribed from an additional promoter (Tynkkynen et al., 1993).

APPLICATION OF THE GENOME SEQUENCING PROJECT: UBIQUITY OF "THREE-COMPONENT" DENSITY-SENSING SYSTEMS?

The GAS serotype M1 genome sequencing project has provided a powerful tool for determining whether homologs to other density-dependent systems exist in GAS (Roe et al., 1997). "Three-component" density-sensing systems (Dunny and Leonard, 1997) are often found in gram-positive bacteria and consist of a gene encoding a peptide signal and a response regulator–histidine kinase protein pair for sensing signal accumulation. The most

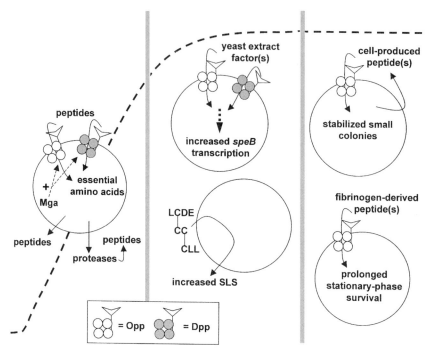

FIGURE 2 Proposed roles of peptides and peptide density-dependent systems in GAS. The dotted line represents the growth curve. Processes occurring during exponential growth are represented in the first panel. Peptides are produced by bacterial cells during growth, either by direct secretion or by protease-mediated breakdown of host or medium proteins. Peptides can be internalized through either the dipeptide permease (Dpp) or the oligopeptide permease (Opp) and subsequently used as sources of essential amino acids. The multiple gene activator Mga, a positive transcriptional activator of virulence factor expression, also up-regulates *dpp* and *opp* transcription. Potential peptide density-dependent processes that occur during the transition from late logarithmic to stationary phase are shown in the middle panel. These include yeast extract factor(s) that are transported through Dpp and Opp and are necessary for *speB* transcription. The complex of three peptides, LCDE, CC, and CLL, linked by disulfide bonds has been shown to increase production of streptolysin S (SLS). Potential peptide density-dependent processes that occur during prolonged stationary-phase incubation (>3 days) are shown in the last panel. The first is the production of a peptide(s) by the cells themselves that leads to a stabilized virulence factor-negative small-colony phenotype. The second is the internalization of a specific fibrinogen-derived peptide(s) that leads to survival in prolonged stationary-phase culture.

striking identity found was to the density-dependent signaling system *cia* involved in competence induction in *S. pneumoniae* (Fig. 3). While competence induction has never been noted in GAS, the presence of these genes in GAS could represent either a similar type of sensing system or an actual competence system. Competence also induces changes in cell wall structure and coregulates

other genes. Therefore, competence inducers could be global regulators of a number of factors, including competence, that allow for survival of GAS when conditions become undesirable.

The GAS *ciaR*-like (*ciaR-L*) response regulator shares 83.9% identity with the corresponding *S. pneumoniae* genes while the histidine kinase (*ciaH-L*) shares 50.7% identity

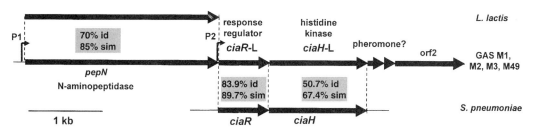

FIGURE 3 Identification of a potential quorum sensor in the GAS serotype M1 genome project database by comparison to known quorum-sensing genes. The *S. pneumoniae cia* operon was used for comparison. Gene sequences are drawn to scale (bar corresponds to 1 kb) and delineated by arrows indicating the direction of transcription. The identity (id) and similarity (sim) scores generated using PC Gene Software are given in the shaded boxes. The presence of corresponding genes in GAS serotypes M2, M3, and M49 was detected by PCR (unpublished observations). "Pheromone?" indicates the presence of a small open reading frame capable of producing a pheromone-like molecule, but no data exist to confirm or deny the transcription, translation, or activity of this region in either GAS or *S. pneumoniae. orf2* has 46% identity and 68% similarity to the *E. coli* Rts pantothenate kinase gene. P1 and P2 indicate the approximate locations of the two promoters for the *cia* operon.

(Guenzi et al., 1994; Zahner et al., 1996). The histidine kinase molecules are expressed at the cell surface and are responsible for the detection of the specific signaling molecule (Håvarstein et al., 1996). The lower degree of identity in the histidine kinase than in the cytoplasmic response regulator may reflect the need for the histidine kinases to specifically interact with distinct peptide signals (Håvarstein et al., 1997). The *cia*-like region is currently being characterized in our laboratories (unpublished observations). The presence of this region can be detected by PCR in serotype M1, M2, M3, and M49 GAS strains. It is transcribed from two promoters, the first resulting in a transcript encompassing *pepN-L* through *ciaH-L* and the second containing just the response regulator–histidine kinase (*ciaR-L, ciaH-L*) genes. The operon is expressed at the highest levels at high cell density and induced by spent medium, suggesting it functions as a legitimate density-regulated system.

The cotranscription of *cia* with an aminopeptidase N–like gene (*pepN-L*) implies that the peptidase has a role in the processing of the *cia* signaling peptide and/or that the aminopeptidase N gene is regulated by *cia*. Aminopeptidase N proteins are present in a wide variety of organisms (Christensen et al., 1995;

Tam et al., 1992) associated with the cytoplasmic side of the membrane (Jensch and Fricke, 1997; van Alen-Boerrigter et al., 1991) and able to cleave N-terminal amino acids of various-length peptides with some preference for certain amino acid residues (Midwinter and Pritchard, 1994; Niven et al., 1995). These peptidases have been shown to be important for the utilization of permease-transported peptides as amino acid sources (Miller and MacKinnon, 1974; Miller and Schwartz, 1978; Mierau et al., 1996).

ciaH-L mutants do not reach the same final optical densities in liquid batch culture as wild-type cells (unpublished observations). This is consistent with the hypothesis that the *cia* mutants are defective in a process that occurs as the cells reach high density. The reduced final density of *cia* mutants could be attributed to loss of the *cia* global regulator or attributed directly to reduced aminopeptidase N production. The loss of aminopeptidase N activity could affect the ability of the cells to scavenge essential amino acids from peptides. This loss of intracellular breakdown of peptides could also affect cell wall recycling (for review see Park, 1995). Consistent with this hypothesis, *cia* mutants of *S. pneumoniae* show a resistance to cell wall-active β-lactams that does not involve an alteration in the penicillin-binding proteins (Zahner et al., 1996). In addition, density-dependent cell wall synthesis regulation has been proposed for the gram-negative *Providencia stuartii* (Rather et al., 1997a,b). In this organism a peptide density signal was found to down-regulate the transcription of aminoglycoside $2'$-N-acetyltransferase upon entry of the cells into stationary phase (Rather et al., 1997a,b). Further characterization of the aminopeptidase N and its regulation by the *cia-L* system should distinguish between these possibilities.

Logarithmic Cell Growth: GAS Survival in Blood

The multiplication of GAS in blood is best mimicked by early- and mid-exponential-phase in vitro batch cultures. Consistent with this idea, a number of important GAS virulence factors that allow for survival and multiplication in blood are expressed during active growth. These factors include the hyaluronic acid capsule and M protein that contribute to phagocytosis resistance (Dale et al., 1996; Moses et al., 1997; for a review on M protein see Kehoe, 1994). Both capsule expression and transcription of the M protein structural gene (*emm*) are the highest during active growth (van de Rijn, 1983; McIver and Scott, 1997). Unlike the capsule synthesis operon *has*, *emm* is coregulated with a number of other virulence factors by the positive transcriptional multiple gene activator Mga (Simpson et al., 1990 [*virR*]; Perez-Casal et al., 1991 [*mry*]). The expression of *mga* is highest during active growth and probably accounts for the higher relative expression of Mga-dependent genes during active growth (McIver and Scott, 1997). Other Mga-dependent factors include fibronectin-binding proteins (serum opacity factor) (Kreikemeyer et al., 1995; Rakonjac et al., 1995), the cysteine protease SpeB, and factors that allow for inactivation of complement (C5a peptidase) (Chen and Cleary, 1990). These Mga-dependent virulence factors are all monocistronically transcribed and can be lo-

cated together either in the core *vir* regulon or in other chromosomal locations (Cleary et al., 1991; Podbielski, 1993; Podbielski et al., 1996b). In addition to being affected by growth phase, Mga-controlled genes are also regulated by pH, temperature, pCO_2, and Fe^{2+} concentrations (Caparon et al., 1992; Cleary et al., 1991; McIver and Scott, 1997; McIver et al., 1995; Podbielski et al., 1992). The combination of the types of virulence factors regulated and the inducing conditions have led investigators to propose that Mga-regulated genes promote bacterial survival in blood and subsequent invasion into deep tissue (Caparon et al., 1992; Lukomski et al., 1998).

GAS are amino acid auxotrophs, unable to synthesize 13 of the 20 amino acids (E, K, L, G, V, I, T, R, S, F, Y, M, H) (Davies et al., 1965). Therefore, they need to obtain essential amino acids by uptake of free amino acid or polypeptides. Through peptide feeding experiments in amino acid-depleted chemically defined medium, it was shown that transport of peptides through the Opp and Dpp can be used to supply the cell with essential amino acids (Podbielski and Leonard, 1998). Since the production of proteins requires amino acids, it is logical that amino acid availability could influence the production of virulence factors. Opp-mediated transport of peptides, therefore, could serve as density signals to coordinate nutritional state and virulence factor production. Both *opp* and *dpp* transcription are increased by the presence of *mga*, which may reflect increased requirements for amino acids for virulence factor synthesis (Podbielski et al., 1996a; Podbielski and Leonard, 1998).

Both the presence of peptides and the ability to transport peptides have been found to be important for expression of virulence factors during active growth. Protein and peptide sources such as Todd-Hewitt yeast extract medium (THY) have been reported to be necessary for the production of M protein (Davies et al., 1967; Davies and Rudd, 1972; Fox, 1961). However, peptide permease mutants showed no change in *emm* transcription (Podbielski et al., 1996a; unpublished observations), suggesting a signaling mechanism or effect distinct from peptide transport. However, mutation in Opp leads to decreased transcription of the capsule synthesis operon *has* during active growth, suggesting a link between peptide transport and capsule synthesis (unpublished observations).

Late-Logarithmic to Early-Stationary-Phase Transition: Growth on and in Tissue

When GAS are growing in blood, it is unlikely that they ever achieve high cell densities. However, when growing attached to eukaryotic cell surfaces or embedded in tissue, the cells will eventually reach high densities. At this point, they will need to either spread or shift to a more quiescent state to survive.

This shift in virulence factor demand at high cell densities is mimicked by virulence factor regulation in in vitro cultures at the transition between late-exponential and stationary-growth phases. Mga and some Mga-dependent genes such as *emm* and *scpA* are expressed in late-logarithmic-growth phase at approximately the same level as they are early in growth and then are shut down once the cells have entered stationary phase (McIver and Scott, 1997). However, there are also a number of virulence factors, both Mga-dependent and Mga-independent, that are up-regulated at high cell densities (>0.8). These genes are subsequently down-regulated after entry into stationary phase. These virulence factors include the mitogenic factor (potentially the same as DNaseB) (Iwasaki et al., 1997), erythrogenic toxin SpeA (McIver and Scott, 1997), and cysteine protease (SpeB) (unpublished observations). Mitogenic factor and SpeA are both superantigens that have a number of effects on the host immune system and are proposed to be important in causing septic shock (for a review see Norgren and Eriksson, 1997). SpeB has been found to cleave, digest, or completely degrade a number of human proteins and streptococcal surface proteins such as M protein and the immunoglobulin-binding proteins (Berge and

Bjoerck, 1995; Herwald et al., 1996; Kapur et al., 1993a,b; Wolf et al., 1994). This activity has been suggested to be important for promoting bacterial dissemination (Berge and Bjoerck, 1995), which could be a response to high cell densities. SpeB appears to be an important component of infection and outcome, since SpeB mutants are significantly reduced in virulence (Lukomski et al., 1997).

ROLE OF PEPTIDES IN ENHANCING VIRULENCE FACTOR EXPRESSION AT HIGH CELL DENSITIES

Peptide transport into the cell has been found to be important for the induction of cysteine protease SpeB production. In the 1960s, Fox (1961) reported that SpeB production required peptides present in the medium. SpeB is transcribed at the highest rate in late logarithmic/early stationary phase when the culture density is over an OD_{600} of 1.0 (unpublished observations). This indicates a need for signaling molecule accumulation. Both oligo- and dipeptide permease mutants have been shown to have significantly reduced *speB* transcription at high cell density (Podbielski et al., 1996a; Podbielski and Leonard, 1998). The involvement of both dipeptide and oligopeptide transport suggests that the accumulated molecule is either nonspecific or, as in the case of down-regulation of protease production in *Lactococcus lactis* (Marugg et al., 1995), due to transport of a specific dipeptide that can be contained within a number of peptides of different lengths.

Expression of the hemolysin streptolysin S (SLS) also has been shown to be dependent on the presence of peptide sources in the medium. The role of peptides for SLS production has been demonstrated by enhancement of SLS production by pretreatment of the cells with protease (Taketo and Taketo, 1984), inhibition of SLS production in the presence of protease inhibitors (Akao et al., 1983), and the purification of an SLS-inducing peptide complex from pronase digests of bovine serum albumin (LCDE, CC, CLL) (Akao et al., 1992). It is unclear whether SLS is specifically induced only by these three peptides connected by disulfide bonds or by any combination of peptides containing a similar disulfide-bonded primary structure (Akao et al. 1988, 1992).

Prolonged Stationary-Phase Survival: Intracellular Survival and Asymptomatic Carriers

The colonization of asymptomatic carriers by GAS has also been reported and suggests that the associated bacteria do not illicit fulminant disease by rapid multiplication in the host and the production of host-damaging virulence factors (Fischetti, 1997). Instead, the bacteria may enter a more "quiescent" state. A number of observations suggest that intracellular GAS may represent such a reservoir. In the 1950s, L-forms of GAS were described that were isolated from eukaryotic cell vacuoles (Freimer et al., 1959; Kagan et al., 1976; Leon and Panos, 1976; Schmidt-Slomska et al., 1972). These L-forms had reduced production of virulence factors such as streptokinase, hyaluronidase, and erythrogenic toxin and were suggested to be in a nonreplicative state (Greco et al., 1995; Kagan, 1968; Leon and Panos, 1976). The down-regulation of virulence factors as a part of intracellular survival could be logical to prevent both eukaryotic cell activation and death and could reflect a possible metabolic inactive state needed for persistence in the human host. The L-forms persisted in a mouse model for up to 52 weeks, where they eventually induced a variety of systemic lesions in tissue (Kagan et al., 1976). More recent studies confirm that GAS can invade eukaryotic cells (Cue and Cleary, 1997; Fluckiger et al., 1998; Greco et al., 1995; LaPenta et al., 1994). An association of intracellular GAS with asymptomatic carriers and patients suffering from recurrent tonsillitis suggests that this intracellular state may be important in vivo (Oesterlund and Engstrand, 1997). Long-term stationary-phase culture has been used as a model for inducing GAS that are able to survive nonreplicatively for long periods of time in enclosed spaces with little fresh nutrient acquisition. This is a possible model for eukaryotic cell vacuole conditions (Leonard et al., 1998).

Studies on nonsporulating gram-negative and gram-positive bacteria have shown that prolonged survival of bacteria in high-density starvation conditions is an induced state (Morton and Oliver, 1994; Reeve et al., 1984a; Watson et al., 1998). Some bacteria, such as *Micrococcus luteus, E. coli,* and *Vibrio* spp., enter a viable but nonculturable state in which the bacteria are metabolically active but unable to undergo sustained cell division (Oliver, 1993; Xu et al., 1982). Other bacteria, such as *S. aureus,* continue "cryptic growth," in which the cells undergo morphological changes but remain metabolically active and culturable (Ryan, 1951; Proctor et al., 1995). The recycling of nutrients, especially proteins, during this cryptic growth (Reeve et al., 1984b; Ryan, 1951) suggests that the peptide environment around the cells can be significantly altered during the course of this prolonged stationary-phase culture.

FIBRINOGEN PEPTIDE(S)-INDUCED PROLONGED STATIONARY-PHASE SURVIVAL: IMPORT OF AN EXOGENOUS PEPTIDE DENSITY SIGNAL?

We have observed that GAS survive for only 3 to 5 days in amino acid-containing CDM as opposed to more than 25 days in complex medium such as THY (unpublished observations). Additionally, Opp mutants show shorter long-term survival when stored at 4°C. Amino acid-depleted chemically defined medium (CDM) supplemented with high-pressure liquid chromatography-prepared crude peptide fractions from THY prolonged the survival of wild-type cells. The peptide signal contained in THY is probably at least semispecific. Specific digests of human proteins, whose bovine homologs are contained in THY and that are known to be bound by GAS, can induce long-term survival. Crude peptide fractions from pepsin digests of human and bovine fibrinogen prolong survival of GAS in CDM. Conversely, papain and trypsin digests of fibrinogen, a number of synthetic peptides, as well as trypsin, papain, or pepsin digests of albumin or casein do not support prolonged survival.

In stationary phase, after depletion of the medium for certain nutrients, cells are still metabolically active. Therefore, the uptake and subsequent breakdown of peptides could be providing the cells in prolonged stationary culture with carbon and energy. However, the fact that only a defined crude peptide preparation prolongs survival strongly indicates that a specific, environmentally derived peptide(s) signaling process is involved in inducing prolonged stationary-phase survival in GAS. The induction of a metabolic state that allows prolonged stationary-phase survival has recently been reported for *S. aureus* (Watson et al., 1998). In these organisms a starvation–survival state could be induced by either glucose starvation or by density-dependent multiple nutrient limitation (Watson et al., 1998), which may be the equivalent of the peptide effects found in GAS.

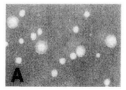

GAS 3 days post-exponential phase growth

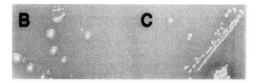

typical and stabilized small colonies (passage 3)

FIGURE 4 Formation of stable atypical small colonies by serotype M49 GAS. GAS serotype M49 were incubated at 37°C in 5% CO_2 for a number of days in THY without addition of fresh media. At various days in postexponential-growth phase, aliquots were withdrawn, plated on THY agar, and incubated at 37°C in 5% CO_2. (A) Colony dimorphism seen in postexponential-phase growth. (B and C) Typical and stabilized small colonies after three consecutive transfers on Mueller-Hinton sheep blood agar plates.

FORMATION OF STABLE SMALL COLONIES DURING PROLONGED STATIONARY-PHASE INCUBATION: AN ENDOGENOUS SENSING SYSTEM?

During prolonged stationary-phase incubation, GAS in both CDM and complex medium (solid and liquid) form stabilized atypical small colonies at a frequency of 5×10^{-1} to 1×10^{-1} after 24 h (Fig. 4) (Leonard et al., 1998). After 3 to 5 days, 20 to 50% of the small colonies remain stable for 132 to 200 generations in fresh medium before complete reversion to typical colonies. The small colony size is not attributable to altered growth of the cells, but instead results from reduced capsule synthetase operon (*has*) transcription. The expression of a number of other virulence factors is also affected, including Mga, a positive virulence regulator that influences transcription of *emm* and *speB* (Simpson et al., 1990 [*virR*]; Perez-Casal et al., 1991 [*mry*]; Podbielski et al., 1996b; Leonard et al., 1998). The decrease in transcription rate is not completely global,

since transcription of some virulence factors such as the streptolysin O hemolysin gene (*slo*) and the housekeeping gene (*recA*) are not decreased. The recovery of virulence factor production after 100 generations indicates that the stabilization is a dynamic phase variation and not due to irreversible mutation in a subpopulation of the cells. The induction of culture dimorphism upon prolonged stationary-phase culture has also been reported for *E. coli*, although unlike in GAS, these cells were found to be mutants (Zambrano et al., 1993).

Owing to the loss of important virulence factors, bacteria that form stabilized small colonies are completely sensitive to phagocytosis (Leonard et al., in press). Cleary et al. (1998) reported reduced eukaryotic cell invasion of unstable smaller, more compact colonies that were reduced in M protein, capsule, and SpeA production. It is formally possible that these unstable smaller, compact colonies are the same as the unstable small-colony variants obtained after 24 h in stationary-phase culture.

Taken together, these data suggest that these small colonies would not survive in a body site exposed to granulocytes and leukocytes. However, recovery of stabilized small colonies from patient samples suggests their persistence in a sequestered site, such as in intracellular vacuoles (unpublished observations in collaboration with A. Oesterlund, Sweden).

The colony variation has been proposed to be a two-step process (Leonard et al., 1998). Onset of the variation occurs shortly after the cells reach stationary phase, while stabilization of the phenotype occurs only after prolonged stationary-phase cultivation. In support of this hypothesis, preliminary data indicate that formation of stabilized small colonies requires cellular oligopeptide permease function (unpublished observations). Oligopeptide permease mutants continue to form small-colony variants, but the phenotype never stabilizes. The involvement of Opp suggests that the formation of stable small colonies results from density-dependent signaling as opposed to depletion of nutrients or buildup of toxic products. Complex medium and CDM, which had supported prolonged growth of GAS, show no significant decrease in free amino acid availability (unpublished observations). This suggests that the cells are not starved for amino acids, which could be provided by peptide transport. The late onset of the phenomenon (≥3 days of postexponential growth) suggests that the observed effects are probably not directly attributable to glucose starvation or buildup of toxic byproducts, since these conditions would also be present at the onset of stationary phase.

A number of changes in peptide availability during prolonged stationary phase have been documented (Reeve et al., 1984b; Ryan, 1951). These changes could be accounted for by either the slow stationary-phase production of a peptide signal or the depletion of a signal formed during exponential growth. As an example of the latter, adding spent growth medium from exponentially growing *M. luteus* cells induces a break in *M. luteus* dormancy, thus suggesting the presence of a pheromone

needed for active growth (Kaprelyants et al., 1994). The stabilization of small colonies occurs in CDM, in the absence of exogenous peptides or proteins, which implies that the peptide(s) required for stabilization is being produced by the cell itself. The turnover of cell wall peptides (for review see Park, 1995), the release of surface proteins by the action of SpeB (Berge and Bjoerck, 1995), and/or the turnover of dying cells (Ryan, 1951) could all be potential sources for unique stationary-phase signaling peptides. With further characterization of the signal(s) using purification methods and continuous culture, it should be possible to dissect the effects of nutrient depletion, toxic product buildup, and cell-density signals in this stationary-phase phenomenon.

OTHER EMERGING FUNCTIONS OF PEPTIDE-DENSITY SIGNALING IN GRAM-POSITIVE COCCI

A number of well-defined functions of peptide-density signaling in gram-positive cocci are presented elsewhere in this volume. The purpose of this section is to describe some interesting observations that have implications for the wide range of processes that may be influenced by density signaling.

Induction of Growth of Specific Bacterial Groups on Solid Surfaces in Mixed Populations

In 1996 Bloomquist et al. described cell density-dependent induction of growth in *Streptococcus sanguis*. They implanted enamel chips onto teeth of human volunteers and then measured growth of the bacteria, primarily *S. sanguis*, in vitro. When cell densities on the chips reached between 2.5×10^6 and 4.0×10^6 bacteria per mm^2, the rate of growth significantly increased. This suggests that in the case of oral-cavity biofilm formation, achievement of a critical bacterial density causes a rapid increase in the growth rate of the bacteria. Presumably (although it has not been tested), this growth rate would again decrease once the bacteria reached very high cell densities. Since these studies were done with forming biofilms, it is not clear whether the effect is species specific or whether this phenomenon may involve induction of several or all bacterial species present. In addition, no further mechanistic details have been published. Nonetheless, it is an interesting observation linking increases in bacterial growth rate to cell density.

Regulation of Adhesins

Successful adherence and survival of bacteria on surfaces such as the intestinal wall, skin, or oral surfaces depends on sensing both the host environment and the number of other bacteria present that may compete for space and nutrients. A possible role for density-dependent signaling in the expression of adhesins in oral streptococci was suggested by studies showing that mutation of the oligopeptide permease systems in *S. gordonii* and *Streptococcus parasanguis* led to a decrease in bacterial cell adhesion (Jenkinson and Easingwood, 1990; Jenkinson et al., 1996; Fenno et al., 1995). It was unclear whether the oligopeptide permease-binding proteins were acting as adhesins themselves or if they were importing signaling molecules. Recently, expression of a major adhesin, *cshA*, in *S. gordonii* was reported to depend on oligopeptide permease (*hpp*)-mediated import of a secreted factor in a density-dependent manner (McNab and Jenkinson, 1998).

Pheromone-Inducible Plasmids: Maintaining Plasmids within Restricted Bacterial Host Populations

Bacteria in nature are often found in mixed cultures. The genetic material encoded on plasmids often helps the bacteria to survive within a particular niche. Therefore, it is important that the plasmids are stable within the target population and that their host range is restricted to allow for a selective advantage. In the case of mobile genetic elements, specific induction of transfer only in the presence of the appropriate species would be a conservation of energy needed for conjugation.

RESTRICTION OF HOST RANGE: A DUAL ROLE OF BACTERIAL CONJUGATIVE PLASMID PHEROMONES

Pheromone-inducible conjugative plasmids are a class of plasmids limited to the enterococci. The transfer of this class of conjugative plasmids to plasmid-free *E. faecalis* can be induced by peptide pheromones secreted by enterococci (see chapter 4 of this volume for details). The mechanism of conjugation induction involves internalization of the pheromone through the cellular oligopeptide permease and its subsequent interaction with intracellular effector molecules (Leonard et al., 1996). Conjugative plasmids can be transferred to a wide variety of bacterial species, including gram-negative bacteria such as *E. coli*, and will be retained provided that an appropriate origin of replication is present (Trieu-Cuout et al., 1988). The pheromone-inducible conjugative plasmids are only retained in *E. faecalis*, where they are extremely stable and can be maintained in >98% of the cells for at least 100 generations in the absence of selection (Dunny et al., 1981; Hedberg et al., 1996). This suggests that maintenance of conjugative plasmids depends on *E. faecalis* specific factor(s).

pCF10 is a pheromone-inducible conjugative plasmid responsive to the pheromone cCF10 (LVTLVFV) that has been used as a model for pheromone-inducible conjugation (for a review see Dunny et al., 1995). Both plasmid-containing and plasmid-free *E. faecalis* cells continue to secrete cCF10 (Nakayama et al., 1994), and the only known producers of cCF10 are enterococci. The host range of pCF10 can be extended to *L. lactis* when the cells are engineered to produce cCF10 using a synthetic gene (Leonard and Dunny, 1996). Like the induction of conjugation, maintenance of pCF10 in *L. lactis* depends on the presence of a functional oligopeptide permease. Since there are no known *E. faecalis* pheromone cCF10 mutants, this cannot be directly tested in *E. faecalis*. However, mutation of the oligopeptide permease and maintenance of pCF10 are mutually exclusive in enterococci. Furthermore, it has been shown that the cCF10 interacts with the putative replication protein (PrgW) of pCF10 (Leonard et al., 1996). Addition of exogenous cCF10 to the bacterial cells does not result in the same stabilization of the pCF10, suggesting that the cCF10 used for plasmid stability comes from an endogenous rather than an exogenous source (unpublished observations). The role of such endogenous peptide signals has also been suggested to be important for sporulation induction in *B. subtilis* (see chapter 16 of this volume).

Taken together, these data led to the testable hypothesis that cCF10 plays a dual role in pCF10 biology (Fig. 5). The first role is to act as an exogenous quorum-sensing molecule, signaling the presence of plasmid-free cells and inducing conjugative transfer of pCF10 (for a review see Dunny et al., 1995). The second role is to function as an endogenous signaling molecule and stably maintain pCF10 within the enterococcal cell. The ability of the cell to use cCF10 for plasmid maintenance without inducing conjugation may be

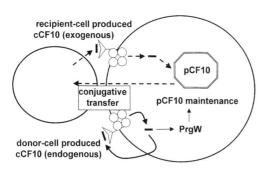

FIGURE 5 Proposed dual role of cCF10 in pCF10 biology. When the pheromone cCF10 is produced by a pCF10-free recipient cell and sensed by a pCF10-containing donor cell, it is internalized through the Opp, interacts with intracellular effector molecules, and induces the conjugative transfer of the plasmid pCF10. The donor cells (pCF10-containing) continue to secrete cCF10. Opp-mediated internalization of donor-secreted cCF10 is necessary for the maintenance of pCF10 within the donor-cell population. The putative pCF10 replication protein PrgW interacts directly with cCF10.

accomplished by spatial or temporal separation of the processes. A better understanding of this dual role awaits further investigation.

PHEROMONE-INDUCIBLE-LIKE SYSTEMS IN OTHER GRAM-POSITIVE MOBILE GENETIC ELEMENTS

In *L. lactis*, the examination of elements needed for a high frequency of conjugative plasmids and transposons suggests the possible existence of a pheromone-inducible conjugative system(s). One feature of *E. faecalis* pheromone-inducible conjugation is the formation of clumps in induced cultures. The clumps are caused by the pheromone induction of an aggregation substance encoded on conjugative plasmids that interacts with a binding substance (lipoteichoic acid) present on all cells (Bensing and Dunny, 1993). The interaction of these molecules causes clumping that is hypothesized to allow for the formation of stable mating pairs and the subsequent transfer of the conjugative plasmid.

In transfer of the *L. lactis* lactose plasmid, involvement of a plasmid-encoded aggregation substance (Clu) and a chromosomal aggregation substance (Agg) has been suggested (van der Lelie et al., 1991). The Agg protein is not present in all *L. lactis* cells. High-frequency transfer of the lactose plasmid occurs only when both Clu and Agg are present. Transfer of the conjugative transposon Tn916 also requires the production of a chromosomally encoded Laff factor for high-frequency transfer in the absence of transposition (Bringel et al., 1992). The Laff factor is proposed to represent a *L. lactis*-specific chromosomal Clu factor. Like the *E. faecalis* aggregation substance, it can enhance transfer of Tn916 regardless of whether it is encoded by the donor or recipient cell (Bringel et al., 1992). This is consistent with the idea that these molecules act as cell adhesins but do not form unidirectional mating channels for mobile element transfer.

Although the Clu structural gene (*clu*) carried by the lactose plasmid is transcribed only after plasmid rearrangement (van der Lelie et al., 1991), to date there is no evidence for pheromone induction of the system.

BACTERIAL PHEROMONE CROSS-TALK?

Skurray and coworkers described the production of an enterococcal pheromone-like activity by *S. aureus* that resulted from the synthesis of a cAD1-like molecule (7 of 8 amino acids identical to cAD1 [Berg et al., 1997; Firth et al., 1994]). The cAD1-like molecule (LFTLVLAG) represents the last eight carboxy-terminal amino acids of the signal sequence for a lipoprotein (TraH) that is found in the conjugation-associated region of the staphylococcal plasmid pSK41 (Firth et al., 1994). The gene *traH* encodes pro-TraH, which is transported to the surface of the cell, where the mature lipoprotein is released. Cleavage of the C-terminal eight amino acids from the remaining signal peptide allows the release of the mature pheromone. While the enterococcal conjugative pheromones are not encoded in the signal sequence of their cognitive lipoprotein receptors, these findings have important implications (Berg et al., 1997). They suggest that, like many of the gram-positive peptide signaling systems in which the peptide is encoded in an operon with the sensor-response proteins, peptides that signal by internalization through an Opp may also be encoded upstream of their specialized OppA-like peptide-binding proteins. Since pSK41 is a conjugative plasmid, and the TraH lipoprotein and cAD1-like molecule are encoded within a region of the plasmid associated with conjugative transfer, these findings also suggest that pheromone involvement in conjugation may not be restricted to the enterococcal plasmids. In addition, it leads to the intriguing possibility that, like homoserine lactone cross-talk in gram-negative bacteria (Gray, 1997), perhaps there is signaling molecule cross-talk between gram-positive species as well.

ACKNOWLEDGMENTS

We acknowledge the numerous intellectual contributions made to this work by the following individuals: Patrick Piggot, Lolita Daneo-Moore, Gerald Shockman, Barbara Bensing, David Mills and Helmut Hirt. In particular, we thank Gary Dunny in whose lab the enterococcal research was initiated. The laboratory of A.P. is supported by DFG grants Po393-3/4 and Po396-2. B.A.B.L. was supported by a NIH postdoctoral fellowship (AI08742) and an Alexander von Humboldt scholarship (IV-USA/1039266).

REFERENCES

Abouhamad, W. N., and M. D. Manson. 1994. The dipeptide permease of *Escherichia coli* closely resembles other bacterial transport systems and shows growth-phase-dependent expression. *Mol. Microbiol.* **14:**1077–1092.

Akao, T., T. Akao, K. Kobashi, and C.-Y. Lai. 1983. The role of protease in streptolysin S formation. *Arch. Biochem. Biophys.* **223:**556–561.

Akao, T., H. Tamei, and K. Kobashi. 1988. The essential factor for streptolysin S production by *Streptococcus pyogenes. Chem. Pharm. Bull.* **36:**3994–3999.

Akao, T., T. Takahashi, and K. Kobashi. 1992. Purification and characterization of a peptide essential for formation of streptolysin S by *Streptococcus pyogenes. Infect. Immun.* **60:**4777–4780.

Alloing, F., P. de Philip, and J. P. Claverys. 1994. Three highly homologous membrane-bound lipoproteins participate in oligopeptide transport by the Ami system of the Gram-positive *Streptococcus pneumoniae. J. Mol. Biol.* **241:**44–58.

Alloing, F., B. Martin, C. Granadel, and J. P. Claverys. 1998. Development of competence in *Streptococcus pneumoniae*: pheromone autoinduction and control of quorum-sensing by the oligopeptide permease. *Mol. Microbiol.* **29:**75–83.

Bensing, B. A., and G. M. Dunny. 1993. Cloning and molecular analysis of genes affecting expression of binding substance, the recipient-encoded receptor(s) mediating mating aggregate formation in *Enterococcus faecalis. J. Bacteriol.* **175:**7421–7429.

Berg, T., N. Firth, and R. A. Skurray. 1997. Enterococcal pheromone-like activity derived from a lipoprotein signal peptide encoded by a *Staphylococcus aureus* plasmid. *Adv. Exp. Med. Biol.* **418:**1041–1044.

Berge, A., and L. Bjoerck. 1995. Streptococcal cysteine proteinase releases biologically active fragments of streptococcal surface proteins. *J. Biol. Chem.* **270:**9862–9867.

Bloomquist, C. G., B. E. Reilly, and W. F. Liljemark. 1996. Adherence, accumulation, and cell division of a natural adherent bacterial population. *J. Bacteriol.* **178:**1172–1177.

Bringel, F., G. L. van Alstine, and J. R. Scott. 1992. Transfer of Tn*916* between *Lactococcus lactis* subsp. *lactis* strains is nontranspositional: evidence for a chromosomal fertility function in strain MG1363. *J. Bacteriol.* **174:**5840–5847.

Caparon, M. G., R. T. Geist, J. Perez-Casal, and J. R. Scott. 1992. Environmental regulation of virulence in group A streptococci: transcription of the gene encoding M protein is stimulated by carbon dioxide. *J. Bacteriol.* **174:**5693–5701.

Chen, C., and P. P. Cleary. 1990. Complete nucleotide sequence of the streptococcal C5a peptidase gene of *Streptococcus pyogenes. J. Biol. Chem.* **265:**3161–3167.

Christensen, J. E., D. L. Lin, A. Palva, and J. L. Steele. 1995. Sequence analysis, distribution and expression of an aminopeptidase N-encoding gene from *Lactobacillus helveticus* CNRZ32. *Gene* **155:** 89–93.

Cleary, P. P., D. LaPenta, D. Heath, E. J. Haanes, and C. Chen. 1991. A virulence regulon in *Streptococcus pyogenes*, p.157–151. *In* G. M. Dunny, P. P. Cleary, and L. L. McKay (ed.), *Genetics and Molecular Biology of Streptococci, Lactococci, and Enterococci.* American Society for Microbiology, Washington, D.C.

Cleary, P. P., L. McLandsborough, L. Ikeda, D. Cue, J. Krawczak, and H. Lam. 1998. High-frequency intracellular infection and erythrogenic toxin A expression undergo phase variation in M1 group A streptococci. *Mol. Microbiol.* **28:**157–167.

Cue, D. R., and P. P. Cleary. 1997. High-frequency invasion of epithelial cells by *Streptococcus pyogenes* can be activated by fibrinogen and peptides containing the sequence RGD. *Infect. Immun.* **65:**2759–2764.

Curtis, N. 1996. Invasive group A streptococcal infection. *Curr. Opin. Infect. Dis.* **9:**191–202.

Dale, J. B., R. G. Washburn, M. B. Marques, and M. R. Wessels. 1996. Hyaluronate capsule and surface M protein in resistance to opsonization of group A streptococci. *Infect. Immun.* **64:**1495–1501.

Davies, H. C., and J. H. Rudd. 1972. Influence of the environment on the growth and cellular content of group A haemolytic streptococci in continuous culture. *J. Appl. Chem. Biotechnol.* **22:**401–403.

Davies, H. C., F. Karush, and J. H. Rudd. 1965. Effect of amino acids on steady-state growth of a group A hemolytic streptococcus. *J. Bacteriol.* **89:** 421–427.

Davies, H. C., F. Karush, and J. H. Rudd. 1967. Synthesis of M protein by group A hemolytic streptococci in completely synthetic media during steady-state growth. *J. Bacteriol.* **95:**162–168.

Dunny, G. M., and B. A. B. Leonard. 1997. Cell-cell communication in Gram-positive bacteria. *Annu. Rev. Microbiol.* **51:**527–564.

Dunny, G. M., C. Funk, and J. Adsit. 1981. Direct stimulation of the transfer of antibiotic resistance by sex pheromones in *Streptococcus faecalis. Plasmid* **6:**270–278.

Dunny, G. M., B. A. B. Leonard, and P. J. Hedberg. 1995. Pheromone-inducible conjugation in *Enterococcus faecalis*: interbacterial and host-parasite chemical communication. *J. Bacteriol.* **177:**871–876.

Fenno, J. C., A. Shaikh, G. Spatafora, and P. Fives-Taylor. 1995. The *fimA* locus of *Streptococcus parasanguis* encodes an ATP-binding membrane transport system. *Mol. Microbiol.* **15:** 849–863.

Firth, N., P. D. Fink, L. Johnson, and R. A. Skurray. 1994. A lipoprotein signal peptide encoded by the staphylococcal plasmid pSK41 exhibits an activity resembling that of *Enterococcus faecalis* pheromone, cAD1. *J. Bacteriol.* **176:**5871–5873.

Fischetti, V. A. 1997. The streptococcus and the host: present and future challenges. *ASM News* **10:** 541–545.

Fluckiger, U., K. F. Jones, and V. A. Fischetti. 1998. Immunoglobulins to group A streptococcal surface molecules decrease adherence to and invasion of human pharyngeal cells. *Infect. Immun.* **66:**974–979.

Fox, E. N. 1961. Peptide requirements for the synthesis of streptococcal proteins. *J. Biol. Chem.* **236:** 166–171.

Freimer, E. H., R. M. Krause, and M. McCarty. 1959. Studies of L forms and protoplasts of group A streptococci. I. Isolation, growth and bacteriologic characteristics. *J. Exp. Med.* **110:**853–873.

Gray, K. M. 1997. Intercellular communication and group behavior in bacteria. *Trends Microbiol.* **5:** 184–188.

Greco R., L. De Martino, G. Donnarumma, M. P. Conte, L. Seganti, and P. Valenti. 1995. Invasion of cultured human cells by *Streptococcus pyogenes. Res. Microbiol.* **146:**551–560.

Guenzi, E., A. M. Gasc, M. A. Sicard, and R. Hakenbeck. 1994. A two-component signal-transduction system is involved in competence and penicillin susceptibility in laboratory mutants of *Streptococcus pneumoniae. Mol. Microbiol.* **12:**505–515.

Håvarstein, L. S., P. Gaustad, I. F. Nes, and D. A. Morrison. 1996. Identification of the streptococcal competence-pheromone receptor. *Mol. Microbiol.* **21:**863–869.

Håvarstein, L. S., R. Hakenbeck, and P. Gaustad. 1997. Natural competence in the genus *Streptococcus*: evidence that streptococci can change pherotype by interspecies recombinational exchanges. *J. Bacteriol.* **179:**6589–6594.

Hedberg, P. J., B. A. B. Leonard, R. E. Ruhfel, and G. M. Dunny. 1996. Identification and characterization of the genes of *Enterococcus faecalis* plasmid pCF10 involved in replication and in negative control of pheromone-inducible conjugation. *Plasmid* **35:**46–57.

Herwald, H., M. Collin, W. Mueller-Esterl, and L. Bjoerck. 1996. Streptococcal cysteine proteinase releases kinins: a novel virulence mechanism. *J. Exp. Med.* **184:**665–673.

Higgins, C. F. 1995. The ABC of channel regulation. *Cell* **82:**693–696.

Iwasaki, M., H. Igarashi, and T. Yutsudo. 1997. Mitogenic factor secreted by *Streptococcus pyogenes* is a heat-stable nuclease requiring His122 for activity. *Microbiology* **143:**2449–2455.

Jenkinson, H. F., and R. A. Easingwood. 1990. Insertional inactivation of the gene encoding a 76-kilodalton cell surface polypeptide in *Streptococcus gordonii* Challis has a pleiotropic effect on cell surface composition and properties. *Infect. Immun.* **58:**3689–3697.

Jenkinson, H. F., R. A. Baker, and G. W. Tannock. 1996. A binding-lipoprotein-dependent oligopeptide transport system in *Streptococcus gordonii* essential for uptake of hexa- and heptapeptides. *J. Bacteriol.* **178:**68–77.

Jensch, T., and B. Fricke. 1997. Localization of alanyl aminopeptidase and leucyl aminopeptidase in cells of *Pseudomonas aeruginosa* by application of different methods for periplasm release. *J. Basic Microbiol.* **37:**115–128.

Ji, G., R. C. Beavis, and R. P. Novick. 1995. Cell density control of staphylococcal virulence mediated by an octapeptide pheromone. *Proc. Natl. Acad. Sci. USA* **92:**12055–12059.

Kagan, G. Y. 1968. Some aspects of investigations of the pathogenic potentialities of L-forms of bacteria, p. 422–443. *In* L. B. Guze (ed.), *Microbial Protoplasts, Spheroplasts, and L-Forms*. The Williams and Wilkins Co., Baltimore.

Kagan, G., Y. Vulfovitch, B. Gusman, and T. Raskova. 1976. Persistence and pathological effect of streptococcal L-forms in vivo. *INSERM* **65:**247–258.

Kaiser, D. 1996. Bacteria also vote. *Science* **272:**1598–1599.

Kaprelyants, A. S., G. V. Mukamolova, and D. B. Kell. 1994. Estimation of dormant *Micrococcus luteus* cells by penicillin lysis and by resuscitation

in cell-free spent culture medium at high dilution. *FEMS Microbiol. Lett.* **115:**347–352.

Kapur, V., M. W. Majesky, L. L. Li, R. A. Black, and J. M Musser. 1993a. Cleavage of IL-1beta by a conserved extracellular cysteine protease from *Streptococcus pyogenes*. *Proc. Natl. Acad. Sci. USA* **90:**7676–7680.

Kapur, V., S. Topouzis, M. W. Majesky, L. L. Li, M. R. Hamrick, R. J. Hamill, J. M. Patti, and J. M. Musser. 1993b. A conserved *Streptococcus pyogenes* extracellular cysteine protease cleaves human fibronectin and degrades vitronectin. *Microbiol. Pathogen.* **15:**327–346.

Kashiwagi, K., Y. Yamaguchi, Y. Sakai, H. Kobayashi, and K. Igarashi. 1990. Identification of the polyamine-induced protein as a periplasmic oligopeptide binding protein. *J. Biol. Chem.* **256:**8387–8391.

Kehoe, M. A. 1994. Cell-wall associated proteins in gram-positive bacteria. *New Comp. Biochem.* **27:**271–261.

Kreikemeyer, B., S. R. Talay, and G. S. Chhatwal. 1995. Characterization of a novel fibronectin-binding surface protein in group A streptococci. *Mol. Microbiol.* **17:**137–145.

LaPenta, D., C. Rubens, E. Chi, and P. P. Cleary. 1994. Group A streptococci efficiently invade human respiratory epithelial cells. *Proc. Natl. Acad. Sci. USA* **91:**12115–12119.

Lazazzera, B. A., J. M. Solomon, and A. D. Grossman. 1997. An exported peptide functions intracellularly to contribute to cell density signaling in *B. subtilis*. *Cell* **89:**917–925.

Leon, O., and C. Panos. 1976. Adaptation of an osmotically fragile L-form of *Streptococcus pyogenes* to physiological osmotic condition and its ability to destroy human heart cells in tissue culture. *Infect. Immun.* **13:**252–262.

Leonard, B. A. B., and G. M. Dunny. 1996. Pheromone regulation of conjugative transfer and replication of pCF10 in *Enterococcus faecalis*, abstr. H210. Abstr. 96th Annu. Meet. Am. Soc. Microbiol. 1996. American Society for Microbiology, Washington, D.C.

Leonard, B. A. B., A. Podbielski, P. J. Hedberg, and G. M. Dunny. 1996. *Enterococcus faecalis* pheromone binding protein, PrgZ, recruits a chromosomal oligopeptide permease system to import sex pheromone cCF10 for induction of conjugation. *Proc. Natl. Acad. Sci. USA* **93:**260–264.

Leonard, B. A. B, M. Woischnik, and A. Podbielski. 1998. The production of stabilized virulence factor negative variants by group A streptococci during stationary phase. *Infect. Immun.* **66:**3841–3847.

Lukomski, S., S. Sreevatsan, C. Amberg, W. Reichardt, M. Woischnik, A. Podbielski, and J. M. Musser. 1997. Inactivation of *Streptococcus pyogenes* extracellular cysteine protease significantly decreases mouse lethality of serotype M3 and M49 strains. *J. Clin. Invest.* **99:**2574–2580.

Lukomski, S., E. H. Burns, Jr., P. R. Wyde, A. Podbielski, J. Rurangirwa, D. K. Moore-Poveda, and J. M. Musser. 1998. Genetic inactivation of the extracellular cysteine protease (SpeB) expressed by *Streptococcus pyogenes* decreases resistance to phagocytosis and dissemination to organs. *Infect. Immun.* **66:**771–776.

Marugg, J. D., W. Meijer, R. vanKranenburg, P. Laverman, P. G. Bruinenberg, and W. M. deVos. 1995. Medium-dependent regulation of proteinase gene expression in *Lactococcus lactis*: control of transcription initiation by specific dipeptides. *J. Bacteriol.* **177:**207–216.

McIver, K. S., and J. R. Scott. 1997. Role of *mga* in growth phase regulation of virulence genes of the group A streptococcus. *J. Bacteriol.* **179:**5178–5187.

McIver, K. S., A. S. Heath, and J. R. Scott. 1995. Regulation of virulence of environmental signals in group A streptococci: influence of osmality, temperature, gas exchange and iron limitation on *emm* transcription. *Infect. Immun.* **63:**4540–4542.

McNab, R., and H. F. Jenkinson. 1998. Altered adherence properties of a *Streptococcus gordonii hppA* (oligopeptide permease) mutant result from transcription effects on *cshA* adhesin gene expression. *Microbiology* **144:**127–136.

Midwinter, R. G., and G. G. Pritchard. 1994. Aminopeptidase N from *Streptococcus salivarius* subsp. *thermophilus* NCDO 573: purification and properties. *J. Appl. Bacteriol.* **77B:**288–295.

Mierau, I., E. R. S. Kunji, K. J. Leenhouts, M. A. Hellendoorn, A. J. Haandrikman, B. Poolman, W. N. Konings, G. Venema, and J. Kok. 1996. Multiple-peptidase mutants of *Lactococcus lactis* are severely impaired in their ability to grow in milk. *J. Bacteriol.* **178:**2794–2803.

Miller, C. G., and K. MacKinnon. 1974. Peptidase mutants of *Salmonella typhimurium*. *J. Bacteriol.* **120:**355–363.

Miller, C. G., and G. Schwartz. 1978. Peptidase-deficient mutants of *Escherichia coli*. *J. Bacteriol.* **135:**603–611.

Morton, D. S., and J. D. Oliver. 1994. Induction of carbon starvation-induced proteins in *Vibrio vulnificus*. *Appl. Environ. Microbiol.* **60:**3653–3659.

Moses, A. E., M. R. Wessels, K. Zalcman, S. Alberti, S. Natanson-Yaron, T. Menes, and E. Hanski. 1997. Relative contributions of hyaluronic acid capsule and M protein in a mucoid strain to virulence of the group A streptococci. *Infect. Immun.* **65:**64–71.

Nakayama, J., R. E. Ruhfel, G. M. Dunny, A. Isogai, and A. Suzuki. 1994. The *prgQ* gene of the *Enterococcus faecalis* tetracycline resistance plasmid pCF10 encodes a peptide inhibitor, iCF10. *J. Bacteriol.* **176:**7405–7408.

Niven, G. W., S. A. Holder, and P. Stroman. 1995. A study of substrate specificity of aminopeptidase N from *Lactococcus lactis* subsp. *cremoris* Wg2. *Appl. Microbiol. Biotechnol.* **44:**100–105.

Norgren, M., and A. Eriksson. 1997. Streptococcal superantigens and their role in the pathogenesis of severe infections. *J. Toxicol.* **16:**1–32.

Oesterlund, A., and L. Engstrand. 1997. An intracellular sanctuary for *Streptococcus pyogenes* in human tonsillar epithelium—studies of asymptomatic carriers and *in vitro* cultured biopsies. *Acta. Otolaryngol.* **117:**883–888.

Oliver, J. D. 1993. Formation of viable but nonculturable cells, p. 239–268. *In* S. Kjelleberg (ed.), *Starvation in Bacteria*. Plenum Press, New York.

Park, J. T. 1995. Why does *Escherichia coli* recycle its cell wall peptides? *Mol. Microbiol.* **17:**421–426.

Perez-Casal, J., M. G. Caparon, and J. R. Scott. 1991. Mry, a trans-acting positive regulator of the M protein gene of *Streptococcus pyogenes* with similarity to the receptor proteins of two-component regulatory systems. *J. Bacteriol.* **173:**2617–2624.

Podbielski, A. 1993. Three different types of organization of the *vir* regulon in group A streptococci. *Mol. Gen. Genet.* **237:**287–300.

Podbielski, A., and B. A. B. Leonard. 1998. The group A streptococcal dipeptide permease (Dpp) is involved in the uptake of essential amino acids and affects the expression of cysteine protease. *Mol. Microbiol.* **28:**1323–1334.

Podbielski, A., J. A. Peterson, and P. P. Cleary. 1992. Surface protein CAT reporter fusions demonstrate differential gene expression in the *vir* regulon of *Streptococcus pyogenes*. *Mol. Microbiol.* **68:**2253–2265.

Podbielski, A., B. Pohl, M. Woischnik, C. Koerner, K.-H. Schmidt, E. Rozdzinski, and B. A. B. Leonard. 1996a. Molecular characterization of group A streptococcal (GAS) oligopeptide permease (Opp) and its effect on cysteine protease production. *Mol. Microbiol.* **21:**1087–1099.

Podbielski, A., M. Woischnik, B. Pohl, and K. H. Schmidt. 1996b. What is the size of the group A streptococcal *vir* regulon? The Mga regulator affects expression of secreted and surface virulence factors. *Med. Microbiol. Immunol.* **185:**171–181.

Proctor, R. A., P. van Langevelde, M. Kristjansson, J. N. Maslow, and R. D. Arbeit. 1995. Persistent and relapsing infections associated with small-colony variants of *Staphylococcus aureus*. *Clin. Infect. Dis.* **20**:95–102.

Rakonjac, J. V., J. C. Robbins, and V. A. Fischetti. 1995. DNA sequence of the serum opacity factor of group A streptococci: identification of a fibronectin-binding repeat domain. *Infect. Immun.* **63**:622–631.

Rather, P. N., K. A. Solinsky, M. R. Paradise, and M. M. Parojcic. 1997a. *aarC*, an essential gene involved in density-dependent regulation of the 2′-N-acetyltransferase in *Providencia stuartii*. *J. Bacteriol.* **179**:2267–2273.

Rather, P. N., M. M. Parojcic, and M. R. Paradise. 1997b. An extracellular factor regulating expression of the chromosomal aminoglycoside 2′-N-acetyltransferase of *Providencia stuartii*. *Antimicrob. Agents Chemother.* **41**:1749–1754.

Reeve, C. A., P. S. Amy, and A. Matin. 1984a. Role of protein synthesis in the survival of carbon-starved *Escherichia coli* K-12. *J. Bacteriol.* **160**:1041–1046.

Reeve, C. A., A. T. Bockman, and A. Matin. 1984b. Role of protein-degradation in the survival of carbon-starved *Escherichia coli* and *Salmonella typhimurium*. *J. Bacteriol.* **157**:758–763.

Roe, B. A., S. Clifton, M. McShan, and J. J. Ferretti. 1997. *Streptococcal Genome Sequencing Project*. University of Oklahoma, Oklahoma City.

Ryan, F. J. 1951. Bacterial mutation in a stationary phase and the question of cell turnover. *J. Gen. Microbiol.* **21**:530–549.

Schmitt-Slomska, J., A. Boue, and R. Caravano. 1972. Induction of L-variants in human diploid cells infected by group A streptococci. *Infect. Immun.* **5**:389–399.

Simpson, W. J., D. LaPenta, C. Chen, and P. P. Cleary. 1990. Coregulation of type 12 M protein and streptococcal C5a peptidase genes in group A streptococci: evidence for a virulence regulon controlled by the virR locus. *J. Bacteriol.* **172**:696–700.

Stevens, D. L. 1996. Invasive group A streptococcal disease. *Infect. Agents. Dis.* **5**:157–166.

Taketo, A., and Y. Taketo. 1984. Enhanced production of cell-bound and extracellular streptolysin S by hemolytic streptococci pretreated with proteases. *Z. Naturforsch.* **40**:166–169.

Tam, P. S., I. J. van Alen-Boerrigter, B. Poolman, R. J. Siezen, W. M. de Vos, and W. N. Konings. 1992. Characterization of the *Lactococcus lactis* pepN gene encoding an aminopeptidase ho-mologous to mammalian aminopeptidase N. *FEBS Lett.* **306**:9–16.

Tam, R., and M. T. Saier. 1993. Structural, functional and evolutionary relationships among extracellular solute-binding receptors of bacteria. *Microbiol. Rev.* **57**:320–346.

Trieu-Cuout, P., C. Carlier, P. Martin, and P. Courvalin. 1988. Conjugative plasmid transfer from *Enterococcus faecalis* to *Escherichia coli*. *J. Bacteriol.* **170**:4388–4391.

Tynkkynen, S., G. Buist, E. Kunji, J. Kok, B. Poolman, G. Venema, and A. Haandrikman. 1993. Genetic and biochemical characterization of the oligopeptide transport system of *Lactococcus lactis*. *J. Bacteriol.* **175**:7523–7532.

van Alen-Boerrigter, I. J., R. Baankreis, and W. M. de Vos. 1991. Characterization and overexpression of the *Lactococcus lactis* pepN gene and localization of its product, aminopeptidase N. *Appl. Environ. Microbiol.* **57**:2555–2561.

van de Rijn, I. 1983. Streptococcal hyaluronic acid: proposed mechanisms of degradation and loss of synthesis during stationary phase. *J. Bacteriol.* **156**:1059–1065.

van der Lelie, D., C. Francisco, G. Venema, and M. J. Gasson. 1991. Identification of a new genetic determinant for cell aggregation associated with lactose plasmid transfer in *Lactococcus lactis*. *Appl. Environ. Microbiol.* **57**:201–206.

Watson, S. P., M. O. Clements, and S. J. Foster. 1998. Characterization of the starvation-survival response of *Staphylococcus aureus*. *J. Bacteriol.* **180**:1750–1758.

Wolf, B. B., C. A. Gibson, V. Kapur, I. M. Hussaini, J. M. Musser, and S. L. Gonias. 1994. Proteolytically active streptococcal pyrogenic exotoxin B cleaves monocytic cell urokinase receptor and releases an active fragment of the receptor from the cell surface. *J. Biol. Chem.* **269**:30682–30687.

Xu, H. S., N. Roberts, F. L. Singleton, R. W. Attwell, D. J. Grimes, and R. R. Colwell. 1982. Survival and viability of non-culturable *Escherichia coli* and *Vibrio cholerae* in the estuarine and marine environment. *Microb. Ecol.* **8**:313–323.

Zahner, D., T. Grebe, E. Guenzi, J. Krauss, M. van der Linden, K. Terhune, J. B. Stock, and R. Hackenbeck. 1996. Resistance determinants for beta-lactam antibiotics in laboratory mutants of *Streptococcus pneumoniae* that are involved in genetic competence. *Microb. Drug Resist.* **2**:187–191.

Zambrano, M. M., D. A. Siegele, M. Almiron, A. Tormo, and R. Kolter. 1993. Microbial competition—*Escherichia coli* mutants that take over stationary phase cultures. *Science* **259**:1757–1760.

THE EMERGENT PROPERTIES OF QUORUM SENSING: CONSEQUENCES TO BACTERIA OF AUTOINDUCER SIGNALING IN THEIR NATURAL ENVIRONMENT

Karen L. Visick and Edward G. Ruby

21

The discovery that *Vibrio fischeri* cells regulate the production of bioluminescence through the use of a cell density-sensing mechanism first termed autoinduction has led to the identification of a common theme among many different species of bacteria, namely, the coordinate control of a set of genes in response to the presence of a quorum of like bacteria (Fuqua et al., 1994). This control occurs through the synthesis of one or more autoinducer (AI) molecules, whose accumulation and recognition by a specific protein receptor/regulator present inside the bacterial cell are responsible for the autoinduction phenomenon. Although this mechanism for sensing the density of sibling cells in a particular environment is apparently not used by all bacteria, there is increasing evidence for its widespread adoption among bacteria found in a variety of biological niches (chapter 19 of this volume; Fuqua et al., 1994). In addition, a number of the AI-producing species carry two or more AI synthase/regulator gene pairs (Kuo et al.,

1994; Pesci et al., 1997). In some of these bacteria, one pair is controlled by another (Pesci et al., 1997), while there is at least a single example in which two autoinduction systems appear to encode redundant regulatory functions (Bassler et al., 1994). Finally, the sets of genes modulated by AI systems are diverse in function, and the mode of regulation, whether positive or negative, also varies.

The best-studied form of AI signaling is the communication between conspecific, or genetically identical, bacterial cells. In the case of many gram-negative bacteria like *V. fischeri*, the signaling molecule is an acyl derivative of homoserine lactone (HSL) [e.g., the *N*-(3-oxohexanoyl)-HSL (3-oxo-C6-HSL), produced by *V. fischeri* (Eberhard et al., 1981)] that freely diffuses through the cell's membranes (Kaplan and Greenberg, 1985). When such an AI is produced by a population of cells in an enclosed space, it accumulates externally, giving rise to an equivalent level (typically in the nanomolar to micromolar range) within the cells. Intracellularly, the AI is believed to bind a receptor protein, which assumes a conformation that stimulates changes in gene expression (Hanzelka and Greenberg, 1995; Schaefer et al., 1996; Sitnikov et al., 1995). An analogous, but distinct, type of cell-density signaling results from the secretion of short

Karen L. Visick, Department of Microbiology and Immunology, Loyola University of Chicago, 2160 South First Avenue, Building 105, Maywood, IL 60153.   *Edward G. Ruby,* Pacific Biomedical Research Center, University of Hawaii, 41 Ahui Street, Honolulu, HI 96813.

Cell-Cell Signaling in Bacteria, Edited by Gary M. Dunny and Stephen C. Winans
©1999 American Society for Microbiology, Washington, D.C.

peptides into the external medium by bacteria such as *Enterococcus faecalis*, *Bacillus subtilis*, and *Staphylococcus aureus* (Kleerebezem et al., 1997). Both HSL-type and peptide-type cues allow bacteria to coordinate expression of genes that are useful and/or required only when a sufficiently large population of cells is present, such as genes that regulate swarming motility (activated by an HSL derivative in *Serratia liquefaciens*) or conjugation (signaled by peptides in *E. faecalis*). For the purposes of this chapter, we will use the term AI only to refer to molecules that are derivatives of HSL.

Recent studies have demonstrated that communication through the use of AI molecules is not limited to recognition among cells of the same species. Although there is often considerable specificity in the recognition of AIs by their receptor proteins, there are examples of cross-species, as well as some cross-genus, regulation by AIs (Bassler et al., 1997; Greenberg et al., 1979; McKenney et al., 1995). Furthermore, many of the genes regulated by these quorum-sensing systems encode important virulence determinants, and evidence is accumulating that AIs can play a significant role in the signaling that occurs between bacteria and their eukaryotic hosts (DiMango et al., 1995; Dunphy et al., 1997; Visick et al., 1996).

The regulation of gene expression by a variety of AIs and their cognate receptors is reviewed at length in other chapters of this volume. However, while the molecular biology and biochemical mechanisms underlying autoinduction have been well studied in culture, the functioning of this signaling mechanism under natural biological conditions has been more difficult to assess (Fig. 1). In this chapter, we focus on what is currently known about the emergent properties of quorum sensing, the properties of a biological system whose existence cannot be predicted simply through a complete description of the components making up an organism (Campbell, 1993). The classic example of emergent properties comes from studies of sickle cell anemia, in which morphologically deformed cells synthesized a human hemoglobin molecule with an altered affinity for oxygen and led biochemists to interpret this condition to be a genetic defect. Only after studies by population biologists, who demonstrated that this trait was maintained in certain human populations, and by ecologists, who linked the distribution of these populations with the co-occurrence of malaria, was it revealed that the "defective" hemoglobin played a crucial protective role against this endemic disease. Thus, only by discovering the presence of the emergent properties, through studies at the population and ecological levels, can we begin to assign an importance to their existence. It seems particularly relevant to address these levels when considering quorum sensing, a mechanism of signaling that, by definition, operates among a population of individual bacterial cells and within a specific subset of environments.

In this chapter we discuss three areas concerning the emergent properties of quorum sensing: (i) the potential consequences to a bacterium of possessing an extracellular signaling system, (ii) the use of AIs in the interactions of quorum-sensing bacteria with genetically distinct organisms, and (iii) some new frontiers in our understanding of AIs in the bacterial cell's extracellular environment.

CONSEQUENCES OF AN EXTRACELLULAR SIGNALING SYSTEM

By their nature, AIs are external signals and, as such, are not without risks as a method of constructive signaling between one bacterium and another. Some of these consequences, their negative and positive effects, and the possible defenses erected by the producing bacterium are outlined in Table 1. For example, alkaline conditions (such as are characteristic of seawater, pH 8.5) potentiate the breakdown of acyl HSL derivatives (Eberhard et al., 1981) and thus could cause attenuation of the cue. While a fast turnover would result in a waste of cellular resources dedicated to AI production for bacteria present in such an en-

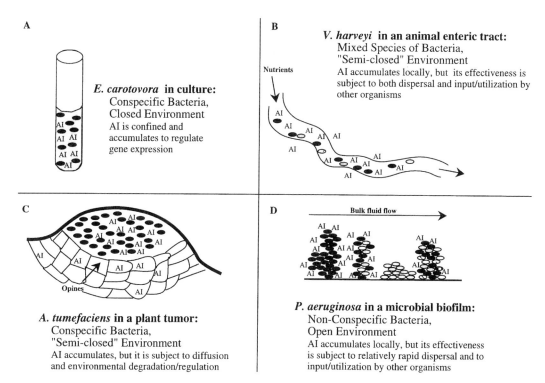

FIGURE 1 Schematic diagram of representative quorum-sensing bacteria in various biological niches: in a laboratory pure culture (A); in the enteric tract of an animal host (B); in a plant tumor (C); and in a biofilm composed of pure and mixed colonies (D). In each panel the quorum-sensing bacteria are depicted as black ovals. The predicted location and relative concentration of autoinducer molecules (AI) are indicated in each panel. Secretion of opines by the plant cells is indicated in panel C.

vironment, it would also allow a more rapid sensing of changes in cell density. Another potential difficulty is that other organisms could secrete similar compounds that function either in competition or as a defense mechanism, thus creating a confusing mixture of "false-positive" or even "false-negative" signals that would lead to unproductive responses. Finally, the AI signal might be recognized and utilized by other organisms. Those organisms that do not invest in creating their own signal, but profit by its production by others, might be considered "cheaters," whose success could lead to a selective disadvantage for producers of the cue. Other organisms may use the AI cue as a means of detecting the presence of the producing bacterium; in some cases, the presence of the external compound may result

in unwanted competition or even predation. AI-excreting bacteria probably face one or more of these challenges and, at least in some cases, may have evolved counteractive strategies to defend against them (Table 1).

One feature of AI systems that may serve as a defense against improper signaling is the specificity of the AI signal recognition process. In some organisms, the receptor protein has a high degree of specificity for its cognate AI and generally will not modulate gene expression in the presence of a different acyl HSL derivative. For example, the *Pseudomonas aeruginosa* receptor protein LasR, together with *N*-3-(oxododecanoyl)-HSL (3-oxo-C12-HSL), regulates the *las* genes by recognizing and binding to upstream regulatory sequences (chapter 10 of this volume; Gray et

TABLE 1 Consequences of using quorum sensing in the natural environment

	Consequences		
	Loss of the signal	Susceptibility to a "false" signal	"Interception" of the signal
Causes	By diffusion into environment By abiotic degradation By competitive removal	Through the secretion of a similar or identical molecule by an unrelated bacterial species	By a signal receptor of an unrelated organism
Negative effects	Creates a continuous energy cost to the cell	Leads to inappropriate gene induction or inhibition	Attracts and/or promotes unwanted competition or predation ("cheater") Initiates host defenses
Positive effects	Prevents unnecessary gene induction Allows a rapid sensing of changes in the environment	Allows input from different organisms in the same niche	Serves to coordinate consortial metabolism among bacterial species Serves to coordinate host and bacterial development Interferes with host defenses
Defenses	Ensure that positive autoregulation occurs only under specific ambient conditions Decrease the threshold concentration of AI required for response	Increase specificity of receptor Increase threshold concentration of AI required for response Use multiple AIs and receptors in overlapping signal pathways	Create community-level responses that are of general selective value

al., 1994; Passador et al., 1993). Interestingly, although the 3-oxo-C12-HSL/LasR complex is also able to recognize similar sequences upstream of the luminescence (*lux*) genes of *V. fischeri* and activate gene expression, LasR in the presence of the *V. fischeri* AI 3-oxo-C6-HSL can activate neither *las* nor *lux* transcription (Gray et al., 1994). Similarly, while the *V. fischeri* regulator LuxR in the presence of the *Pseudomonas* AI (3-oxo-C12-HSL) (Pearson et al., 1994) cannot activate either *lux* or *las* transcription, the LuxR/3-oxo-C6-HSL couple can (Gray et al., 1994). Thus, despite the fact that the two autoinducers are structurally similar, their respective receptor proteins have a high degree of specificity for the native signaling molecule. Recent results suggest that both the length and substituents of the acyl chain are responsible for this specificity (Eberhard et al., 1986; McClean et al., 1997; Passador et al., 1996; Schaefer et al., 1996; Zhang et al., 1993).

Other receptor proteins appear to be less specific in the recognition of their cognate AIs. It is possible that this lack of specificity occurs only in the artificial conditions of the test tube, since the biologically relevant concentration of AIs in nature has not been well investigated. Alternatively, the lack of specificity may indicate an adaptation of the organism to accept signaling input from nonconspecific bacteria. For example, a *swrI* mutant of *S. liquefaciens*, which is defective for the synthesis of its own two AIs (*N*-butanoyl-L-HSL [C4-HSL] and *N*-hexanoyl-L-HSL [C6-HSL]), is unable to achieve swarming motility (Eberl et al., 1996). Restoration of swarming motility requires either the addition of synthetic C4-HSL (150 nM) or the addition of C6-HSL (900 nM). However, swarming motility can also be restored by the addition of a 5- to 50-fold higher concentration of certain other AIs, such as *N*-3-oxobutanoyl-HSL (6 μM) or 3-oxo-C6-HSL (9 μM) (Eberl et al., 1996). Thus, higher concentrations of a nonnative signaling molecule are capable of activating gene expression through the native AI regulator, although it is not known whether these concentrations would be present in the natural environment of *S. liquefaciens*.

In addition, the threshold of detection of the same native AI may be different for different organisms. For example, the closely related species *Vibrio logei* and *V. fischeri* both produce and respond to 3-oxo-C6-HSL (Greenberg, personal communication); however, the induction of AI-regulated genes in *V. logei* requires higher levels of VAI-1 than those required for induction of their homologs in *V. fischeri* (Boettcher and Ruby, 1990). These differences in minimal effective dosages may reflect the levels of AI achieved in the niches typically occupied by these organisms.

Another mechanism by which bacteria achieve a greater signal specificity may be the encoding of multiple, interacting AI systems within a single organism. *V. fischeri* cells carry two AI systems, one of which is the well-studied prototype LuxR/3-oxo-C6-HSL, which is involved in the control of bioluminescence, and the other is the more recently described *N*-(octanoyl-L-/-HSL)-synthase, AinS (Gilson et al., 1995; Kuo et al., 1994). While C8-HSL is capable of stimulating luminescence in the presence of LuxR, its natural role appears to be one of competitive inhibition (Kuo et al., 1996). A mutant defective in *ainS* induces luminescence at a lower cell density, and, once induced, luminescence production increases more rapidly in this mutant (Kuo et al., 1996). The addition of C8-HSL (75 nM) either to the *ainS* mutant or to the wild-type parent causes a delayed onset of luminescence (Kuo et al., 1996). These results suggest that the inhibition by C8-HSL may provide a level of "fine-tuning" that prevents premature gene induction under environmental conditions that are inappropriate for luminescence. Such a mechanism would also be useful in the hypothetical situation in which other organisms emit a signal that falsely simulates the presence of a quorum; the secretion of a second, inhibitory, signaling molecule that competes for the same receptor protein may function as a defense mechanism to decrease the possibility of false signaling, as well as premature gene expression. C8-HSL could

also have a distinct role in the control of other genes, but such a function has not yet been established.

In summary, there are multiple means by which bacteria may defend against false signaling in the environment. They have evolved receptors that are fairly specific to their cognate AIs and that are responsive to a certain minimal concentration of signal. The presence of multiple AI systems in a single bacterium may also serve as a defense. However, while cell-to-cell communication through co-occurring AI signals in nature can lead to nonproductive signaling, in some cases stimulation of autoinduction by the presence of other species of bacteria may represent an effective mechanism for a bacterium to recognize and respond to additional biological characteristics of its environment.

QUORUM SENSING WITHIN AND BETWEEN POPULATIONS

The natural environments in which AI-producing bacteria are found are diverse in character and in many cases contain multiple bacterial taxa (Fig. 1). In other cases, such as certain well-studied bacteria-host associations (Ruby, 1996), the bacterial population is present as a pure or monospecific culture for at least a portion of its life cycle. In addition, new AI-producing strains from other environments are currently being identified using global screening methods (Shaw et al., 1997; Taylor et al., 1996). Thus, when determining the role of AIs in nature, the synergistic effects of the surrounding environment, including other bacteria and/or host tissue, must be considered. In the following sections, we discuss the influence that other organisms may have on AI-mediated gene expression in a quorum-inducible bacterium, as well as the consequences that this consortial or group behavior of bacteria may exert on other organisms.

Quorum Sensing between Bacterial Taxa

One of the earliest studies to consider the role of AIs in the natural environment demonstrated that autoinduction of bioluminescence in the marine organism *Vibrio harveyi* could be stimulated by other marine bacteria (Greenberg et al., 1979). Since that time, it has been established that *V. harveyi* encodes two quorum-sensing systems for the control of bioluminescence, sensor system 1 and sensor system 2 (chapter 17 of this volume). The two systems converge to stimulate regulation of the *lux* genes through a single regulatory protein (Showalter et al., 1990). Sensor system 1 produces a typical AI/receptor complex, while the nature of the second quorum-sensing signal, believed not to be an HSL derivative, remains undetermined (Bassler et al., 1997).

Bassler et al. (1997) have clarified the original *V. harveyi* study and added depth to our understanding of the interactions of different bacteria with each other. Although the two sensor systems initially appeared to be redundant in the control of bioluminescence, it now seems likely that the presence of two systems allows *V. harveyi* to respond to input signals from either conspecific or nonconspecific bacteria. The majority of the other species of microbes that showed the ability to stimulate bioluminescence of *V. harveyi* did so by acting through system 2. These microbes included the closely related *Vibrio* species *V. cholerae*, *V. anguillarum*, and *V. natriegens*, as well as the more distantly related *Yersinia enterocolitica*. Although we do not yet know whether the AI activities produced by these bacteria are identical to that of *V. harveyi*, there are examples of two unrelated bacteria making identical AIs (Fuqua et al., 1996). Only one of the organisms that was examined signaled *V. harveyi* luminescence via system 1 (Bassler et al., 1997). These data indicate that the quorum-sensing mechanism of *V. harveyi* regulates bioluminescence genes by integrating conspecific input (obtained specifically through system 1) with input from other organisms (through system 2).

This ability to receive and respond to a signal from other organisms may be an important factor in communication in certain environments, such as the enteric tracts of fishes, which are typically colonized by a diverse population of bacteria, often including *V.*

harveyi (Fig. 1) (Ruby and Morin, 1979). The potential for external control of luminescence suggests that light emission is useful when *V. harveyi* is present in a group of cells, even of nonconspecific bacteria. It is not clear why the expression of bioluminescence genes might be important to *V. harveyi* and others in such an environment, but a survival advantage may be linked to the high affinity that luciferase has for oxygen (Lloyd et al., 1985). The ability to lower the ambient oxygen tension, and thereby decrease potential oxidative stress or competition from aerobes, may confer a survival advantage on the facultatively anaerobic portion of the population, including *V. harveyi* (Nealson and Hastings, 1979). An alternative explanation that has been suggested is that the fecal pellets produced by marine animals, if bioluminescent, may be preferentially consumed by other organisms, thus returning the excreted fecal cohort to an enteric environment where they can continue to propagate (Andrews et al., 1984; Nealson and Hastings, 1979).

In addition to the level of bioluminescence, the production of poly-3-hydroxybutyrate, a fatty acid storage product, is also induced in *V. harveyi* by the accumulation of AI (β-hydroxy-C4-HSL), although the significance of this connection between luminescence and carbon metabolism is unclear (Sun et al., 1994). Additional sets of *V. harveyi* genes may be controlled by one or both of the sensing systems, and the expression of these other genes may be important under specific environmental conditions. The identification of additional genes controlled by either of these sensing systems, and the determination of the environmental conditions optimal for their expression, will help elucidate whether the systems are truly redundant or whether the different signals result in distinct physiological responses. This work has revealed an example of a bacterium that has evolved a regulatory network based on signaling input from other species of apparently cooperative bacteria, resulting in a community-level communication process.

Interspecies communication through the use of AIs has been proposed as a possible mechanism by which the pathogenicity of *Burkholderia cepacia* is enhanced (McKenney et al., 1995). *B. cepacia* is associated with fatalities in patients with cystic fibrosis, but in most of these cases this organism was found to be co-colonizing tissue with *P. aeruginosa*. The addition of the *P. aeruginosa* AI (3-oxo-C12-HSL) to cultures of *B. cepacia* potentiated their production of virulence factors, including a sevenfold increase in siderophore synthesis and more than a twofold increase in protease excretion (McKenney et al., 1995). A small elevation in virulence factor production (threefold in the case of siderophores) was also obtained by the addition of a concentrate of spent medium obtained from *B. cepacia* cultures, suggesting that *B. cepacia* may also encode its own quorum-sensing system (McKenney et al., 1995). The supernumerary colonization of lungs by the opportunistic *B. cepacia*, following infection by other microbes such as *P. aeruginosa*, may be aided by the presence of 3-oxo-C12-HSL molecules (McKenney et al., 1995). If this is in fact the case, *B. cepacia* would represent an example of a "cheater" organism that exploits a microenvironment through the use of another organism's cellular investment in signal production (Table 1).

We have described two examples of AI signaling that may occur between distantly related bacteria that are present in the same niche. While the biological relevance has yet to be confirmed, it is likely that, with the widespread nature of quorum-sensing systems, such interactions are occurring in the environment. The development of specific signaling molecules that can be used to sense not only conspecific bacteria but also certain nonconspecific bacteria present in specific niches is a clever mechanism for community-level regulation of gene expression.

Quorum Sensing in a Bacterial-Host Symbiosis

Many quorum-sensing bacteria are found in high-cell-density niches inside eukaryotic hosts. *V. fischeri* cells, unlike those of *V. harveyi*, are known to occur in dense, monospecific populations in nature where an environmental importance for bioluminescence has been established. Certain species of marine squids and fishes possess specialized light-emitting organs that contain as many as 10^{10} *V. fischeri* cells. This bacterial culture produces the light that is used in predatory and/or antipredatory behaviors of the host (Nealson and Hastings, 1979). The symbiosis between *V. fischeri* and the Hawaiian sepiolid squid *Euprymna scolopes* has been established as a model system for studying the signaling interactions between bacteria and their hosts (reviewed by Ruby, 1996).

Newly hatched *E. scolopes* juveniles are rapidly colonized by symbiosis-competent *V. fischeri* cells that are present in the surrounding seawater. Colonization is accompanied by the onset of the AI-controlled (chapter 18 of this volume) luminescence activity within 8 to 10 h of the initiation of the association (Fig. 2). The amount of light emitted per cell, or specific luminescence, also increases over 100-fold (Ruby and Asato, 1993). The maximal level of colonization in the juvenile squid is between 10^5 and 10^6 *V. fischeri* cells and occurs extracellularly in a series of crypts that have a total volume of about 750 pl (Montgomery and McFall-Ngai, 1993). The result is an effective concentration of greater than 10^9 bacteria per ml, a level that is 10-fold higher than that necessary to fully induce luminescence in culture (Boettcher and Ruby, 1990).

In addition to providing an environment suitable for both rapid colonization and the production of bioluminescence by *V. fischeri* cells, the squid host plays an active role in controlling the level of their bioluminescence emission. This is achieved by a combination of at least three mechanisms. The first and simplest is the presence of several accessory tissues, including a muscle-controlled ink sac that the animal can use as a moveable shutter to either conceal or reveal the bioluminescence of its symbiotic partner (Fig. 3) (McFall-Ngai and Montgomery, 1990).

FIGURE 2 Diagrammatic representation of the daily rhythm of luminescence and symbiont cell expulsion by the squid *E. scolopes*. The thick black line represents the level of luminescence emitted by the juvenile squid and is roughly proportional to the level of colonization by *V. fischeri* cells. This luminescence begins during the first 12 h after inoculation and subsequently cycles up during the night (black boxes across top strip) and down during the day (white boxes). The onset of the drop in luminescence corresponds to the peak of bacterial cell expulsion from the light organ crypts into the surrounding seawater during the periods indicated by the arrows. Modified from Boettcher et al. (1996) and Lee and Ruby (1994).

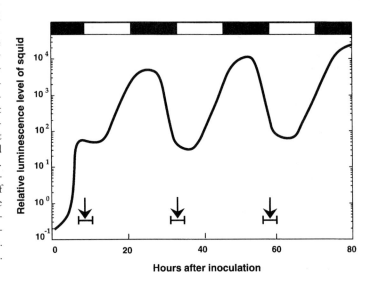

The other two light-modulating mechanisms are linked to a diurnal phenomenon that illustrates the host's ability to control both the number of cells in the bacterial population and the level of luminescence of these cells. Each morning, over 90% of the population of bacterial symbionts is expelled from the light organ crypts, with a concomitant drop in bioluminescence (Fig. 2) (Boettcher et al., 1996; Lee and Ruby, 1994). Luminescence subsequently increases to a maximal level by nightfall, owing in part to the regrowth of the remaining bacteria in the crypts. This cyclic pattern of bioluminescence is not observed in squid that are held in either constant dim illumination or constant dark conditions, demonstrating that the primary cue for the control of bioluminescence is external to both the bacterium and the squid (Boettcher et al., 1996) and is linked to the environmental light conditions.

In addition, regardless of the number of symbiont cells present, the host is also able to directly modulate the specific activity of light emission by the bacteria (Boettcher et al., 1996). Boettcher et al. (1996) compared the average level of luminescence produced by

bacterial cells in the intact juvenile squid light organ (their "actual luminescence"), with that of these same cells immediately after their release from the light organ (their "potential luminescence"), at various times during the normal day/night cycle. There was a significant similarity between the actual and potential levels of luminescence (per cell) during the "peak" period of luminescence (that is, near the onset of nightfall) (Fig. 2). However, during daylight or "non-peak" periods, the actual luminescence level was about 10-fold below that of the potential luminescence. These data suggested that the animal host actively represses the bioluminescence of its symbiotic partner during a significant portion of each 24-h cycle. One mechanism for repression may be through the control of the rate that oxygen, a substrate for the bioluminescence reaction, is provided by the host to the symbiont population (Boettcher et al., 1996). Another hypothetical mechanism is that luminescence is repressed in part through the stimulation of synthesis of C8-HSL, which may competitively inhibit LuxR stimulation of *lux* gene expression at certain times of the day. In any case, the multiple levels of control

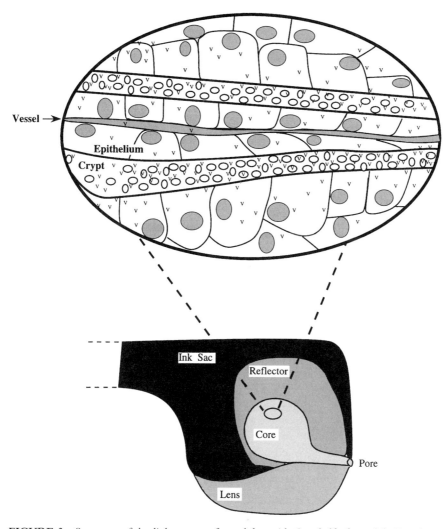

FIGURE 3 Structure of the light organ of an adult squid. One half of an adult *E. scolopes* light organ is depicted in the lower drawing. The central core is composed of a series of crypt spaces in which the bioluminescent bacteria are located, flanked by host epithelial cells. The core is surrounded by three tissues: the reflector, lens, and ink sac. The crypts of the light organ communicate with the surrounding seawater through a lateral pore. A small segment of the core is enlarged to show the location of the symbiotic *V. fischeri* (indicated by small white ovals) in the crypt spaces. The arrangement of epithelial cells and the vascularization of the tissue are also indicated. The small v's represent 3-oxo-C6-HSL and C8-HSL and their diffusion into the host tissue (Boettcher and Ruby, 1995), although there are as yet no data to confirm the presence of C8-HSL in the light organ. Modified from McFall-Ngai and Montgomery, 1990.

exerted by both the squid and the bacterium strongly suggest that the ability to modulate bioluminescence is of considerable importance to these organisms.

Interestingly, when grown in culture, bioluminescent strains of *V. fischeri* isolated from the light organ of *E. scolopes* emit only low levels of light that are not visible to the human eye (measured at 0.8 quantum per s per cell with a sensitive photometer), but are 1,000-fold brighter immediately after release from the light organ, achieving a level of about 800 quanta per s per cell (Boettcher and Ruby, 1990). This low level of in-culture luminescence is not seen in other *V. fischeri* strains isolated from the light organs of other symbiotic squid or fish associations. For example, MJ-1, a strain of *V. fischeri* isolated from the fish *Monocentris japonicus* (Ruby and Nealson, 1976), produces over 1,000 times higher luminescence in culture than do *E. scolopes* isolates such as strain ES114 (Boettcher and Ruby, 1990). Studies of strain ES114 in culture have revealed that these *V. fischeri* cells are underproducers of 3-oxo-C6-HSL (Gray and Greenberg, 1992). The amount of this AI produced in culture by ES114 is about 10-fold less than that of *V. fischeri* cells isolated from the related squid *Euprymna morsei*, although the concentrations of AI present in the respective squid associations are approximately the same (Boettcher and Ruby, 1995). These data suggest that the *E. scolopes* light organ association provides a special environment that promotes a sufficiently high concentration of *V. fischeri* cells to result in the maximal induction of luminescence even in these AI-underproducer isolates. Whether this induction involves the presence of conditions or molecules other than 3-oxo-C6-HSL remains to be determined. (Studies describing the amount and diffusion of 3-oxo-C6-HSL in the squid light organ are discussed further later in this chapter.)

Mutants of *V. fischeri* that are defective for either of the luminescence regulators, LuxR or LuxI, do not show an increase in specific activity of luminescence during colonization of the squid light organ and, in fact, produce no detectable light at all in the early stages of the association, despite the relatively high level of residual luminescence observed in these strains in laboratory culture (Fig. 4) (Dunlap and Kuo, 1992; Visick et al., 1996). These data support the hypothesis that the mechanism of induction of bioluminescence in the squid has the same requirement for the presence of the *luxR* and *luxI* regulatory genes as is seen in culture. In addition, these data suggest that C8-HSL, despite having some ability to initiate luminescence (Kuo et al., 1996), is not, by itself, sufficient for symbiotic induction of luminescence (Visick et al., 1996).

In addition to the diminished capacity to luminesce in the squid, both *luxI* and *luxR* mutants exhibit between a 3- and 10-fold decrease in the level of colonization of the juvenile light organ within 48 h of the initiation of the association (Visick et al., 1996). Curiously, a similar colonization defect is seen for a mutant lacking the AI-regulated gene, *luxA*, which encodes a subunit of bacterial luciferase, the enzyme that catalyzes the luminescence reaction. It is unclear at this time whether the colonization deficiency of all three *lux* mutant strains is due to an inability of these dim or dark cells to survive certain conditions in the light organ or to a recognition of the luminescence defect by the squid, leading to their expulsion in favor of light-emitting strains.

In summary, the squid–*V. fischeri* symbiosis has proved to be an effective model system for studying autoinduction at the community level and has revealed that AI-controlled genes and their products are subject to multiple levels of control in nature. Luminescence induction inside the host is higher than the maximal level obtained in culture and is dependent on the same regulatory control genes, *luxR* and *luxI*. The absence either of these quorum-sensing components or of luciferase itself results in a defect in the colonization of the juvenile squid, further supporting the importance of luminescence in the symbiotic association. Continued investigation of the role of bioluminescence should yield a better understanding of the consequences of AI-controlled gene regulation in the natural environment provided by the host light organ.

Quorum Sensing during Bacterial Virulence

It is rapidly becoming clear that both AI/receptor regulatory pairs and the genes that they

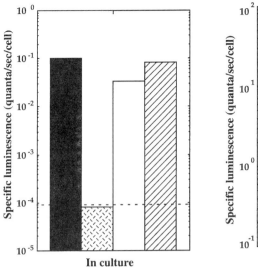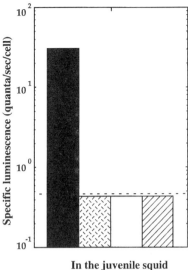

FIGURE 4 Specific luminescence of wild-type and *lux* mutants of *V. fischeri*, both in culture and in the juvenile squid light organ. The maximum specific activities of luminescence of the parent strain (solid box) and the *luxA* (hatched box), *luxI* (open box), and *luxR* (striped box) mutants are plotted. The horizontal dashed line at the bottom of the graph represents the lower limit of luminescence detection by the photometers used in each experiment.

control play an important role in the interactions of bacteria with a number of higher organisms. There are several examples of plant and animal pathogens whose virulence is attenuated by mutations in their AI synthase or receptor genes (Beck von Bodman and Farrand, 1995; Tang et al., 1996). While the vast majority of host-associated bacteria are commensal or beneficial, the examination of a number of pathogenic associations has provided an important source of information on the role of quorum sensing in the ecology of bacteria that colonize host tissues.

AI-production mutants of the plant pathogens *Erwinia carotovora* and *Erwinia stewartii*, bacteria that cause soft rot and Stewart's wilt, respectively, are avirulent (chapter 7 of this volume; Beck von Bodman and Farrand, 1995; Pirhonen et al., 1993). The avirulence is presumably due to the inactivity of the AI-controlled genes, encoding either cell wall-degrading enzymes (in the case of *E. carotovora*) or heteropolysaccharide capsule production

(in *E. stewartii*). When the mutant bacteria are preincubated with their species-specific AI and inoculated onto plant leaves, virulence is at least partially restored, but this restoration is dependent upon additional application of the AI to the leaf. In both cases, continued treatment of the plant leaves with added AI was required to sustain the spread of the bacterial infection. The addition of the AI alone to mock-inoculated leaf tissue did not induce any disease symptoms (Beck von Bodman and Farrand, 1995), suggesting that it is not the AI itself that is responsible. Interestingly, in the case of *E. stewartii*, external application of AI on the leaves was insufficient for the mutant to elicit all of the disease symptoms that the wild-type strain produced. These latter results may simply indicate either that the AI fails to persist on the leaves or that the levels of AI added were not comparable to those resulting from a natural infection and thus were insufficient to completely complement the virulence defect.

It is interesting that restoration of virulence to the AI mutants of both of the *Erwinia* species required exogenous addition of AI, even though the mutants had been preincubated with AI. Thus, AI is required not only for *Erwinia* species to initiate the infection, but also for the infection to persist. This is indicative of a need for the continual communication of the bacteria with each other during the infection. Neither study (Beck von Bodman and Farrand, 1995; Pirhonen et al., 1993) reported the concentration of AI at the application site or the minimal level required for successful virulence. Thus, it appears that the study of plant pathogenesis could benefit from an investigation of the rate at which AIs can penetrate plant cells and a description of the concentration and stability of AI present in a natural infection.

A mutant strain of the opportunistic human pathogen *P. aeruginosa*, defective for the 3-oxo-C12-HSL receptor LasR, has a decreased ability to bind host epithelial cells and is significantly impaired in virulence when tested in a mouse infection model (Tang et al., 1996). These defects in virulence may result from an underexpression of the genes that LasR regulates (chapter 10 of this volume), whose products include the virulence factors elastase, endotoxin A, and alkaline protease, as well as those involved in the general secretion pathway (Chapon-Hervé et al., 1997). The relative consequences on virulence due to the direct and indirect effects of a mutation in *lasR* await further evaluation.

A role for AIs in virulence has also been demonstrated in the insect pathogen *Xenorhabdus nematophilus*, which forms a cooperative symbiosis with the nematode *Steinernema carpocapsae* (Forst and Nealson, 1996). *S. carpocapsae* invades susceptible insect larvae and delivers a fatal inoculation of *X. nematophilus*. Direct injection of a wild-type *X. nematophilus* strain into insects causes a 50% mortality within 9.6 h, while insects injected with an AI mutant (defective for the production of *N-β*-hydroxybutanoyl HSL [*β*-hydroxy-C4-HSL]) survive for over 2 weeks (Dunphy et al.,

1997). Infections carried out in the presence of 1 μg of added AI decrease the time necessary to achieve 50% mortality both for the wild-type strain (to 8.6 h) and for the AI mutant (to less than 2 days) (Dunphy et al., 1997). Although the results are highly suggestive of a direct role for AI in the virulence of this bacterium, it should be noted that the experimental injection of the bacterium and AI is not the natural avenue of infection and that the concentration of AI present in the natural infection is as yet unknown. Future studies that ask whether the AI-synthesis mutation has an effect not only on infection of the insect but also on the bacterium's ability to form the natural cooperative association with the nematode will yield useful information regarding the roles of AI in the cooperative and pathogenic phases of the ecology of *X. nematophilus*.

NEW FRONTIERS IN AI FUNCTIONS

The environment includes not only other bacteria and host cells, but also other chemical cues. We discuss in this section what is currently known about the consequences of AI diffusion into a bacterium's surroundings, the influence of other molecules on AI-regulatory circuits, as well as the potential conflicts posed by AI mimicry in the environment.

Effect of AI Diffusion into the Surrounding Environment

While much is known about the activity of AIs inside the cell, other characteristics, such as the diffusion of these molecules through the bacterial cell envelope, across abiotic surroundings (water and soil) and into other cells are only beginning to be studied. As yet, the ability of acyl HSL derivatives to freely diffuse through the bacterial envelope has been experimentally demonstrated only for 3-oxo-C6-HSL (Kaplan and Greenberg, 1985). Although it is likely that all acyl HSL derivatives can permeate the cell membrane, the diversity of acyl adjuncts suggests that the rates of diffusion may be different for different molecules. Similarly, the nature of the bacterial (or host) membrane may present a barrier of differential effectiveness. For example, the concentration of C4-HSL required to achieve induction of a reporter construct in *P. aeruginosa* is only one-tenth that required for the activation of the same construct in *Escherichia coli* (Pearson et al., 1997). While these results may be due simply to a higher expression of LasR in *E. coli*, the authors speculate that C4-HSL may pass more freely through the *P. aeruginosa* membrane, although they acknowledge that these results could also be explained by the absence of other specific factors in *E. coli*. Clearly, the rates of diffusion of various acyl HSL derivatives through different bacterial membranes could have a profound influence on both cell-cell communication and gene induction, because the relative levels of these molecules appear to be a significant factor in at least some systems (Passador et al., 1996; Schaefer et al., 1996).

Once the AI leaves the cell, the nature of its environment becomes important. Recent studies have shown that AI signaling between cells is able to occur within the rhizosphere of wheat seedlings. In co-inoculation studies, the level of expression of AI-controlled genes in an AI-synthase mutant strain increased with increasing ratios of complementing wild-type to mutant cells in the mixed population (Wood et al., 1997). Studies of this type provide biologically relevant information about the bacterial communication in natural settings.

The consequence of AI diffusion through eukaryotic tissues must also be considered, because most of the AI-producing bacteria identified to date are known to interact with the cells of specific eukaryotic host species (chapter 19 of this volume). In principle, the chemical properties of AIs should make their diffusion into host cells impossible to prevent. However, in only one bacteria-host association do we have data on the amount and flux of an AI in host tissue. The amount of 3-oxo-C6-HSL autoinducer present in an adult *E. scolopes* light organ has been estimated to be 100 to 200 pg of extractable AI activity, based on a luminescence bioassay (Boettcher and

Ruby, 1995). This level would result in an effective concentration of at least 100 nM of 3-oxo-C6-HSL in the whole light organ. Further experiments showed that this AI diffuses into and through the epithelial cells surrounding the crypt space of the light organ where the bacteria are housed (Fig. 3). The epithelium is surrounded by two other tissues, the reflector and lens (McFall-Ngai and Montgomery, 1990). Because of the chemical and structural nature of these tissues, AIs are less likely to penetrate them freely. Thus, these tissues may effectively confine the AI, promoting the high level of luminescence induction characteristic of the bacterial symbionts (Ruby and Asato, 1993). It is less clear how AI molecules would accumulate within the light organ of a juvenile squid, in which the reflector and lens are less developed (Montgomery and McFall-Ngai, 1993). Similarly, the possible effect that the vascularization of the crypt epithelium (Fig. 3) might have on the flux of AI out of the crypts and into the rest of the animal is unknown. The biochemical consequences of the AI molecules diffusing into the epithelial cells of the adult squid remains to be defined, although, by itself, AI is incapable of initiating the program of host tissue apoptosis (Doino and McFall-Ngai, 1995) that normally ensues after colonization of the nascent juvenile light organ (Montgomery and McFall-Ngai, 1994).

The first evidence that AIs may provide a signal to eukaryotic cells has been obtained during experiments with the *P. aeruginosa* autoinducer 3-oxo-C12-HSL. The addition of synthetic 3-oxo-C12-HSL was shown to be sufficient to elicit interleukin-8 production in a respiratory epithelial cell line (DiMango et al., 1995), a response that is characteristic of infection of these cells with *P. aeruginosa*. Because interleukin-8 is a chemoattractant for neutrophils (Huber et al., 1991), it appears unlikely that this effect would benefit the infecting *P. aeruginosa* cells; rather, AI production in this case may be an inadvertent signal that alerts the host to the pathogen's presence. Thus, this phenomenon might be an example of a host adaptation to take advantage of a chemical cue excreted by the bacterium and to utilize it for an entirely different purpose.

A more extensive study of the role of 3-oxo-C12-HSL in eliciting an immune response has yielded data that support a role for this molecule in immunomodulation (Telford et al., 1998). Levels of IL-12 and tumor necrosis factor-α are decreased by the addition of 3-oxo-C12-HSL, although the latter effect required approximately 40 μM AI, a level that may or may not be achieved in the host tissue (Telford et al., 1998). These data suggest that the AI produced and secreted by *P. aeruginosa* may by itself function as a virulence determinant by modulating the host immune response.

AIs have also been shown to influence host development in a bacterial-plant symbiosis, suggesting that they may be used as host signals in cooperative bacterial associations as well. Inoculation of bean seedlings with an AI-synthase mutant, *raiI*, of *Rhizobium etli* resulted in a twofold increase in nodule formation in the plant relative to the number produced by inoculation with the wild-type strain (Rosemeyer et al., 1998). The same effect was not seen with a *raiR* mutant defective for the putative receptor protein, suggesting that the AI is functioning with another receptor, perhaps in the host. *R. etli* produces seven AIs, two of which are differentially expressed by the *raiI* and *raiR* mutants (lower in the *raiI* mutant). These data implicate a role for one or both of these AIs, or for genes downstream of these regulatory factors, in the restriction of host nodule number (Rosemeyer et al., 1998).

Taken together, these studies provide evidence that AIs may function not only to signal cell density and coordinate gene expression in the bacterium, but also to provide a cue that is of significance to the host (Table 1). Clearly, much remains to be learned about the consequences of AIs in nature; for instance, the turnover rates of AIs in any natural environment, including host tissues, have yet to be measured. It will also be exciting to discover

the identity of target molecules in or on the eukaryotic cells that recognize and respond to the presence of AI or of the targets of the genes that they may control. Whether the targets of AI in eukaryotic cells are the same in hosts for pathogenic bacteria as in hosts for symbiotic bacteria is clearly another important and useful line of inquiry.

Selective Advantage of AI Regulatory Circuits

The results of many genetic studies have shown that quorum sensing is an effective mechanism for coordinating the regulation of a set of genes that may be useful only when a high density of cells is present. What has been less well documented is how AI-regulated genes may confer a selective advantage on the bacterium that encodes them. Such a selective advantage must manifest itself by an increased fitness or competitiveness in the organism's natural environment. Earlier, we described the importance of luminescence induction to the maintenance of a symbiotic light organ population. Below we discuss other organisms for which evidence is emerging for the use of AI-controlled genes to confer a selective advantage.

In the pathogenic association between *Agrobacterium tumefaciens* and its plant host, it is clear that the signaling process follows a two-way path: the T-DNA that is transferred from the bacterium into the plant cell encodes proteins that cause plant cells to produce and secrete a class of compounds called opines; these opines are in turn recognized by the bacterium, resulting in an induction of both the opine-catabolizing genes and the AI-regulated genes required for conjugation and transfer of plasmid DNA (chapter 8 of this volume; Dessaux et al., 1992; Fuqua and Winans, 1994; Piper et al., 1993; Zhang et al., 1993). In nature, the ability to utilize opines as a source of nutrients may confer a selective advantage on the bacteria. It has been shown that in a mixed infection, bacteria that have the ability to catabolize opines can successfully outcompete bacteria that cannot utilize them, even though both bacterial strains colonized the plant roots with the same kinetics when inoculated individually (Savka and Farrand, 1997). In addition, the soil environment of plants that have been engineered to produce a particular opine become enriched for bacteria that can catabolize that compound (Oger et al., 1997). Because the genes encoding the advantageous opine catabolism enzymes are carried by the conjugative plasmid, the AI-controlled conjugation behavior is therefore beneficial to the community of conjugation-competent recipients.

AIs are also important in biofilms. When compared to liquid cultures, cells in biofilms are known to have an increased resistance to such stresses as antibiotics and starvation (Costerton et al., 1995). Researchers have determined that the recovery of ammonia-oxidizing bacteria from ammonia starvation in liquid culture is significantly enhanced (i.e., the lag phase is significantly decreased) by the addition of 3-oxo-C6-HSL (Batchelor et al., 1997). They speculate that the higher cell densities achieved in a biofilm result in a greater concentration of AI, which in turn may induce genes whose products promote a more rapid recovery from ammonia starvation (Batchelor et al., 1997). The demonstration of the presence of AIs in naturally occurring biofilms (McLean et al., 1997) further supports this hypothesis.

Additional evidence of a role for AIs in the development of biofilms has been found using an AI-synthase mutant of *P. aeruginosa* (defective for *lasI*). The *lasI* mutant produced a biofilm that was thinner and contained denser layers of cells than that achieved by the wild-type strain (Davies et al., 1998), a phenotype that was rescued by the topical addition of the *P. aeruginosa* AI 3-oxo-C12-HSL. Unlike the biofilms produced by the wild-type strain, those of the *lasI* mutant were susceptible to the addition of detergent. Growth of the *lasI* biofilm in the presence of 3-oxo-C12-HSL restored the detergent-resistant phenotype displayed by the wild-type strain (Davies et al., 1998). These data provide the first direct evidence of a role for AIs not only in the structural development of a biofilm, but also in the establishment of at least one of the resistance qualities known to be characteristic of cells in biofilms.

Dunlap (1997) has speculated that AI-mediated control of antibiotic production by *E. carotovora* may provide a selective advantage for this organism in its natural environment of a plant infection. The production of cell wall-degrading exoenzymes and the consequent release of nutrients could result in competition from other soil bacteria. To minimize this, *E. carotovora* has adapted the strategy of placing antibiotic biosynthetic genes under the same regulatory system, thereby eliminating competition from other soil bacteria for the nutrients released during the infection (Dunlap, 1997). Antibiotic production is also under the control of AIs in *Pseudomonas aureofaciens* and *Pseudomonas fluorescens*, where these molecules may also bestow a competitive advantage over other bacteria (Cook et al., 1995; Pierson et al., 1994; Wood and Pierson, 1996). Although it is difficult to distinguish whether the functions of AI-controlled genes are necessary, or merely advantageous, it seems likely that examples of key ecological functions for these molecules will be plentiful, given the widespread nature of the AI signal among gram-negative bacteria.

Production of AI Antagonists in the Environment

Because of the ubiquitous distribution of AI-producing bacteria in the natural environment, it would not be surprising if other organisms have evolved competitor molecules as a defense against microbial colonization. This evolutionary strategy appears to be functioning in the seaweed *Delisea pulchra*, which produces compounds called furanones that are structurally similar to AIs and appear to be involved in the inhibition of bacterial growth at the growing tip of the organism (Steinberg et al., 1997). While the basis for this inhibition is still under investigation, it may be related to the loss of the colonizing organism's swarming motility. Both *S. liquefaciens* and *Proteus mirabilis* are able to swim either as individual cells or in a multicellular swarm. The addition of either of two furanones (designated 1 and 2) to *S. liquefaciens* cells inhibits the swarming but not the swimming behavior of these cells (Givskov et al., 1996). This effect can be overcome by the addition of 0.2 μM *N*-butanoyl-L-HSL (C4-HSL), one of the known *S. liquefaciens* AIs. *P. mirabilis* swarming can also be inhibited by the addition of another furanone (designated 4), but in this case, the ability to swarm is not restored by either C4-HSL or an extract from *P. mirabilis* cells (Gram et al., 1996). In addition, the ability of *P. mirabilis* cells to come into close contact, which is required for effective consolidation and swarming, is abolished by a crude preparation extracted from *D. pulchra* cells but not by addition of any of the individual furanones tested, suggesting that an additional inhibitor may be present in this alga that has a further effect on bacterial behavior (Gram et al., 1996).

Although direct proof that furanones are used specifically to inhibit bacterial growth awaits further investigation, it is not unlikely that eukaryotic cells have developed mechanisms for controlling bacterial colonization of their surfaces by interference with AI signaling. The work with *D. pulchra* appears to be the first description of a eukaryotic organism creating a substance(s) that antagonizes a bacterial signaling system by mimicking an AI. Further studies will help delineate the array of eukaryotic defenses against bacteria that may occur via AI molecules or AI antagonists.

CONCLUDING REMARKS

Quorum sensing is only one of a number of known bacterial-sensing systems. Individual bacterial cells have the ability to coordinate their own gene expression using sigma factors, antisigma factors, and other global protein response regulators. By themselves they can sense changes in the environment and relay that information through methylation and phosphorylation relay pathways. They are able to produce surface or exported proteins and to use those proteins to bind to other bacteria and even to eukaryotic hosts. Specific communication with eukaryotic hosts can proceed through pathways stimulated by attachment, or even through the insertion of proteins into their hosts. Given all of these other mechanisms, one can ask: why have quorum-sensing systems evolved, and why are they so broadly distributed in gram-negative bacteria?

With perhaps one exception (Puskas et al., 1997), AIs are found in bacteria that are known to exist in high cell densities in nature (chapter 19 of this volume). Although microbiologists generally think about selection at the cellular level, it is important to remember that AI systems typically function at the population or community level and may in some cases represent a mechanism for survival of the group rather than of the individual. Perhaps AIs are unique in providing a low-energy mechanism by which to signal across distances and through host tissue. In these natural settings, the accumulation of an AI may be easier to accomplish than the probability of achieving direct cell-to-cell contact. Regardless of the reason, it is clear not only that AIs are an effective method of promoting bacterial unity in the coordination of gene expression, but also that the existence of these chemical signals in the environment will surely translate into a reciprocal response by other bacteria and eukaryotic cells to achieve interorganismal communication. As the field of quorum sensing continues to grow and mature in the coming years, the development of our knowledge of the biological emerging properties of this signaling strategy will be an increasingly important area of discovery.

ACKNOWLEDGMENTS

We thank the following individuals for their review of the manuscript: Margaret McFall-Ngai, Frank Aeckersberg, Eric Stabb, Spencer Nyholm, and Jonathan Visick. We also thank E. Peter Greenberg and Staffan Kjelleberg for communicating their results in advance of publication. This work was funded by National Science Foundation grant IBN96-01155 to M. McFall-Ngai and E.G.R, and by National Institutes of Heath grant RR12294 to E.G.R and M. McFall-Ngai. K.L.V was funded by National Institutes of Health Research Service Award 1F32GM174724-01A1.

REFERENCES

Andrews, C. C., D. M. Karl, L. F. Small, and S. W. Fowler. 1984. Metabolic activity and bioluminescence of oceanic faecal pellets and sediment trap particles. *Nature* **307:**539–541.

Bassler, B. L., M. Wright, and M. R. Silverman. 1994. Multiple signalling systems controlling expression of luminescence in *Vibrio harveyi*: sequence and function of genes encoding a second sensory pathway. *Mol. Microbiol.* **13:**273–286.

Bassler, B. L., E. P. Greenberg, and A. M. Stevens. 1997. Cross-species induction of luminescence in the quorum-sensing bacterium *Vibrio harveyi*. *J. Bacteriol.* **179:**4043–4045.

Batchelor, S. E., M. Cooper, S. R. Chhabra, L. A. Glover, G. S. A. B. Stewart, P. Williams, and J. I. Prosser. 1997. Cell density-regulated recovery of starved biofilm populations of ammonia-oxidizing bacteria. *Appl. Environ. Microbiol.* **63:**2281–2286.

Beck von Bodman, S., and S. K. Farrand. 1995. Capsular polysaccharide biosynthesis and pathogenicity in *Erwinia stewartii* require induction by an N-acylhomoserine lactone autoinducer. *J. Bacteriol.* **177:**5000–5008.

Boettcher, K. J., and E. G. Ruby. 1990. Depressed light emission by symbiotic *Vibrio fischeri* of the sepiolid squid *Euprymna scolopes*. *J. Bacteriol.* **172:** 3701–3706.

Boettcher, K. J., and E. G. Ruby. 1995. Detection and quantification of *Vibrio fischeri* autoinducer from symbiotic squid light organs. *J. Bacteriol.* **177:** 1053–1058.

Boettcher, K. J., E. G. Ruby, and M. J. McFall-Ngai. 1996. Bioluminescence in the symbiotic squid *Euprymna scolopes* is controlled by a daily biological rhythm. *J. Comp. Physiol.* **179:**65–73.

Campbell, N. A. 1993. *Biology.* The Benjamin/Cummings Publishing Company, Inc., Redwood City, Calif.

Chapon-Hervé, V., M. Akrim, A. Latifi, P. Williams, A. Lazdunski, and M. Bally. 1997. Regulation of the *xcp* secretion pathway by multiple quorum-sensing modulons in *Pseudomonas aeruginosa*. *Mol. Microbiol.* **24:**1169–1178.

Cook, R. J., L. S. Thomashow, D. M. Weller, D. Fujimoto, M. Mazzola, G. Bangera, and D.-S. Kim. 1995. Molecular mechanisms of defense by rhizobacteria against root disease. *Proc. Natl. Acad. Sci. USA* **92:**4197–4201.

Costerton, J. W., Z. Lewandowski, D. E. Caldwell, D. R. Korber, and H. M. Lappin-Scott. 1995. Microbial biofilms. *Annu. Rev. Microbiol.* **49:** 711–745.

Davies, D. G., M. R. Parsek, J. P. Pearson, B. H. Iglewski, J. W. Costerton, and E. P. Greenberg. 1998. The involvement of cell-to-cell signals in the development of a bacterial biofilm. *Science* **280:**295–298.

Dessaux, Y., A. Petit, and J. Tempe. 1992. Opines in *Agrobacterium* biology, p. 109–136. *In* D. P. S. Verma (ed.), *Molecular Signals in Plant-Microbe Communications,* CRC Press, Ann Arbor, Mich.

DiMango, E., H. J. Zar, R. Bryan, and A. Prince. 1995. Diverse *Pseudomonas aeruginosa* gene products stimulate respiratory epithelial cells to produce interleukin-8. *J. Clin. Invest.* **96:**2204–2210.

Doino, J. A., and M. J. McFall-Ngai. 1995. A transient exposure to symbiosis-competent bacteria induces light organ morphogenesis in the host squid. *Biol. Bull.* **189:**347–355.

Dunlap, P. V. 1997. N-acyl-L-homoserine lactone autoinducers in bacteria: unity and diversity, p. 69–106. *In* J. A. Shapiro and Dworkin, M. (ed.), *Bacteria as Multicellular Organisms,* Oxford University Press, New York.

Dunlap, P. V., and A. Kuo. 1992. Cell-density modulation of the *Vibrio fischeri* luminescence system in the absence of autoinducer and LuxR protein. *J. Bacteriol.* **174:**2440–2448.

Dunphy, G., C. Miyamoto, and E. Meighen. 1997. A homoserine lactone autoinducer regulates virulence of an insect-pathogenic bacterium, *Xenorhabdus nematophilus* (Enterobacteriaceae). *J. Bacteriol.* **179:**5288–5291.

Eberhard, A., A. L. Burlingame, C. Eberhard, G. L. Kenyon, K. H. Nealson, and N. J. Oppenheimer. 1981. Structural identification of autoinducer of *Photobacterium fischeri* luciferase. *Biochemistry* **28:**2444–2449.

Eberhard, A., C. A. Widrig, P. McBath, and J. B. Schineller. 1986. Analogs of the autoinducer of bioluminescence in *Vibrio fischeri*. *Arch. Microbiol.* **146:**35–40.

Eberl, L., M. K. Winson, C. Sternberg, G. S. A. B. Stewart, G. Christiansen, S. R. Chhabra, B. Bycroft, P. Williams, S. Molin, and M. Givskov. 1996. Involvement of N-acyl-L-homoserine lactone autoinducers in controlling the multicellular behaviour of *Serratia liquefaciens*. *Mol. Microbiol.* **20:**127–136.

Forst, S., and K. Nealson. 1996. Molecular biology of the symbiotic-pathogenic bacteria *Xenorhabdus* spp. and *Photorhabdus* spp. *Microbiol. Rev.* **60:**21–43.

Fuqua, W. C., and S. C. Winans. 1994. A LuxR-LuxI type regulatory system activates *Agrobacterium*

Ti plasmid conjugal transfer in the presence of a plant tumor metabolite. *J. Bacteriol.* **176:**2796–2806.

Fuqua, W. C., S. C. Winans, and E. P. Greenberg. 1994. Quorum sensing in bacteria: the LuxR-LuxI family of cell density-responsive transcriptional regulators. *J. Bacteriol.* **176:**269–275.

Fuqua, C., S. C. Winans, and E. P. Greenberg. 1996. Census and consensus in bacterial ecosystems: the LuxR-LuxI family of quorum-sensing transcriptional regulators. *Annu. Rev. Microbiol.* **50:**727–751.

Gilson, L., A. Kuo, and P. V. Dunlap. 1995. AinS and a new family of autoinducer synthesis proteins. *J. Bacteriol.* **177:**6946–6951.

Givskov, M., R. De Nys, M. Manefield, L. Gram, R. Maximilien, L. Eberl, S. Molin, P. D. Steinberg, and S. Kjelleberg. 1996. Eukaryotic interference with homoserine lactone-mediated prokaryotic signaling. *J. Bacteriol.* **178:**6618–6622.

Gram, L., R. de Nys, R. Maximilien, M. Givskov, P. Steinberg, and S. Kjelleberg. 1996. Inhibitory effects of secondary metabolites from the red alga *Delisea pulchra* on swarming motility of *Proteus mirabilis. Appl. Environ. Microbiol.* **62:**4284–4287.

Gray, K. M., and E. P. Greenberg. 1992. Physical and functional maps of the luminescence gene cluster in an autoinducer-deficient *Vibrio fischeri* strain isolated from a squid light organ. *J. Bacteriol.* **174:**4384–4390.

Gray, K. M., L. Passador, B. H. Iglewski, and E. P. Greenberg. 1994. Interchangeability and specificity of components from the quorum-sensing regulatory systems of *Vibrio fischeri* and *Pseudomonas aeruginosa. J. Bacteriol.* **176:**3076–3080.

Greenberg, E. P. Personal communication.

Greenberg, E. P., J. W. Hastings, and S. Ulitzur. 1979. Induction of luciferase synthesis in *Beneckea harveyi* by other marine bacteria. *Arch. Microbiol.* **120:**87–91.

Hanzelka, B. L., and E. P. Greenberg. 1995. Evidence that the N-terminal region of the *Vibrio fischeri* LuxR protein constitutes an autoinducer-binding domain. *J. Bacteriol.* **177:**815–817.

Huber, A. R., S. J. Kunkel, R. F. Tod, and S. J. Weiss. 1991. Regulation of transendothelial neutrophil migration by endogenous interleukin-8. *Science* **254:**99–102.

Kaplan, H. B., and E. P. Greenberg. 1985. Diffusion of autoinducer is involved in regulation of the *Vibrio fischeri* luminescence system. *J. Bacteriol.* **163:**1210–1214.

Kleerebezem, M., L. E. N. Quadri, O. P. Kuipers, and W. M. de Vos. 1997. Quorum sensing by peptide pheromones and two-component signal-transduction systems in gram-positive bacteria. *Mol. Microbiol.* **24:**895–904.

Kuo, A., N. V. Blough, and P. V. Dunlap. 1994. Multiple N-acyl-L-homoserine lactone autoinducers of luminescence in the marine symbiotic bacterium *Vibrio fischeri. J. Bacteriol.* **176:**7558–7565.

Kuo, A., S. M. Callahan, and P. V. Dunlap. 1996. Modulation of luminescence operon expression by N-octanoyl-L-homoserine lactone in *ainS* mutants of *Vibrio fischeri. J. Bacteriol.* **178:**971–976.

Lee, K.-H., and E. G. Ruby. 1994. Effect of the squid host on the abundance and distribution of symbiotic *Vibrio fischeri* in nature. *Appl. Environ. Microbiol.* **60:**1565–1571.

Lloyd, D., C. J. James, and J. W. Hastings. 1985. Oxygen affinities of the bioluminescence systems of various species of luminous bacteria. *J. Gen. Microbiol.* **131:**2137–2140.

McClean, K. H., M. K. Winson, L. Fish, A. Taylor, S. R. Chhabra, M. Camara, M. Daykin, J. H. Lamb, S. Swift, B. W. Bycroft, G. S. A. B. Stewart, and P. Williams. 1997. Quorum sensing and *Chromobacterium violaceum*: exploitation of violacein production and inhibition for the detection of N-acylhomoserine lactones. *Microbiology* **143:**3703–3711.

McFall-Ngai, M., and M. K. Montgomery. 1990. The anatomy and morphology of the adult bacterial light organ of Euprymna scolopes Berry (Cephalopoda:Sepiolidae). *Biol. Bull.* **179:**332–339.

McKenney, D., K. E. Brown, and D. G. Allison. 1995. Influence of *Pseudomonas aeruginosa* exoproducts on virulence factor production in *Burkholderia cepacia*: evidence of interspecies communication. *J. Bacteriol.* **177:**6989–6992.

McLean, R. J. C., M. Whiteley, D. J. Stickler, and W. C. Fuqua. 1997. Evidence of autoinducer activity in naturally occurring biofilms. *FEMS Microbiol. Lett.* **154:**259–263.

Montgomery, M. K., and M. McFall-Ngai. 1993. Embryonic development of the light organ of the sepiolid squid *Euprymna scolopes* Berry. *Biol. Bull.* **184:**296–308.

Montgomery, M. K., and M. McFall-Ngai. 1994. Bacterial symbionts induce host organ morphogenesis during early postembryonic development of the squid *Euprymna scolopes. Development* **120:**1719–1729.

Nealson, K. H., and J. W. Hastings. 1979. Bacterial bioluminescence: its control and ecological significance. *Microbiol. Rev.* **43:**496–518.

Oger, P., A. Petit, and Y. Dessaux. 1997. Genetically engineered plants producing opines alter their biological environment. *Nat. Biotechnol.* **15:** 369–372.

Passador, L., J. M. Cook, M. J. Gambello, L. Rust, and B. H. Iglewski. 1993. Expression of the *Pseudomonas aeruginosa* virulence genes requires cell-to-cell communication. *Science* **260:**1127–1130.

Passador, L., K. D. Tucker, K. R. Guertin, M. P. Journet, A. S. Kende, and B. H. Iglewski. 1996. Functional analysis of the *Pseudomonas aeruginosa* autoinducer PAI. *J. Bacteriol.* **178:** 5995–6000.

Pearson, J. P., K. M. Gray, L. Passador, K. D. Tucker, A. Eberhard, B. H. Iglewski, and E. P. Greenberg. 1994. Structure of the autoinducer required for expression of *Pseudomonas aeruginosa* virulence genes. *Proc. Natl. Acad. Sci. USA* **91:**197–201.

Pearson, J. P., E. C. Pesci, and B. H. Iglewski. 1997. Roles of *Pseudomonas aeruginosa las* and *rhl* quorum-sensing systems in control of elastase and rhamnolipid biosynthesis genes. *J. Bacteriol.* **179:** 5756–5767.

Pesci, E. C., J. P. Pearson, P. C. Seed, and B. H. Iglewski. 1997. Regulation of *las* and *rhl* quorum sensing in *Pseudomonas aeruginosa. J. Bacteriol.* **179:**3127–3132.

Pierson, L. S., III, V. D. Keppenne, and D. W. Wood. 1994. Phenazine antibiotic biosynthesis in *Pseudomonas aureofaciens* 30-84 is regulated by PhzR in response to cell density. *J. Bacteriol.* **176:** 3966–3974.

Piper, K. R., S. Beck von Bodman, and S. K. Farrand. 1993. Conjugation factor of *Agrobacterium tumefaciens* regulates Ti plasmid transfer by autoinduction. *Nature* **362:**448–450.

Pirhonen, M., D. Flego, R. Heikinheimo, and E. T. Palva. 1993. A small diffusible signal molecule is responsible for the global control of virulence and exoenzyme production in the plant pathogen *Erwinia carotovora. EMBO J.* **12:**2467–2476.

Puskas, A., E. P. Greenberg, S. Kaplan, and A. L. Schaefer. 1997. A quorum-sensing system in the free-living photosynthetic bacterium *Rhodobacter sphaeroides. J. Bacteriol.* **179:**7530–7537.

Rosemeyer, V., J. Michiels, C. Verreth, and J. Vanderleyden. 1998. *luxI-* and *luxR-* homologous genes of *Rhizobium etli* CNPAF512 contribute to synthesis of autoinducer molecules and nodulation of *Phaseolus vulgaris. J. Bacteriol.* **180:**815–821.

Ruby, E. G. 1996. Lessons from a cooperative, bacterial-animal association: the *Vibrio fischeri*-*Euprymna scolopes* light organ symbiosis. *Annu. Rev. Microbiol.* **50:**591–624.

Ruby, E. G., and L. M. Asato. 1993. Growth and flagellation of *Vibrio fischeri* during initiation of the sepiolid squid light organ symbiosis. *Arch. Microbiol.* **159:**160–167.

Ruby, E. G., and J. G. Morin. 1979. Luminous enteric bacteria of marine fishes: a study of their distribution, densities, and dispersion. *Appl. Environ. Microbiol.* **38:**406–411.

Ruby, E. G., and K. H. Nealson. 1976. Symbiotic association of *Photobacterium fischeri* with the marine luminous fish *Monocentris japonica*: a model of symbiosis based on bacterial studies. *Biol. Bull.* **151:** 574–586.

Savka, M. A., and S. K. Farrand. 1997. Modification of rhizobacterial populations by engineering bacterium utilization of a novel plant-produced resource. *Nat. Biotechnol.* **15:**363–368.

Schaefer, A. L., B. L. Hanzelka, A. Eberhard, and E. P. Greenberg. 1996. Quorum sensing in *Vibrio fischeri*: probing autoinducer-LuxR interactions with autoinducer analogs. *J. Bacteriol.* **178:** 2897–2901.

Shaw, P. D., G. Ping, S. L. Daly, C. Cha, J. E. J. Cronan, K. L. Rinehart, and S. K. Farrand. 1997. Detecting and characterizing N-acyl-homoserine lactone signal molecules by thin-layer chromatography. *Proc. Natl. Acad. Sci. USA* **94:** 6036–6041.

Showalter, R. E., M. O. Martin, and M. R. Silverman. 1990. Cloning and nucleotide sequence of *luxR*, a regulatory gene controlling bioluminescence in *Vibrio harveyi. J. Bacteriol.* **172:**2946–2954.

Sitnikov, D. M., J. B. Schineller, and T. O. Baldwin. 1995. Transcriptional regulation of bioluminescence genes from *Vibrio fischeri. Mol. Microbiol.* **17:**801–812.

Steinberg, P. D., S. Rene, and S. Kjelleberg. 1997. Chemical defenses of seaweeds against microbial colonization. *Biodegradation* **8:**211–220.

Sun, W., C. J.-G., K. Teng, and E. A. Meighen. 1994. Biosynthesis of poly-3-hydroxybutyrate in the bioluminescent bacterium, *Vibrio harveyi*, and regulation by the *lux* autoinducer, N-(3-hydroxylbutanoyl) homoserine lactone. *J. Biol. Chem.* **269:**20785–20790.

Tang, H. B., E. DiMango, R. Bryan, M. Gambello, B. H. Iglewski, J. B. Goldberg, and A. Prince. 1996. Contribution of specific *Pseudomonas aeruginosa* virulence factors to pathogenesis of pneumonia in a neonatal mouse model of infection. *Infect. Immun.* **64:**37–43.

Taylor, A. M., K. H. McClean, S. R. Chhabra, M. Camara, M. K. Winson, M. Daykin, B. W. Bycroft, G. S. A. B. Stewart, and P. Williams. 1996. Exploitation of violacein production and inhibition in *Chromobacterium violaceum* for the detection of *N*-acyl homoserine lactones, abstr. H-206, p. 519. Abstr. 96th Gen. Meet. Am. Soc. Microbiol. 1996. American Society for Microbiology, Washington, D.C.

Telford, G., D. Wheeler, P. Williams, P. T. Tomkins, P. Appleby, H. Sewell, G. S. A. B. Steward, B. W. Bycroft, and D. I. Pritchard. 1998. The *Pseudomonas aeruginosa* quorum-sensing signal molecule *N*-(3-oxododecanoyl)-L-homoserine lactone has immunomodulatory activity. *Infect. Immun.* **66:**36–42.

Visick, K. L., V. O. Orlando, and E. G. Ruby. 1996. Role of the *luxR* and the *luxI* genes in the *Vibrio fischeri-Euprymna scolopes* symbiosis, abstr. H-150, p. 509. Abstr. 96th Gen. Meet. Am. Soc. Microbiol. 1996. American Society for Microbiology, Washington, D.C.

Wood, D. W., and L. S. Pierson III. 1996. The *phzI* gene of *Pseudomonas aureofaciens* 30-84 is responsible for the production of a diffusible signal required for phenazine antibiotic production. *Gene* **168:**49–53.

Wood, D. W., F. Gong, M. M. Daykin, P. Williams, and L. S. Pierson III. 1997. *N*-acyl homoserine lactone-mediated regulation of phenazine gene expression by *Pseudomonas aureofaciens* 30-84 in the wheat rhizosphere. *J. Bacteriol.* **179:**7663–7670.

Zhang, L., P. J. Murphy, A. Kerr, and M. E. Tate. 1993. *Agrobacterium* conjugation and gene regulation by *N*-acyl-L-homoserine lactones. *Nature* **362:**446–448.

INDEX

353